Basic Electricity And DC Circuits

Ralph A. Oliva, Ph.D.
Texas Instruments Learning Center

Charles W. Dale, Ed.D.
Dallas County Community College District
Eastfield College

With Contributions by
David Clemens
Ross Wise
Donald W. Taylor

LEARNING
CENTER

SECOND EDITION

This book is part of the BASIC ELECTRICITY SERIES
from the TEXAS INSTRUMENTS LEARNING CENTER consisting of:

Basic Electricity and DC Circuits
Basic AC Circuits

It is designed for use either as a basic stand-alone text,
or as an integral part of the TI course package *Basic Electricity and DC Circuits.*

TEXAS INSTRUMENTS

P.O. BOX 225012, MS-54 • DALLAS, TEXAS 75265

This book was developed by:

The Staff of the Texas Instruments Learning Center
P.O. Box 225012 MS-54
Dallas, Texas 75265

For marketing and distribution inquire to:

James B. Allen
Marketing Manager
P.O. Box 225012, MS-54
Dallas, Texas 75265

THIRD PRINTING

ISBN 0-89512-034-8
Library of Congress Catalog Number: 79-92192

Important Note to the Student and Instructor

This is a large-looking book of nearly 1000 pages, but it is important to note that **about half of this book is devoted to WORKED THROUGH EXAMPLES with step-by-step solutions.** Many students learn about electricity and dc circuits most easily from problems of this sort. Students must often buy an additional book to obtain extra worked through examples; however, a wide variety of problems with solutions is included in the body of this text for you, and in a large set at the end of each lesson.

In addition, each text section uses a new **special layout** which has several distinct features:

1. Quite a bit of blank space has been deliberately left on the text pages so that there is ample room for highlighting, extra notes, questions, etc.

2. There are two "pathways" through the text material. One is a fast path through the essential required materials. The other path includes extra examples and detailed discussion in areas where questions are commonly raised.

The page layout and type faces are your keys to the use of these two pathways. The explanation of their use is a simple one, as is shown on the next page. It's as if you are on a railroad train and you have the option of selecting an "express route", or taking a tour through a new "sidings" of interest. Here's how it works.

- **Basic Concept #1**
- Supplemental Material
- **Basic Concept #2**

Basic Concept #1 — Basic materials are presented on the left side of printed text pages in type like this. (Figure A) The main body of the course will be presented using this layout. Also, if you are watching a portion of this course on videotape, all *taped* material is printed here for you in this way. In key course areas, supplemental materials are inserted for you. At these points, you have the option of swinging your attention to the right side of the page, or moving straight down to the next basic concept, skipping the supplemental material.

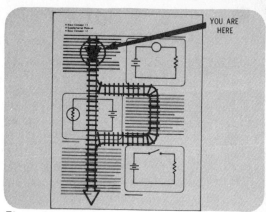

Figure A

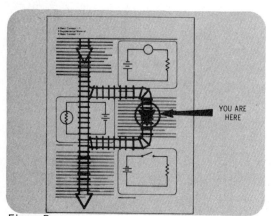

Figure B

Supplemental Material — If you are interested in "digging in" a little deeper for an extra example or more detailed discussion, swing your attention to the right, where supplemental materials are printed in a slightly lighter type like this. On these "sidings" you will hopefully find a question answered, a helpful extra example, or extra details in an area of interest to you.

Basic Concept #2 — If you move straight down the left side of the textbook pages, from basic concept to basic concept, you will have no problems with the continuity of the text or flow of material. If you have some background in basic electricity, or wish a shorter "express" pathway through the material, this is the path for you. In addition (as mentioned), if you are watching the TI videotape course, *Basic Electricity and DC Circuits*, all of the basic concepts presented on the tape are printed on this express path. *Each figure shown on the TV screen is reproduced for you in sequence to the right of the text material.*

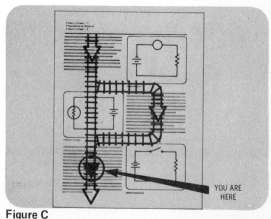

Figure C

Features of This Book

As a Textbook

This book is designed primarily for the entry level student. No sophisticated math background or previous knowledge of electricity is assumed, and no matter what your background you can learn the basic concepts that have enabled man to harness and control dc electricity.

There are several features of this book specifically designed to increase its efficiency and aid you in grasping the principles of predicting and controlling dc electricity.

1. There is plenty of blank space near each text passage and figure for your notes, calculations or sketches. Keep your notes here in the book so they can serve as a handy reference.

2. At the beginning of each lesson, on a shaded page, detailed OBJECTIVES are listed for you. These objectives state, step-by-step, what new things you should be able to do upon successful completion of each chapter. (The things you will actually be doing are printed in *italics; write, sketch, calculate, predict, construct, explain, etc.)*

3. At the end of each lesson there are three types of examples for your use:

   First, in addition to those examples in the text, there is a set of examples with detailed step-by-step solutions. These WORKED THROUGH EXAMPLES *apply* the theory in each lesson to typical situations involving dc electricity. In this way you are led from *knowledge* of the concepts you need to know, to the *application* of those concepts in circuits you will build and handle.

   Second, you are provided with a complete set of PRACTICE PROBLEMS with answers on the back of each page. These problems will give you the opportunity to *try* your new skills and *test* your accuracy.

   Finally, a set of questions without solutions or answers is arranged as a two-page QUIZ, and will take you less than an hour to complete. Your quiz will serve as an indicator of areas where you need further review of key concepts and principles.

   The quizzes and the worked through problems were taken from the most asked questions in TI's instructor conducted course on Basic Electricity and DC Circuits.

For Use With Videotape

Although this book is designed to be a stand alone text, it is also an essential part of the Texas Instruments Learning Center course, *Basic Electricity and DC Circuitry*. When using this book along with the videotape, you can increase your efficiency, retention, and enjoyment of the material by keeping the following points in mind.

LISTEN FAST AND WATCH FAST. Video taped lectures typically cover two to three times more material as a live instructor would in the same length of time. The steady pace of videotape gives you a highly concentrated message and leaves no time for asking questions, day dreaming or chatting. If you sneeze, you might miss something important. If the TV lecture raises any question in your mind, jot it down quickly and get clarification after the tape, but put it out of your mind while you absorb the rest of the lecture.

TAKE VERY FEW NOTES. This book contains every figure you will see on the screen, including formulas and related information. It is also a printed record of the speaker's remarks. Limit your notes to brief comments that will help you study later, and write them in the book.

All of the material covered in the videotape is printed on the left hand side of each text page, with all TV figures reproduced in sequence, on the right. Supplemental material not included in the tape is also printed on the right-hand side but in a lighter type face. As you watch the tape, you can follow along in your book by looking at the left-hand column for text and the right-hand column for figures. At the completion of the tape you can then go back and review any supplemental material that you feel you need.

KEEP YOUR PRIMARY ATTENTION ON THE TV SCREEN. Open your book to the section being covered to allow you to take quick notes or get a closer look at the figures; but *do not try to follow the speaker's words in the text book*. Special media techniques are used to link the speaker's voice and visual aids to your thinking, allowing quicker coverage and better retention. Take full advantage of this by keeping your primary attention on the tape.

RELAX. The TV lecturer will not be interrupted with questions. Nor will he ask for your opinions. You will be freed from such distractions so you can open your consciousness to absorb the material.

ENJOY IT. Education using videotape coupled with special texts such as this one is the wave of the future brought to you today. Over the years, attendees at videotaped seminars from TI's Learning Center have found them to be informative, effective, and enjoyable. We hope you do too.

Contents

Lesson 1

An Introduction to Electricity

This lesson defines electricity through a discussion of atomic theory. The structure of the atom is examined in detail. All new terms such as *proton, electron, ion, valence*, to name a few, are defined and discussed. The first law of electrostatics is explained, illustrated, and demonstrated. Many important concepts such as *potential difference* and the complete *circuit* are fully explained and illustrated.

LESSON 1. AN INTRODUCTION TO ELECTRICITY

● Objectives

At the end of this introductory lesson, the student should be able to:

1. *Write* a simple definition of electricity.

2. *Sketch* the structure of a simple atom, labeling all its parts.

3. *State* the first law of electrostatics, and *sketch* with lines of force the appearance of the electric field between two attracting and two repelling bodies.

4. *Write* a definition of a free electron, including a description of how electrons can become free.

5. *List* the major characteristics of a substance that govern whether or not it is a good conductor.

6. *Draw and label* the schematic symbol for a battery.

7. *Write a description*, with sketches, of how a potential difference may be created, and what the effects of applying a potential difference to a conductor might be.

8. *Write a description*, with sketches of what a circuit is.

9. *Identify, define or sketch* the definition of the following key terms needed in the discussion and explanation of electricity:

   1. Electricity
   2. Atom
   3. Element
   4. Nucleus
   5. Electron
   6. Proton
   7. Neutron
   8. Positive charge
   9. Negative charge
   10. Electrically neutral
   11. Orbit
   12. Shell
   13. Valence shell
   14. Coulomb
   15. "Free electron"
   16. Ion
   17. Potential difference
   18. Circuit
   19. Open circuit

To help you see how well you have achieved these objectives, a short quiz has been provided at the end of each lesson. The quiz is designed to point out for you in which areas further detailed study should be directed.

- **TI-50 Calculator**
- **Approach to the Course**

TI-50 Calculator — In recent times man is learning to harness electricity to act for him in more and more complex and astonishing ways. One of the most remarkable achievements of the "electronic revolution" is the integrated circuit, which is the heart and brain of hand-held calculators such as the one in Figure 1.1. These little devices, remarkable as they may be, are rapidly becoming "taken for granted" as are many other devices which use electricity in our everyday lives. In our home, automobile, business and industry, electricity is used to handle information, as well as to provide the power to do many jobs. This course begins at the most basic level with an introduction to basic electricity and dc circuits. The innovative approach to the course material uses several media which are made possible only with electricity itself.

Figure 1.1

Approach to the Course — As shown in Figure 1.2, it is assumed that you're a student who is just beginning the study of electricity. Very little or no prior knowledge of electricity or mathematics will be required as you begin and proceed through the course. All new concepts are introduced and explained as they are needed and where they logically fit into the course material. This printed lesson guide has been specially prepared and organized for this course and will be an extremely valuable aid particularly if you are viewing the course on videotape. At the beginning of each lesson, detailed goals or objectives will be set down. These will indicate what new things you can expect to be able to do after successfully

- NO PRIOR KNOWLEDGE REQUIRED
- SPECIALLY PREPARED LESSON SUMMARY
- DETAILED OBJECTIVES
- QUIZ AT THE END OF EACH LESSON

Figure 1.2

completing each lesson. A quiz is included for you at the end of each lesson, and your performance on this quiz will indicate how well you have achieved the objectives of the lesson.

The complete multimedia coursework package available for this course includes a video laboratory package. The laboratory manual that forms a part of that package includes lab experiments which are directly coordinated with the lecture information presented in each lesson of this course. In this way and others, you will be provided with the opportunity to make practical applications of the theoretical concepts presented in the videotape series.

Course Objectives — Listed in Figure 1.3 is a very brief condensation of the overall objectives for this entire course on "Basic Electricity and DC Circuits." Generally speaking, upon successful completion of this course, you will understand the concepts of voltage, current, and resistance. You will be able to use simple dc circuit laws to predict the behavior of a variety of dc circuits when given a schematic of the circuit. Then, you will be able to identify and select the actual working components necessary to build typical circuits, and, using the proper test equipment, safely measure all current, voltage, and resistance values. You will also be able to predict the behavior of dc circuits that contain special components called capacitors and inductors.

COURSE OBJECTIVES
- UNDERSTAND VOLTAGE, CURRENT, AND RESISTANCE
- USE SIMPLE DC CIRCUIT LAWS
- IDENTIFY AND SELECT COMPONENTS
- MEASURE CURRENT, VOLTAGE AND RESISTANCE
- FUNCTIONS OF CAPACITORS AND INDUCTORS IN DC CIRCUITS

Figure 1.3

- **Objectives — Lesson 1**
- **Definition of Electricity**
- **Definition of the Atom**

Objectives — Lesson 1 — As shown in Figure 1.4, this first lesson begins with "an introduction to basic electricity." When you complete this lesson, you should have a clear understanding of what electricity is and a mental picture of how it may be generated, as well as an understanding of its basic characteristics. To begin — consider the meaning of the word electricity itself.

OBJECTIVES — LESSON 1

- WHAT IS ELECTRICITY?
- HOW IS IT GENERATED?
- WHAT ARE ITS BASIC CHARACTERISTICS?

Figure 1.4

Definition of Electricity — Electricity (as shown in Figure 1.5) may be defined as "the flow of *electrons* through simple materials and devices." An obvious question that you may be asking is: "What is an electron?" The answer to that question is very important to your study of electricity, so it is necessary that it be discussed in detail. To understand what an electron is, it is important to investigate first what an *atom* is, because electrons are basic parts that help build atoms.

ELECTRICITY IS THE FLOW OF ELECTRONS THROUGH SIMPLE MATERIALS AND DEVICES

Figure 1.5

Definition of the Atom — As stated in Figure 1.6, an atom is the smallest part into which an *element* can be divided. Elements are the basic building blocks of all matter. Some common elements include silver, copper, gold, tin, hydrogen, and oxygen. An atom is the smallest piece of an element that will still act like that element. The atom as shown in Figure 1.7 is the smallest particle into which an element can be reduced that will still retain the characteristics of the element. To date, man has discovered 106 different elements and therefore, there are 106 different kinds of atoms.

THE ATOM

SMALLEST PART INTO WHICH AN ELEMENT CAN BE DIVIDED

ELEMENTS

BASIC BUILDING BLOCKS OF ALL MATTER

Figure 1.6

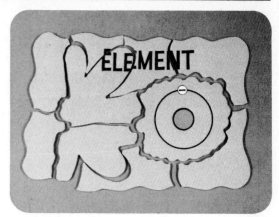

Figure 1.7

Elements — So far you have seen quite a chain of definitions. Electricity is the flow of electrons; electrons are parts of atoms; atoms make up elements. Focus your attention on these words for a minute. All around you in the environment are all sorts of materials and objects that make up the physical world. In order to visualize where electricity comes from, you have to get deep inside matter itself. For example, visualize yourself shrinking in size, smaller and smaller, until you become so small that you could get deep inside one sheet of paper on which this book is printed. As you become small enough, you will see that the paper is made up of a complex array of different types of bodies, these are called atoms. Virtually everything is composed of atoms. In a substance like paper, many different types of atoms are present, all joined together in complex ways to form what are called compounds.

Scientists have examined many, many substances and have found that in nature there is not an extremely large number of different *types* of atoms. There are only about 100 different distinct types of atoms (106 have been discovered at the time this was written). Substances and objects you find around you in everyday life are arrays of countless billions of atoms. Some objects and substances are made up of a huge array of the *same* types of atoms. These substances found in nature, such as gold, copper, hydrogen, and oxygen, are called elements. If you were able to get inside a copper penny, you would see that it is made up of an array of countless atoms, all of the same type, copper atoms. Now, if you took a copper penny and cut it in half, each half would still exhibit the properties of copper, and you could identify each as a piece of copper easily. If you could keep dividing each half smaller and smaller, the smallest piece still retaining the properties of copper would be a single copper atom standing alone. That would be some job of slicing — to get down to one atom, but even that is not enough to get to the key to electricity. You would have to go one step smaller and get inside the atom itself. So now you need to see how atoms are put together.

The Atom — As shown in Figure 1.8 all atoms consist of two basic parts: a body at the center of the atom, called the *nucleus*, and at least one other very important body called an *electron*, orbiting around the nucleus. Atoms may have more than one orbiting electron, but each atom contains only one nucleus.

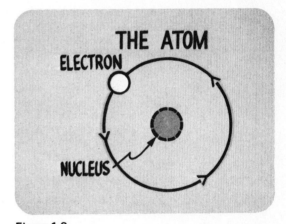

Figure 1.8

Electrostatic Force — One of the greatest mysteries of nature is that the nucleus and the electron attract one another. This attraction is given the name *electrostatic force*, and it is this force that holds the electron in orbit. This force may be illustrated with *lines* as shown in Figure 1.9. Without this electrostatic force, the electron, which is whirling around the nucleus at high speed, would fly away. Bodies that attract each other in this special electrostatic way are described as *charged* objects (Figure 1.10). The electron has what is called negative charge, while the nucleus has what is called positive charge.

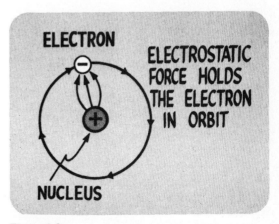

Figure 1.9

CHARGED OBJECTS

● ELECTRONS HAVE A NEGATIVE CHARGE

● NUCLEUS HAS A POSITIVE CHARGE

Figure 1.10

The Why of Electrostatic Force — A question you may be asking yourself is "why?" *Why* do the nucleus and electrons attract each other? The *real* reason for this attraction is a mystery. There are theories that attempt to explain the *why* of electrostatic force, but all that is known about it are its *effects*, that is, what it *does*. Names and simple rules have been developed to help describe these effects, and these will be used in later discussions of electricity and electronics. Again, the force acting between the nucleus and electrons in an atom is called the *electrostatic* force. Bodies that exhibit electrostatic forces are called charged bodies, and by common agreement the charge of an electron is called a *negative charge*, and the charge of the nucleus is a *positive charge*. Now you might wonder what goes on inside the nucleus and how it gets its positive charge.

Protons — The positive charge of the nucleus is actually due to particles called *protons*, which are inside the nucleus and have a positive charge equal to the electron's negative charge. The position of the protons in a simple nucleus is illustrated in Figure 1.11. In the normal atom, the number of protons in the nucleus *equals* the number of orbiting electrons. This means that the atom has an equal number of positive and negative charges. When there is an equal number of electrons and protons in the atom, the atom is said to be electrically *neutral* (Figure 1.12).

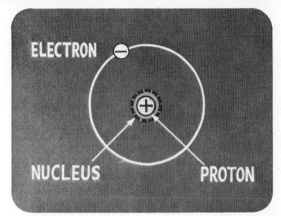

Figure 1.11

ATOMS HAVING AN EQUAL NUMBER OF
POSITIVE AND NEGATIVE CHARGES ARE
ELECTRICALLY NEUTRAL

Figure 1.12

First Law of Electrostatics — Focus your attention on another intriguing fact about electrical charges. It has already been stated that inside the atom the *protons* and *electrons* attract each other. In nature *unlike* charges (like the positive protons and negative electrons) attract each other. As it turns out *like* charges *repel* each other, that is, electrons repel other electrons and protons repel other protons. These facts are put together in the first law of electrostatics which simply states: Like charges repel each other, and unlike charges attract each other. (Figure 1.13) Remember this law, it's one of the vital concepts in electricity.

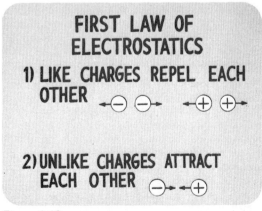

FIRST LAW OF
ELECTROSTATICS
1) LIKE CHARGES REPEL EACH OTHER
2) UNLIKE CHARGES ATTRACT EACH OTHER

Figure 1.13

What Holds the Nucleus Together — Another question may come to mind here. If protons repel other protons, why do protons stay together in the nucleus of an atom? The reason is that the protons are held in the nucleus by another very strong force called the nuclear binding force, or nuclear strong force. This force is also quite mysterious in origin and is strong enough to hold the protons together, even though their electrostatic repulsion is trying to push them apart.

Demonstration of Laws of Electrostatics — An important point to remember concerning charge is that an atom or even an object or body which contains billions of atoms is *normally electrically neutral*. This means it has an equal number of positive protons and negative electrons and since there is no excess of any one type of charge, it is neither attracted nor repelled by other similar objects. These facts can be illustrated with the glass rod and pith balls (Figure 1.14). Pith balls are very lightweight balls with a thin metal cover. When the glass rod and pith balls are electrically neutral, they can be physically touched and nothing happens, that is, they neither attract nor repel each other. However, it is possible to change the electrical charge on the glass rod and balls. If the glass rod is rubbed with silk cloth for a few seconds, electrons (which are negatively charged) are removed from the glass rod through the action of friction. Since negatively charged electrons are removed from the rod, this leaves the rod with a positive charge. When the positively charged glass rod is brought close to the pith balls, two interesting effects are seen. First the balls are attracted to the glass rod for a short period of time. What is happening here? The pith balls originally are neutral, that is, they contain an equal number of positive protons and negative electrons. When the positively charged glass rod is brought near the balls, the negative electrons in the balls move as close as they can to the rod, giving one side of each ball a high

Figure 1.14

concentration of negative charges close to the rod. This concentration of negative charge is closer to the rod than the remaining positive charges in the balls and for this reason is more strongly affected by the rod. The result is that the negative electrons in the balls are strongly attracted to the charged rod and these charges, and the whole ball move towards it. A second effect is seen after the balls actually touch the glass rod. The tremendous lack of electrons on the glass rod pulls electrons off of the balls. Thus, the balls *both* develop a *positive charge* (remember that a positive charge is just a *lack* of electrons), and hence after the balls have touched the positive rod, they *repel each other*. This occurs because they both have a positive charge and *like charges repel*. This demonstration again illustrates that like charges repel and unlike charges attract.

Lines of Force Illustrating an Electrostatic Field — This demonstration has shown that electrically charged objects are attracted or repelled depending on the types of charge they possess. Some special force is acting between these charged bodies, and although you can't see the force, you can see its effects. Forces of this type are the result of what is called an *electrostatic field* which exists around each charged object. This *electrostatic field* and the force it creates may be illustrated with lines which are called *lines of force*. These lines can be drawn around charged bodies to help you visualize the patterns of force acting between them.

The lines of force representing the field around a positively charged body are illustrated in Figure 1.15. Notice that the arrows on the lines are drawn by convention to point *away* from a positively charged body. You will be seeing the phrase "by convention" in your study of various concepts of electricity. The phrase "by convention" refers to a common agreement or set of rules set up by some group of people to avoid confusion. The aim of these conventions or rules is to enable people writing or describing some topic to use the same basic terms. Various rules or conventions will be discussed throughout this course.

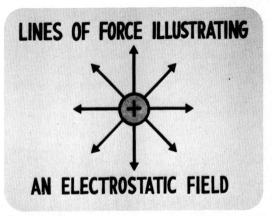

Figure 1.15

- **Like Charges Repel**
- **Electrostatic Field Around the Negative Charge**
- **Unlike Charges Attract**

Like Charges Repel — If one positively charged body is brought near another positively charged body, it will be repelled. Each charged body has its own field and the two fields interact as shown in Figure 1.16. As you can see, the two fields almost seem to act like springs. You can almost see the lines of force pushing against one another, repelling the two charged bodies.

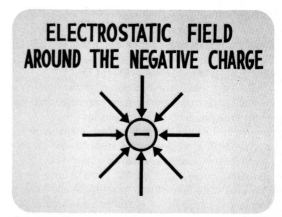

Figure 1.16

Electrostatic Field Around the Negative Charge — The field around a negatively charged object is illustrated with lines of force in Figure 1.17. It has the same general "shape" as that near the positive body, but in this case the arrows point inward.

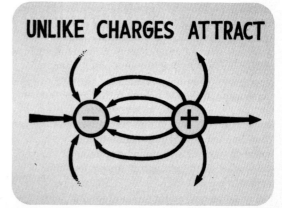

Figure 1.17

Unlike Charges Attract — When positive and negative bodies come into close contact, their fields interact as shown in Figure 1.18. In this case, the lines of force actually *connect* the two objects, starting on the positive and ending on the negative body. Here you can almost feel these lines pulling these two bodies together. The result of these fields interacting is that *unlike charges attract each other.*

Figure 1.18

Electron in Orbit — To get back to the discussion of charged objects, consider again the structure of the atoms shown in Figure 1.19. Since the electrons in orbit possess a negative charge, and the protons in the nucleus possess a positive charge, they attract each other. You may ask, why aren't the two pulled together, so that they hit each other?

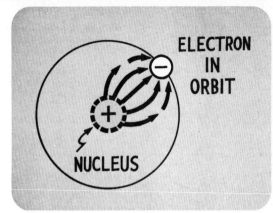

Figure 1.19

Electron in Motion — The answer is illustrated by Figure 1.20. The electrons are in very rapid motion around the nucleus. So, while there is an electrostatic force trying to pull the nucleus and electron together, the electron is constantly in motion and keeps trying to pull away. These two effects balance. There is a balance of electrostatic force attracting, and the force of the electron trying to pull away, which keeps the electron in its orbit and hence gives the atom its general appearance, like a little sun with orbiting planets.

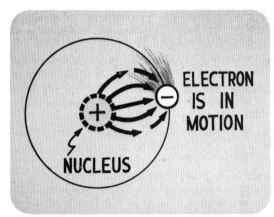

Figure 1.20

The Coulomb — The discussions of electricity that follow will involve many, many electrons that come from the billions of atoms that make up materials such as copper wire. So it will often be convenient to discuss large amounts of charge instead of one electron at a time. Scientists recognizing this, assigned a special name to a *large quantity of electrons* that is convenient in working with electricity. It was named after a French physicist, Charles A. Coulomb, who made many of the first discoveries involving electrostatic force. As shown in Figure 1.21, *one coulomb* is an extremely large number of electrons: 6.25 billion billion, and the *coulomb* is the common unit used in specifying how big or small a given charge is.

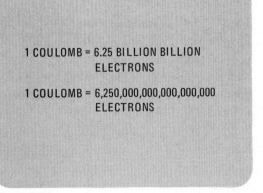

1 COULOMB = 6.25 BILLION BILLION ELECTRONS

1 COULOMB = 6,250,000,000,000,000,000 ELECTRONS

Figure 1.21

The coulomb is an important and realistic unit to use when considering charge and its effects. In general, you won't ever be able to count electrons one by one, or even detect the effects of a single electron acting alone, its charge and effects are too small. However, the effects of *6.25 billion billion, or 1 coulomb* of electrons, are easily detectable and measurable. The coulomb will be more meaningful when the effects of charges are discussed in real-world situations.

Free Electrons — As mentioned earlier, electricity deals with electrons flowing through simple materials and devices. In order for electrons to flow, they somehow have to get free from their atoms to move through the material (Figure 1.22). As it turns out, certain atoms can free one of their electrons easily. These atoms, and the elements which contain these atoms, are called *good conductors* of electricity. Good conductors contain many free electrons which can be made to move (Figure 1.23). There are other elements, however, which have very few free electrons and these are called *nonconductors* or *insulators* (Figure 1.24). Therefore the number of available free electrons a material contains is really a key to its electrical properties. Of the roughly 100 different atoms man knows about, only a few have the ability to easily free electrons.

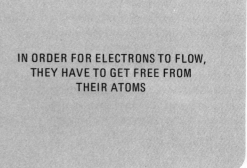

IN ORDER FOR ELECTRONS TO FLOW, THEY HAVE TO GET FREE FROM THEIR ATOMS

Figure 1.22

GOOD CONDUCTORS HAVE MANY FREE ELECTRONS

Figure 1.23

NONCONDUCTORS OR INSULATORS HAVE
VERY FEW FREE ELECTRONS

Figure 1.24

Electrical Properties — As indicated by
Figure 1.25, this property depends very strongly
on the number of electrons an atom has rotating
around its nucleus, and *how these electrons are
arranged*. Consider how the structure of an atom is
related to its electrical properties.

As mentioned, the number of protons in the
nucleus of a normal atom equals the number of
orbiting electrons. The total positive charge in the
normal atom equals the total negative charge, and
so the atom is considered *electrically neutral.*

ELECTRICAL PROPERTIES DEPEND VERY
STRONGLY ON THE NUMBER OF
ELECTRONS ROTATING AROUND
THE NUCLEUS.

Figure 1.25

Neutrons in the Atom — The nucleus may contain
other particles called *neutrons*, as shown in
Figure 1.26. However, as their name implies, these
particles have no charge, and therefore they will
not be important in the discussions of electricity.

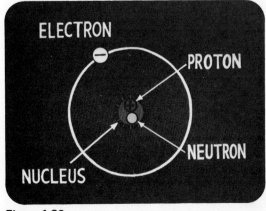

Figure 1.26

- **Lithium Atom**
- **Hydrogen Atom**

Lithium Atom — One example of a typical electrically neutral atom is the lithium atom which has three positive protons in its nucleus and three negative electrons orbiting the nucleus (Figure 1.27).

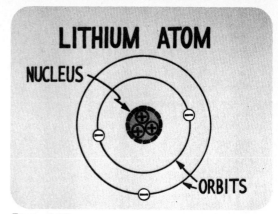

Figure 1.27

Hydrogen Atom — As another example, Figure 1.28 illustrates the hydrogen atom which is the simplest atom. It contains only *one proton* in its nucleus, and only one orbiting electron.

Note that all electrons are just like all other electrons, all protons are like all other protons, and all neutrons are alike. What makes the difference between the 106 different elements and all of the different materials found in nature, is the number and arrangement of protons, neutrons, and electrons within the individual atoms. Atoms with more and more protons, neutrons and electrons become increasingly complex. More complex atoms, such as copper have many more protons, neutrons, and electrons than simple atoms like hydrogen or lithium; however, the parts of these complex atoms don't combine in just any old way. Definite interesting periodic behaviors are seen in atomic structure. The structure of some of these more complex atoms and some of the additional features about how atoms are put together are important to your study of electricity.

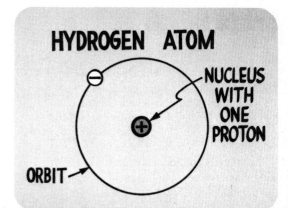

Figure 1.28

Copper Atoms — Consider the copper atom in Figure 1.29. This atom is different from the hydrogen or the lithium atom only in the number of particles in its nucleus and the number of orbiting electrons. The nucleus of the copper atom contains 29 protons with a positive charge and balancing out this positive charge, it has 29 orbiting electrons. Notice that the 29 electrons in the copper atom do not drift aimlessly. The electrons travel in specific well-defined orbits, and this fact is true for every atom.

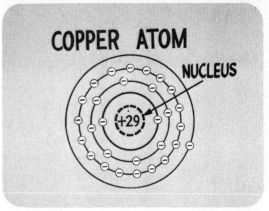

Figure 1.29

Electron Shells — The electron's orbits are also called shells. These shells nest inside one another and surround the nucleus. The nucleus is the center of all the shells. The shells are lettered, by convention, beginning with the shell nearest the nucleus: K, L, M, N, O, P, and Q. Another important fact about the electron shells of an atom is that each shell has a *maximum number of electrons it can hold*. For example, the K shell will hold a maximum of two electrons. The next shell, the L shell, will hold a maximum of eight electrons. As shown in Figure 1.30, each shell has a specific number of electrons it will hold for a particular atom.

An important point is that when the outer shell of an atom contains eight electrons, the atom becomes very *stable* or very resistant to changes in its structure. This means an atom with one or two electrons in its outer shell can lose electrons much more easily than one which has a full outer shell.

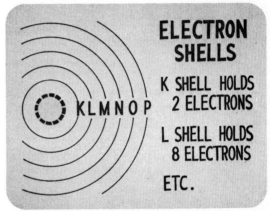

Figure 1.30

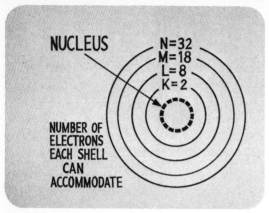

Figure 1.31

The Shell Structure of an Atom — The shell structure of an atom explains a great deal of the electrical and chemical properties of substances and materials you encounter in everyday life. This description of the atom was first set down by the Danish scientist Niels Bohr, and is often called the "Bohr model." Scientists studying matter in greater detail divide the shells into subshells to explain certain phenomena. In your study, however, a brief discussion of only the main shells will be necessary. Several important points have been made about atomic shells that can be discussed in a little more depth. First of all, in each specific atom, each shell has a maximum number of electrons it can accommodate. The shell closest to the nucleus, the K shell, can contain two electrons. The next succeeding shells can, in general, accommodate more electrons depending on their position away from the nucleus: the L shell, 8; the M shell, 18; the N shell, 32; etc. (Figure 1.31).

There are two simple rules concerning electron shells which make it possible to predict the electron distribution of any element:

1. The maximum number of electrons that can fit into the outermost shell of any atom cannot be more than eight.
2. The maximum number of electrons that can fit into the next to the outermost shell of any atom cannot be more than 18.

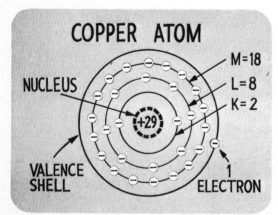

Figure 1.32

Copper Atom — As an example, you can use these rules to determine how the electrons in a copper atom are distributed among their various shells. The copper atom has 29 electrons. Filling the shells from the inside out (Figure 1.32) gives you two electrons in the first (or K shell), eight electrons in the second (or L shell), 18 electrons in the third (M) shell, and only one electron in the outermost shell. This outermost shell, and the number of electrons in it, is one of the key factors determining the electrical properties of an atom.

Valence Shell — The outermost shell of any atom is given the special name *valence shell*, and the electrons in this valence shell are what cause electricity (Figure 1.33).

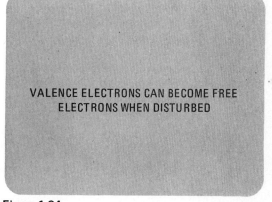

THE OUTERMOST SHELL OF ANY ATOM
IS CALLED THE VALENCE SHELL

Figure 1.33

Valence Electrons — *The key point in this lesson is that the outermost, or valence, shell in some atoms contains electrons which can easily break free of the atom when disturbed* (Figure 1.34). Electrons which have broken free from the atom's valence shell are called *free electrons*. These negatively charged free electrons are what cause electricity.

VALENCE ELECTRONS CAN BECOME FREE
ELECTRONS WHEN DISTURBED

Figure 1.34

Valence Electrons — Because the valence shell and the electrons contained in it are so important in determining the electrical and chemical properties of an atom, you may want to know more about these subjects. The number of electrons in the valence shell (called aptly enough valence electrons) determine the atom's ability to gain or lose electrons. The ability of an atom to gain or lose electrons determines the atom's electrical and chemical properties. If an atom is lacking only one or two electrons in its valence shell, that is if its valence shell is almost full, it will be very difficult to remove any electrons from this shell and make them free electrons. Freeing electrons from atoms is also difficult in atoms whose valence shell (remember that's the outermost shell) is exactly full. If, however, the atom's valence shell contains only one or two electrons, in general it is quite easy to remove these from the atom. For this reason, as will be shown, atoms having only one or two valence electrons generally make the best *conductors* of electricity. Now you may ask another question, how do these electrons get "free"? What causes an atom to lose an electron and what happens to that atom after the loss?

Three Ways an Electron can be Disturbed — As shown in Figure 1.35 an atom can lose an electron if it is disturbed in one of three ways: by heat, by light, or by an applied field or electrical disturbance. If an electron is disturbed in one of these three ways and acquires enough energy, it can break free of the influence of its atom and become a free electron.

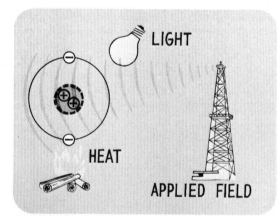

Figure 1.35

Positive Ion — Notice that when an electron is freed, the *atom* left behind is no longer electrically neutral. It now has more positive protons in its nucleus than it has orbiting negative electrons. Therefore, this atom will now have a net positive charge. Unbalanced atoms such as this are called *positive ions* (Figure 1.36).

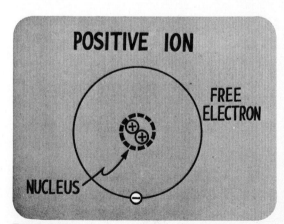

Figure 1.36

- **Balanced Lithium Atom**
- **Lithium Ion**
- **Negative Ion**

Balanced Lithium Atom — Go back and consider as an example of an electrically neutral atom, the lithium atom, which has three protons in the nucleus and three orbiting electrons (Figure 1.37). The net electrical charge for the atom is zero, and thus this balanced atom is electrically neutral.

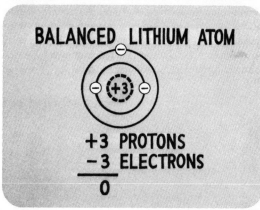

Figure 1.37

Lithium Ion — If some heat, light, or applied electrical disturbance were to cause one of the orbiting electrons to break away, the lithium atom would still retain the +3 charge in its nucleus for the three protons (Figure 1.38). Its negative charge, however, would now be only a −2. This leaves the atom with a net charge of +1. The atom has now become a positive ion.

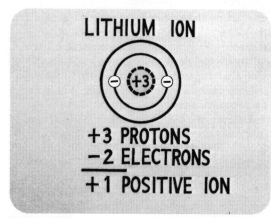

Figure 1.38

Negative Ion — In contrast, if a balanced atom should gain an additional electron, instead of losing one, it would then have a net charge of −1. As shown in Figure 1.39, it would then be called a negative ion.

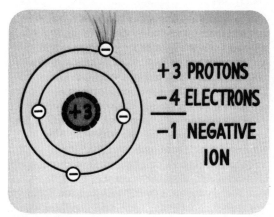

Figure 1.39

- Positive or Negative Ions
- The Ability to be a Good Conductor
- Periodic Table

Positive or Negative Ions — An *ion* is, therefore, an *electrically charged atom*, and it can have either a positive or negative net charge. An atom becomes a positive ion if it gives up electrons or a negative ion if it acquires electrons (Figure 1.40).

AN ATOM BECOMES

1. A POSITIVE ION IF IT GIVES UP ELECTRONS

2. A NEGATIVE ION IF IT ACQUIRES ELECTRONS

Figure 1.40

The Ability to be a Good Conductor — As has been mentioned, different elements have varying abilities to give up and to acquire electrons. This ability depends largely on how many electrons each atom has in its valence shell. This is what determines the ability of an element or object to be a good conductor of electricity. *Good conductors* of electricity are materials which have *many free electrons*.

Periodic Table — Conductors and nonconductors can be better understood by examining how elements and their valence shell structures are classified on the *periodic table of elements*. The shell structure of atoms shows a great deal of order. All of the elements found in nature fall into a position on this table as shown in Figure 1.41.

On the vertical left-hand side of the table are numbers which indicate *how many shells each atom contains*. This number is called the *period* of the element. The vertical columns have numbers at the top which indicate the *number of valence electrons* in the atoms within each column. Each of these vertical columns is called a *group*. You may want to know more about the various elements, so

Periodic Table

INERT GASES

GROUP 0 →		I	II	III	IV	V	VI	VII	VIII
PERIOD 1	(He) HELIUM 2	HYDROGEN 1 (H)							
2	(Ne) NEON 10	LITHIUM 3 (Li)	BERYLLIUM 4 (Be)	BORON 5 (B)	CARBON 6 (C)	NITROGEN 7 (N)	OXYGEN 8 (O)	FLUORINE 9 (F)	
3	(Ar) ARGON 18	SODIUM 11 (Na)	MAGNESIUM 12 (Mg)	ALUMINUM 13 (Al)	SILICON 14 (Si)	PHOSPHORUS 15 (P)	SULFUR 16 (S)	CHLORINE 17 (Cl)	
4	(Kr) KRYPTON 36	POTASSIUM 19 (K) (Cu) COPPER 29	CALCIUM 20 (Ca) (Zn) ZINC 30	SCANDIUM 21 (Sc) (Ga) GALLIUM 31	TITANIUM 22 (Ti) (Ge) GERMANIUM 32	VANADIUM 23 (V) (As) ARSENIC 33	CHROMIUM 24 (Cr) (Se) SELENIUM 34	MANGANESE 25 (Mn) (Br) BROMINE 35	IRON 26 COBALT 27 NICKEL 28
5	(Xe) XENON 54	RUBIDIUM 37 (Rb) (Ag) SILVER 47	STRONTIUM 38 (Sr) (Cd) CADMIUM 48	YTTRIUM 39 (Y) (In) INDIUM 49	ZIRCONIUM 40 (Zr) (Sn) TIN 50	NIOBIUM 41 (Nb) (Sb) ANTIMONY 51	MOLYBDENUM (Mo) 42 (Te) TELLURIUM 52	TECHNETIUM (Tc) 43 (I) IODINE 53	RUTHENIUM 44 RHODIUM 45 PALLADIUM 46
6	(Rn) RADON 86	CESIUM 55 (Cs) (Au) GOLD 79	BARIUM 56 (Ba) (Hg) MERCURY 80	57 – 71* (Tl) THALLIUM 81	HAFNIUM 72 (Hf) (Pb) LEAD 82	TANTALUM 73 (Ta) (Bi) BISMUTH 83	TUNGSTEN 74 (W) (Po) POLONIUM 84	RHENIUM 75 (Re) (At) ASTATINE 85	OSMIUM 76 IRIDIUM 77 PLATINUM 78
7		FRANCIUM 87 (Fr)	RADIUM 88 (Ra)	ACTINIUM 89 (Ac)	THORIUM 90 (Th)	PROTACTINIUM (Pa) 91	URANIUM 92 (U)	NEPTUNIUM 93 (Np)	PLUTONIUM 94 AMERICIUM 95 CURIUM 96 96 – 106†

*LANTHANUM AND THE LANTHANONS   † ACTINIDE SERIES

Figure 1.41

consider the table and look at a few of the important groups.

Listed in Group 0 on the left-hand side of the periodic table are the inert gases, which will not combine chemically with any other element. These elements are very reluctant to give up any electrons. In fact, they act as if they had no valence electrons to give up at all.

Listed in Group I is hydrogen, symbolized with an H. As mentioned before, hydrogen has one orbiting electron, which is considered its valence electron. Elements in Group I have only one valence electron, which can be removed easily; therefore, elements in this group are good conductors. The good conductors in this group are copper, symbolized Cu; silver (Ag); and gold (Au); which have one valence electron each.

Listed in Group VIII are atoms with eight valence electrons. These atoms generally are more reluctant to give up electrons because their outer shells are full. These elements, therefore, are considered poor conductors.

Between these two extremes of good conductors on the left and poor conductors on the right, are the *semiconductors* in the central group, Group IV. These include elements such as silicon (Si) and germanium (Ge), the materials from which transistors are made. You might be familiar with the fact that transistors and other components manufactured with silicon or germanium are commonly called semiconductor devices.

Random Movement of Electrons — In a good conductor like copper wire, there are millions and millions of free electrons available. These electrons are free of their atoms and hence, if left alone, drift randomly throughout the wire.

Free Electrons in Copper Atoms — Each copper atom has one electron in its valence shell which will quite easily become free. When many copper atoms are joined together to form a piece of wire, the atoms interact with each other in what is known as a *metallic* bond. In the process of interacting with each other, each atom gives up its valence electron, creating a huge cloud of free electrons that roams randomly through the body of the wire.

Controlled Movement of Electrons — If the movement of these free electrons in the wire can be controlled so that they do not drift aimlessly and randomly, *electricity* can be produced; that is, the electrons can be made to flow in an orderly and prescribed manner (Figure 1.42).

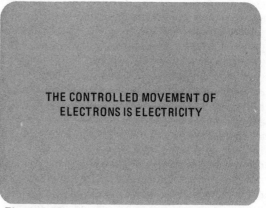

THE CONTROLLED MOVEMENT OF ELECTRONS IS ELECTRICITY

Figure 1.42

Definition of Electricity — At this point in your work you should understand fairly well the definition of electricity. Electricity is the flow of electrons through simple materials and devices. What electrons are, where they are commonly located in nature, and how they become free to move have been discussed. The "simple devices" mentioned are really devices that contain free electrons capable of conducting electricity. These devices can include wires, switches, light bulbs, and others. The study of electron flow through more complex devices is known as *electronics*. This area of study includes devices that *amplify, process*, or *transmit* electrical signals, such as transistors, tubes, or integrated circuits.

In this coursework, you are starting at the beginning with simple devices.

Opposite Charges — How can you create a flow of electrons through a copper wire rather than just a random motion? The answer to this question goes back to the earlier discussion of charged objects. If you recall, it was demonstrated that two oppositely charged bodies always attract each other, and two objects with the same charge repel each other. This was summed up in the first law of electrostatics: like charges repel and unlike charges attract. Also recall that the two kinds of charges found in nature have been named *positive* and *negative*. A body with a negative charge has an excessive amount of electrons; a body with a positive charge has a lack of electrons. So, if a negatively charged and a positively charged object are brought close to each other, they will attract one another (Figure 1.43).

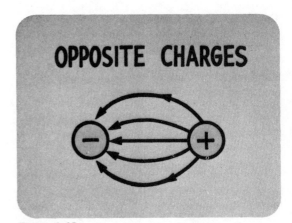

Figure 1.43

Potential Difference — Here's an important *new* term. Whenever this type of electrostatic force exists between two objects, it is said that a *potential difference* exists between them. *Potential difference* is the term used to describe how large the force acting between the two bodies will be, this will be discussed more fully in the next lesson. For now, focus your attention on this idea: if a charged body were to be placed between two objects having a potential difference, this charged body would try to move in one direction or the other. Figure 1.44 shows what would happen if an electron were put between a positively charged and a negatively charged body. The action of the potential difference between the two bodies would be to push the electron from left to right. The electron, being negative itself, is repelled from the negative body and attracted to the positive body.

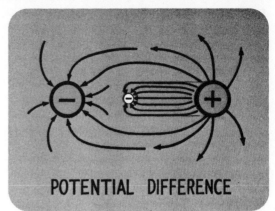

Figure 1.44

Copper Wire — If a copper wire were placed between two oppositely charged bodies, as shown in Figure 1.45, a potential difference would exist between the ends of the wire. All the negatively charged free electrons in the wire will feel a force pushing them from the negative to the positive charge (left to right), opposite to the conventional direction of the electric field lines.

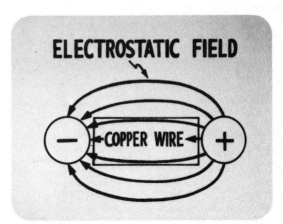

Figure 1.45

Drift of Electrons from Negative to Positive — If you actually touch the charged balls to the ends of the copper wire, as shown in Figure 1.46, what will happen to the free electrons in the wire? The free electrons near the *negatively charged ball*, because they possess a negative charge, and since like charges repel, are repelled and drift through the wire. These same negative electrons feel the effects of the positively charged ball at the other end and are attracted to it. Thus, *the electrons move through the conductor under the influence of a potential difference.* Instead of drifting randomly, these electrons move from the negative end to the positive end. This movement or flow continues until the two charged balls have equalized and possess the same charge. In other words, what's happening is that electrons leave the negative ball and move through the wire to the positive ball. When enough electrons have moved from the negative to the positive ball, the system will be electrically neutral (assuming the positive and negative charges on the balls were of the same size). When this happens, the electrons in the wire will return to their random drifting motion as shown in Figure 1.47. At this point, the potential difference no longer exists between the ends of the wire.

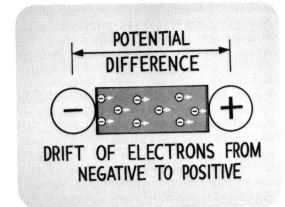

Figure 1.46

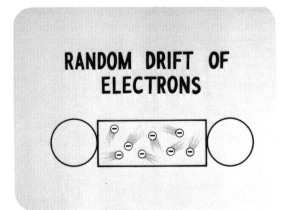

Figure 1.47

Electron Motion — While the electrons are moving under the influence of the potential difference, they don't just scoot through the wire as if they were in a peashooter. They still bounce around in irregular paths, similar to random motion, except that the force acting on them makes them drift in a specific direction. As the electrons flow, they undergo collisions with various obstacles in the wire and don't really proceed in one neat, orderly line from end to end. The chances are very small that even one single electron can make it through a small piece of wire without stopping. As a matter of fact, if you could sit on one electron and ride it through the wire, you would probably be surprised at what a long, bumpy ride it would be. Although it takes a long time for one electron to go all the way through the wire, some electrons are already near the end. So when the potential difference is applied, electrons move out the end of the wire almost instantly, even though each electron is actually moving fairly slowly through the wire.

Now, a device could be constructed which would *maintain a potential difference*, a continuous flow of electrons could be maintained through a copper wire.

A Battery — One such device that most of you are familiar with is called a *battery*. A battery can maintain a potential difference between the ends of a wire by a chemical action. The battery will be discussed more fully in the next lesson, but for now all you need to know is that a battery creates a potential difference and maintains a negative charge at its negative terminal and a positive charge at its positive terminal. You might think of a battery as a pump which can cause charge to move. All you must do is place the charges between the potential difference at the terminals.

Throughout your study of electricity, simple line diagrams will be used to illustrate the paths of electron flow. These diagrams are called *schematic diagrams*, and they use straight lines to indicate wires and special standard symbols to illustrate the various devices used to control the flow of electrons. The symbol for a battery in these schematic type diagrams is shown in Figure 1.48. The positive terminal is always shown connected to the long plate and the negative to the short plate on the schematic symbol. More complete discussions of schematic diagrams and symbols will be included as you proceed through this course.

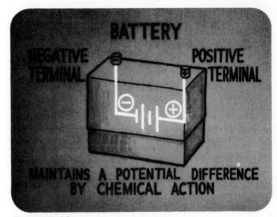

Figure 1.48

A Simple Circuit — As this lesson nears completion, consider another important point. If a copper wire is placed across the terminals of a battery as shown in Figure 1.49, it will provide a *complete path* through which electrons can flow. This is another key concept. In order for electrons to flow at all, they must have a complete path to flow through. This *complete path* is called a *circuit*. When the copper wire is connected, electrons will flow through the circuit because they are forced by the potential difference of the battery. The copper wire across the battery provides a complete circuit.

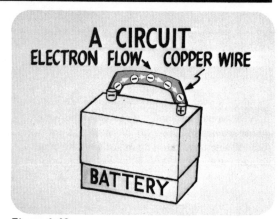

Figure 1.49

An Open Circuit — As shown in Figure 1.50, if the wire is disconnected from the terminal of the battery, the flow will stop. Even though the force on the electrons is still present, they have nowhere to go. Thus, charges build up in the wire until there is no potential difference between the *ends* of the wire, and no more electron flow exists in the wire. This is called an *open circuit*. It is very important to realize that *without a complete path or circuit, there is no flow of electrons*. As you might imagine, the concept of a circuit is an important one in your study and will be discussed further as the course continues.

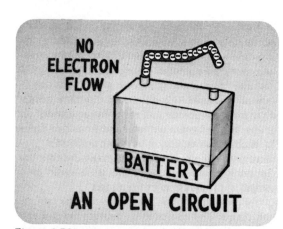

Figure 1.50

Basic Electricity — In this lesson some of the fundamental concepts of electricity, as listed in Figure 1.51, were examined. The structure of an atom and the laws that govern it were described. The laws of electrostatics were discussed, and it was shown how like electrical charges repel and unlike electrical charges attract. It was explained that a normal atom is electrically neutral and that certain atoms can be made to change from their neutral state when disturbed by heat, light, or an applied field. These atoms can either give up or acquire free electrons. Then the normal motion of free electrons was shown to be random. Finally, the concept of a potential difference between two oppositely charged bodies was introduced, and

BASIC ELECTRICITY

- STRUCTURE OF AN ATOM
- LAWS OF ELECTROSTATICS
- FREE ELECTRONS
- POTENTIAL DIFFERENCE
- FLOW OF ELECTRONS

Figure 1.51

how a potential difference can cause electrons to flow in an orderly and prescribed manner was discussed.

Electron Motion and Resistance: Lab Exercise 1 — The actual picture of electron motion was briefly discussed. The electrons in motion through a wire or device under the influence of a potential difference actually do a great deal of bouncing around, as they drift from minus to plus. In other words, they encounter *opposition* to their flow. How much opposition an electron will encounter as it moves through a substance depends on what that substance is, and the conditions under which the electron is trying to get through it. Because different substances and devices offer different amounts of opposition to electron flow, a uniform way is needed to measure and describe this opposition. This brings up several important new ideas which are discussed in your first laboratory exercise. The *opposition to electron flow* that electrons feel when they pass through a wire or device is called *resistance*. The units used to measure the resistance of a device or wire are called *ohms*. The more ohms of resistance a device has, the more difficulty electrons will have flowing through it. These terms are introduced for you in laboratory exercise 1, along with an important device called an *ohmmeter*, which allows you to easily measure resistance in the laboratory. Here's an extremely important note: The concepts of resistance measured in ohms and electron flow, or electron *current* as it's called, are major topics discussed in Lesson 2. Be sure to relax and enjoy your first laboratory exercise, and if some of the new concepts you are working with appear unclear, you will find them fully discussed in Lesson 2 which covers voltage, current, and resistance.

LESSON 1. AN INTRODUCTION TO ELECTRICITY

● **Worked Through Examples**

The basic concepts and definitions presented in this lesson provide the groundwork for your later studies in electricity and electronics. The following examples are designed to set you thinking in a little more detail about some of the concepts and terms that have been discussed.

1. Sketch the structure of a sodium atom showing the position of its 11 protons, 11 electrons, neutrons, nucleus, shells, valence shell, and show which electron would most likely become a free electron if the atom were disturbed. Would you expect sodium to be a good conductor?

   Solution: Following the picture of atomic structure given in this lesson, the atom would appear as follows with all its parts properly labeled:

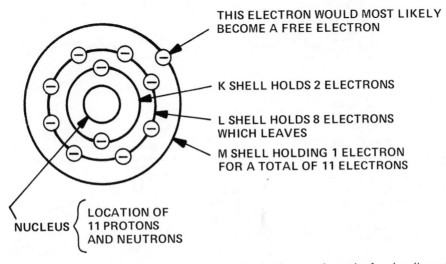

   THIS ELECTRON WOULD MOST LIKELY BECOME A FREE ELECTRON

   K SHELL HOLDS 2 ELECTRONS

   L SHELL HOLDS 8 ELECTRONS WHICH LEAVES

   M SHELL HOLDING 1 ELECTRON FOR A TOTAL OF 11 ELECTRONS

   NUCLEUS { LOCATION OF 11 PROTONS AND NEUTRONS

   Since there is only one electron in the outermost shell of this atom, it can be freed easily and you would expect sodium to be a good conductor.

2. A negatively charged rod is brought near one end of an isolated (originally neutral) copper bar as shown. Describe what happens to the free electrons in the bar.

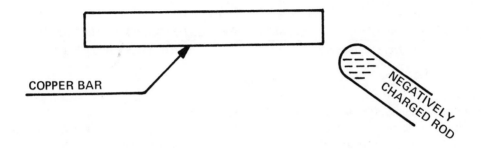

   COPPER BAR

   NEGATIVELY CHARGED ROD

Solution: The free electrons in the bar that are closest to the charged rod will be repelled from the rod to the other side of the bar; therefore, electrons will flow in the bar from right to left. Flow will continue until there is a negative charge on the left-hand side and positive charge on the right-hand side of the bar. Eventually, the opposing charges on the opposite ends of the bar will build up to a point where charge flow will stop. At the point where charge flow has ceased, the charge buildup in the bar is pushing electrons to the right, just balancing the charged rod's push to the left, as shown here.

Electrons flow for a short while

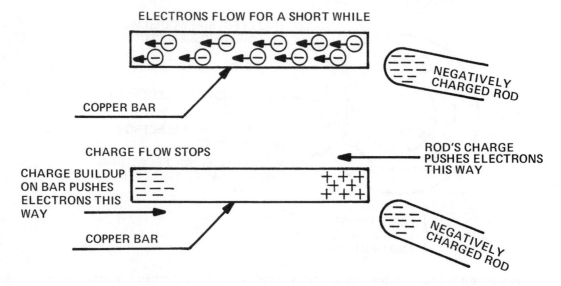

3. A copper penny contains roughly 29,000,000,000,000,000,000,000 atoms. Approximately how many coulombs of free electrons does it contain?

   Solution: Since copper has one valence electron, for each atom you can assume that one free electron is released into the penny. If you also assume that one coulomb equals 6.25 billion billion electrons, then you may calculate:

$$\frac{\text{No. of electrons/penny}}{\text{No. of electrons/coulomb}} = \frac{\text{No. of coulombs}}{\text{penny}} = \frac{29{,}000{,}000{,}000{,}000{,}000{,}000{,}000}{6{,}250{,}000{,}000{,}000{,}000{,}000}$$

$$= 4{,}640 \text{ coulombs/penny}$$

   Notice that handling numbers this large can be quite cumbersome, with many zeros to keep track of, and many opportunities to make a mistake. Later on in this course, in Lesson 3, you'll learn the use of scientific notation (or powers of ten notation) which will make handling numbers this size a much easier task.

4. Two fixed metal spheres are given charges as shown, and an electron is placed between them.

    a. Draw in the lines of force near these bodies
    b. Describe the motion of the electron

    Solution:

    a.

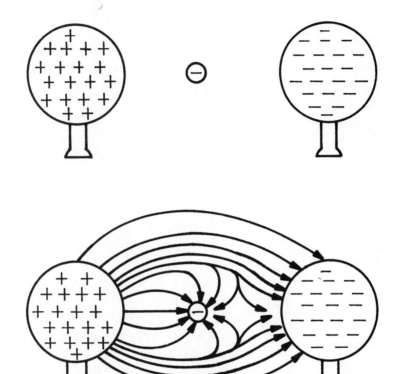

    b. The electron is pushed from right to left in the direction of the positive sphere.

LESSON 1. AN INTRODUCTION TO ELECTRICITY

• **Practice Problems**

To help you review some of the important definitions introduced in this lesson, answer these questions by placing the letter of the correct answer in the column provided. Fold over page to check your answers.

Fold Over

Answer

1. Like charges ___ each other.
   A. attract
   B. repel
   C. destroy
   D. neutralize
   E. none of the above

   1.

2. When a balanced atom loses electrons, it becomes a ___ ion.
   A. positive
   B. negative
   C. neutral
   D. potential
   E. none of the above

   2.

3. A balanced atom is one which:
   A. contains one neutron for every electron
   B. contains even numbered electrons
   C. has one electron for every proton
   D. has one proton for every neutron
   E. none of the above

   3.

4. Objects containing more positive ions than negative ions have a ___ net charge.
   A. positive
   B. neutral
   C. negative
   D. ion
   E. none or all of the above

   4.

5. If two objects both have an excess of electrons and are brought together electrons will flow to the object with the ___ negative charge.
   A. greater
   B. same
   C. lesser
   D. makes no difference — it is a 50-50 chance
   E. none or all of the above

   5.

Answers

1. B

2. A

3. C

4. A

5. C

	Answer	Fold Over

6. If a material lacks one coulomb of electrons, it has a ____ charge.
   A. positive
   B. neutral
   C. negative
   D. ion
   E. none of the above

6.

7. Electrons always flow toward the more ____ charge.
   A. ionized
   B. negative
   C. neutral
   D. positive
   E. greater

7.

8. The outermost shell of any atom is called its ____ .
   A. orbit
   B. M shell
   C. valence shell
   D. complete path
   E. circuit

8.

9. The name given to the common unit of charge, equivalent to 6.25 billion-billion electrons is:
   A. one ion
   B. one ampere
   C. one coulomb
   D. one neutron
   E. one bundle

9.

10. Which of the following neither attract nor repel each other?
    A. 2 electrons
    B. an electron and a neutron
    C. a proton and an electron
    D. 2 protons
    E. a proton and a positive ion

10.

Answers

6. A

7. D

8. C

9. C

10. B

1. All atoms have:

   a. The same three types of particles
   b. The same number of protons
   c. A nucleus and orbiting electrons
   d. Protons that attract positive charges
   e. Electrons that are attracted by a positive charge
   f. a, b, c and e above
   g. a, c, d, above
   h. c and e above

2. Atoms are:

   a. The smallest part into which an element can be divided
   b. Basic building blocks of all materials
   c. Always have a neutral charge
   d. Have at least one nucleus and one electron
   e. All of the above
   f. a, b and d above

3. The first law of electrostatics states that:

   a. Neutrons will repel each other
   b. Electrons must stop before proceeding
   c. Like charges repel each other
   d. Unlike charges attract each other
   e. a and b above
   f. c and d above
   g. all of the above

4. Lines of Force Illustrating an Electrostatic Field:

   a. Exist around a charged object
   b. Point away from a positively charged body
   c. Point toward a negatively charged body
   d. Start on a positive charge and end on a negative charge when unlike charges attract each other
   e. Start on a positive charge and end on a negative charge when unlike charges attract each other
   f. All of the above

5. Electricity, which is defined as electrons flowing through simple materials and devices:

   a. Depends on free electrons for its flow
   b. Depends on good conductors which have free electrons
   c. Flows easily through materials that are called insulators
   d. Depend on valence electrons to become free electrons
   e. Can flow easily in materials that have completely full valence shells.
   f. a, c and e
   g. a, b and d

6. An atom is more likely to lose an electron if it has:

   a. One electron in its valence shell
   b. One proton in its valence shell
   c. Six electrons in its valence shell
   d. Eight electrons in its valence shell

7. The terms "orbit" and "shell" both describe

   a. Protons and electrons
   b. Protons only
   c. Electrons only
   d. Protons, electrons and neutrons

8. An electrically balanced atom that loses one or more electrons becomes:

   a. A positive ion
   b. A negative proton
   c. A positive electron
   d. A negative ion

9. If two objects both have an excess of electrons and are brought together, electrons will flow to the object with the:

   a. Greater negative charge
   b. Greater number of electrons
   c. Lesser number of neutrons
   d. Lesser negative charge

10. In an open circuit:

    a. Maximum electron flow occurs
    b. There is no electron flow
    c. There is maximum proton flow
    d. There cannot be any potential difference anywhere in the circuit

11. Good conductors can be identified:

    a. By examining the periodic table
    b. By the free electrons available in the material
    c. By the negative ions in the material
    d. Are usually in Group I and II of the periodic table
    e. a, b and c above
    f. a, b and d above

12. A positively charged body has:

    a. An excess of electrons
    b. An excess of neutrons
    c. An excess of protons
    d. A lack of electrons
    e. All of the above
    f. c and d above
    g. d only

13. An object that has a negative charge has:
    a. An excess of electrons
    b. An excess of neutrons
    c. An excess of protons
    d. A lack of electrons
    e. All of the above

14. In a closed circuit with electricity flowing:
    a. All switches are open
    b. All materials used are good insulators
    c. Electrons move through the circuit conductors under the influence of a potential difference
    d. Positive charges are flowing in the conventional way

15. A battery is:
    a. A device that opens a circuit
    b. A device that causes electron flow to stop
    c. A device that maintains a status quo
    d. A device without terminals
    e. A device that maintains a potential difference by chemical action

16. In a closed circuit:
    a. Electrons flow because there is a potential difference
    b. There is a complete path
    c. The potential difference pumps the free electrons
    d. All of the above

17. A Coulomb is:
    a. A unit of charge containing 6.25 x 10^{18} electrons
    b. A unit charge containing 6.26 x 10^9 protons
    c. A unit of charge containing 6.25 x 10^{18} neutrons
    d. None of the above

18. Insulators:
    a. Are materials that have very few free electrons
    b. Cause only a very small electron flow when placed across a potential difference
    c. Are usually found in Group VIII of the periodic table
    d. Have their outer valence shell full of electrons
    e. All of the above

19. Ions are formed by atoms:
    a. Gaining a nucleus
    b. Gaining a neutron
    c. Losing a nucleus
    d. Gaining or losing electrons
    e. None of the above

20. If a closed circuit is placed between two oppositely charged bodies:
    a. Electrons will flow from the positive charge to the negative charge
    b. Protons will flow from the positive charge to the negative charge
    c. Electrons will flow from the negative charge to the positive charge
    d. Atoms will flow from one charge to another
    e. All of the above

Lesson 2

Voltage, Current, and Resistance

This lesson introduces, explains, and illustrates the key concepts of *voltage, current, and resistance. Conventional* current and *electron* current are discussed and shown to be equivalent. A discussion of *resistance* and *resistors* includes the factors that affect resistance, the resistor *color code*, and *wattage ratings* of resistors.

LESSON 2. VOLTAGE, CURRENT, AND RESISTANCE

● **Objectives**

This lesson continues studying the basics of electricity by introducing the key concepts of voltage, current and resistance. At the end of this lesson the student should be able to:

1. *Write* an explanation of voltage, current, and resistance including the units used to measure each and how they act together in circuits.

2. *Write* a detailed explanation of what a circuit is, including sketching a simple schematic diagram showing a battery, wires, and a resistor.

3. *Write* a description of *conventional* current, how it differs from *electron* current.

4. *List* the 4 factors that determine the resistance of a conductor.

5. Given a color code chart, *identify* the resistance and tolerance of any given resistor.

6. *Write* an explanation of the wattage rating of a resistor and why it's important.

- **Electricity**
- **The Atom**
- **First Law of Electrostatics**

Electricity — In the previous session you were introduced to the basic forces at work in creating and moving electricity. This discussion begins with a brief review of the basic topics covered in that lesson. First, recall that electricity was *defined* as the *flow of electrons through simple materials and devices* (Figure 2.1).

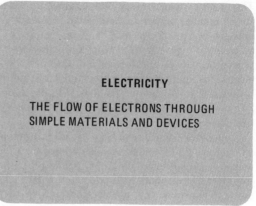

Figure 2.1

The Atom — The origins of electricity were discussed including a picture of the structure of the *atom*. Recall that atoms are the basic building blocks of elements, and that they are made up of a positively charged nucleus, surrounded by orbiting negatively charged electrons, as shown in Figure 2.2.

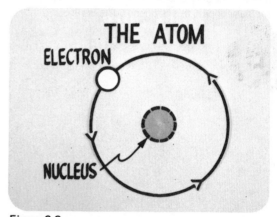

Figure 2.2

First Law of Electrostatics — Then you saw that charged particles such as electrons and protons, exhibit an *electrostatic force* between them. The action of this force is described by the first law of electrostatics which as shown in Figure 2.3 states:

Like charges repel, unlike charges attract.

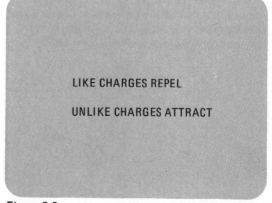

Figure 2.3

The Electric Field Between Two Charged Bodies —
Recall that the pattern of electrostatic forces
between charged bodies may be illustrated by
diagrams of the electric *field* surrounding charged
objects using *lines of force*. An illustration of the
electric field between two charged bodies is shown
in Figure 2.4. Notice that the lines of force *seem to
pull* the unlike charges together and to push the
like charges apart.

Since all electrons are alike and all protons are
alike, the only difference between atoms is the
makeup of their nucleii and the number of orbiting
electrons they contain. It's important in your
study to remember that the electrons in the
outermost, or *valence*, shell of some atoms can be
easily removed and become "free" electrons. In
atoms with only one electron in the valence shell,
such as copper, this happens quite easily.

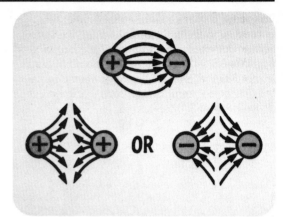

Figure 2.4

Conductors and Insulators — The availability of
free electrons makes materials such as copper good
conductors of electricity. Materials with *very few*
free electrons, on the other hand, are called
insulators (Figure 2.5).

MATERIALS WITH MANY FREE ELECTRONS
ARE GOOD CONDUCTORS

MATERIALS WITH FEW FREE ELECTRONS
ARE INSULATORS

Figure 2.5

- **Electricity**
- **The Coulomb**
- **Objectives for Lesson 2**

Electricity — Then you learned that if a conductor is placed between two charged bodies, the free electrons inside it will feel a force and begin to flow. When electrons are flowing through a *simple* material or device, this is electricity (Figure 2.6).

ELECTRICITY IS THE FLOW OF ELECTRONS THROUGH SIMPLE MATERIALS OR DEVICES.

Figure 2.6

The Coulomb — Remember, since the electron is a very small quantity of charge, it is often convenient to speak of large groups of electrons in the study of electricity (Figure 2.7). The coulomb will be a convenient unit of charge for you to use, and is defined as 6.25 billion billion electrons.

A COULOMB EQUALS 6.25 BILLION BILLION ELECTRONS

Figure 2.7

Objectives for Lesson 2 — This lesson goes on to further discuss electricity. As the objectives state, you'll be introduced to some important new concepts as summarized in Figure 2.8. Among the most important of these basic objectives is the introduction and discussion of the terms voltage, current, and resistance and the *units* used to measure these quantities.

Remember also that a short quiz is included at the end of each lesson. If the objectives of the lesson have been satisfactorily achieved, you should score well on this quiz. If you do not score well, the quiz can serve as a guide to show you where review of the lesson may be in order.

OBJECTIVES FOR LESSON 2

STUDENT UNDERSTANDING OF:

- VOLTAGE
- CURRENT
- RESISTANCE
- CIRCUIT
- CONVENTIONAL CURRENT
- ELECTRON CURRENT
- COMPONENTS
- RESISTOR COLOR CODE

Figure 2.8

- **Voltage**
- **Electron Flow**

Voltage — Consider the first of these important new concepts: *Voltage*. In the previous lesson you learned how free electrons are produced in a conductor and how these electrons can be made to move by applying a *potential difference* or *voltage* across the conductor. As you proceed through this lesson, you'll see that the terms *voltage* and *potential difference* are often used interchangeably.

Now as shown in Figure 2.9, one way this potential difference could be applied would be to place charged bodies on either side of the conductor. The electrons in the conductor would all feel a force, due to the field set up by these bodies, and they would begin to move from the negative to the positive end of the wire. The electrons are repelled from the negatively charged body, and attracted to the positively charged body.

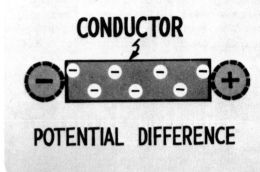

Figure 2.9

Electron Flow — In the term *"potential difference"*, the word "difference" comes about because more positively charged bodies are said to be at *higher potential* than more negatively charged bodies. In any situation where there are free electrons, and a path exists which they can flow through, when two *differently* charged objects are placed at either end, electrons will begin to flow. The charged objects provide the force or push that moves these electrons.

The amount and type of electrical charge on a given object determines what is called its electric *potential*. The more charge an object has, the more *potential* it has to create electron movement. Now in any complete electrical path, *two* charges will be involved, one at either end of the path. So actually, it's the *difference in potential*, or potential difference between the 2 charged objects that provides the force that moves the electrons. (Notice that if both objects at either end of the conductor had the exact same charge and exact same potential, the push they exerted on the electrons would cancel out and there would be no electron flow.)

So whenever a potential difference is created across a conductor, the electrons will flow from low to high potential (Figure 2.10).

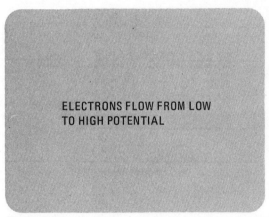

Figure 2.10

Rapid Flow of Electrons — For example, if the charged bodies discussed previously are touched to the conductor as shown in Figure 2.11, electrons quickly flow from the negatively charged body which is at a lower potential, through the wire to the positively charged one, which is at a higher potential, until both bodies have no excess charge.

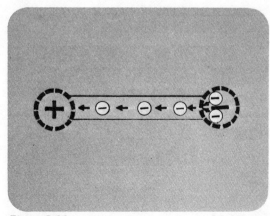

Figure 2.11

Random Motion of Electrons — At this point, both of the bodies are at the *same potential*, and the electrons resume their normal random motion as shown in Figure 2.12. Notice that as long as there is no *potential difference* between the two objects, electrons between them will feel no force, and there will be no electron flow.

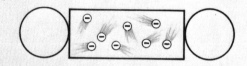

Figure 2.12

Figure 2.13

Figure 2.14

Potential Differences Between Like Charges — Notice that electrons will flow anytime a potential difference exists between the ends of a conductor. This potential difference can be created by *two positively charge bodies,* or *two negatively charged bodies;* as well as *two bodies with different types of charge.* For example, if one body has a much higher positive potential than another, and these bodies are placed on either side of a conductor, electrons will flow from the lower to the higher positive potential (Figure 2.13) until there's no potential difference between the bodies. Similarly electrons will feel the potential difference between bodies with different negative potentials and they will flow from the body with the greater negative charge to the body with the lesser negative charge (Figure 2.14), again from low to high potential.

In all the simple examples discussed thus far, charged bodies were used to create the potential difference that "pushes" on the electrons and causes them to flow.

Actually, these bodies would not be very practical for sustaining electron flow because as charge flows between them they rapidly reach a condition where *no* potential *difference* exists between them.

Battery — Another way to create a potential difference or voltage and to maintain it across a conductor such as a wire is to connect a *battery* across it as shown in Figure 2.15. A battery relies on the chemical action which occurs between two different metals called electrodes which are immersed in either a salt or acid solution called an electrolyte. As the chemicals react, the electrodes accumulate opposite charges. If a conductor is connected between the two electrodes, electrons will leave the negative terminal and flow through the conductor to the positive terminal. This is called *discharging* the battery and the ultimate result would be its destruction.

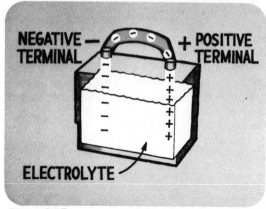

Figure 2.15

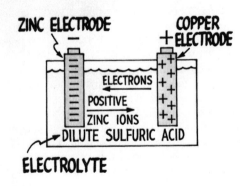

Figure 2.16

The battery (or *cell*) was originally discovered by an Italian scientist, Allessandro Volta. The construction of one type of simple "Voltaic cell" is shown in Figure 2.16. Basically, 2 electrodes, one of zinc and one of copper, are immersed in a dilute solution of *sulfuric acid* called an electrolyte. Through the action of the electrolyte several interesting effects occur. The acid acting on the zinc electrode causes some of it to dissolve. Many *positive* zinc ions leave the zinc electrode. Since the zinc *loses* positive charges, this leaves the zinc electrode with an abundance of negative charges, making it the negative terminal of the battery. Meanwhile, the acid acting in the electrolytic solution creates an abundance of positive hydrogen ions. These come in contact with the copper electrode and pull many electrons from it, leaving it with a *positive* charge. Thus the copper electrode becomes the *positive* terminal. If a conductor is placed across these terminals, electrons will flow from the negative to the positive terminal until the cell is destroyed. This type of cell cannot be "recharged" or restored to its original condition and is called a "primary cell".

Dry Cell — A battery that is more convenient to use, but employs similar principles is the dry cell shown in Figure 2.17. This is another common form of the simple primary voltage cell. The electrolyte, however, is a damp paste instead of a liquid. The action is identical to the simple battery just discussed, but this battery is sealed to prevent leakage and drying out of the electrolyte, and it can be used in any position.

Figure 2.17

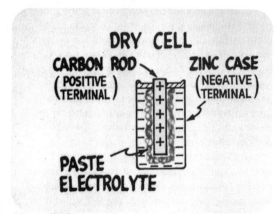

Figure 2.18

The internal parts of a dry cell are enclosed in a cylindrical zinc container which actually serves as the cell's negative electrode (Figure 2.18). This container is lined with a non-conducting material, such as blotting paper, which separates the zinc from the paste electrolyte. A carbon or high-carbon metal center electrode runs right into the electrolyte paste. The paste itself is made up of a variety of materials (depending on the manufacturer) and usually includes ground carbon, manganese dioxide, aluminum chloride, and zinc chloride. The electrode material reacts with the paste so that electrons are forced from the center terminal to the outer case, making the case the negative battery terminal and the center electrode positive. The actual negative terminal of the cell may be connected to the can anywhere, such as at the bottom in a typical flashlight cell, or at the top in the dry cell shown in Figure 2.19. These batteries are also called primary cells; once they're discharged, they can't be recharged again.

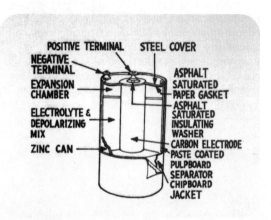

Figure 2.19

The typical automobile battery is called a *lead acid* battery. This type of battery has the advantage of being recharged when electrons are forced through it in the reverse direction.

Rechargeable batteries are called *secondary* batteries and are commonly constructed from lead electrodes in an acid electrolyte or a nickel-iron-alkali or nickel-cadmium-alkaline combination. The common automotive battery (or wet cell) uses lead dioxide as the positive electrode and spongy lead as the negative plate in a sulfuric acid electrolyte. The strength of the electrolyte is measured in terms of its *specific gravity*. (Specific gravity is the ratio of the weight of a given volume of electrolyte to a given volume of water.) By measuring the specific gravity of the electrolyte, you can determine the state of charge of the battery.

In the secondary wet cell, the action of the acid reacting with the spongy lead plates gives them a negative charge and deposits a chemical, lead sulfate, on them. The acid acting with the lead dioxide plates makes them positive and creates lead sulfate on them. As the battery is *discharged*, the acid in the battery gets weaker becoming less and less concentrated and *lead sulfate* builds up on the plates until the battery gradually becomes inoperative. If at this point electrons are forced to flow through the battery in reverse, the lead sulfate will eventually be removed from the plates, restoring the acid content to the battery's electrolyte. At this point the battery will be recharged and ready for use again.

DC Generator — Another common method used to produce voltage is the dc generator shown in Figure 2.20. It *converts mechanical energy to electrical energy* and can be used in automobiles, boats, airplanes, and fixed sites where a source of potential difference is required. The theory of operation for a dc generator is beyond the scope of this course and will not be necessary in the discussions that follow. In this course, a battery or dc power supply will usually suffice as a source of voltage and current.

Figure 2.20

Potential Difference — Voltage, or potential difference, can be produced by batteries or generators, but what is voltage and what does it do? The simple key to voltage is that a greater potential difference across a conductor produces a greater force on the electrons and as a result, a greater rate of electron flow. Potential differences of many different "sizes" will be discussed in this course, so a unit is needed to easily specify the *size*. The *size* of potential difference is measured in volts. As shown in Figure 2.21, the *unit of measure for potential difference is the volt*.

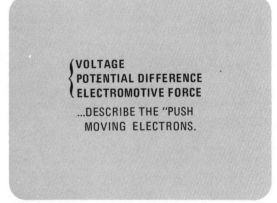

Figure 2.21

Voltage — Volts or voltage is a measure of the *push* on each electron, and it's this voltage, this pressure or push, that makes the electrons move. So the terms *potential difference* and voltage are often used interchangeably to mean the "push" on the electrons (Figure 2.22). Other words are also used to describe this "push", and so you may see the term electromotive force (EMF), or just the word *potential*, used to describe the electron push in certain instances.

Figure 2.22

Here's another important point. Often when writing voltages in electrical formulas, the letter E is used to symbolize voltage or potential difference, while the letter V is more commonly used to abbreviate the unit of voltage, *volts*. Instead of "the potential difference is 5 volts" on a schematic diagram as shown in Figure 2.23, you would more likely see "E = 5V".

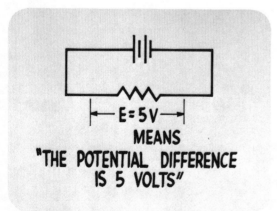

Figure 2.23

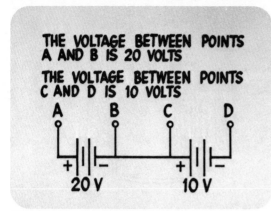

Figure 2.24

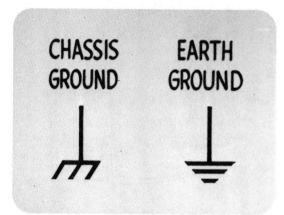

Figure 2.25

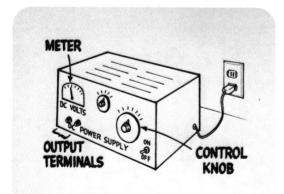

Figure 2.26

Potential Difference Between Two Points and Ground — It's important to recognize at this point that voltages or potential *differences* always involve *two points*. If a potential difference exists between two points and a conductor is placed between them, current will flow. The voltage is always measured or stated as existing between points "A and B", or from "point A to point B". The strength of the potential difference is measured in volts; typically you may say that "the voltage between points A and B is 20 volts", or "the voltage between points C and D is 10 volts", etc. (Figure 2.24).

Often it is convenient to pick one point and use it as a *reference* point. Then voltages are measured at other points with respect to that reference point. This is particularly true in complex electronic equipment where voltages between many pairs of points have to be specified. In cases like this, the metal case or *chassis* in which the equipment is wired is usually considered to be the reference point. Voltages are then specified between any specific point and the chassis. When used as the reference point, the chassis is said to be at *ground potential* and the chassis is referred to as a *chassis ground*. Often the earth itself or water pipes which are buried in the earth, are used as the reference point for measuring voltage. A reference point connected to the earth is called an *earth ground*. In schematic diagrams, chassis and earth grounds are indicated by the symbols shown in Figure 2.25. Grounds and grounding will be covered more thoroughly later in the course.

Power Supplies — In a variety of applications, particularly in the laboratory, you will need a convenient, easy-to-use source of dc voltage. For this purpose, a wide variety of dc *power supplies* is available. A power supply is basically a device which is usually plugged into a wall outlet and can replace a battery in many applications by providing a known potential difference between two convenient terminals (Figure 2.26). The power supply may be variable with a knob or similar adjustment which allows you to adjust the voltage, or potential difference, between the terminals, and may include meters to indicate this voltage.

- **Water Pressure**
- **Water Flow**
- **Battery**

Water Pressure — Because electricity is actually a flow of electrons "under pressure", electricity is often compared to water flow. As shown in Figure 2.27, a water faucet holds back water under pressure. This pressure is similar to the voltage in a battery — a measure of how hard the water is pushed. Water pressure is often measured in "pounds per square inch" while electron pressure is measured in volts.

Figure 2.27

Water Flow — As you have seen, a push or voltage is needed to create electron flow, but that's not all that is needed. Consider another important concept, as illustrated in Figure 2.28. If a water faucet is turned on, water flows. There is a push on it, and *opening the faucet gives the water a place to go*.

Figure 2.28

Battery — The water faucet is similar to a battery. In a battery, as shown in Figure 2.29, the potential difference or push also exists, but *no current flows unless a path is provided* through which the electrons can move. Without this path, no electron flow can occur.

Figure 2.29

- **Circuit**
- **Open Circuit**
- **Electron Current**

Circuit — Placing a wire across the terminals of the battery as shown in Figure 2.30 provides a complete path and electrons then flow from the negative terminal through the wire to the positive terminal. This important concept, the complete path needed for current flow, is called a closed or complete *circuit*.

Figure 2.30

Open Circuit — You might wonder what happens when the wire is removed. If the wire is disconnected from one terminal, or it is broken at any point, there is no longer a complete path. This is now called an *open circuit* (Figure 2.31). In open circuits, even though the voltage or push may be there, there is *no electron flow*.

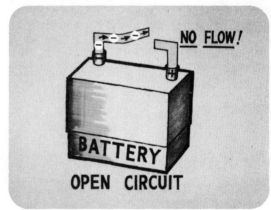

Figure 2.31

Electron Current — As mentioned earlier, if a voltage exists across a completed conducting path, or circuit, electrons flow in this circuit from low to high potential. This flow of electrons is given the name "electron current" (Figure 2.32), and the investigation of this electron current flow will be an important part of your study. Often it will be important to specify *how much electron current* is flowing in a specific conductor at a specific time.

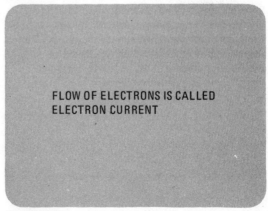

FLOW OF ELECTRONS IS CALLED ELECTRON CURRENT

Figure 2.32

The Ampere — When specifying an amount of current flow, you can't count actual electrons as they go by one point in the circuit. Electrons and their effects are usually too small to detect at a one-at-a-time basis, as mentioned earlier. Recall from the previous lesson that a *coulomb* is a very large *group* of electrons. So when specifying current flow, you examine one point in a circuit and specify how many *coulombs* of electrons go by in one second. This unit of measure for current is the *ampere* (named after Andre Ampere, an early pioneer in the study of electricity). As shown in Figure 2.33, *one ampere* (or amp) of current is flowing past a point in a circuit if *one coulomb of electrons passes by the point in one second.*

1 AMPERE (AMP) EQUALS
1 COULOMB PASSING ANY POINT
IN 1 SECOND

Figure 2.33

Abbreviations and Shorthand — In the study of electricity, you'll find many shortcuts and shorthand types of notation have developed to make the writing and handling of electrical quantities easier. For example, electron current is often referred to as "amperage", and when writing down a quantity of *amperes or amps*, it's usually abbreviated with the letter *A*. Another shortcut is often used when writing current. In circuit diagrams and the equations, that will be discussed later, *current* will most often be designated by the letter *I*. For example, you would find the expression "the current flow is 6 amperes" written more simply as I = 6A.

Simple Circuit — In simple circuits of the type discussed so far in this course, the current *has only one path* to follow, as shown in Figure 2.34. The voltage is provided by the battery, the wire provides a complete path, and so the voltage "push" creates the current flow.

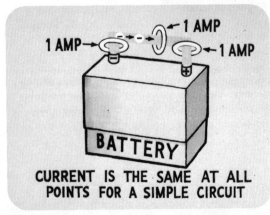

CURRENT IS THE SAME AT ALL
POINTS FOR A SIMPLE CIRCUIT

Figure 2.34

Open Circuit — Another important point is that voltage can exist *without* a current flow. This occurs when there is no complete circuit across the potential difference. A current, however, cannot exist without a voltage. Current flows when a voltage or potential difference exists across a complete path, as shown in Figure 2.35.

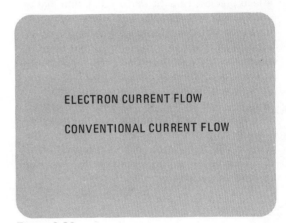

VOLTAGE CAN EXIST WITH OUT CURRENT FLOW

FOR A CURRENT TO EXIST, THERE MUST BE A VOLTAGE.

Figure 2.35

Two Conventions for Current Flow — Up to this point whenever current flow has been discussed, it's been called *electron* current and has meant the motion of electrons through conducting materials. In the study of current flow in science or engineering textbooks, there are two ways of looking at current flow, or two *conventions* for current flow in use today. One of these is electron current flow, and the other is called *conventional current* flow (Figure 2.36). Electron and conventional current are each perfectly adequate and equivalent ways to describe electricity, and both are in use today in different textbooks. A review of the history of these two conventions will help you feel more familiar with them as you see them in your future work.

ELECTRON CURRENT FLOW

CONVENTIONAL CURRENT FLOW

Figure 2.36

Conventional Current — Ben Franklin, one of America's first scientists, studied electricity and was one of the first to actually describe current flow. Unfortunately, although he knew that some type of charge was flowing, he did not know *what type*. Whether it was actually the positive or negative charge that flowed was not discovered until much later. So Ben made some guesses. As it turns out, he guessed that positive charges were flowing in conductors, and in his system they flowed from high to low potentials or from positive to negative as shown in Figure 2.37. Note that this type of flow is just the opposite to the flow of electrons from low to high potential, or from negative to positive, which again is called electron current.

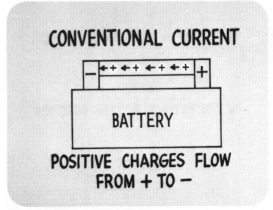

CONVENTIONAL CURRENT

BATTERY

POSITIVE CHARGES FLOW FROM + TO −

Figure 2.37

Electron Current and Conventional Current — As stated in Figure 2.38, all of the electrical effects of electron flow from low to high potential are exactly the same as those that would be created by a flow of positive charges in the opposite direction. Because of this, the two conventions work equally well in describing electricity. Since Franklin's convention was the first one used, it was adopted in textbooks and is still used widely today. Because it is now known that it is the negative electrons that actually move through conductors, the electron flow convention is also in widespread use. Therefore, it is important to realize that both conventions are in use, and that they're essentially equivalent. It makes no difference which convention is used in describing electricity; all the effects predicted are the same. You'll find that as you become more familiar with electricity, you'll have little trouble understanding and using either convention. The terms "current, current flow, electron current, electron flow," etc., are often haphazardly applied to different phenomena. In your work when the word "current" is used, actually *either* electron or conventional current may be substituted in its place with no change in the substance of the discussion. If, however, you want to talk about electron current instead of conventional current, be sure to specify "electron current" and vice versa.

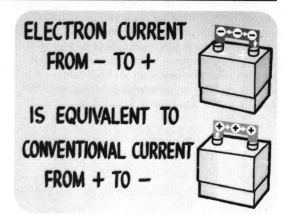

Figure 2.38

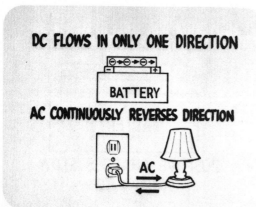

Figure 2.39

DC and AC — Generally, electric current can be classified as one of two general types: *direct* current (dc), or *alternating* current (ac). A direct current flows continuously in the same direction; while alternating current (common house current) periodically reverses its direction (Figure 2.39). Throughout this course, *direct current* only will be discussed. (Alternating current will be introduced and fully covered in other courses.)

Resistance — Now that the concepts of voltage and current have been introduced and discussed, consider a third key concept in this lesson: resistance. Simply stated, *resistance*, as its name implies, is the *opposition to current flow* (Figure 2.40). Resistance is a lot like friction; they both act to oppose motion and generate heat.

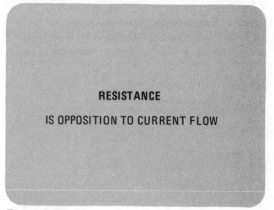

RESISTANCE

IS OPPOSITION TO CURRENT FLOW

Figure 2.40

The amount of opposition to current flow a material will produce depends on the amount of available free electrons it contains, and what types of "obstacles" the electrons encounter as they attempt to move through the material. When electrons try to travel through a substance, under the influence of a potential difference, they often collide with atoms of the conducting material (which are vibrating due to heat in the material), impurities, defects in the material's structure, etc. All these things affect the resistance of a given device or material.

Conductors — As mentioned earlier, some materials have many more free electrons than others and offer relatively few obstacles to electron flow. As a result, current flows through these materials much more easily than in those with few free electrons. In general, materials with many free electrons have *low resistance* (Figure 2.41) and hence are called good conductors.

● MANY FREE ELECTRONS
● LOW RESISTANCE
● GOOD CONDUCTORS

Figure 2.41

- **Insulators**
- **Resistance vs. Current**
- Other Examples of Resistance

Insulators — On the other hand, materials with very few free electrons and many obstacles to electron flow (Figure 2.42) have high resistance, and are called *nonconductors or insulators*.

- FEW FREE ELECTRONS
- HIGH RESISTANCE
- NONCONDUCTORS (INSULATORS)

Figure 2.42

Resistance vs. Current — If a voltage is applied to a conductor, current flows. Just *how much* current flows depends on the resistance of the conductor. As you can see in Figure 2.43, a lower resistance allows a greater flow of current and a higher resistance allows less current to flow.

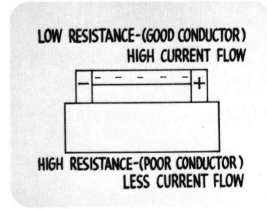

LOW RESISTANCE-(GOOD CONDUCTOR) HIGH CURRENT FLOW

HIGH RESISTANCE-(POOR CONDUCTOR) LESS CURRENT FLOW

Figure 2.43

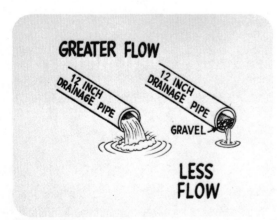

GREATER FLOW

12 INCH DRAINAGE PIPE

12 INCH DRAINAGE PIPE

GRAVEL

LESS FLOW

Figure 2.44

Other Examples of Resistance — It should appear logical to you that the greater the resistance a material has, the less current flow it allows. For example, water will flow quite readily through a 12-inch drainage pipe, as shown in Figure 2.44. If the pipe were filled with gravel, the flow would decrease. The number of obstacles the water must overcome (the pipe resistance) has been increased. In your car, for example, good air flow is maintained into your engine as long as the air filter is clean (Figure 2.45). As the air filter clogs up, the holes in the paper element through which the air passes become clogged, and the *resistance* it offers to the air flow is increased. As a result, less air gets into your engine, and it runs less efficiently. So in

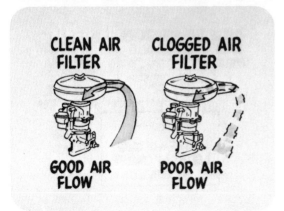

Figure 2.45

circuits, more resistance allows less current to flow. This can be a very desirable result if you want to control or limit the amount of current flowing in a circuit, or undesirable if you wish to deliver a great quantity of current in a given application.

The Ohm — As you will see, resistance is a property of every electrical component. Since, as has been mentioned, some devices will have much more resistance than others, a unit is needed to specify resistance. The unit used to specify just how much resistance is present in a circuit is called the *ohm* (Figure 2.46). The ohm is a unit related closely to the volt and the ampere.

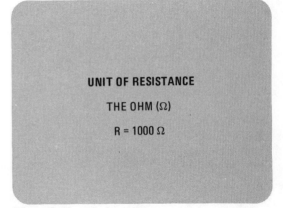

UNIT OF RESISTANCE

THE OHM (Ω)

R = 1000 Ω

Figure 2.46

Shorthand for Resistance — As with other units, shorthand notation is often used when referring to the resistance of a specific device in a circuit. When referring to a specific quantity of ohms, the Greek letter Ω (capital omega) is used to signify *ohms*. In electrical formulas, the word *resistance* is usually abbreviated with the letter *R*. So in the study of electricity, the statement "the resistance is 1000 ohms", may be abbreviated R = 1000 Ω.

The Relationship of Voltage, Current, and Resistance — George Simon Ohm was an early scientist who worked with electricity. He discovered that for simple materials or circuits of the type shown in Figure 2.47, the relationship between the resistance of current flowing through; and voltage across a circuit was quite simple.

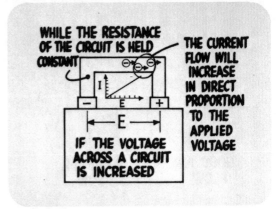

Figure 2.47

Ohm's Basic Relationship — Since this relationship is one of the most important in the fields of electricity and electronics, the unit of resistance, as well as the relationship, was named after Ohm. Basically, Ohm's discovery, simply stated, says that if the voltage across a circuit is *increased* while the resistance is held constant, the *current* will increase in direct proportion to the applied voltage. Materials, devices, and circuits that follow this behavior are said to "follow Ohm's law", or are called *ohmic* devices.

Ohm's Law — Ohm's law may be stated in formula form as shown in Figure 2.48. The formula states that the number of volts across a circuit, equals the number of amps through the circuit, times the circuit resistance in ohms.

The symbols that have been introduced are put together in an important formula to show their relationship: E for voltage or potential difference (in volts), equals I for current (in amps), times R for resistance (in ohms). So, simply stated, the Ohm's law formula reads E = I X R. Since law is so important, it will be the topic of detailed discussion in later lessons. The intent here is simply to expose you to Ohm's law, and you are *not* expected to fully understand all aspects of the law at this point.

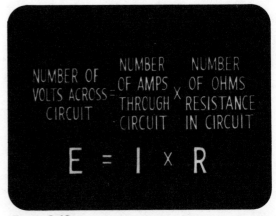

Figure 2.48

Resistors — Often it is desirable to control the amount of current in circuits by adding a known amount of resistance. To do this, special components called resistors are manufactured and are readily available. As shown in Figure 2.49, the symbol for a resistor is a zig-zag line in "schematic" diagrams and as has been mentioned, the Greek letter Ω (omega) is used as the symbol for ohm, and each resistor's value is usually written right next to each resistor symbol.

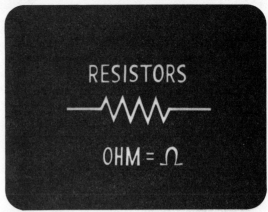

Figure 2.49

A Simple Schematic — Consider the schematic for a simple circuit is shown in Figure 2.50. The basic, simple circuit consists of a 10-volt battery, a 5-ohm resistor, and connecting wires making contact between the two.

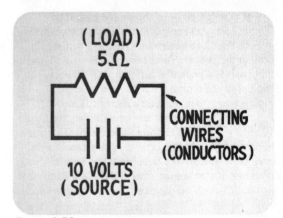

Figure 2.50

Fundamental Circuit Structure and Short Circuits — The basic circuit diagram, shown in Figure 2.50, illustrates the schematic for a typical *circuit* of the type that will be the subject of much study in later lessons. Most all of the circuits discussed will consist of the three basic elements shown in this schematic: a *source*, a *load*, and *conductors*. The source is a device, such as a battery or dc power supply, which supplies the potential difference and electrical energy to the circuit. The electrical energy is carried to the load by means of electrons flowing in the wire conductors. The load is a device such as a resistor, light bulb or motor, which receives the electrical energy.

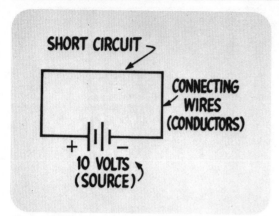

Figure 2.51

If the load device were omitted from this circuit and a wire alone were connected across the terminals of the battery as shown in Figure 2.51, the circuit would then have very little or no resistance. The result would be an extremely high current flow in the wire — enough to discharge the battery quickly and often heat or melt the wire. Circuits such as this with little or no resistance are called *short circuits*, and, in general, the term short circuit refers to a circuit drawing an unsafe amount of current. Except in special circumstances, creating such circuits should be avoided. Adequate resistance should always be placed in a circuit to limit the current to a suitable level.

Use of Ohm's Law — In the simple circuit of Figure 2.50, you can now use Ohm's law as shown in Figure 2.52 to determine the current that will flow in the circuit. You know that E = I X R and that there is a potential difference of 10 volts applied to the 5-ohm resistor. As shown in Figure 2.52, there must be a current flow of 2 amps in the circuit for the equation to hold true: 10 volts = 2 amps X 5 ohms. The use of Ohm's law to determine voltage, current, or resistance is covered in much more detail in later lessons. However, if you are interested, there are additional worked-through examples at the end of this lesson.

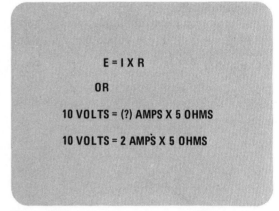

Figure 2.52

Factors that Determine Resistance — To get back to the topic of resistors, resistors are deliberately manufactured with known, and sometimes very large, resistance values. In contrast, the wires used for connections have very little resistance. In most applications, the low resistance of the wires is desirable, since this allows the total resistance of the circuit to be accurately determined by inserting resistors.

The four factors that determine resistance are as follows:

1. The *material* itself determines the resistance of a conductor. Silver is a better conductor than copper, and copper is better than aluminum

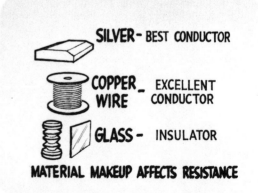

Figure 2.53

(Figure 2.53). The atomic structure plays a role in the resistance.

2. Another factor which affects the resistance of a conductor is its *cross-sectional area*. If you cut a piece of wire, as shown in Figure 2.54, and look at the end, you will see what is called the cross-sectional area. The resistance is inversely proportional to the cross-sectional area. This means if the cross-sectional area of the conductor is increased, its resistance will decrease.

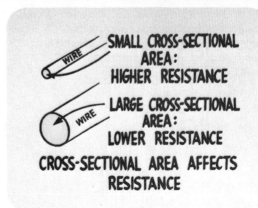

Figure 2.54

3. The third factor which determines resistance is *length*. Resistance is directly proportional to the length of a conductor; that is, a longer conductor has a greater resistance, as shown in Figure 2.55.

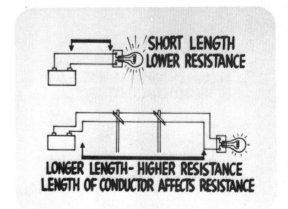

Figure 2.55

4. *Temperature* affects the resistance of a conductor (Figure 2.56). As the temperature of a material increases, the atoms in the material increase their activity which causes the free electrons to undergo more collisions and hence encounter more obstacles. This thermal agitation causes an increase in the resistance of most common materials. For common conducting materials, the higher the temperature, the higher the resistance.

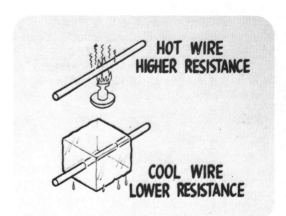

Figure 2.56

Notice that in Figure 2.57 these four factors are listed so that the first letters spell MALT. This memory key will help you to remember the four factors that determine the resistance of a conductor.

MATERIAL

AREA (CROSS-SECTIONAL)

LENGTH

TEMPERATURE

DETERMINE RESISTANCE

Figure 2.57

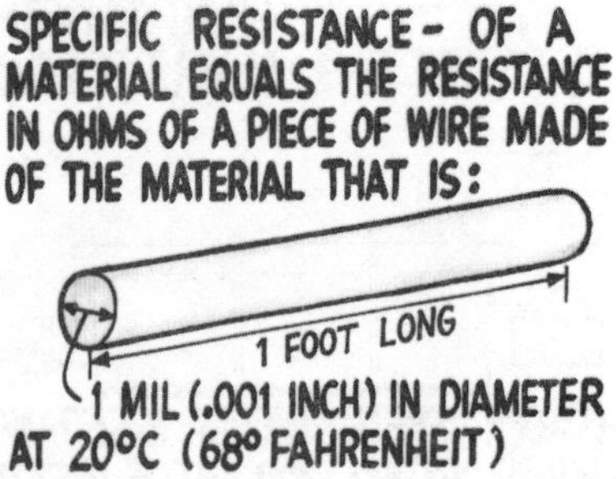

SPECIFIC RESISTANCE - OF A MATERIAL EQUALS THE RESISTANCE IN OHMS OF A PIECE OF WIRE MADE OF THE MATERIAL THAT IS:

1 FOOT LONG

1 MIL (.001 INCH) IN DIAMETER AT 20°C (68° FAHRENHEIT)

Figure 2.58

MATERIAL	SPECIFIC RESISTANCE (OHMS) RESISTANCE OF A CONDUCTOR 1 MIL IN DIAMETER, 1 FOOT LONG AT 20°C
SILVER	9.9
COPPER (DRAWN)	10.4
GOLD	14.7
ALUMINUM	17.0
BRASS	42.1

Figure 2.59

Specific Resistance — Of the four factors that affect the resistance of any device, the material it consists of is by far the most important factor. Because different materials vary greatly with respect to their resistance, a convenient way to describe this property is needed. One common way to express how *material makeup* affects resistance is in terms of what is called the *specific resistance* of the material. The specific resistivity of different materials is determined by measuring the resistance of different wires of the same size and shape, but made of the different materials to be tested. As shown in Figure 2.58, the standard size and shape most commonly used is a piece 1 foot long and 1 mil (0.001 inch) in diameter. When wires of this length and diameter are made from different materials, they will each have different resistance values. The specific resistance of a substance is the resistance (in ohms) of a piece that is 1 foot long and 1 mil in diameter. Because temperature also affects resistance, the specific resistance is specified with the wire held at 20° centigrade (or Celsius), which is 68° Fahrenheit or normal room temperature. Figure 2.59 is a table of specific resistances of common materials. A more complete table of material specific resistances is included in the Appendix.

Once you know the specific resistance of a material, you can use it to calculate the approximate resistance of any actual wire with the following formula:

$$R = \rho \text{ (specific resistance)} \; \frac{\text{(length)}}{\text{(cross section area)}}$$

If the cross section area is expressed in *circular mils* this becomes:

$$R = \rho \text{ (specific resistance)} \; \frac{l}{d \cdot d} \text{ or}$$

$$R = \rho \text{ (specific resistance)} \; \frac{l}{d^2}$$

[Note: A circular mil is defined such that the cross sectional area of a piece of wire equals its (diameter in mils)^2.]

where

R = the resistance of the wire in ohms

ρ = (the Greek letter "Rho") stands for the resistance of the material

l = the length of the wire in feet

d = the diameter of the wire in mils (thousandths of an inch)

In the second formula, d^2 is just a shorthand notation meaning "d times d" and is said "d squared". More on this type of shorthand will be discussed in later lessons.

As an example, if you needed to know the resistance of 100 feet of copper wire that has a diameter of 0.002 inch (2 mils), you'd perform the following calculation:

R (ohms) =

$$\frac{\rho \text{ (specific resistance from table)} \times l \text{ (length in feet)}}{d \text{ (diameter in mils)} \times d}$$

$$R = \frac{(10.4) \times 100}{(2) \times (2)}$$

$$= \frac{10.4 \times 100}{4} = \frac{1040}{4}$$

$$= 260 \; \Omega$$

Note that the Greek letter, ρ, used to represent the *specific resistance* of a material, is also used to represent what is generally referred to as the *resistivity* of a material. The *resistivity* of a material is the resistance in ohms that would be measured across one-meter block of the material. This measure of relative resistance is normally used to compare the conducting properties of materials not normally used in making wire (nonconductors, etc.). The formula for finding the resistance of objects of various shapes is given by:

$$R = \rho \text{ (resistivity) } \times \frac{L \text{ (length of the object in meters)}}{A \text{ (cross-sectional area in square meters)}}$$

The *difference* between the specific resistance of a material and what is generally called the resistivity of a material, is the size and shape of the standard object that is considered.

The *specific resistance* is the resistance of a piece of wire 1 mil in diameter, 1 foot long; and the *resistivity* is generally expressed as the resistance that would be measured across the faces of a one-meter cube.

Temperature Effects on Resistance — For most common materials, the hotter they get, the more resistance they will have if used as a conductor. As has been mentioned, as the temperature increases, the atoms making up these materials vibrate more and collide with electrons as they try to travel through the material. Materials such as this, whose *resistance increases as their temperature increases*, are said to have a *positive temperature coefficient*. In some other materials more heat in the material frees more electrons to take part in current flow. In these materials (notably the semiconductors), the *higher* the temperature is, the *lower* the resistance will be; and hence, they are said to have a *negative temperature coefficient*. Certain special materials (notably the alloys named constantan and maganin) are compounded especially so that changes in temperature hardly affect their resistance at all. These materials are said to have a *zero* or *nearly zero temperature coefficient*.

AWG Table — In building actual circuits the most common material used in making the circuit connections will be copper wire. Since copper wire is manufactured by many different companies, a method is needed to allow the user some freedom of choice as to manufacturer. In the U.S., manufacturers conform to a standard, known as the American Wire Gauge (AWG), which standardizes the cross-sectional dimensions of wire. As shown in Figure 2.60, the gauge numbers run from 0000 for the largest wire to 40 for the smallest. A more complete AWG table is included for your reference in the Appendix.

GAUGE NUMBER	DIAMETER (MILS)	CROSS SECTION CIRCULAR (MILS)	OHMS PER 1000 FT. 25°C (=77°F.)
0000	460.0	212,000.0	.0500
2	258.0	66,400.0	.159
6	162.0	26,300.0	.403
10	102.0	10,400.0	1.02
14	64.0	4,110.0	2.58
18	40.0	1,620.0	6.51
22	25.3	642.0	16.5
26	15.9	254.0	41.6
30	10.0	101.0	105.0
36	5.0	25.0	423.0
38	4.0	15.7	673.0
40	3.1	9.9	1,070.0

- STANDARD ANNEALED SOLID COPPER WIRE

Figure 2.60

Resistor Types — As mentioned earlier, the common components used to regulate current flow in circuits are called resistors. Resistors are manufactured with a wide variety of known values of resistance, for use where required. Resistors are available in many sizes and shapes depending on their function in an electric circuit. As you can see in Figure 2.61, all resistors can be grouped into two basic types, fixed resistors or variable resistors. These two classifications can be divided into two more subgroups depending on the material and construction of the resistor, as either composition or wirewound resistors.

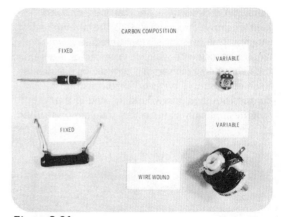

Figure 2.61

Carbon Composition Resistor — The resistor shown in Figure 2.62 is a carbon composition resistor. (These can usually be identified by the colored bands painted around them and the wire leads coming out of the center of each end.) This device usually contains a mixture of granulated carbon, which is a conductor, and a synthetic resinous binder, which is an insulator. The percentage of carbon to binder will determine the resistance of the device.

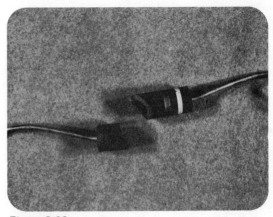

Figure 2.62

Carbon composition resistors are the most common type you will encounter because they're easy and inexpensive to manufacture, they're rugged, and they have accurate enough resistance values for most applications. The major disadvantages of carbon composition resistors are that they tend to change their resistance as they age; and because they are small, they can't handle much power, as will be discussed in a later lesson.

Wirewound Resistor — Another type of fixed resistor shown in Figure 2.63 is the wirewound resistor. This resistor is constructed of special, high-resistance wire which is wrapped on a ceramic tube and then enclosed in a ceramic or plastic case. Each of these two types of fixed resistors, composition and wirewound, is used in the specific applications for which they are best suited.

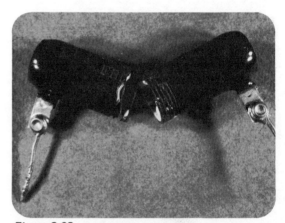

Figure 2.63

Because wirewound resistors are wound on a ceramic case, which is often hollow, they can withstand higher currents and heat and, hence, higher power than carbon composition types. Their resistance value doesn't change as much with age as carbon types, since the wire they consist of is much more stable. Wirewound resistors can also be made with much more accurate values than carbon types. All of these advantages, however, cost more, and so the main disadvantage of wirewound resistors is high cost.

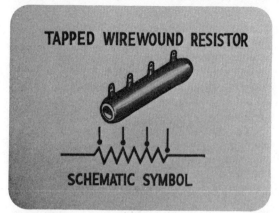

Figure 2.64

Wirewound fixed resistors are often *tapped* at several points. Connections are made at one or more points along its length so that various values of resistance are available at different intervals. The schematic symbol for a typical tapped resistor is shown in Figure 2.64.

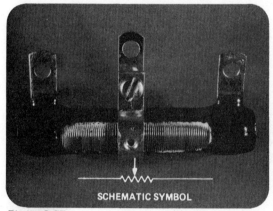

Figure 2.65

Another type of wirewound resistor commonly used is the sliding contact resistor shown in Figure 2.65. The resistance between the taps may be adjusted to any desired value. Usually with this type of resistor a resistance value is selected and then the tap is locked with a screw to hold it in this position.

Variable Resistors — Other types of resistors found in electric circuits are variable resistors. Variable resistors are made of the same materials used for the fixed value resistors, either carbon composition or wire winding. The major difference is that a variable resistor provides a way of adjusting the amount of resistance desired.

The schematic symbols for variable resistors are shown in Figure 2.66. A variable resistor is called by many different names, such as potentiometer or pot, rheostat, and others. In this course, it will be referred to as a potentiometer, or the shortened version — pot.

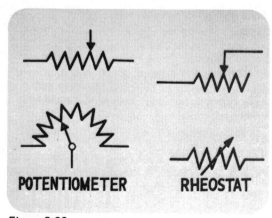

Figure 2.66

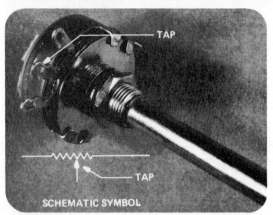

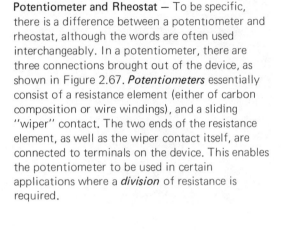

Figure 2.67

Potentiometer and Rheostat — To be specific, there is a difference between a potentiometer and rheostat, although the words are often used interchangeably. In a potentiometer, there are three connections brought out of the device, as shown in Figure 2.67. *Potentiometers* essentially consist of a resistance element (either of carbon composition or wire windings), and a sliding "wiper" contact. The two ends of the resistance element, as well as the wiper contact itself, are connected to terminals on the device. This enables the potentiometer to be used in certain applications where a *division* of resistance is required.

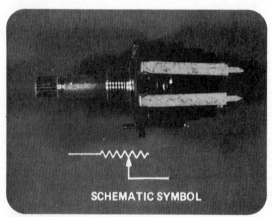

Figure 2.68

A *rheostat*, in particular, has only two external contacts; one contact is at the end of the resistance and the variable contact is a tap as shown in Figure 2.68. A *variable resistance* is set up between the contacts, but the rheostat can't be used in applications requiring a divided resistance. Notice then that a potentiometer can be used as a rheostat but not vice versa.

Other Factors Concerning Resistors — To complete this basic discussion of resistors, several additional key topics need to be discussed. Usually in applying a resistor in a circuit, you need to be aware of its value in ohms, its tolerance, and its power rating. Also, in certain applications and circuit analysis situations, knowledge of what's known as a resistor's *conductance* will be helpful. Consider these factors one at a time (Figure 2.69).

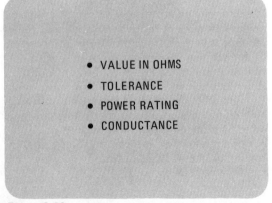

- VALUE IN OHMS
- TOLERANCE
- POWER RATING
- CONDUCTANCE

Figure 2.69

Resistor Color Code — As has been discussed, the unit of measure of resistance is the ohm, abbreviated with the Greek letter Ω. One ohm is the amount of resistance which allows an ampere of current to flow through it when one volt is applied to it. In almost any resistor application, you'll need to be able to select a resistor of some predetermined ohmic value to insert into a circuit. On some resistors that are physically large enough, the value is simply labeled on the side of the resistor. Since carbon composition resistors are quite small physically, an alternate standard method has been devised. The system uses colored bands painted on the body of the resistor to indicate its resistance. This allows the resistor's value to be easily read, no matter what position it has in the circuit. A standard *color code* is used to indicate the amount of resistance in ohms. In this code, each color represents a number. The colors and the numbers they represent are shown in Figure 2.70. Since you must memorize this code, a memory key has been devised to help you. If you remember the sentence, "*Bad Beer Rots Our Young Guts But Vodka Goes Well*" and notice that the first letter of each word in this sentence is also the first letter of each color in the code, you'll be able to read resistance color bands. As an example, the B in Bad stands for B in black. It's the *first* word in the sentence and represents the number *zero*, which is the *first* number in the sequence of numbers from zero to nine.

RESISTOR COLOR CODE		
BLACK	0	BAD
BROWN	1	BEER
RED	2	ROTS
ORANGE	3	OUR
YELLOW	4	YOUNG
GREEN	5	GUTS
BLUE	6	BUT
VIOLET	7	VODKA
GRAY	8	GOES
WHITE	9	WELL

Figure 2.70

An Example of Reading the Resistor Color Code — When reading the color bands, it's easiest to turn the resistor so that the band closest to the end is on your left. The color bands are then read from left to right. The color band nearest the end indicates the first digit of the numerical value in ohms. For example consider the resistor in Figure 2.71; the first band is brown. If you refer to the color chart, you'll note that brown is the color for one. The second band indicates the second digit of the resistor's ohmic value; it's indicated as gray in the figure, the color for 8. The third color band is the *decimal multiplier* which indicates how many zeros should be behind the second number. In the figure, the third band or multiplier is indicated as

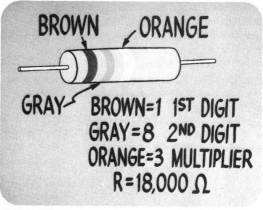

Figure 2.71

orange, which says there should be three zeros
after the 8. So, when the number is completed, the
value of this resistor is seen to be 18,000 ohms.

**Another Example of Reading the Resistor Color
Code** — Another resistor is shown in Figure 2.72.
The first band is brown; so again the first digit is
one. The second band is green, the color for 5, so
the second digit is 5. With a black third band, you
might think this is a 150 ohm resistor, but that is
not quite right. Remember the multiplier band
indicates the number of zeros behind the second
digit, and *black is the color for zero*. This means
that there are *zero zeros* behind the second digit.
Thus, this is a 15-ohm resistor.

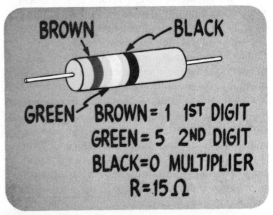

Figure 2.72

Fractional Multipliers — As shown in Figure 2.73,
there are resistors available which have a value of
less than 10 ohms. On these resistors, the third
color band (fractional multiplier) will be gold or
silver. When gold is used, the first two digits are
multiplied by 0.1 or one-tenth. When silver is used,
the first two digits are multiplied by 0.01, or one
one-hundredth.

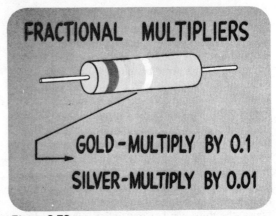

Figure 2.73

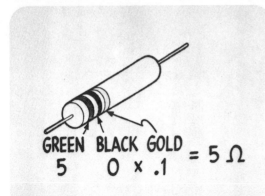

Figure 2.74

An example of a resistor with a fractional multiplier is shown in Figure 2.74. The first band is green, which is the color for five, and the second is black, the color for zero. The first two digits then are 50. The third band is gold, so you multiply 50 by 0.1, or one tenth, and you see that the resistance value is 5 ohms.

Tolerance — The first three bands indicate the value of the resistor in ohms, but some deviation from the exact value can be tolerated in most electrical circuits. Because carbon composition resistors are mass-produced with low cost as a prime factor, each resistor isn't exactly the resistance value it's labeled. In general, the more accuracy with which a resistor is manufactured, the higher its cost will be. To indicate how accurately a resistor has been manufactured, a fourth band is added to the resistor to indicate the *tolerance* of the ohmic value of the resistor. The *tolerance* of a resistor indicates how much a resistor's actual ohmic value can vary above or below its marked value (Figure 2.75). These tolerances are expressed as a percentage of the resistor value. Normal resistor tolerances available include 20%, 10%, or 5%. As has been mentioned, lower resistor tolerance increases the price of the resistor.

TOLERANCE

HOW MUCH A RESISTOR MAY VARY ABOVE OR BELOW ITS MARKED VALUE

Figure 2.75

- **Color Code for the Tolerance Band**
- **Example of Tolerance**

Color Code for the Tolerance Band — The color code for the fourth band, the tolerance band, is shown in Figure 2.76. A gold fourth or tolerance band indicates a tolerance of ±5%; a silver band indicates ±10%, and no fourth band at all indicates ±20%.

4TH BAND:	TOLERANCE
NO BAND	± 20%
SILVER	± 10%
GOLD	± 5%

Figure 2.76

Example of Tolerance — Consider an example of how to determine the tolerance of a resistor, Figure 2.77. Suppose you have a resistor marked with only three bands — brown, black, brown. Using the color code, color by color, you can see that the first digit is one, the second is zero, and the multiplier or third band is brown, meaning that you should add one additional zero after the first two digits. This resistor would have a resistance of 100 ohms. Its tolerance would be ±20%, since it has *no fourth* band.

If a 100-ohm resistor has a tolerance of plus or minus 20%, this means that the resistor's actual value may be as much as 20% of 100 ohms (or 20 ohms) above or below its marked value of 100 ohms. The resistor's actual value is in the range from 80 to 120 ohms.

$100\Omega \pm 20\%$

20% OF 100Ω IS 20Ω

LOWER LIMIT = $100\Omega - 20\Omega = 80\Omega$

UPPER LIMIT = $100\Omega + 20\Omega = 120\Omega$

RESISTANCE RANGE = $80\Omega - 120\Omega$

Figure 2.77

Ranges of Resistance for Various Tolerances — As shown in Figure 2.78, the actual value of the resistor is somewhere between 80 and 120 ohms. Likewise, if you had a 100-ohm resistor with a silver fourth band, its tolerance would be ±10%. This means that its actual value is within ±10% of its marked 100-ohm value. Ten percent of 100 ohms is 10 ohms. Therefore a resistor with a brown, black, brown, silver code, would have a true resistance anywhere between 100 ohms minus 10 ohms and 100 ohms plus 10 ohms (anywhere between 90 and 110). If the resistor had a gold fourth band, its resistance would be within ±5% of its marked value, in this case, between 95 and 105 ohms.

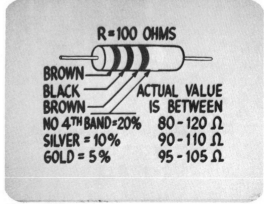

Figure 2.78

Military Failure Rate Band — On some resistors, you may also find a fifth band, called the military failure rate band as shown in Figure 2.79. This band is next to the tolerance band, on the right side of it. These resistors are part of a batch that have been tested for reliability, and the code gives a description of how reliable the resistor will be. In certain special applications this band will be of importance.

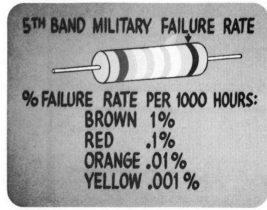

Figure 2.79

TEMP CHARACTERISTIC
TOLERANCE
RB09A12501F
STYLE
RESISTANCE VALUE
MILITARY SPECIFICATION NUMBERS

Figure 2.80

Military Specification Numbers — Some carbon composition resistors you may encounter may be coded with what is called the MIL SPEC (military specification) number system. In this system, a sequence of numbers and letters is stamped on the body of the resistor as shown in Figure 2.80. You may encounter resistors with these markings in "military surplus" electronic supply stores. This code contains quite a bit of information about the resistor in addition to its ohmic value. For your purposes, all you really need to know is how you would determine the resistance of a resistor marked in this way. In this system, the *first four* symbols give information about the resistor's construction; the *fifth symbol* gives information

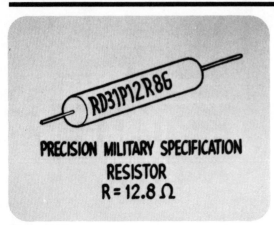

PRECISION MILITARY SPECIFICATION RESISTOR
R = 12.8 Ω

Figure 2.81

concerning its temperature characteristic. The remaining numbers specify the resistor's value in ohms; the final letter at the end tells you its tolerance. The actual resistance value may be indicated by three, four, or five digits. In each case the *last digit*, the one right before the final letter, indicates the number of zeros to be placed behind the first digits. In this figure, the number 12501 tells you that the resistor's value is 12,500 ohms.

If the resistor is a precision resistor that needs a decimal point to specify it, the letter R is used in the code to represent the decimal point. As shown in Figure 2.81, the letter R is used in the portion of the code that indicates the resistor's ohmic value. The R indicates the location of the decimal point and the final digit denotes the decimal fraction rather than the multiplier. Therefore, the resistance of the resistor shown in the figure would be 12.8 ohms.

Power Rating of a Resistor — A third important factor concerning resistors is power. As has been discussed, when electrons flow through a resistor, they meet an opposition to their passage. To overcome this opposition, a certain amount of power is expended and this power shows up in the form of *heat* in the resistor. Excessive heat will eventually damage resistors, so they are rated by the amount of power they can safely give off or dissipate into the air. This power rating is expressed in units called *watts*. The wattage rating of a resistor is the amount of power it can safely handle without damage (Figure 2.82).

POWER RATING OF A RESISTOR

HOW MUCH POWER IT CAN DISSIPATE (GIVE OFF) SAFELY IN THE FORM OF HEAT (IN WATTS).

Figure 2.82

Wattage Sizes of Resistors — The wattage rating, in general, is not marked on carbon composition resistors. The *size of the resistor* is most commonly used to indicate a resistor's wattage capacity. Generally, with two resistors of equal value, the larger resistor will have the greater wattage rating. For applications in standard use, standard resistor body sizes indicate the wattage rating. These resistor sizes and their ratings are shown in Figure 2.83 and listed in the table below:

Wattage Rating (watts)	Diameter (inches)	Length (inches)
1/4	3/32	3/8
1/2	1/8	13/32
1	1/4	23/32
2	1/4	1-1/2

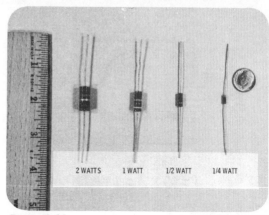

Figure 2.83

Note that there are exceptions to these standard sizes. In particular, a variety of custom-made resistors is available for applications in miniaturized equipment that will deviate from the standard in the table.

In general, where a resistor capable of handling greater than 2 watts Is needed, a wire-wound resistor is the best choice. The ohmic value and wattage rating are usually stamped directly on wire-wound resistors.

Basic Formula for Power — How do you know how much power a resistor will have to handle in a circuit? As it turns out, this calculation is fairly easy. A basic formula for the calculation of how much power a resistor will dissipate while acting in a circuit is shown in Figure 2.84. The power the resistor has to handle (watts), equals the voltage across the resistor, times the number of amps flowing through it. Like Ohm's law, power and power formulas will be discussed in much more detail in later lessons.

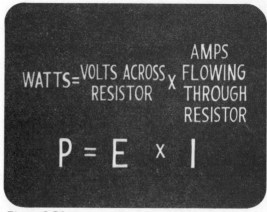

Figure 2.84

- Example: Power Dissipated by a Resistor
- Another Formula for Power
- Example: $P = I^2 \times R$

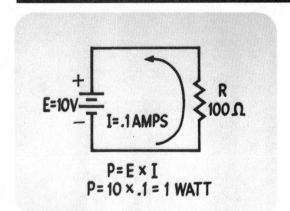

Figure 2.85

Example: Power Dissipated by a Resistor —
Consider the resistor shown in the simple circuit in
Figure 2.85. The figure shows that this resistor has
10 volts across it and is carrying 0.1 amp. How
much power will it have to dissipate in watts?
Using the formula: $P = E \times I$, you can see that P
equals 10 volts times 0.1 amp, and 10×0.1 is one.
Therefore, 1 watt is being dissipated in this
resistor.

Another Formula for Power — If the equivalent of
E from Ohm's law is substituted in the power
equation, then another useful formula for
calculating the power dissipated by a resistor can
be found. As shown in Figure 2.86, the original
formula reads $P = E \times I$. Ohm's law states that
$E = I \times R$. So replace the E in the power formula
with $I \times R$, and you'll see that power can also be
calculated as $I \times I \times R$ or I^2R. The formula
$P = I^2 \times R$ will be useful in calculating the power
dissipated by a resistor in cases where the current
flowing through it is known.

$$P = E \times I$$
SUBSTITUTE
$$E = I \times R$$
$$P = (I \times R) \times I$$
$$P = I \times I \times R$$
$$P = I^2 \times R$$

Figure 2.86

Example: $P = I^2 \times R$ — In the example circuit of
Figure 2.85, it is shown that 0.1 amp is flowing
through the 100-ohm resistor. The power
dissipated may then be calculated:
$$P = 0.1 \times 0.1 \times 100 = 0.01 \times 100 = 1 \text{ watt}$$

Resistor Selection —Here's an important point concerning the selection of a resistor for a circuit application. For most low-power applications, select a resistor with a wattage rating that is *two times* the wattage it will have to handle (Figure 2.87). So if you calculate that the resistor will have to handle 1/2 watt of power, you should use a 1-watt resistor; if you calculate 1 watt of power dissipation, use a 2-watt resistor, etc. This ensures that the resistor will stay fairly cool during operation and will extend its working life. The wattage rating of a resistor does not affect its ohmic value.

RESISTOR SELECTION

USE A RESISTOR WITH **TWICE** THE RATING OF THE POWER YOU CALCULATE

Figure 2.87

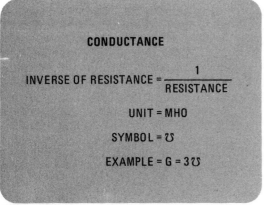

CONDUCTANCE

$$\text{INVERSE OF RESISTANCE} = \frac{1}{\text{RESISTANCE}}$$

UNIT = MHO

SYMBOL = ℧

EXAMPLE = G = 3℧

Figure 2.88

Conductance — Another important characteristic of resistance and resistors as shown in Figure 2.88 is called *conductance*. Conductance is just the inverse of resistance; that is, the *ability of a material to conduct current*. Conductance is found by taking the reciprocal of the resistance, or just one divided by the resistance. The unit used to specify conductance is called *mho*, which is just ohm spelled backwards. The symbol for the mho is the Greek letter omega inverted, and in formulas the letter G is used to indicate conductance. As you will see in a later lesson, conductance is an important concept in the study of certain types of circuits.

Resistor Preferred Values — Before finishing this lesson one more topic needs mentioning concerning resistors. As you begin working in the laboratory you will find that resistors are not commonly available having every single value in ohms from a fraction of an ohm to several million ohms. Here's why. As has been discussed, resistor *tolerance* allows any resistor to vary from its labeled ohmic value — plus or minus the amount specified by its fourth color band. This means that any resistor will actually cover a *range* of resistance values within the limits of its tolerance. So, resistors of a given tolerance are manufactured in a *set of values*, called preferred values that allow you

PREFERRED RESISTOR VALUES

A		B		C
20% TOLERANCE		10% TOLERANCE		5% TOLERANCE
PREFERRED VALUES	RANGE OF RESISTANCES	PREFERRED VALUES	RANGE OF RESISTANCES	
10	8 RANGE OF RESISTANCE VALUES COVERED BY 10-Ω, 20% RES.	10		10
			11 RANGE OF RESISTANCE VALUES COVERED BY 10% TOLERANCE 12-Ω RESISTOR	11
	12 RANGE OF 20% TOLERANCE 15-Ω RESISTOR	12		12
			13	13
15		15		15
			16	16
	18	18		18
			20	20
22		22		22
			24	24
	27	27		27
			30	30
33		33		33
			36	36
	39	39		39
			43	43
47		47		47
			51	51
	56	56		56
			62	62
68		68		68
			75	75
	82	82		82
			91	91
100		100		100

Figure 2.89

to always choose a resistor that will be within tolerance limits of any value you need, and minimize unnecessary "overlap" in manufacture.

For example, as shown in column A of Figure 2.89, the set of preferred resistor values for resistors with a 20% tolerance will contain the numbers 10, 15, 22, 33, 47, 68, and 100. So at your lab bench or in electrical equipment you'll find that 20% resistors will have typical values such as 100 ohms, 470 ohms, 33,000 ohms, etc. You will not commonly find 20% resistors having values that do not contain the numbers or decimal multiples of the numbers listed in the figure. If you examine the top entry in Figure 2.89 you will see why. A 20% tolerance, 10-ohm resistor can vary plus or minus 20% from its labeled value of 10 ohms. This means it may be as low as 8 ohms, (10 ohms minus 20%) or as high as 12 ohms (10 ohms plus 20%) and still function normally in 20% tolerance applications. So if you need a resistance of 8, 9, 10, 11, or 12 ohms, and required only a 20% tolerance, a 10-ohm, 20% tolerance resistor may be used.

Notice in the second entry of Figure 2.89, column A, that a 15-ohm, 20%-tolerance resistor may have a value as low as 12 or as high as 18 ohms. As you look down the line, notice how the preferred resistance values interlock to cover all possible resistance values. Again note that resistors are commonly available in these listed values, as well as *decimal multiples* of these values: Ten times preferred, 100 times, 1000 times, etc. If your calculations showed that you needed a 4620-ohm resistor, but the application was not critical (a 20% resistor would be acceptable), you'd select the preferred resistor value that's closest to the one you need — in this case a 4700-ohm resistor. Notice that a 47 is a preferred resistor value in the 20% column and a 4700-ohm resistor would be used to cover resistance values from 3900 to 5600 ohms if only 20% accuracy is required.

If you desire a little more accuracy, you can select and use 10% tolerance resistors having a silver fourth band. Since 10% tolerance resistors will not vary as much as 20% tolerance resistors, all possible resistance values will not be covered if the 20% tolerance preferred values are used. As you can see from Figure 2.89 column B, 10% tolerance resistors are commonly manufactured in *twice as many preferred values* as 20% tolerance resistors. Notice that half of these values are just the 20% tolerance preferred values. The new numbers are simply placed halfway between the 20% tolerance values.

Similarly, the preferred resistance values for 5% tolerance resistors will occur in a pattern like that observed for the 10% resistors. There will be twice as many preferred values for 5% tolerance resistors as there are for 10% tolerance resistors, and the new values will fall into the spaces between the existing numbers as shown in Figure 2.89, column C.

Other types of resistors are available commonly called "precision resistors". These resistors are typically of 1 % tolerance, and are usually specially manufactured for specific applications. As you might expect, they cost considerably more than standard tolerance resistors, which are manufactured in the preferred values in large quantities.

Selecting Resistors for Circuit Application — At this point you should know enough about resistors to be able to select them for use in the laboratory. Usually a schematic diagram will tell you the ohmic value of the resistor you need and your lab manual should tell you what wattage size and tolerance is necessary. (Most labs exercises you will be performing will probably use 1/2 watt, 10% tolerance resistors for the majority of applications.) You will usually be given resistor values in ohms (for example 3300 Ω); and asked to find a 3300-ohm resistor from your bench. The process is simple: first, convert the number of ohms into the color code; and then scan through your supply of resistors until you find one that matches. The resistor's wattage rating is indicated by its physical size and can be checked by placing the resistor on the watt size chart you've been provided. The tolerance is indicated by the fourth band.

To find a 3300-ohm, 10%-tolerance resistor — first (in your mind or on paper if necessary), write down its color code. To find the code from a resistance value, use the color chart in reverse. The color for 3 is orange. (Remember the little memory key, *Bad Beer Rots Our* . . . correspond to the numbers 0, 1, 2, and 3. The "O" in "Our" corresponds to the "O" in orange. So the first color band on your resistor should be orange. Since the second figure of the resistance is 3 also, the second band on the resistor should be orange. Now you need 33*00* ohms; so *two* zeros must follow the 33, and the color representing 2 is red. Finally for the 10% tolerance you should look for a silver fourth band. So the resistor you look for is one whose bands are orange, orange, red, and silver.

One more point to remember is that carbon composition resistors are commonly manufactured in the preferred values only. If a resistor value is called for such as 34 ohms, and the tolerance specified is 10%; examine the preferred value 10% resistors and you'll find one that comes within 10% of the value you need. (In this case a 33-ohm, 10%-resistor will do the job.)

Summary — Figure 2.90 briefly lists the topics covered in this lesson. There are three basic concepts important in the study of electric circuits.

1. *Voltage* is the pressure or potential difference which makes electrons move in a conductor.
2. *Current* is charge flowing in a conductor as a result of a potential difference across the conductor. Both electron current (which flows from negative to positive potential) and conventional current (which flows from positive to negative potential) were discussed. Either picture of the current is perfectly

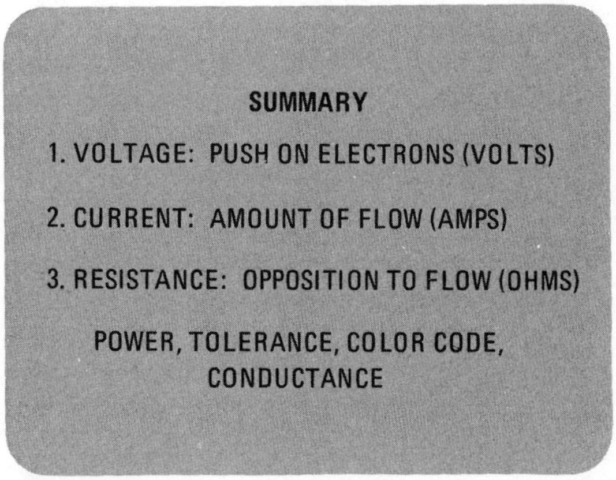

SUMMARY

1. VOLTAGE: PUSH ON ELECTRONS (VOLTS)

2. CURRENT: AMOUNT OF FLOW (AMPS)

3. RESISTANCE: OPPOSITION TO FLOW (OHMS)

POWER, TOLERANCE, COLOR CODE,
CONDUCTANCE

Figure 2.90

adequate. The unit of measure of current is the ampere or amp.

3. *Resistance*, which is the opposition to current flow, is a third basic concept in the study of electric circuits. Resistors are devices manufactured to have desired values of resistance. The two most basic types have been discussed: *wire-wound* and *carbon composition*. Also discussed was the way these two types can be manufactured as either fixed value or variable resistors.

Also reviewed were the resistor color code for reading ohmic values, the concept of resistor tolerance, and resistor power rating.

Conductance, which is the ability of a device to conduct current, was discussed. Conductance is the reciprocal of resistance.

LESSON 2. VOLTAGE, CURRENT, AND RESISTANCE

● **Worked Through Examples**

The following examples are designed to help you become more familiar with the new concepts presented in this lesson. You'll see that there are examples dealing with some key definitions, the resistor color code, Ohm's law, resistor power or wattage rating, and power dissipation in a resistor. Ohm's law and the formulas for power were only briefly introduced in this lesson, but as you'll see, they will be examined and discussed in much greater detail in Lesson 4. Therefore, the examples and practice problems presented here will be very simple and straightforward, and they will not involve calculations with very large or small quantities. The techniques that you'll need to handle these large or small quantities are presented in Lesson 3 and their use with Ohm's law and the power formulas is thoroughly illustrated in Lesson 4.

1. A copper wire is placed between two charged bodies as shown. Explain what happens to the free electrons in the wire.

The electrons in the wire feel the effects of the potential difference created by the two charged bodies. Since electrons are negative, they are repelled by the negatively charged body and attracted by the positively charged body. So, they flow through the wire under the influence of the potential difference, from negative to positive. If the two charged bodies are touching the wire, the electrons will continue to flow through the wire until both bodies have no excess charge. At this point the electrons will resume their random motion in the wire.

2. What is the definition for voltage, what are the other terms used to describe it, and what is its unit of measure?

Voltage is a measure of the push on each electron and it is this push that makes electrons move. Other terms used to describe voltage are "potential difference" and "electromotive force (EMF)", and the unit of measure is the *volt*.

3. Define and compare electron current and conventional current, and give the unit of measure.

Electron current is the *flow of electrons* through a material from low to high potential or from negative to positive. Conventional current is another way of considering current flow. In the conventional current scheme, positive charges are considered to flow through a material from high to low potential or from positive to negative. Both of these conventions for current are adequate and equivalent ways of describing electricity. The unit of measure is the *ampere* or *amp*.

4. Define resistance, list the four factors that determine the resistance of a conductor and give the unit of measure.

Resistance is the *opposition* to current flow, and its unit of measure is the *ohm*. The factors that determine the resistance of a conductor are the material itself, the area (cross-sectional), length, and temperature of the material.

5. In the circuit shown, 5 amps of current is flowing through a 4 ohm resistor. Using the formula for Ohm's law introduced in this lesson, calculate the voltage applied to the circuit.

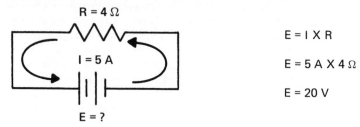

$$E = I \times R$$

$$E = 5\,A \times 4\,\Omega$$

$$E = 20\,V$$

6. In the diagram shown, indicate the power rating of the four resistors.

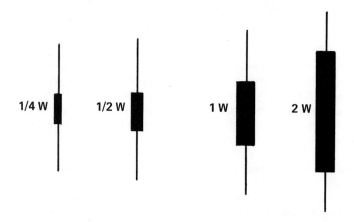

7. In the circuit of example 5, a current of 5 amps is flowing through a 4-ohm resistor with 20 volts applied to the circuit. Use each of the power formulas introduced in this lesson to calculate the power dissipated by the resistor.

$$P = E \times I$$
$$P = 20\,V \times 5\,A$$
$$P = 100 \text{ watts}$$

$$P = I^2 \times R$$
$$P = I \times I \times R$$
$$P = 5\,A \times 5\,A \times 4\,\Omega$$
$$P = 25 \times 4\,\Omega$$
$$P = 100 \text{ watts}$$

8. Review Example: Reading the Resistor Color Code

			Tolerance	
0	Black	Bad		
1	Brown	Beer	Gold	5%
2	Red	Rots	Silver	10%
3	Orange	Our	No band	20%
4	Yellow	Young		
5	Green	Guts		
6	Blue	But	Fractional Multipliers	
7	Violet	Vodka	Gold X	.1
8	Gray	Goes	Silver X	.01
9	White	Well		

The resistor color code is a very important part of electronics. You must learn the code and how to apply it to a resistor. You will experience little difficulty with the code if you follow these steps:

1. Turn the resistor so that the band closest to the end of the resistor body is to your left. The blank end of the resistor body is to your right.

START READING HERE

2. The first band indicates a number. Look up the number on the color code chart that corresponds to the color of the band and write it down.

RED ➔ 2

For example:
If the first band is red, the first digit is 2.

3. The second band also indicates a number. Go to the chart again and write down the number that corresponds with the color of the second band.

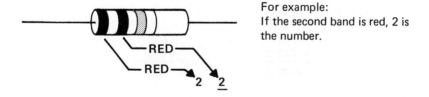

RED
RED ➔ 2 2

For example:
If the second band is red, 2 is the number.

4. The third band of the resistor is the *multiplier band*. This band simply tells you how many zeros to write after the first two digits. If, as in this case, this band is also red, you should write two zeros after the first two digits.

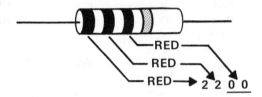

RED
RED
RED→ 2 2 0 0

If the third band is silver or gold, a "fractional multiplier" is indicated. A gold band indicates that the first two digits of the resistor's value should be multiplied by .1. A silver band tells you to multiply the first two digits by .01. Remember, these fractional multipliers are to be used only when the *third* band is gold or silver.

5. The fourth band is called the "tolerance" band. This band indicates the range of values the resistor may have and still be considered "good". If this resistor has a silver tolerance band, according to the chart this means that the resistor's value is supposed to be within plus or minus 10% of its indicated value. The indicated value of the resistor is 2200 ohms. In order to find what 10% of 2200 is, you must multiply 2200 by .1 (you get point one by moving the decimal point two places to the left like this . . .10% = .10. (You may remember how to do this by remembering that the two zeros in the percent sign stand for the two places you must shift when converting from a percentage to a decimal fraction.) Now, to find the actual ohmic value of 10%, you multiply 2200 by .1.

$$
\begin{array}{r}
2200 \\
\times \quad .1 \\
\hline
220.0
\end{array}
$$

Since there is only one decimal place in the numbers to be multiplied, the decimal point in the answer is shifted one place, giving you 220 ohms.

To find the actual ohmic range of the resistor, you must add 220 ohms to 2200 ohms, and subtract 220 ohms from 2200 ohms.

$$
\begin{array}{r}
2200\ \Omega \\
+\ 220\ \Omega \\
\hline
2420\ \Omega
\end{array}
\qquad
\begin{array}{r}
2200\ \Omega \\
-\ 220\ \Omega \\
\hline
1980\ \Omega
\end{array}
$$

This 2200-ohm resistor is within tolerance if its value is between 1980 and 2420 ohms.

9.   Review Example: Selecting a resistor from its ohmic value.

In your work with electricity, not only will you have to identify a resistor's value by reading its color code, you will often have to pick out a resistor from a supply cabinet, given its value in ohms. For this reason, it is also important to learn how to determine a resistor's color code, given its value in ohms. The following problems are designed to illustrate this process.

Given the following resistor values, specify the color code you would look for in selecting them from a supply cabinet.

a.   680 Ω ± 10%

1)   Using either the chart, memory key, or more hopefully just recall; determine the color for the first digit: 6. If you use the memory key (Bad, Beer, Rots, etc.) remember that the little saying corresponds to the numbers *zero* through 9. (Black is the color representing zero, NOT one; this is a commonly forgotten item.) The color for 6 is blue so that should be your first band.

2)   The color for 8 is gray; so using the same methods as above you determine that your second band should be gray.

3)   The 6 and 8 must be followed by ONE zero to give you 680. The third band should then be the color for one (which is brown), representing one zero.

4)   A 10% tolerance resistor will have a silver fourth band.

So the colors, in the correct order, that you should look for are: blue, gray, brown, silver.

b.   1,000,000 Ω ± 5%   Brown Black Green Gold

1)   The first digit you need is a one; the color for one is brown (first band).

2)   The second digit is zero; the color for zero is black (second band).

3)   You now need to follow the 10 for your first two digits with 5 more zeros to yield 1 million Ω. The color for 5 is green (third band).

4)   Finally, for a 5% tolerance you'll need a gold fourth band.

The colors you should look for are: brown, black, green, gold.

LESSON 2. VOLTAGE, CURRENT, AND RESISTANCE

● **Practice Problems**

Fold Over

The following problems are designed to begin to familiarize you with circuits, Ohm's law in very basic cases, and to provide lots of practice in interpreting the resistor color code. Fold over the page to check on your results.

1. Draw in the space provided below a schematic diagram of a complete circuit containing:

   a. a 10-volt battery
   b. a 100-ohm resistor
   c. all necessary connections (label all components as necessary)

2. In the following circuit a current of 1/10 amp is flowing. Using the basic Ohm's law and power formulas calculate:

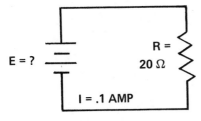

E = ? R = 20 Ω I = .1 AMP

   a. The value of the applied voltage

   E = ____2____ volts.

   b. The power dissipated by the resistor in this circuit

   P = ____.2____ watts

   c. If you were building this circuit, what power rating resistor would you select?

   ____.4____ watt.

Answers

1.

E = 10 V

R =
100 Ω

2.a. E = 2 volts

2.b. P = .2 watt

2.c. .4 watt
      you would probably
      select a 1/2 watt
      carbon resistor

3. In the following circuit 1 amp is flowing through the resistor.

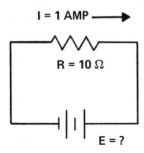

I = 1 AMP ⟶

R = 10 Ω

E = ?

a. Calculate the applied voltage E = ___10___ volts

b. If you wanted 2 amps to flow in this circuit with this same applied voltage, what should the resistance be changed to?

R = ___5___ ohms

c. If you wanted 2 amps to flow through the 10-ohm resistor you had initially, what should the voltage be changed to?

E = ___20___ volts

4. In a certain circuit situation you need a 91-ohm 5%-tolerance resistor that is set up to carry .01 amp of current.

a. What is the color code of this resistor?

    White Brown Black Gold

b. What voltage must be applied to it to yield a .01 amp current flow?   .91

c. Draw a complete schematic of the circuit needed to apply the above voltage to this resistor yielding the .01 amp of current flow?

R= .91Ω

E= .91 V

Answers

3.

3.a. E = 10 volts

3.b. R = 5 Ω

3.c. E = 20 volts

4.

4.a. white, brown, black, gold

4.b. .91 volt

4.c.

Fold Over

4. d. What power rating should the resistor you select have in this circuit? ,0182

5. To really learn the resistor color code, there's no substitute for the old standby, PRACTICE. Shown below are the codes for common resistors you'll be using in the laboratory. When you can write any resistor's value from its code without looking at your chart, you'll have achieved a major objective of this lesson. Fold over the page to check your answers.

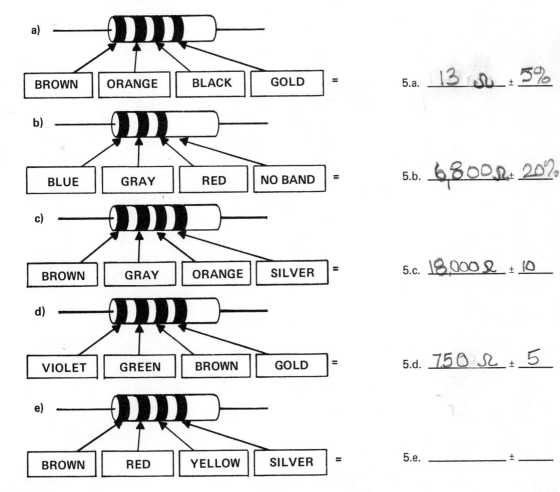

a) | BROWN | ORANGE | BLACK | GOLD | =

5.a. 13 Ω ± 5%

b) | BLUE | GRAY | RED | NO BAND | =

5.b. 6,800 Ω ± 20%

c) | BROWN | GRAY | ORANGE | SILVER | =

5.c. 18,000 Ω ± 10

d) | VIOLET | GREEN | BROWN | GOLD | =

5.d. 750 Ω ± 5

e) | BROWN | RED | YELLOW | SILVER | =

5.e. _____ ± _____

Answers

4.d.   The resistor needs
to dissipate .0091 watt.
        Two times this is
.0182 watt.
        Select a 1/4 watt
resistor.

5.

5.a. 13 Ω ± 5%

5.b. 6800 Ω ± 20% ✓

5.c. 18,000 Ω ± 10%

5.d. 750 Ω ± 5%

5.e. 120,000 Ω ± 10%

Fold Over

5. f)

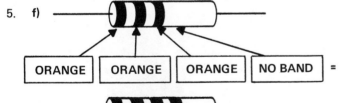

| ORANGE | ORANGE | ORANGE | NO BAND | =

5.f. _____ ± _____

g)

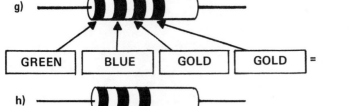

| GREEN | BLUE | GOLD | GOLD | =

5.g. _____5.6_____ ± _5%_

h)

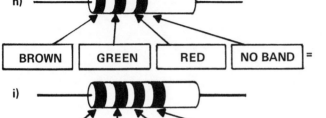

| BROWN | GREEN | RED | NO BAND | =

5.h. _____ ± _____

i)

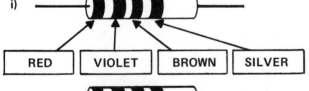

| RED | VIOLET | BROWN | SILVER | =

5.i. _____ ± _____

j)

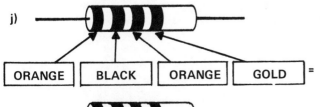

| ORANGE | BLACK | ORANGE | GOLD | =

5.j. _____ ± _____

k)

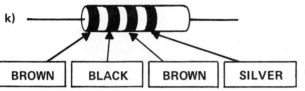

| BROWN | BLACK | BROWN | SILVER | =

5.k. _____ ± _____

Answers

5.f. 33,000 Ω ± 20%

5.g. 5.6 Ω ± 5%

5.h. 1500 Ω ± 20%

5.i. 270 Ω ± 10%

5.j. 30,000 Ω ± 5%

5.k. 100 Ω ± 10%

Fold Over

5. l)

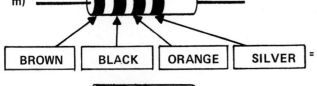

| BROWN | BLACK | RED | NO BAND | = |

5.l. _____ ± _____

m)

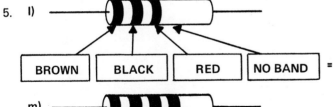

| BROWN | BLACK | ORANGE | SILVER | = |

5.m. _____ ± _____

n)

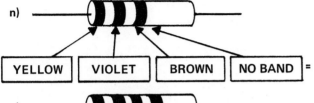

| YELLOW | VIOLET | BROWN | NO BAND | = |

5.n. _____ ± _____

o)

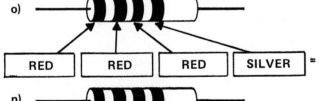

| RED | RED | RED | SILVER | = |

5.o. _____ ± _____

p)

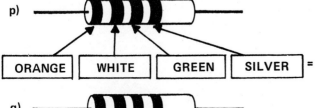

| ORANGE | WHITE | GREEN | SILVER | = |

5.p. _____ ± _____

q)

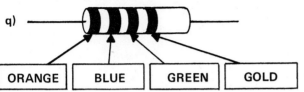

| ORANGE | BLUE | GREEN | GOLD | = |

5.q. _____ ± _____

Answers

5.l.   1000 Ω ± 20%

5.m. 10,000 Ω ± 10%

5.n. 470 Ω ± 20%

5.o. 2200 Ω ± 10%

5.p. 3,900,000 Ω ± 10%

5.q. 3,600,000 Ω ± 5%

Fold Over

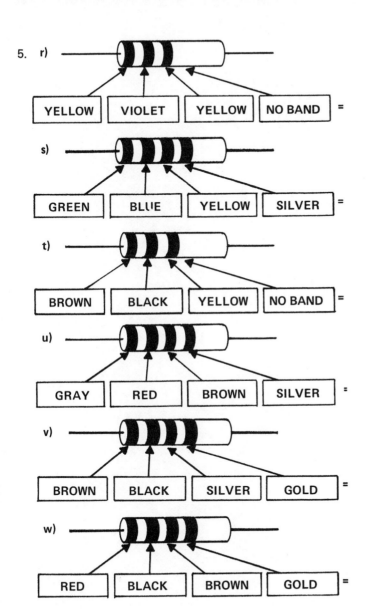

5. r)

YELLOW | VIOLET | YELLOW | NO BAND =

5.r. _____ ± _____

s)

GREEN | BLUE | YELLOW | SILVER =

5.s. _____ ± _____

t)

BROWN | BLACK | YELLOW | NO BAND =

5.t. _____ ± _____

u)

GRAY | RED | BROWN | SILVER =

5.u. _____ ± _____

v)

BROWN | BLACK | SILVER | GOLD =

5.v. _____ ± _____

w)

RED | BLACK | BROWN | GOLD =

5.w. _____ ± _____

Answers

5.r. 470,000 Ω ± 20%

5.s. 560,000 Ω ± 10%

5.t. 100,000 Ω ± 20%

5.u. 820 Ω ± 10%

5.v. .1 Ω ± 5%

5.w. 200 Ω ± 5%

• Practice Problems

Fold Over

6. As mentioned it will be important for you to be able to select resistors for use in circuits, given their value in ohms. For each of the resistor values and tolerances indicated below label the color bands that the resistor would have. Fold over the page half way to check your answers.

6.

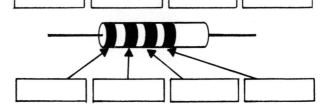

a. 1500 Ω ± 5%   6.a.

b. 330,000 Ω ± 20%   6.b.

c. 7,500,000 Ω ± 10%   6.c.

d. 22 Ω ± 5%   6.d.

e. 470 Ω ± 10%   6.e.

Answers

6.

6.a. brown, green, red, gold

6.b. orange, orange, yellow, no band

6.c. violet, green, green, silver

6.d. red, red, black, gold

6.e. yellow, violet, brown, silver

Fold Over

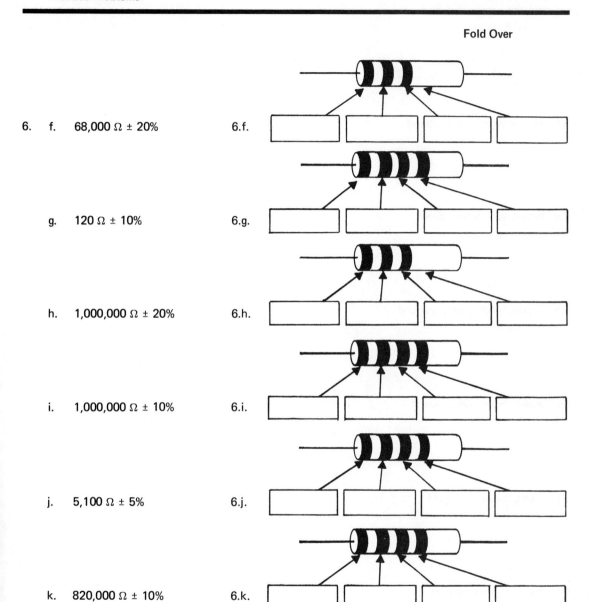

6.   f.   68,000 Ω ± 20%   6.f.

     g.   120 Ω ± 10%   6.g.

     h.   1,000,000 Ω ± 20%   6.h.

     i.   1,000,000 Ω ± 10%   6.i.

     j.   5,100 Ω ± 5%   6.j.

     k.   820,000 Ω ± 10%   6.k.

Answers

6.f.   blue, gray, orange, no band

6.g.   brown, red, brown, silver

6.h.   brown, black, green, no band

6.i.   brown, black, green, silver

6.j.   green, brown, red, gold

6.k.   gray, red, yellow, silver

Fold Over

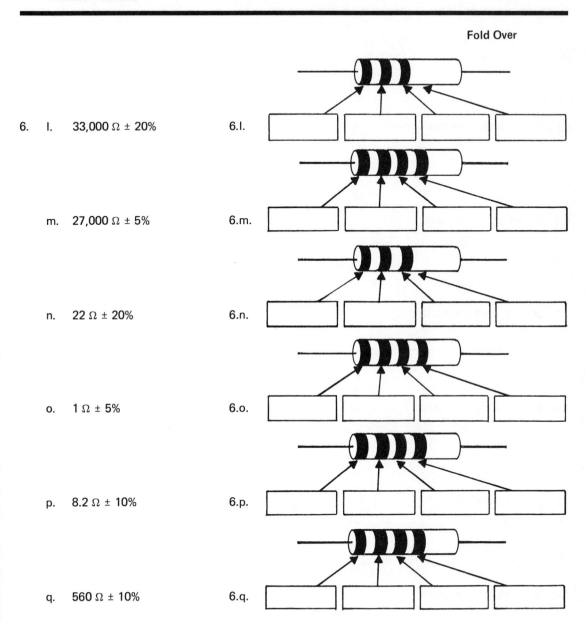

6. l. 33,000 Ω ± 20% 6.l.

m. 27,000 Ω ± 5% 6.m.

n. 22 Ω ± 20% 6.n.

o. 1 Ω ± 5% 6.o.

p. 8.2 Ω ± 10% 6.p.

q. 560 Ω ± 10% 6.q.

Answers

6.l. orange, orange, orange, no band

6.m. red, violet, orange, gold

6.n. red, red, black, no band

6.o. brown, black, gold, gold

6.p. gray, red, gold, silver

6.q. green, blue, brown, silver

1. When a potential difference produced by charged bodies exists in a closed circuit:
   a. More positively charged bodies are said to be at a higher potential than more negatively charged bodies
   b. If free electrons are available, they will flow from the negative to the positive charged bodies
   c. The charged objects provide the push or force required for the flow
   d. If free electrons flow, they will flow until each body has the same charge
   e. all of above
   f. a and c above

2. Potential difference between two charged bodies at each end of conductor can exist:
   a. When one charged body has a positive charge and the other has a negative charge
   b. When the charged bodies have different positive charges
   c. When the charged bodies have different negative charges
   d. When both charged bodies have equal charge
   e. When both charged bodies are neutral (have no charge)
   f. a only
   g. a, b and c above
   h. All of above

3. Voltage is a measure of the:
   a. The electron flow in a circuit
   b. Push that makes electrons move in a closed circuit.
   c. The amount of charge on an object.
   d. All of the above

4. Voltage is also known by:
   a. Potential
   b. Electromotive force
   c. Potential difference between two points on a circuit
   d. All of the above
   e. None of the above
   f. a only

5. The unit of measure of a potential difference is:
   a. The coulomb
   b. The resistor
   c. The volt
   d. A battery

6. Even though there is an electromotive force in a circuit, electrons will not flow until the circuit is:
   a. Opened
   b. Closed
   c. Bypasses
   d. Neutralized
   e. None of the above

7. If one coulomb of electron passes by a point in a circuit in a second, the electron flow is:
   a. One ampere
   b. One volt
   c. One ohm
   d. In the wrong direction
   e. Too small to measure

8. In simple circuits with only one current path between the voltage source:
   a. The voltage source is open
   b. The voltage is the same at all points in the circuit
   c. The current flow is the same at all points in the circuit
   d. The current flows in the wrong direction

9. Conventional and electron current flow:
   a. Can be used to predict the effects of electricity
   b. Flow in opposite directions in a circuit
   c. Both describe the flow of charge in a circuit
   d. Are equivalent if the direction of flow is taken into account
   e. All of above

10. The opposition to current flow in a closed circuit is called:
    a. Resistance
    b. A volt
    c. An ampere
    d. A coulomb
    e. All of above

11. The amount of current flow in a simple circuit depends on the:
    a. Electromotive force in the circuit
    b. Resistance in the circuit
    c. Ohm's Law
    d. c only
    e. a and b above
    f. a, b and c above

12. The unit used to specify the amount of resistance in a circuit is the:
    a. Volt
    b. Ohm
    c. Ampere
    d. Coulomb
    e. Potential

13. Resistance of a conductor is dependent on:
    a. The type of conductor material
    b. The cross-section area
    c. The length of the conductor
    d. The temperature of the material
    e. All of the above
    f. None of the above

14. A carbon composition resistor with Brown, Gray, Orange and Gold bands has a resistance of:
    a. 1800 ohms ±10%
    b. 183 ohms
    c. 18,000 ohms ± 5%
    d. 1830 ohms ± 10%
    e. 18.3 ohms

15. A resistor with Red, Violet, Gold, Gold bands has a resistance of:
    a. 27 ohms ± 5%
    b. 2.7 ohms ± 5%
    c. 270,000 ohms ± 5%
    d. 27,000 ohms ± 5%
    e. 27 ohms ± 10%

16. A resistor with Blue, White, Green, Silver bands has a resistance of:
    a. 6,900,000 ohms ± 10%
    b. 69,500 ohms ± 10%
    c. 6.95 ohms
    d. 690,000 ohms
    e. 6950 ohms ± 10%

17. A composition resistor having a resistance of 4,700 ohms ± 5% has the following color code:
    a. Yellow, violet, yellow, silver
    b. Yellow, violet, red, gold
    c. Orange, blue, red, gold
    d. Green, gray, red, gold
    e. Yellow, violet, gold, gold

18. A 200 ohm resistor is connected across a 20 volt battery. The current is:
    a. 1 amp
    b. 0.01 amp
    c. 10 amps
    d. 0.1 amp
    e. None of the above

19. The power dissipated in the resistor of the above circuit is:
    a. 20 watts
    b. 2 watts
    c. 200 watts
    d. 4000 watts
    e. 4 watts

20. A 1.2 watt light bulb is placed across a 12 volt battery. How much resistance does it have?
    a. 10 ohms
    b. 12 ohms
    c. 144 ohms
    d. 1.2 ohms
    e. 120 ohms

Lesson 3

Scientific Notation and Metric Prefixes

This lesson introduces and examines scientific notation, "powers of ten" notation, and metric prefixes used with electrical quantities. Included also is a discussion of conversion from one form of notation to another. Addition, subtraction, multiplication, and division with powers of ten are introduced and examined in detail.

LESSON 3. SCIENTIFIC NOTATION AND METRIC PREFIXES

• Objectives

At the completion of this lesson you should (with pencil and paper) be able to:

1. *Convert* any number into its:

   a. Decimal

   b. Scientific notation ("powers of ten")

   c. Metric prefixed form

   rounding off where appropriate, given the number in any one of these three forms.

2. *Convert* numbers from any nonstandard ("incorrect") powers of ten or metric prefixed form into standard ("correct") form.

3. *Add* and *subtract* numbers expressed in powers of ten form.

4. *Correctly manipulate the exponents* as required in the multiplication and division of numbers expressed in powers of ten form.

- Useful Tools in the Study of Electricity
- Voltage
- Electron Current

Useful Tools in the Study of Electricity — The goals that will be discussed in this lesson (Figure 3.1), are the metric prefixes used with electrical quantities, and scientific notation, which is more commonly called "powers of ten" notation. With the aid of these tools, the extremely large or small quantities you will often encounter in electronic calculations can be handled with greater ease.

Before these tools are discussed, it will be helpful to refresh your memory on the fundamental concepts discussed thus far. Essentially, three basic ideas concerning electric circuits have been discussed.

Metric Prefixes

SCIENTIFIC NOTATION

Figure 3.1

Voltage — The first concept discussed is voltage, electromotive force, or potential difference (Figure 3.2). This is the pressure which causes electrons to move in a conductor. In electrical formulas and equations, you will see voltage symbolized with a capital E, while on laboratory equipment or schematic diagrams the voltage is often represented with a capital V.

VOLTAGE
ELECTROMOTIVE FORCE
POTENTIAL DIFFERENCE
...THE PRESSURE WHICH CAUSES
ELECTRONS TO MOVE...

Figure 3.2

Electron Current — The second concept, shown in Figure 3.3, is electron current, or amperage, which is the movement of free electrons through a conductor. In electrical formulas, current is symbolized with a capital I, while in the laboratory or on schematic diagrams, it is common to use a capital A to indicate amps or amperage.

ELECTRON CURRENT

...THE MOVEMENT OF FREE ELECTRONS
THROUGH A CONDUCTOR

Figure 3.3

- **Resistance**
- **Resistor Color Code**
- **Large and Small Quantities**

Resistance — Resistance is the third basic property found in an electric circuit and in Figure 3.4 it is defined as the opposition to current flow. *Resistors* are manufactured components which have specific resistances. As you have seen, the schematic symbol for a resistor is —\/\/\/— ; in electrical formulas and equations, a capital R is used to indicate resistance. On schematic diagrams and in the laboratory, the Greek letter omega (Ω) is used to indicate the unit of resistance, the ohm.

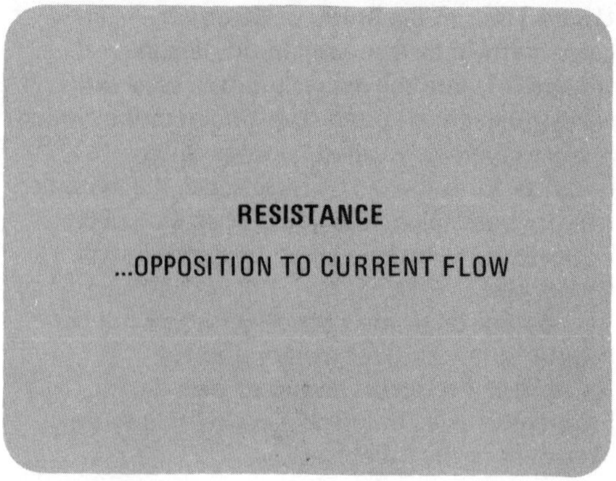

RESISTANCE

...OPPOSITION TO CURRENT FLOW

Figure 3.4

Resistor Color Code — The resistor color code, listed in Figure 3.5, is used to determine the ohmic value of a composition resistor from the color bands painted on it. The memory key included in the figure will help you to remember the code.

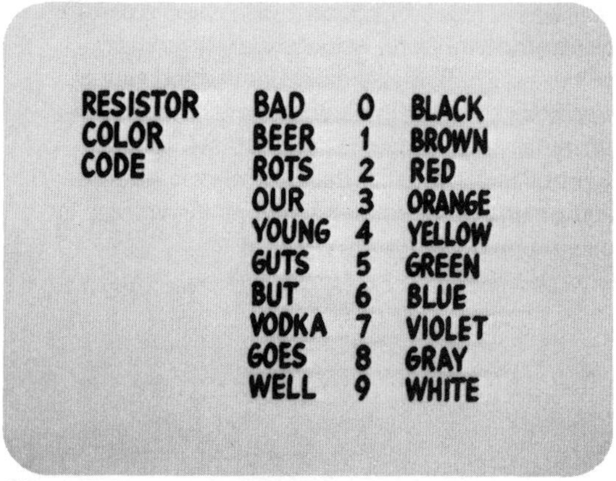

RESISTOR COLOR CODE	BAD	0	BLACK
	BEER	1	BROWN
	ROTS	2	RED
	OUR	3	ORANGE
	YOUNG	4	YELLOW
	GUTS	5	GREEN
	BUT	6	BLUE
	VODKA	7	VIOLET
	GOES	8	GRAY
	WELL	9	WHITE

Figure 3.5

Large and Small Quantities — Getting into the topic of this lesson, recall that in Lesson 2, Ohm's law was introduced and it was stated that resistors sometimes have very large values. Because of the large range of values encountered when working with electricity and electrical components, a method for easier handling of extremely large and extremely small quantities is needed. As has been stated, this lesson deals with methods that make writing and handling these quantities much simpler and easier.

Ohm's Law — An example illustrating some of the reasons for learning these methods, is shown in Figure 3.6. In this figure, a circuit with an unknown voltage is forcing 10 amps of current through 10,000 ohms of resistance.

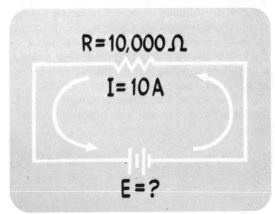

Figure 3.6

To find this unknown voltage, you would use the form of Ohm's law shown in Figure 3.7, which states that voltage equals current times resistance. When the 10 amps of current in the circuit are multiplied by the 10,000 ohms of resistance, you find the required voltage to be 100,000 volts.

This is just one example of a sizable quantity you might find when carrying out your own electrical calculations. An easier way to handle large quantities such as this, as well as the very small quantities, is called scientific notation or "powers of ten" notation.

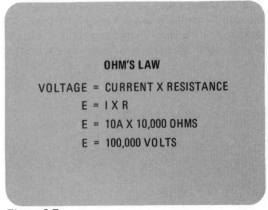

Figure 3.7

Metric Chart — To aid in your study of these important tools, examine Figure 3.8 which illustrates what is called a "metric chart". This chart is designed to help you understand and use scientific notation, and what are called metric prefixes.

The vertical line on Figure 3.8 represents the location of the decimal point for ordinary, nonprefixed units such as volts, amps, or ohms. Thus, the three zeros written on the chart are labeled as units, and in the decimal system they represent the position of the ones, tens, and hundreds units.

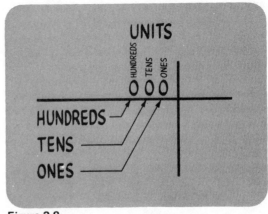

Figure 3.8

Decimal 2 — To illustrate how numbers are written on the chart, consider the simple number 2, with no decimal point written next to it. As shown in Figure 3.9, when a number has no decimal point, it is important to realize that the decimal point *is always understood to be to the right of the last digit.* Thus, the simple number 2, written by itself, may be considered to have a decimal point on its right.

Figure 3.9

2 on the Metric Chart — When correctly placed on the metric chart as shown in Figure 3.10, the number 2 falls under the ones unit column, and so the number 2 represents two "ones" on the chart.

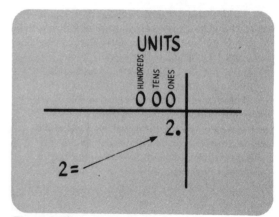

Figure 3.10

452 on the Metric Chart — Another example of how numbers are written on the chart is shown in Figure 3.11. Here the number 452 has been placed on the metric chart. Note that the 4 falls under the hundreds column, the 5 falls under the tens column and the 2 falls under the ones column. Remember, this number has a decimal point to the right of the 2. So, the number 452 represents 4 one hundred units, 5 ten units, and 2 one units, and is commonly called four hundred and fifty-two.

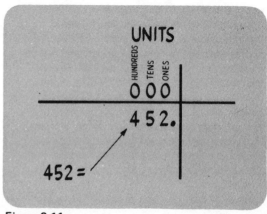

Figure 3.11

Scientific Notation — The important use of the metric chart will not be in analyzing these common numbers, but in helping you to correctly use scientific notation and the metric prefixes. So before the chart is discussed further, an introduction to scientific notation and some discussion of how it works will be helpful.

Basically, scientific notation is simply a type of shorthand which is used to keep track of decimal places. To show you some of the key parts of scientific notation, consider the number 100, shown in Figure 3.12. As shown, the number 100 can be written as 10 to the +2 "power". Now focus your attention on the special positioning and significance of this "power".

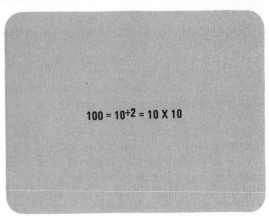

$$100 = 10^{+2} = 10 \times 10$$

Figure 3.12

The Power or Exponent — In Figure 3.13, this power, or *exponent* as it is sometimes called, is shown as a number written above and to the right of another number called the *base*. Now here's a key point: The power is a number which simply indicates how many times the base is multiplied by itself. Thus, in Figure 3.12, ten to the power two, or ten to the second power, equals 10 times 10, which is 100. In this course, only *powers of ten are used*. That is, ten is the base number commonly used for writing numbers in scientific notation. It is for this reason that scientific notation is often referred to as "powers-of-ten" notation. You will find these terms used interchangeably in most discussions on handling numbers in electronics.

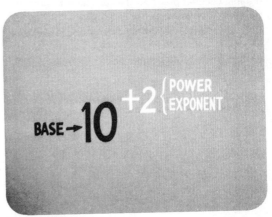

$$\text{BASE} \rightarrow 10^{+2} \begin{cases} \text{POWER} \\ \text{EXPONENT} \end{cases}$$

Figure 3.13

Rule for Powers of Ten — When a number is put into "powers of ten" form, a standard or "correct" method of writing the number is commonly used. It is easiest to write any number in standard "powers of ten" form by remembering the rule illustrated in Figure 3.14. The number being placed into standard "powers of ten" form should first be written in the form: digit, point, digit, digit.

RULE FOR POWERS OF TEN

THE NUMBER MUST BE WRITTEN
DIGIT, POINT, DIGIT, DIGIT.

Figure 3.14

452 in Standard Powers of Ten Form — In Figure 3.15, the number 452 is again placed on the metric chart, this time to show how it may be converted to standard powers of ten form. To put this number into digit, point, digit, digit form, the decimal point must be moved two places to the left from its original position until it is between the 4 and the 5. This puts 452 into digit, point, digit, digit form: 4.52. To complete writing the number in scientific notation, it is written as 4.52 times 10 to some *positive* power, which you will write in the position of the asterisk, once you determine its value.

    Here's an important procedure: You *count* the number of spaces you moved the decimal point. That is, the number of digits between the original decimal point on the vertical line, and the new one between the first two digits.

    This number of places moved becomes the exponent, and so the number 452 is correctly written in standard powers of ten form as $4.52 \times 10^{+2}$.

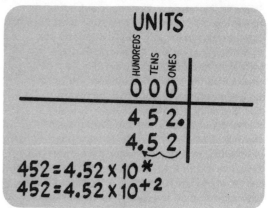

Figure 3.15

Significant Digits — You may wonder how to proceed if you have to write a number in scientific notation that has *more* than three digits, such as the number in Figure 3.16. To write a number in the digit, point, digit, digit form for scientific notation or powers of ten, only what are called the *first three most significant digits* are used.

Figure 3.16

Three Most Significant Digits — What is meant by the most significant digit of a number? Consider this example. Suppose you were out shopping for a new car and the price of the car you wanted was $3,965.95 (Figure 3.17). Ask yourself which of these numbers is the most significant to you? Well, the 3 of course, because it represents $3000. The 9 is the next most significant digit because it represents $900. The 6 is less significant than the 9 because it represents $60. So, of the first three digits, the 6 is called the *least significant digit*. It is obvious that the last figures are not as significant as the first ones. Likewise, in the study of electricity or electronics, the first three significant digits are usually the most important since most components indicate values with only three significant digits.

Figure 3.17

Rounding Off — If you are starting with a number that has more than three digits, you can reduce it to a number with only three significant digits by a procedure called *rounding off*. The procedure for rounding off, shown in Figure 3.18, is as follows:

1. Keep the first three most significant digits of the number and look at the fourth one.
2. Increase the third significant digit by one if the fourth significant digit is 5 or more.
3. Leave the three most significant digits as they are if the fourth significant digit is less than 5.
4. Drop any remaining digits beyond the third (replacing them with zeros up to the decimal point).

ROUNDING OFF

- KEEP THE 3 MOST SIGNIFICANT DIGITS.
- IF THE 4TH DIGIT IS 5 OR MORE, INCREASE THE 3RD DIGIT BY 1.
- IF THE 4TH DIGIT IS LESS THAN 5, LEAVE THE FIRST 3 AS THEY ARE.

Figure 3.18

An Example of Rounding Off — In the example shown in Figure 3.19, the car sticker price discussed earlier, $3965.95, when correctly rounded off, becomes $3970. Since the fourth significant digit was 5, the third significant digit is increased by one. To go one step further and write this number in scientific notation, the decimal point must first be moved three places to the left to put the number in standard digit, point, digit, digit form. Recall that since the decimal point is moved *three places to the left*, the power of ten (or exponent) is 3, and 3970 can be written as $3.97 \times 10^{+3}$.

$$3965.95 = 3970$$
$$3970. = 3.97 \times 10^{+3}$$

Figure 3.19

"Kilo" on the Metric Chart — With this introduction to scientific notation, the metric chart will be expanded so that the role and function of the *metric prefixes* can be introduced. More detail and examples covering "powers of ten" notation will be included along the way.

Figure 3.20 shows an expansion of the metric chart with three more zeros to the left of the units group. This group of decimal places represents thousands, and in the metric system, a prefix called *kilo* is used to represent thousands and is abbreviated k. Note that under the kilo group there is a one k column, a ten k column, and a hundred k column. It is important to realize that when the metric prefix kilo or k is used, a decimal point must be placed between the units group and the k group. This new decimal point is called the decimal point for *k*.

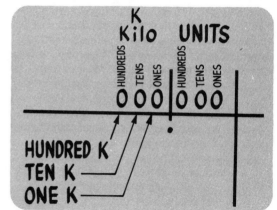

Figure 3.20

- **45,000 on the Metric Chart**
- Use of Prefixes with Electrical Units
- **45 k in Powers of Ten Form**

45,000 on the Metric Chart — Figure 3.21 shows how 45,000 is placed on the metric chart. Notice carefully that the 45 appears under the k group. This is interpreted as 4 ten k and 5 one k; so in abbreviated form, this number can be called 45 k which means the same as 45 thousand.

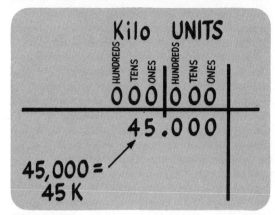

Figure 3.21

Use of Prefixes with Electrical Units — If this 45 k were the result of a calculation for a resistance in ohms, you would then have 45 kilohms or 45 kΩ. The k would be written right in front of the symbol for ohms and that is why it is called a metric *prefix*. In other calculations, you may see results in kilovolts (abbreviated kV) or even kiloamps (abbreviated kA).

45 k in Powers of Ten Form — For some additional practice, follow through the steps that would be needed to write 45 k in powers of ten form. First, a decimal point must be placed between the most significant digit and the second most significant digit as shown in Figure 3.22. This puts the number in the standard digit, point, digit, digit form. Then the number is written as 4.5 times 10 to some power. The next step is simply to count over from the original decimal point to the new decimal point — four decimal places — so the exponent of ten is 4. The number 45 k in correct or standard scientific notation is then $4.5 \times 10^{+4}$.

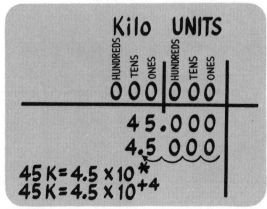

Figure 3.22

You may ask why scientific notation or powers of ten notation is used. One obvious reason is that it saves a lot of inconvenience and possible error when writing very large or small numbers. The number 4.79×10^{12} is much easier to write than 4,790,000,000,000. Another important reason is that extremely large or small numbers can be multiplied or divided much easier when the numbers are written in powers of ten form. In particular, it is much easier to make quick estimates and to keep track of the decimal point in multiplication or division when the numbers are first put in powers of ten form. The ins and outs of multiplication and division of numbers using scientific notation will be discussed later. For now, consider some additional features of the metric chart and some additional examples of how it is used.

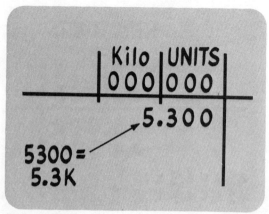

Figure 3.23

Additional Examples — Figure 3.23 shows how 5,300 is placed on the metric chart. Notice that the 5 appears under the one k column and the 3 is under the hundred unit's column. Notice also that the decimal point for kilo is between the 5 and the 3; thus, you can say that 5,300 equals 5.3 k.

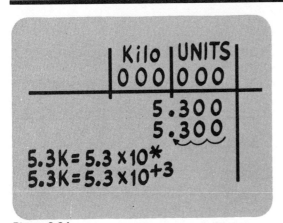

Figure 3.24

Figure 3.24 shows how you can convert 5.3 k to scientific notation. In the number 5.3 k, the decimal point is already in the proper location, that is, between the most significant and the second most significant digit. Then write the number as 5.3 times 10 to some positive power. Next count over from the original decimal point, which, as you recall is always represented by the central vertical line on the chart, to the location of the new decimal point. This is 3 decimal places to the left, so the exponent or power of ten is +3. So, 5.3 k in standard scientific notation is $5.3 \times 10^{+3}$.

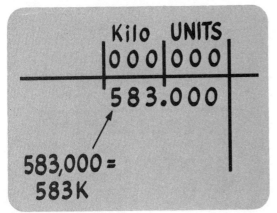

Figure 3.25

As another example, Figure 3.25 shows how 583,000 is placed on the metric chart. Check carefully and notice that the 5 falls under the hundred k column, the 8 falls under the ten k column, and the 3 falls under the one k column. So, 583,000 is interpreted as 5 hundred k, 8 ten k, and 3 one k. In metric prefix form, it is called 583 k, which means the same as (or equals) 583,000.

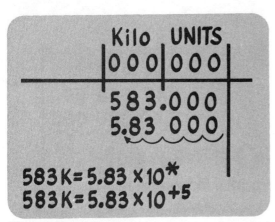

Figure 3.26

Figure 3.26 shows how you can convert 583 k to powers of ten notation. First, place a decimal point between the 5 and the 8, that is, between the most significant and the second most significant digit. Then write the number as 5.83 times 10 to some positive power. Next, simply count over from the original decimal point (the vertical line) to the location of the new decimal point. You should count five decimal places to the left. Thus, 583,000 equals $5.83 \times 10^{+5}$ in standard scientific notation.

"Mega" on the Metric Chart — Figure 3.27 shows how the metric chart can be further expanded to accommodate millions. This is done by simply adding three more zeros to the left of the k group.

Note that a new decimal point appears between the k group and the millions group and this is called the decimal point for *mega*. Mega, or meg, is the metric prefix representing millions and is abbreviated capital *M*.

This new group takes on the same characteristics as the two preceding groups, and you will see that there is a column for one mega, ten mega, and hundred mega.

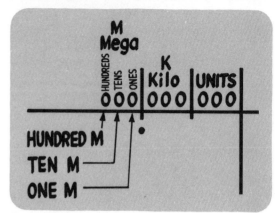

Figure 3.27

"742 Million" on the Metric Chart — To practice using the chart for numbers in this new group, consider Figure 3.28, which shows the number 742 million in its correct position. Notice that it appears as 7 one hundred mega, 4 ten mega, and 2 one mega. So, this number can be simply abbreviated as 742 M.

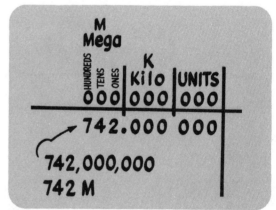

Figure 3.28

742 Million in Powers of Ten Notation — Figure 3.29 shows how 742 M can be written in scientific notation. By now the steps should be getting fairly familiar to you. First a decimal point is placed between the most significant digit and the second most significant digit. The second step is to write 7.42 times 10 to some power.

As usual, the third step is to count the number of decimal places that the original decimal point was moved to get to the new decimal point between the 7 and the 4. This is eight decimal places to the left, and so +8 is used as the exponent in the powers of ten form. The result is $7.42 \times 10^{+8}$.

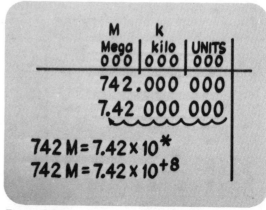

Figure 3.29

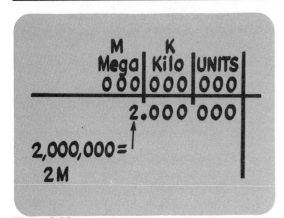

Figure 3.30

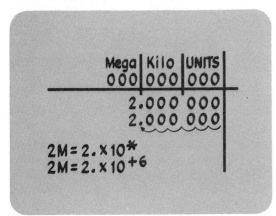

Figure 3.31

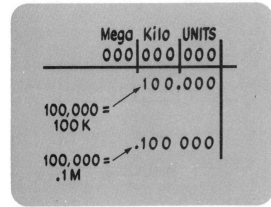

Figure 3.32

Additional Examples — Here are some additional examples illustrating how numbers in the mega group are handled using the metric chart. Figure 3.30 shows 2 million on the chart, and notice that it appears as 2 one mega. This is abbreviated simply as 2 M.

Figure 3.31 shows how you can convert 2 M to standard scientific notation. First, notice that the decimal point for mega falls just to the right of the 2, or between the most significant and the second most significant digit, which in this case is just a zero. Next, count over from the original decimal point (the vertical line) to the new decimal point, which is six decimal places to the left. So the exponent is +6. Thus, 2,000,000 equals $2 \times 10^{+6}$ in standard scientific notation.

Notice an important point. Often when writing numbers in scientific notation, if the least significant digit (or digits) is a zero, the zero (or zeros), and often the decimal point are left off the number. You may see a number written as $2 \times 10^{+6}$, or $2.00 \times 10^{+6}$, either means the same thing.

Figure 3.32 shows 100,000 on the chart. How would you write this in abbreviated form? As you will see there are two ways to do it, and both of them are acceptable. Normally, you would say that since the 1 falls under the hundred k column, 100,000 should equal 100 k. That is the standard form that you have been using and it is perfectly all right, but there is another way to abbreviate 100,000. If you use the decimal point in the mega group, it falls just to the left of the 1. So, you could also write 100,000 as *.1 M*, which equals 100 k.

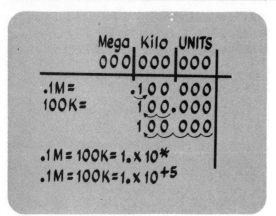

Figure 3.33

Figure 3.33 illustrates how you can convert .1 M or 100 k to standard scientific notation. The key is to use the metric chart to correctly keep track of the decimal places. Here is a list of the steps you need to complete the process.

1. If the number is expressed as .1 M, first locate the decimal point for mega (M) on the metric chart and place a 1 in the column immediately to the *right* of it — for .1 M.

or 2. If the number is expressed as 100 k, locate the decimal point for the kilo group and write 100 on the chart immediately to the left of it — for 100 k.

3. In either case, the 1 should be in the hundred k column. If you fill in all the places between the 1 and the vertical line (representing the position of the original decimal point), you see that the number again just equals 100,000.

4. Place a decimal point in the digit, point, digit, digit position on this number.

5. Count the number of places between the vertical line and the position of this new decimal point — here there are five places.

6. This *number of places* then becomes the power or exponent of the number in correct powers of ten form: $1.00 \times 10^{+5}$.

"Giga" on the Metric Chart — To handle the largest numbers usually encountered in the study of electricity, the metric chart must be expanded to include one more group. Figure 3.34 shows how the chart can be expanded to include quantities that fall in the billions range. The metric prefix for billion is called giga and is abbreviated capital *G*.

The decimal point between the mega group and the giga group is called the giga decimal point. The giga group has the same characteristics as the three preceding groups. That is, it has a one giga column, a ten giga column, and a hundred giga column, from right to left.

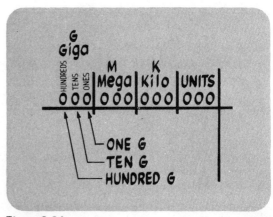

Figure 3.34

"9.456 Billion" on the Metric Chart — As shown in Figure 3.35, the number 9.456 billion can be written in abbreviated form as 9.46 giga. Notice again that in your study of electricity, you will normally only use the first three significant digits of any number. In this case, the fourth significant digit is greater than 5, so when the number is rounded off the third significant digit is increased by one.

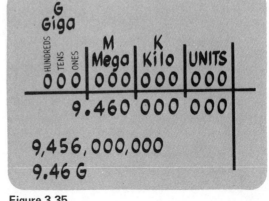

Figure 3.35

9.46 Giga in Scientific Notation — Figure 3.36 shows how 9.46 giga can be written in powers of ten form. All the steps used in the prior examples are used, and the result is $9.46 \times 10^{+9}$.

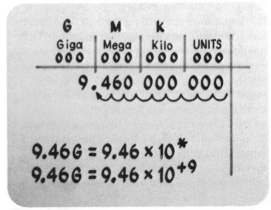

Figure 3.36

To follow through the steps needed to complete this, first locate the decimal point for giga on the chart (right between the giga and mega groups). Write the 9.46 giga on the chart so that its decimal point is right on decimal point for giga position. (If this is done correctly then the 9 will be in the one giga column, the 4 in the hundred mega column, and the 6 in the ten mega column.) Next, fill in all the places left empty between the 9.46 and the vertical line with zeros. (Remember that each of the units, kilo, mega, and giga columns has *three* places.) The number is already written in digit, point, digit, digit form, therefore, write it as 9.46 times 10 to some power. Count the number of places between the vertical line and the decimal point in the digit, point, digit, digit position, there are nine places. Plus 9 then becomes the exponent of 9.46 giga in scientific notation $9.46 \times 10^{+9}$.

".05 M" on the Metric Chart — Up to this point, this lesson has dealt with numbers in their decimal form, then showed how they can be written with metric prefixes and put in powers of ten form. An example of a number that is abbreviated in a nonstandard way or "incorrectly" abbreviated might be .05 mega. This number can be written more correctly with a different prefix.

To convert the number to a more standard form, first write it on the metric chart as shown in Figure 3.37. With the chart as an aid, you can see that .05 M actually represents 5 ten k and correctly abbreviated in standard form it becomes 50 k.

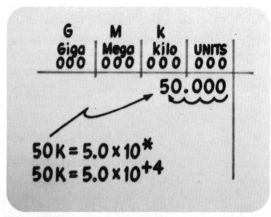

Figure 3.37

50 k in Scientific Notation — Figure 3.38 shows how 50 k can be written in powers of ten form. Following the procedures outlined earlier, the exponent is found to be +4 and 50 k is written $5 \times 10^{+4}$. Notice this number is not usually written $5.00 \times 10^{+4}$. As you have seen, when there are only zeros to the right of the decimal point, the zeros and the decimal point are usually omitted.

Figure 3.38

5,560 k Equals 5.56 M — In Figure 3.39 is another example of a number that can be expressed more simply with a different metric prefix. The number 5,560 k does not fit into what has been called standard form. A better way to express this number is 5.56 M. When written in correct powers of ten form, it is $5.56 \times 10^{+6}$.

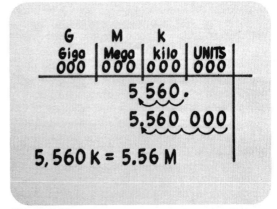

Figure 3.39

Additional Shorthand: 10^0 = ? — Consider a few important additional topics before proceeding. Often, in many texts, you will find that when a number greater than one is expressed in scientific notation, the plus sign in front of the exponent is left off. This means that a number such as $3.00 \times 10^{+7}$ may be written as 3×10^7. (Recall, also, that it is common to omit least significant digits when they are zeros; and, in some instances, the decimal point is omitted when both of the least significant digits are zeros.) In this text, the standard form will be adhered to for the most part to minimize confusion.

Consider one other item, the number 5.678. How might this be expressed in scientific notation? The decimal point is already in the correct digit, point, digit, digit position, round the number off and it becomes 5.68. Since the decimal point is not moved at all, or actually it is moved zero places, this number could also be expressed as 5.68×10^0. Here's a key point: 10^0 equals one; and in mathematics, *any base number* raised to the zeroth power equals one.

Now turn from the use of scientific notation and metric prefixes in handling very large quantities to see how these techniques might be used to handle the *smaller* quantities commonly encountered in the study of electricity.

- Quantities Less than One on the Metric Chart
- ".001" on the Metric Chart

Quantities Less than One on the Metric Chart — So far the metric chart has been developed for use with numbers that are one or greater. Next quantities are considered that are some fractional part of a whole number, that is, numbers *less than one*. It is important to learn how to put these in powers of ten form, and also how to use their metric prefixes, because electronic calculations can often result in extremely small quantities, as well as the extremely large ones you have been shown so far.

You will find that the procedure to follow for handling numbers less than one is very similar to the one used in dealing with large quantities. As a first step, the metric chart is expanded to begin handling these quantities.

In Figure 3.40, three zeros have been added to the *right* of the original decimal point line on the metric chart. The *prefix* used to represent this new group is called *milli*, and it is abbreviated with a small *m*. *Be certain not to confuse this with the capital M used for mega numbers.*

This group has characteristics similar to the larger quantities discussed previously, and among them, it requires a new decimal point. The new decimal point for milli is placed to the right of the group as shown. The milli group also has a one milli column, a ten milli column, and a hundred milli column. Note that the one milli column is just to the left of the decimal point for milli and that the ten milli and hundred milli columns are arranged to the left in the same way the columns were arranged for numbers larger than one.

".001" on the Metric Chart — In Figure 3.41, the number .001 is placed on the chart, where you can see that it can also be called one milli, and is abbreviated 1 m.

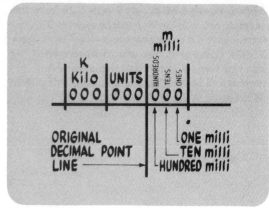

Figure 3.40

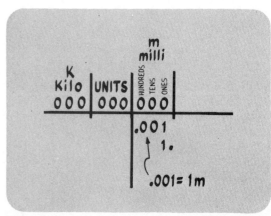

Figure 3.41

.001 in Powers of Ten Form — In order to change .001 to powers of ten form, as usual a decimal point must first be placed between the most significant digit and the second most significant digit. In this case, it simply goes to the right of the 1, because there is no second most significant digit.

Another key point regarding powers of ten: When working with numbers larger than one, the number of places that the decimal point was moved to the *left* becomes the exponent. When a number *less than one* is written in scientific notation, the places that the decimal point is moved to the *right* from the vertical line becomes the exponent or power of ten. A *minus sign* is put in front of the power to indicate a quantity that is less than one. Thus, in Figure 3.42, .001 or one milli is written as 1×10^{-3}.

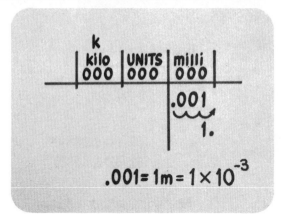

Figure 3.42

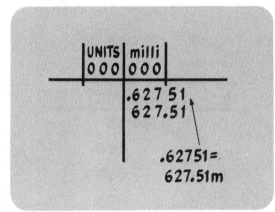

Figure 3.43

Additional Examples — As another example, Figure 3.43 shows how the number .62751 can be converted to the metric prefix form, and then Figure 3.44 shows how it can be written in correct powers of ten form. First, write the number on the metric chart; notice that there are 6 hundred millis, 2 ten millis, and 7 one millis. An important point to notice here deals with rounding off numbers less than one. The most significant digit in this number is the 6 because it is closest to the original decimal point line (the vertical line), and hence represents the largest, most significant quantity. The next two significant digits are the 2 and the 7 in that order. Now, the 5 must also be considered in rounding off this number. So the number .62751 is correctly

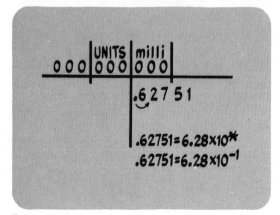

Figure 3.44

expressed as 628 milli. To put this number into standard powers of ten form, first move the decimal until the number is in digit, point, digit, digit form with the decimal point between the most significant and second most significant digit. Notice again that you move the decimal point to the right one place to put the number in standard form. Since you move the decimal point one place to the right, the exponent is *minus* one. Thus .62751 equals 6.28×10^{-1} in standard powers of ten notation.

Micro on the Metric Chart — As the chart is expanded to take in even smaller numbers, keep in mind that numbers farther to the right from the original decimal point line become smaller and smaller, and that the number farthest from original decimal point line is the least significant digit.

Figure 3.45 shows a new group called the *micro group*, and its abbreviation is the Greek letter μ (mu). As before, this group also has a one micro column, a ten micro column, and a hundred micro column. The decimal point for micro is to the right of the one micro column.

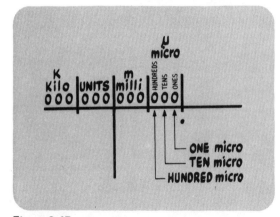

Figure 3.45

".00058" on the Metric Chart — How can the number .00058 be written with a metric prefix? If this number is placed on the chart (Figure 3.46), you will see that it represents 5 one hundred micro and 8 ten micro. So in abbreviated form, it would be called 580 micro (five hundred and eighty micro).

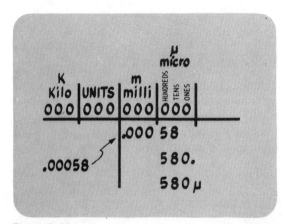

Figure 3.46

580 Micro in Scientific Notation — To write 580 micro in scientific notation (Figure 3.47), a decimal point is placed between the most significant digit and the second most significant digit which is between the 5 and the 8.

Remember, numbers that are less than one must have a negative exponent. The number of places from the original decimal point line to the new decimal point is used as the exponent. Thus 380 micro becomes 5.8×10^{-4}.

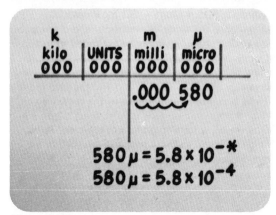

Figure 3.47

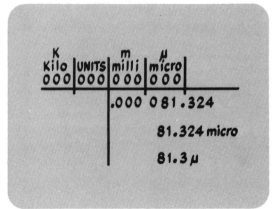

Figure 3.48

Additional Examples — Figure 3.48 shows .000,081,324 on the metric chart. The 8 falls under the ten micro column, and the 1 falls under the one micro column. Since the decimal point for micro is between the 1 and the 3, this number can be written as 81.324 micro. To round off this number you keep the three most significant digits and look at the fourth significant digit which is 2. Since 2 is less than 5, drop the 2 and the 4, which leaves 81.3 micro, when rounded off.

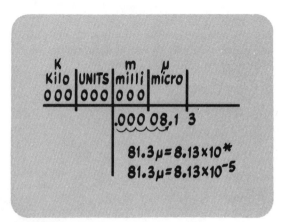

Figure 3.49

Figure 3.49 shows how you can convert 81.3 micro to standard scientific notation. First move the decimal point so that it is between the 8 and the 1. Now count over from the original decimal point (the vertical line) to the new decimal point. You should count five decimal places to the right, which means the power of ten is −5. Thus, 81.3 micro equals 8.13×10^{-5} in standard powers of ten form.

Complete Metric Chart — For most practical purposes in the study of electricity and electronics, the metric chart need only be extended to include two more groups. Figure 3.50 shows the complete metric chart including the last two groups you will probably need, the nano group and the pico group. These two groups are handled in the same way as the others. It's important to note that "pico" is a relatively new term that replaces the older term "micro micro", but you may still see the micro micro notation in use.

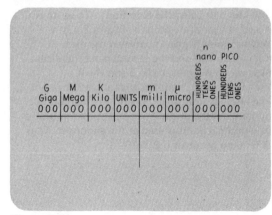

Figure 3.50

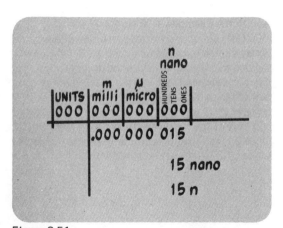

Figure 3.51

Nano and Pico Group in Detail — The nano group, which is to the right of the micro group, has a one nano column, a ten nano column, and a hundred nano column. The abbreviation for nano is a small n. The decimal point for nano is to the right of the one nano column.

Figure 3.51 shows .000,000,015 on the metric chart. The 1 falls under the ten nano column and the 5 falls under the one nano column; so this number is abbreviated 15 n (15 nano).

Figure 3.52 shows how you can convert 15 nano to standard scientific notation. Try this one on your own. If you follow the same procedure as before, you should get 1.5×10^{-8}.

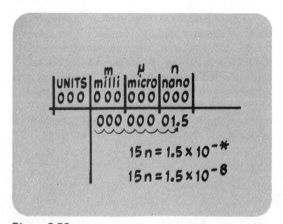

Figure 3.52

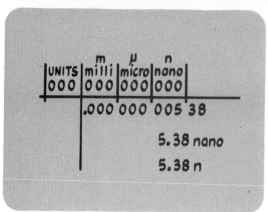

Figure 3.53

Figure 3.53 shows .000,000,005,38 on the metric chart. The 5 falls under the one nano column and the nano decimal point is to the right of the 5, so this number is 5.38 n in abbreviated form.

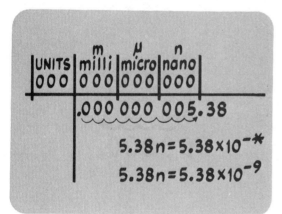

Figure 3.54

Figure 3.54 shows the conversion of 5.38 nano to 5.38×10^{-9}. Remember, the exponent has a minus sign because the decimal point is moved to the right of the vertical line.

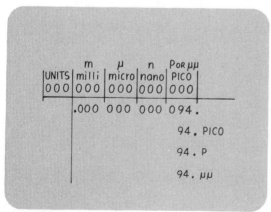

Figure 3.55

The next group on the right in Figure 3.55 is the pico group. Pico is abbreviated with a small p or, as mentioned before, it is sometimes called micro micro and abbreviated $\mu\mu$. The pico group has a one pico column, a ten pico column, and a hundred pico column. The decimal point for pico is to the right of the one pico column.

Figure 3.55 shows .000,000,000,094 on the metric chart. The 9 falls under the ten pico column, and the 4 falls under the one pico column, so this number is abbreviated 94 pico.

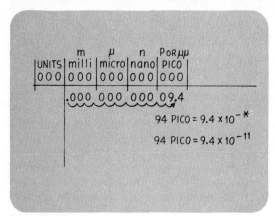

Figure 3.56

Figure 3.56 shows the conversion of 94 pico to 9.4×10^{-11}.

Converting a Number from Scientific Notation to a Correctly Prefixed Number — So far this lesson has shown how to write numbers with the correct metric prefix, and then how to write them in powers of ten form. You may wonder how to reverse this process. How can a number be converted *from scientific notation to a correctly prefixed number*? An example illustrating how to do this is shown in Figure 3.57. To convert $5.3 \times 10^{+4}$ to its correctly prefixed metric form, first write 5.3. Remember the *positive* exponent represents the number of places the decimal point was moved to the *left* from the original decimal point line. So the procedure can be *reversed* by moving the decimal point back four places to the *right*. Next the blank spaces to the right of the 3 are filled with zeros, which puts the number in its decimal form. Now convert it to metric prefixed form, you simply put it on the metric chart.

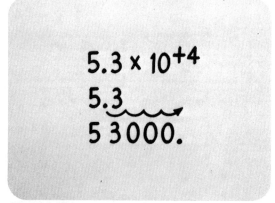

Figure 3.57

- "53,000" on the Metric Chart
- Converting Numbers with Negative Exponents to their Decimal Form

"53,000" on the Metric Chart — As you can see in Figure 3.58, when 53,000 is placed on the chart, it is equivalent to 53 k, because the 5 is in the ten k column and the 3 is in the one k column.

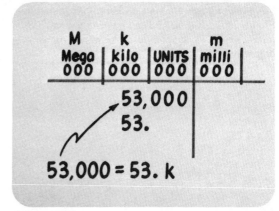

Figure 3.58

Converting Numbers with Negative Exponents to their Decimal Form — Numbers with a negative exponent are handled basically in the same way as positive exponent numbers. Consider the number 75.3×10^{-7}. Notice that this number isn't written in the standard digit, point, digit, digit form, yet as you will see, converting to the correct metric prefixed form will be easy. When a number *less than one* is originally written into powers of ten form, remember that the decimal point is moved from the original decimal point line to the *right*. As shown in Figure 3.59 this procedure can be *reversed* for this number by moving the decimal point seven places to the *left*.

When the blank spaces between the 7 and the new placement of the decimal point are filled with zeros, the number is back in its decimal form. To convert this number to its metric prefixed form, simply place it on the metric chart.

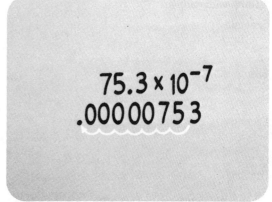

Figure 3.59

"000,007,53" on the Metric Chart — As shown in Figure 3.60, .000,007,53 can be changed to 7.53 micro, since the decimal point for micro appears between the 7 and the 5.

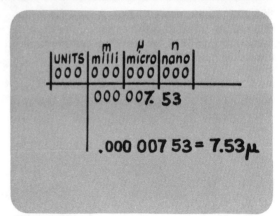

Figure 3.60

Additional Examples

Example 1

$8.27 \times 10^{+1}$
$8.27 = 82.7$ (units)

Move the decimal point one place to the right

Example 2

$1 \times 10^{+1}$
$1. = 10.$ (units)

Move the decimal point one place to the right
Fill in the space with a zero

Example 3

$.82 \times 10^{+3}$
$.82 = 820.$ (units)

Move the decimal point three places to the right
Fill in the space with a zero

Example 4

1×10^{-1}
$1. = .1$ (units)

Move the decimal point one place to the left
On the metric chart this number equals 100 milli

Example 5

857.3×10^{-2}
857.3
8.57 (units)

Move the decimal point two places to the left
Round off by dropping the 3

Example 6

$.93 \times 10^{-4}$
$.93$

$.000093$ (units)

Move the decimal point four places to the left
Fill in the four spaces to the left of the 9 with zeros
On the metric chart this number equals 93 micro

Example 7

1×10^{-8}
$1.$

$.000,000,01$ (units)

Move the decimal point eight places to the left
Fill in the seven spaces to the left of the 1 with zeros
On the metric chart this number equals 10 nano

Adding with Scientific Notation — In your study of electricity you will find that there are many times when you need to add two quantities in powers of ten form. The only rule you must remember for adding numbers in scientific notation is shown in Figure 3.61. *The power of ten or exponent for each number must be the same before the numbers are added.* When the exponents are not the same, they must first be changed to the same power before adding.

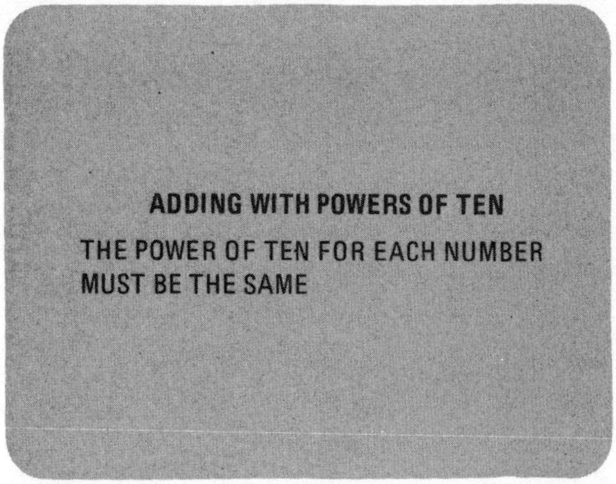

ADDING WITH POWERS OF TEN

THE POWER OF TEN FOR EACH NUMBER MUST BE THE SAME

Figure 3.61

Example of Addition — As shown in Figure 3.62, $5.2 \times 10^{+2}$ is 520 in decimal form. To add this number to $5 \times 10^{+1}$, which is 50 in decimal form, the exponents must be the same.

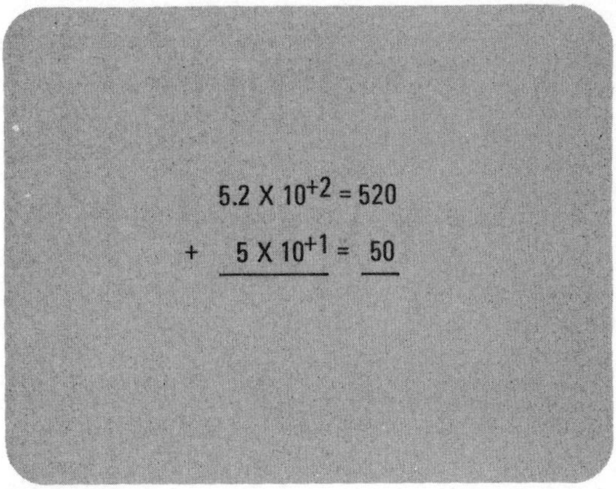

$$5.2 \times 10^{+2} = 520$$
$$+ \quad \underline{5 \times 10^{+1} = 50}$$

Figure 3.62

CHANGING THE VALUE OF THE POWER OR EXPONENT

WHEN YOU MOVE THE DECIMAL POINT ONE PLACE TO THE LEFT, INCREASE THE POWER BY 1. $3.0 \times 10^{+2} = .3 \times 10^{+3}$

WHEN YOU MOVE THE DECIMAL POINT ONE PLACE TO THE RIGHT, DECREASE THE POWER BY 1. $3.0 \times 10^{+2} = 30 \times 10^{+1}$

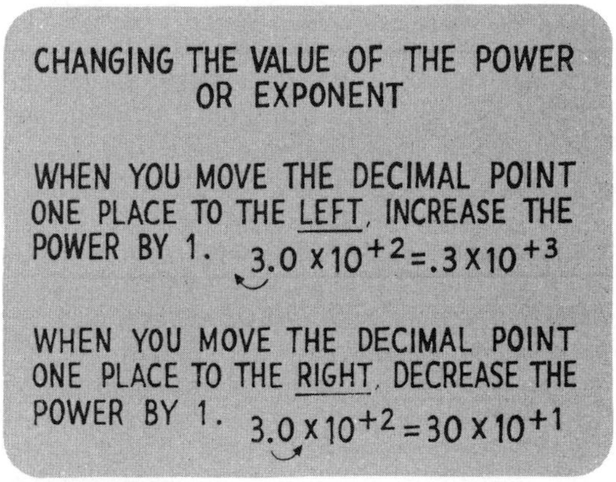

Figure 3.63

Changing the Value of the Power or Exponent — In order to add numbers in scientific notation, then, it is often necessary to change a number written in the standard form of scientific notation to another equivalent number with a different exponent. In the example above, the number 50 or $5.0 \times 10^{+1}$ must be rewritten so that its exponent is +2 before it can be added to $5.2 \times 10^{+2}$. Changing the exponent of a number in scientific notation is really just an extension of what you have learned this far, and the process is illustrated in Figure 3.63. Basically, it works like this. The total value of a number written in scientific notation is actually a ***product*** of two numbers: the leading number times the base raised to a certain power. In the standard form, the leading number is always in digit, point, digit, digit form. For the same number, the **power** can be changed by moving the position of the decimal point in the leading number. Every time you move the decimal point one place to the *left*, algebraically add +1 to the exponent. Every time you move the decimal point one place to the *right*, algebraically subtract one from the exponent. This way you can take a number in standard scientific notation and change it to an *equal* number having any power you may require to perform an addition.

$$5.1 \times 10^{+1}$$

$$.51 \times 10^{+1+1}$$

$$.51 \times 10^{+2}$$

Figure 3.64

Example — In Figure 3.64 the steps for rewriting $5.1 \times 10^{+1}$ into a number with a power of two are shown. The power must increase by one, so move the decimal point one place to the *left* and raise the power by one. Thus $5.1 \times 10^{+1}$ is equivalent to $.51 \times 10^{+2}$.

$$5.1 \times 10^{+1} = 510. \times 10^{-1}$$
$$+ 3.9 \times 10^{-1} = 3.9 \times 10^{-1}$$
$$\underline{\phantom{+ 3.9 \times 10^{-1}}}$$
$$? \qquad 513.9 \times 10^{-1}$$

Figure 3.65

Now suppose you were adding $5.1 \times 10^{+1}$ to another number in scientific notation whose exponent was -1, such as 3.9×10^{-1}. You would first want to change the number $5.1 \times 10^{+1}$ to an equivalent one whose exponent is -1. This means you have to lower the exponent by two, since +1 minus 2 equals -1. To lower the exponent, move the decimal point two places to the *right*, as shown in Figure 3.65. Each time you move the decimal point one place to the right, you *decrease* the exponent by one. Then $5.1 \times 10^{+1}$ equals 510×10^{-1}. Using this number, you could then complete the addition.

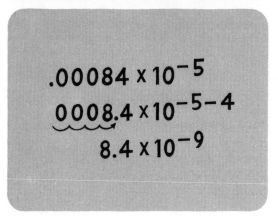

Figure 3.66

Notice also that the same procedures may be followed in reverse to convert a number in a nonstandard form of scientific notation back into standard form. Just move the decimal point to the digit, point, digit, digit position and carefully count the number of places and the direction you moved it. For each place you move it to the *left, raise* the exponent by one; for each place you move it to the *right, lower* the exponent by one. For example, in Figure 3.66 a number is shown in nonstandard form: $.00084 \times 10^{-5}$. To put this number back into standard form just move the decimal point to the digit, point, digit, digit position. This requires that you move it *four places* to the *right*; algebraically *subtract* 4 from the exponent, and the number becomes 8.4×10^{-9} in standard form. With these procedures in mind, you can get back into the discussion of adding with powers of ten.

Adding with the Same Exponents — In Figure 3.67, $5 \times 10^{+1}$ has been changed to an equivalent form, or $.5 \times 10^{+2}$. Now the powers are the same for both numbers. After the addition is performed, the answer is $5.7 \times 10^{+2}$. Note that the *exponents are not added* and that the sum has the same exponent as the two numbers being added. If the sum is converted to decimal form, the answer may be expressed as 570.

$$5.2 \times 10^{+2} = 520$$
$$\underline{.5 \times 10^{+2} = 50}$$
$$5.7 \times 10^{+2} = 570$$

Figure 3.67

Examples of Addition

Note: In addition either exponent (or both) may be changed until all powers are the same; then the addition is carried out.

Example 1

Add: $7.5 \times 10^{+3} + 356.0$

$356.0 = 356. \times 10^{0}$
$356.0 = .356 \times 10^{+3}$

$7.5 \times 10^{+3}$
$\underline{.356 \times 10^{+3}}$
$7.856 \times 10^{+3}$
$7.86 \times 10^{+3}$

1. Change the power of 356 to +3, that is, raise it by 3.
2. Move the decimal point three places left, raise the power by 3.
3. Carry out addition: add numbers, assign same exponent to answer.
4. Round off

Example 2

Add: $9.1 \times 10^{-6} + 1.23 \times 10^{-5}$

$1.23 \times 10^{-5} = 1.23 \times 10^{-5-1}$
$1.23 \times 10^{-5} = 12.3 \times 10^{-6}$

12.3×10^{-6}
$\underline{+ 9.1 \times 10^{-6}}$
21.4×10^{-6}
2.14×10^{-5}

1. Change powers
2. Move the decimal point one place to the right; lower the exponent by 1.

3. Add

4. Convert to standard form; move decimal one place to the left, add +1 to the exponent

The rules for addition with powers of ten can be used to add numbers expressed in prefixed notation.

Example 3

Add: $33 k + 4.7 k$

Kilo	Units
0 0 0	0 0 0
3 3 0	0 0
4 7	0 0

$33 k = 3.3 \times 10^{4}$
$4.7 k = .47 \times 10^{4}$

3.3×10^{4}
$\underline{.47 \times 10^{4}}$
3.77×10^{4}

1. Convert to scientific notation with same power of power of ten for each number

2. Use metric chart

3. Add

Example 4

Add: $50\,\mu + 12\,p$

milli	micro	nano	pico
0 0 0	0 0 0	0 0 0	0 0 0

0 . 0 0 0 5 0

0 0 0 0 0 0 0 0 0 1 2

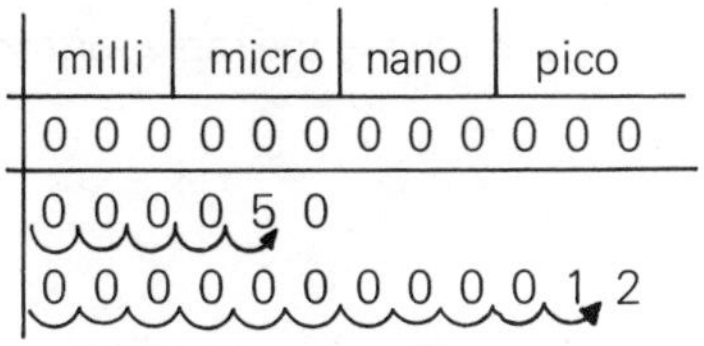

$50\,\mu = 5 \times 10^{-5}$

$12\,p = 1.2 \times 10^{-11}$

$.0000012 \times 10^{-11}$

$12\,p = .0000012 \times 10^{-5}$

$\begin{aligned} 5.0 \qquad &\times 10^{-5} \\ .0000012 &\times 10^{-5} \\ \hline 5.0000012 &\times 10^{-5} \end{aligned}$

5×10^{-5}

$50\,\mu + 20\,p = 50\,\mu$

1. Convert to scientific notation with same power of ten for each number (use metric chart)

1. Raise power by 6; move decimal six places to the left.

2. Add

3. Round off
(Note: In cases such as this, where the two numbers being added have values that are far apart, the smaller number may be dropped and the sum simply equals the larger number.)

Subtraction with Powers of Ten — Subtraction with powers of ten follows the same rules as addition as shown in Figure 3.68. Before you subtract two numbers in powers of ten form, you must take steps to be sure that both have the same power of ten. Once they do, subtract the ordinary numbers and assign the same power of ten to the answer. When the powers of ten are not the same, first change them to the same power before performing the subtraction.

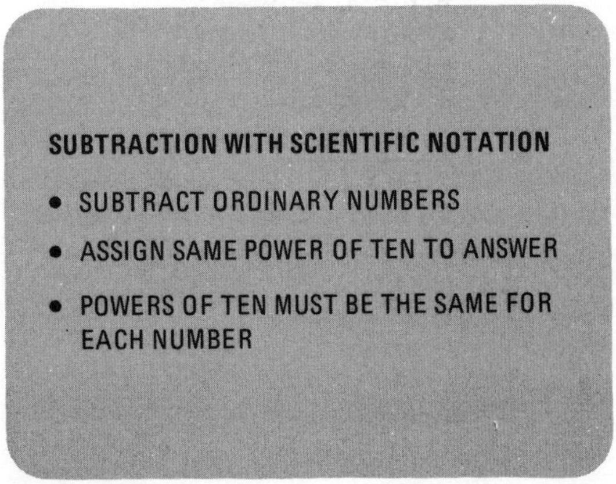

SUBTRACTION WITH SCIENTIFIC NOTATION

- SUBTRACT ORDINARY NUMBERS

- ASSIGN SAME POWER OF TEN TO ANSWER

- POWERS OF TEN MUST BE THE SAME FOR EACH NUMBER

Figure 3.68

Subtraction Examples

Example 1

Subtract: $3.5 \times 10^{+3} - 1.4 \times 10^{+3}$

$$\begin{array}{r} 3.5 \times 10^{+3} \\ -1.4 \times 10^{+3} \\ \hline 2.1 \times 10^{+3} \end{array}$$

Note: The powers of ten are the same for each number
Subtract

Example 2

Subtract: $7.5 \times 10^{+4} - 400$

$400 = \underset{\curvearrowleft}{400 \times 10^{0}}$

$400 = .04 \times 10^{4}$

$$\begin{array}{r} 7.5 \times 10^{+4} \\ - \ .04 \times 10^{+4} \\ \hline 7.46 \times 10^{+4} \end{array}$$

1. Convert one number so that both have equivalent powers of ten
2. Move decimal point four places left; raise power of 10 by four

3. Subtract the numbers; assign same exponent to answer

With the use of the subtraction rules and the metric chart, subtraction of numbers expressed in metric form, may also be carried out.

Example 3

Subtract: $830\ m - 52\ m$

Milli
0 0 0
8 3. 0
5. 2

$830\ m = 83.0 \times 10^{-2}$
$-52\ m = -5.2 \times 10^{-2}$
$\overline{\qquad\qquad 77.8 \times 10^{-2}}$
7.78×10^{-1}

1. Convert to scientific notation

2. Subtract; numbers have same exponent in answer
3. Convert to standard form

Example 4

Subtract: $10 - 20\ m$
$10 = 10 \times 10^{0}$
$20\ m = 20 \times 10^{-3}$
$20 \times 10^{-3} = .\underset{\curvearrowleft}{20} \times 10^{-3+3}$
$20 \times 10^{-3} = .02 \times 10^{0}$
$10. \times 10^{0}$
$\underline{-.02 \times 10^{0}}$
$9.98 \times 10^{0} = 9.98$ (since $10^{0} = 1$)

1. Convert to scientific notation

2. Change powers until they're equal
3. Move decimal three places left, raise exponent by three

4. Subtract

- **Multiplying with Scientific Notation**
- **Multiplication Example**
- Multiplication Examples

Multiplying with Scientific Notation — As shown in Figure 3.69, multiplication with scientific notation is somewhat different. The powers of ten do not have to be the same for all numbers. When multiplying with powers of ten, the simple rule to remember is: *multiply the leading numbers and algebraically add the exponents*. It is important to recall when handling the exponents that adding a negative number to something is equivalent to subtracting it.

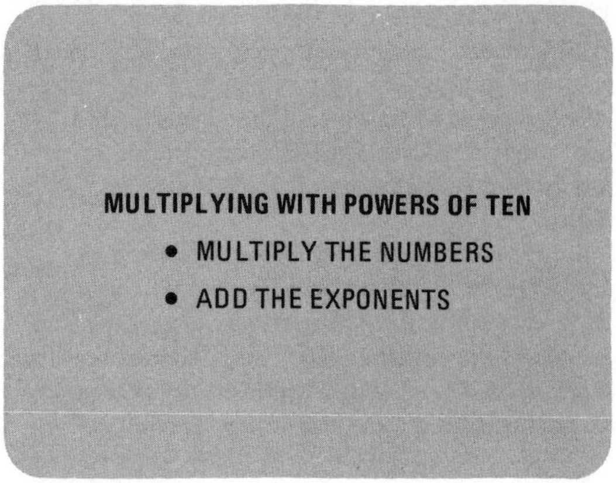

MULTIPLYING WITH POWERS OF TEN
- MULTIPLY THE NUMBERS
- ADD THE EXPONENTS

Figure 3.69

Multiplication Example — As an example of multiplication with scientific notation, multiply 22 kilohms by 1 milliamp as shown in Figure 3.70. First convert the numbers from prefixed quantities to scientific notation. In powers of ten form, 22 k = 2.2×10^4 and 1 milli is 1×10^{-3}. Following the rule step by step, perform the multiplication. First, multiply 2.2 by 1 to get 2.2. Next, combine the exponents which are +4 and —3. In this case, 3 is actually *subtracted* from 4 to get +1, which is the power of ten for the answer. Thus, 22 k times 1 milli equals $2.2 \times 10^{+1}$. Expressed in decimal form, this is 22.

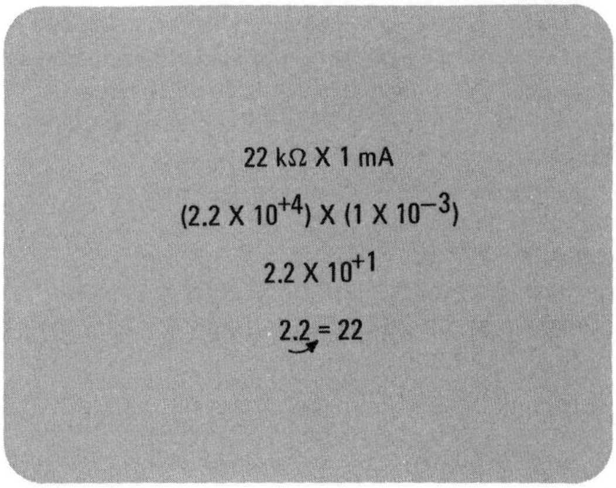

$$22 \text{ k}\Omega \times 1 \text{ mA}$$
$$(2.2 \times 10^{+4}) \times (1 \times 10^{-3})$$
$$2.2 \times 10^{+1}$$
$$2.2 = 22$$

Figure 3.70

Multiplication Examples

Example 1

Multiply: $10 \text{ k} \times 3 \text{ m}$
$1 \times 10^{+4}$
$\times 3 \times 10^{-3}$
—————————
$3 \times 10^1 = 30$

1. Convert to powers of ten form

2. Multiply the leading numbers, and add the exponents algebraically

Example 2

Multiply: $150 \text{ k} \times 50 \mu$
$1.5 \times 10^{+5}$
$\times \ 5 \times 10^{-5}$
—————————
$7.5 \times 10^0 = 7.5$

1. Convert to powers of ten form

2. Multiply the leading numbers and add the exponents algebraically
3. Convert to standard form

Division with Powers of Ten — In working problems dealing with electricity, you often need to divide with powers of ten. Figure 3.71 illustrates how to divide with powers of ten. In this case, 6 kilovolts is divided by 300 ohms. In powers of ten form, 6 k volts equals $6 \times 10^{+3}$ and 300 is $3 \times 10^{+2}$.

In division when numbers are written in fractional form, the bottom number is always divided into the top. When dividing numbers expressed in powers of ten form, a special new step is required. *The bottom exponent is first brought across the division line and its sign is changed.*

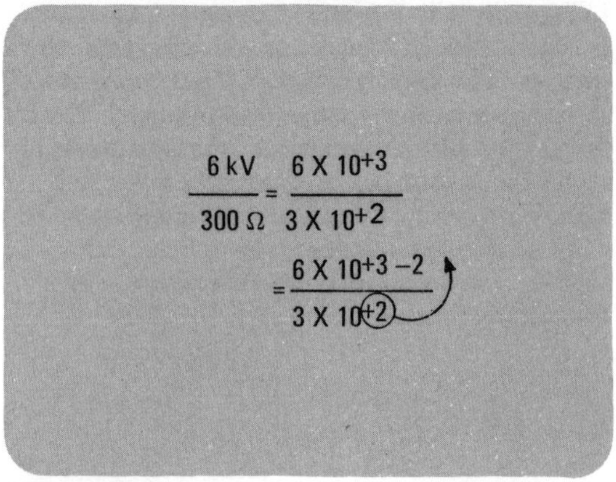

$$\frac{6\,kV}{300\,\Omega} = \frac{6 \times 10^{+3}}{3 \times 10^{+2}}$$

$$= \frac{6 \times 10^{+3\,-2}}{3 \times 10^{+2}}$$

Figure 3.71

Once both exponents are on top (remember the sign of what used to be the bottom exponent is changed), the *second step is to combine the exponents*. In Figure 3.72, a +3 and a −2 equal +1. This is now the exponent of the answer.

The third step is to divide the leading numbers. In this case, 3 divides into 6 which equals 2. The final answer is $2 \times 10^{+1}$ which in decimal form is 20. Thus, 6 kilovolts divided by 300 ohms equals 20 amps.

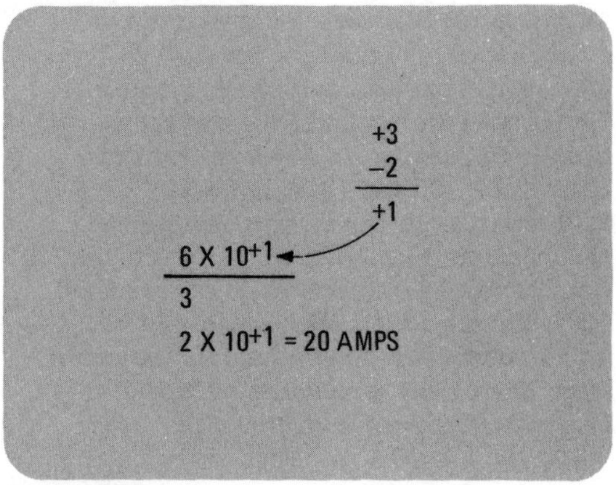

$$\begin{array}{r} +3 \\ -2 \\ \hline +1 \end{array}$$

$$\frac{6 \times 10^{+1}}{3}$$

$$2 \times 10^{+1} = 20 \text{ AMPS}$$

Figure 3.72

Division Examples

Example 1

Divide: $10 \div 3.4\,m$

$$\frac{1 \times 10^{+1}}{3.4 \times 10^{-3}}$$

$$\frac{1 \times 10^{+1\,+3}}{3.4 \times 10^{-3}} = \frac{1 \times 10^{+1+3}}{3.4}$$

$$\frac{1 \times 10^{+4}}{3.4}$$

$$.294 \times 10^{+4}$$

$$2.94 \times 10^{+3}$$

1. Convert to scientific notation

2. Bring the bottom exponent up to the top and change its sign

3. Combine the exponents

4. Divide the leading numbers

5. Convert to standard form

Example 2

Divide: 24 k ÷ 3.3 k

$$\frac{2.4 \times 10^{+4}}{3.3 \times 10^{+3}}$$

1. Convert to scientific notation

$$\frac{2.4 \times 10^{+4} \,\,(-3)}{3.3 \times 10\,(+3)} = \frac{2.4 \times 10^{+4-3}}{3.3}$$

2. Bring the bottom exponent up to the top and change its sign

$$\frac{2.4 \times 10^{+1}}{3.3}$$

3. Combine the exponents

$.727 \times 10^{+1}$

4. Divide the leading numbers

7.27×10^{0}

5. Convert to standard form

Example 3

Divide: 25 k ÷ 5 k

$$\frac{2.5 \times 10^{+4}}{5 \times 10^{+3}}$$

1. Convert to scientific notation

$$\frac{2.5 \times 10^{+4} \,\,(-3)}{5 \times 10\,(+3)} = \frac{2.5 \times 10^{+4-3}}{5}$$

2. Bring the bottom exponent up to the top and change its sign

$$\frac{2.5 \times 10^{+1}}{5}$$

3. Combine the exponents

$.5 \times 10^{+1}$

4. Divide the leading numbers

$5. \times 10^{0} = 5$

5. Convert to standard form

The topic of division is also explained in detail in Lesson 4.

Summary — Figure 3.73 contains a summary of the topics covered in this lesson. To really learn how to use powers of ten and metric prefixes, practice is essential. The practice problems following this lesson are designed to provide the practice you will need. Further detailed examples covering multiplication and division, arising out of electrical examples, are covered in Lesson 4.

SUMMARY

- OHM'S LAW PROBLEM
- METRIC PREFIXES
- SCIENTIFIC NOTATION
- EXPONENTS, SIGNIFICANT NUMBERS, ROUNDING OFF
- ADDITION, SUBTRACTION, MULTIPLICATION, DIVISION WITH POWERS OF TEN

Figure 3.73

- **Worked Through Examples**

1. Express the following numbers in correct, standard metric prefixed form, and in powers of ten form.

    a.   6138.95

    b.   3427126

    c.   .0000004134

Solution:

    a.   6138.95 — Use the metric chart

	Kilo	Units
	0 0 0	0 0 0
6138.95 =	6 . 1 3 8	9 5

You see that 6 represents 6 ones k and the decimal point for kilo follows it. So, this number is 6.13895 k. To round off, keep the first three most significant digits, that is, the 6, the 1 and the 3. Then examine the fourth significant digit, the 8 (which is greater than 5), so you increase the third digit by one, and then drop the remaining digits.

So 6.13895 k equals 6.14 in standard metric prefixed form.

To put the number in powers of ten form.

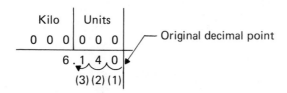

Place a decimal point between the first two significant digits, and write the number in the standard digit, point, digit, digit form, which is then multiplied by ten to some power: 6.14 X 10* (where the asterisk indicates the position of the exponent). Count over from the position of the original decimal point (vertical line) to the new one (three places to the left). This number is the exponent and since the decimal point was moved to the *left*, the exponent is *positive*. In correct scientific notation, the number may then be expressed as $6.14 \times 10^{+3}$.

b.   3427126 – Place the number on the chart,

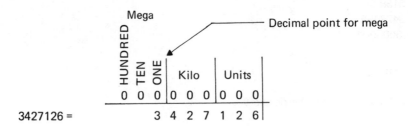

3427126 =

and you see that it represents 3.427126 mega. To round off, keep the 3, 4, and 2. The fourth digit is 7 (greater than 5), so increase the third significant digit by one (the two becomes a three) and the answer is 3.43 M.

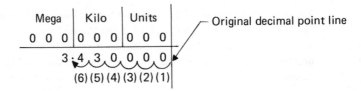

To convert this to powers of ten form, place a decimal point between the first two most significant digits and count over from the original decimal point line to find the exponent. This is six places to the left, so the exponent is +6, and the number is written $3.43 \times 10^{+6}$.

c.   .0000004134 – Write this on the metric chart

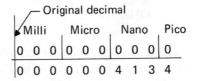

and find that it represents:
        3 one nanos
        1 ten nano
        4 hundred nanos
or 413.4 nano, or 413.4 n.

To round off, drop the 4. Since the fourth significant digit, 4, is less than 5, don't change the 3. Your answer is then 413 nano or 413 n.

To place this in powers of ten form, first put a decimal point between the first and second most significant digits.

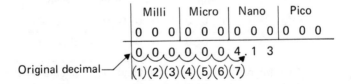

Count over from the original decimal point line to the position of the new decimal point, seven places to the right. When the decimal point is moved to the right (whenever handling numbers less than one), the exponent is negative. The result is 4.13×10^{-7}.

2. Express the following quantities in correct decimal form and in scientific notation:

   a. 25 MV (mega volts)
   b. 35 pA (pico amps)

Solution:

a. Put the number on the metric chart in its correct location:

Mega	Kilo	Units
0 0 0	0 0 0	0 0 0
	2 5	0 0 0 0 0 0

Fill in the spaces with zeros, and the number in decimal form is 25,000,000. volts.

To put it into scientific notation, place a decimal point between the first two significant digits

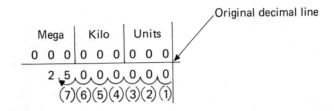

and count over from the original decimal point line to the position of the new decimal point, seven places to the left. This again becomes your exponent and the number correctly written in powers of ten form is $2.5 \times 10^{+7}$ volts.

b. Put 35 pA on the metric chart in its correct position.

Milli	Micro	Nano	Pico
0 0 0	0 0 0	0 0 0	0 0 0

0 0 0 0 0 0 0 0 0 ,3 5

Then fill in the digits between it and the original decimal with zeros. The number in decimal form is then .000,000,000,035 amps.

To put the number in standard powers of ten form, place a decimal point between the two most significant digits, and count the number of places between it and the original line.

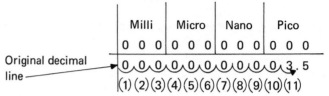

	Milli	Micro	Nano	Pico
	0 0 0	0 0 0	0 0 0	0 0 0

Original decimal line → 0 0 0 0 0 0 0 0 0 0 3 . 5
(1) (2) (3) (4) (5) (6) (7) (8) (9) (10) (11)

This is 11 places to the right, so the exponent is −11. The number in correct form is then 3.5×10^{-11} amps.

3. Place the following numbers in standard "correct" metric prefixed form:

   a. 94267 μA (microamps)
   b. .00241 kV (kilovolts)

Solution:

   a. Again, just place this number on the metric chart, aligning the one, ten, and hundred micros.

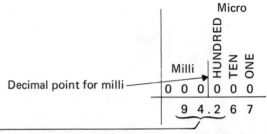

Decimal point for milli ──

Milli	Micro (HUNDRED TEN ONE)
0 0 0	0 0 0

9 4 . 2 6 7

You can see from the chart that this number is more correctly abbreviated 94.267 milli. Rounding off, you should get 94.3 milliamps.

b. Place .00241 kilovolts in its correct position on the chart,

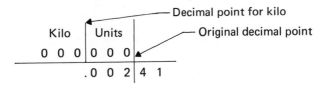

and you can see it is better written as 2.41 volts.

4. Place the following numbers in "correct" or standard scientific notation.

a. $21432681. \times 10^{-9}$ 2.14×10^{-2}
b. $.00314 \times 10^{+16}$ 3.14×10^{13}

Solution:

a. Place a decimal point between the first two digits of the number (putting it in digit, point, digit, digit form.

$$2.1432681. \times 10^{-9}$$
$$2.1432681 \times 10^{-9}$$
$$7654321$$

Count over the number of places moved from the decimal point you had at first. For each place you move to the *left*, add +1 to the exponent; for each place you moved to the *right*, subtract 1 (or add −1) to the exponent. The decimal point is moved 7 places to the *left*, so the number becomes:

$$2.1432681 \times 10^{-9+7} \text{ or}$$
$$2.1432681 \times 10^{-2}$$

Round off to get 2.14×10^{-2}.

b. Place a decimal point in the digit, point, digit, digit spot.

$$.00314 \times 10^{+16}$$
$$00314 \times 10^{+16}$$
$$123$$

Count the number of places between it and the one you had at first. The decimal point is moved to the right three places, so you subtract 3 from the exponent.

$$3.14 \times 10^{+16-3}$$

The final answer is then $3.14 \times 10^{+13}$.

5. Add the following:

   a. $2.41 \times 10^{+10}$
       $+3.18 \times 10^{+8}$ 2.44×10^{10}

   b. $2.38 \times 10^{+9}$
       $+1.19 \times 10^{-6}$

Solution:

a. $2.41 \times 10^{+10}$
    $+3.18 \times 10^{+8}$

Remember the rules for adding numbers in powers of ten form:
- the powers of ten must be the same
- add the leading numbers
- the sum has the same power of ten as the numbers you are adding

1. $3.18 \times 10^{+8} = 3.18 \times 10^{+8+2}$

    1. First convert the $3.18 \times 10^{+8}$ to a number whose whose exponent is +10

2. $3.18 \times 10^{+8} = .0318 \times 10^{+10}$

    2. Move the decimal two places left and increase the exponent by 2

3. $.0318 \times 10^{+10}$
    $2.41 \times 10^{+10}$
    _____
    $2.4418 \times 10^{+10}$

    3. Now add

4. $2.44 \times 10^{+10}$

    4. Round off

b. $2.38 \times 10^{+9}$
    $+1.19 \times 10^{-6}$

By examination, you can see that $2.38 \times 10^{+9}$ is so much larger than 1.19×10^{-6} that the 1.19×10^{-6} is negligible by comparison. (or converting it to a +9 exponent, you would get: $.00000000000000119 \times 10^{+9}$.)

So the answer is just $2.38 \times 10^{+9}$ and you neglect the 1.19×10^{-6} completely.

6. Subtract

   a. 8.13×10^{-3}
      -6.25×10^{-2}

   b. $8.95 \times 10^{+8}$ 8.95×10^8
      $-3.21 \times 10^{+5}$

The rules for subtraction are essentially similar to those for addition:

- The powers of ten of the numbers being subtracted must be the same
- Subtract the leading numbers
- The result has the same power of ten as the numbers being subtracted

Solution:

a. 1. 8.13×10^{-3}
         -6.25×10^{-2}

1. Change powers of ten so that powers of both numbers involved are the same. (Either one may be changed; examine 6.25×10^{-2}.)

6.25×10^{-2}

Move the decimal point one place to the right   Subtract 1 from the exponent

You need to change 10^{-2} to 10^{-3} for this problem, so move decimal one place to the *right*.

$62.5 \times 10^{-2} = 62.5 \times 10^{-3}$

   2. 8.13×10^{-3}
         -62.5×10^{-3}
         -54.37×10^{-3}

2. Subtract

   3. $= -5.44 \times 10^{-2}$

3. Convert to standard form and round off

b. 1. $8.95 \times 10^{+8}$
         $-3.21 \times 10^{+5}$

1. Change powers of ten until powers of both numbers are the same

$3.21 \times 10^{+5}$

Move decimal three places left   Add 3 to exponent

You need to change 10^{+5} to 10^{+8}, so move decimal three places left

$-3.21 \times 10^{+5} = - .00321 \times 10^{+8}$

2.   8.95   X 10^{+8}
      − .00321 X 10^{+8}
      _____
      8.94679 X 10^{+8}

2. Subtract

3.   8.95 X 10^{+8}

3. Round off

(Note, in this case, the number .00321 X 10^{+8} is so much smaller than 8.95 X 10^{+8} that subtracting it has little effect on the result.)

7.   Multiply

a.   3.18 X 10^{-6} 8.30×10^{2}
      X2.61 X 10^{+8}
      _____

b.   8.92 X 10^{+4} 28.10×10^{-5}
      X3.15 X 10^{-9}
      _____

Recall the rules for multiplying numbers in powers of ten form:
• Multiply the leading numbers
• Algebraically add the exponents
(Recall that adding a negative number to something is equivalent to *subtracting* it.)

Solution:

a.   3.18 X 10^{-6}
      X2.61 X 10^{+8}
      _____

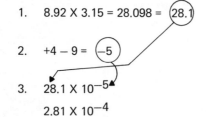

1.   3.18 X 2.61 = 8.2998 = ⟨8.30⟩

1. Multiply the leading numbers (round off). (This gives the leading numbers of the result.)

2.   −6 + 8 = ⟨+2⟩

      8.30 X 10^{+2}

2. Algebraically add the exponents. (This gives the exponent or power of the result.)

b.   8.92 X 10^{+4}
      X3.15 X 10^{-9}
      _____

1.   8.92 X 3.15 = 28.098 = ⟨28.1⟩

1. Multiply the leading numbers (round off). (This gives leading number of result.)

2.   +4 − 9 = ⟨−5⟩

2. Algebraically add exponents. (This gives exponent exponent of result.)

3.   28.1 X 10^{-5}

      2.81 X 10^{-4}

3. Convert to standard form

8. Divide

a. $$\frac{3.97 \times 10^{-6}}{2.31 \times 10^{-4}}$$

b. $$\frac{2.65 \times 10^{+3}}{3.33 \times 10^{-3}}$$

Recall the rules for dividing numbers expressed in powers of ten form:

- Divide the leading numbers
- Bring the exponent of the number in the denominator up across the division line, and *change its sign*
- Algebraically add the exponents to find the exponent of the result

Solution:

a. $$\frac{3.97 \times 10^{-6}}{2.31 \times 10^{-4}}$$

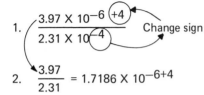

1. $\dfrac{3.97 \times 10^{-6} \; \boxed{+4}}{2.31 \times 10^{-4}}$ Change sign

    1. Divide the leading numbers

2. $\dfrac{3.97}{2.31} = 1.7186 \times 10^{-6+4}$

    2. Bring lower exponent above the division line and change its sign

3. $\quad\quad = 1.72 \times 10^{-6+4}$

    3. Round off

4. $\quad\quad = 1.72 \times 10^{-2}$

    4. Algebraically add the exponents

b. $$\frac{2.65 \times 10^{+3}}{3.33 \times 10^{-3}}$$

1. $\dfrac{2.65}{3.33} = .79579 = .796$

    1. Divide the leading numbers and round off

2. $\dfrac{.796 \times 10^{+3} \; \boxed{+3}}{10^{-3}}$ Change sign

    2. Bring lower exponent to the top and change its sign

3. $\quad = .796 \times 10^{+3+3}$

    3. Algebraically add the exponents

4. $\quad = .796 \times 10^{+6}$

$\quad\quad = 7.96 \times 10^{+5}$

    4. Convert to standard form

●   **Practice Problems**

To really tackle the problems of predicting how electricity will act, or of controlling its operation, a thorough knowledge of powers of ten notation and the metric prefixes is needed. Try these problems on your own. Then fold the page to check your results.

Fold Over

1. Convert the following electrical quantities to "correct" standard powers of ten and metric prefixed form. (V = volts, A = amps, Ω = ohms)

1.

Powers of Ten

Metric Prefix

a.   .00031568 A

1.a. $3.16 \times 10^{-4} A$

b.   160,000 Ω

1.b.

c.   3.94 V

1.c.

d.   4000 X 10^{+8} Ω

1.d.

e.   .0000000002578 A

1.e.

f.   8951763214 V

1.f.

g.   895176.3214 Ω

1.g.

h.   .8951763214 A

1.h.

i.   160,000 X 10^{-3} V

1.i.

j.   75000 μA

1.j.

k.   8500 mA

1.k.

l.   .00005 X 10^7 V

1.l.

m.   .12856 X 10^{-3} A

1.m.

n.   63.7 X 10^{+3} Ω

1.n.

o.   819 X 10^{+3} Ω

1.o.

p.   819 X 10^{-6} A

1.p.

Answers

1.

Powers of Ten	Metric Prefix
1.a. 3.16×10^{-4} A	$316\,\mu$A
1.b. 1.6×10^{5} Ω	160 kΩ
1.c. 3.94×10^{0} or 3.94 V	3.94 V
1.d. $4.0 \times 10^{+11}$	400 GΩ
1.e. 2.58×10^{-10} A	258 pA
1.f. $8.95 \times 10^{+9}$ V	8.95 GV
1.g. $8.95 \times 10^{+5}$ Ω	895 kΩ
1.h. 8.95×10^{-1} A	895 mA
1.i. $1.60 \times 10^{+2}$ V	160 V
1.j. 7.5×10^{-2} A	75 mA
1.k. 8.5×10^{0} A or 8.5 A	8.5 A
1.l. 5.0×10^{2} V	500 V
1.m. 1.29×10^{-4} A	$129\,\mu$A
1.n. $6.37 \times 10^{+4}$ Ω	63.7 kΩ
1.o. $8.19 \times 10^{+5}$ Ω	819 kΩ
1.p. 8.19×10^{-4} A	$819\,\mu$A

	Powers of Ten	Metric Prefix

q. 213×10^{-3} A   1.q.

r. $.00001561 \times 10^{+5}$ V   1.r.

s. $.0003343 \times 10^{-3}$ A   1.s.

t. 161×10^{-9} A   1.t.

2. Add the following numbers, expressing the result in correct powers of ten form.   2.

     a. $\begin{aligned} 3.196 &\times 10^{+8} \\ +2.165 &\times 10^{+7} \end{aligned}$ 2.a.

     b. $\begin{aligned} 2.1426 &\times 10^{+6} \\ +2.1426 &\times 10^{+4} \end{aligned}$ 2.b.

     c. $\begin{aligned} 2.891 &\times 10^{-5} \\ +1.891 &\times 10^{-3} \end{aligned}$ 2.c.

     d. $\begin{aligned} 2.53 &\times 10^{+6} \\ +2.53 &\times 10^{-6} \end{aligned}$ 2.d.

     e. $\begin{aligned} 2.53 &\times 10^{+3} \\ +3000 \end{aligned}$ 2.e.

3. Subtract the following numbers, placing the answer in standard powers of ten form.   3.

     a. $\begin{aligned} 3.985 &\times 10^{+3} \\ -2000 \end{aligned}$ 3.a.

     b. $\begin{aligned} 9.81 &\times 10^{-6} \\ -3.45 &\times 10^{-4} \end{aligned}$ 3.b.

Answers

	Powers of Ten	Metric Prefix
1.q.	2.13×10^{-1} A	213 mA
1.r.	1.56×10^{0} V or 1.56 V	1.56 V
1.s.	3.34×10^{-7} A	334 nA
1.t.	1.61×10^{-7} A	161 nA

2.

2.a. $3.41 \times 10^{+8}$

2.b. 2.16×10^{6}

2.c. 1.92×10^{-3}

2.d. $2.53 \times 10^{+6}$ (Hint: This was primarily a rounding-off problem, since one number was so small as to be negligible compared to the other.)

2.e. $5.53 \times 10^{+3}$

3.

3.a. $1.99 \times 10^{+3}$

3.b. -3.35×10^{-4}

Fold Over

c.   128000
      $-4.5 \times 10^{+4}$

3.c.

d. 3.9×10^{-7}
      -2.1×10^{3}

3.d.

e. $8.15 \times 10^{+3}$
      $-2.13 \times 10^{+2}$

3.e.

4.   Multiply the following numbers, placing the results in correct "standard" scientific notation.

4.

a. $6.75 \times 10^{+6}$
      $\times 2.38 \times 10^{-5}$

4.a.

b. 3.18×10^{-6}
      $\times 2.75 \times 10^{-5}$

4.b.

c. $2.1864 \times 10^{+5}$
      $\times 3.25 \times 10^{+5}$

4.c.

d. $7.95 \times 10^{+2}$
      $\times 3.0 \times 10^{-7}$

4.d.

e. 8.88×10^{-5}
      $\times 3.33 \times 10^{-2}$

4.e.

Answers

3.c. $8.3 \times 10^{+4}$

3.d. $-2.1 \times 10^{+3}$

3.e. $7.94 \times 10^{+3}$

4.

4.a. 1.61×10^{2}

4.b. 8.75×10^{-11}

4.c. $7.11 \times 10^{+10}$

4.d. 2.39×10^{-4}

4.e. 2.96×10^{-6}

Fold Over

5. Divide the following numbers, placing the result in "correct" standard scientific notation.

    5.

    a. $\dfrac{7.98 \times 10^{-5}}{2.01 \times 10^{+4}}$

    5.a.

    b. $\dfrac{3000}{2.31 \times 10^{+4}}$

    5.b.

    c. $\dfrac{2.31 \times 10^{+4}}{3000}$

    5.c.

    d. $\dfrac{8.85 \times 10^{-12}}{3.14 \times 10^{+10}}$

    5.d.

    e. $\dfrac{3 \times 10^{2}}{2 \times 10^{2}}$

    5.e.

Answers

5.

5.a. 3.97×10^{-9}

5.b. 1.30×10^{-1}

5.c. 7.7×10^{0} or 7.7

5.d. 2.82×10^{-22}

5.e. 1.5×10^{0} or 1.5

1. Scientific notation is:
   a. A way of working a math problem
   b. Another name for powers of ten notations.
   c. A simple type of shorthand to keep track of decimal points
   d. Taking notes in scientific terms
   e. b and c above
   f. a, b and d above

2. 680,000 is written as_____in powers of ten or scientific notation:
   a. 6.8×10^5
   b. 6.8×10^4
   c. 680×10^3
   d. 68×10^4
   e. 680K

3. Kilo is an abbreviation for_____when working with electrical quantities:
   a. Tens
   b. Hundreds
   c. Thousands
   d. Units
   e. Millions

4. 670 Milliamperes is how many amperes in decimal form:
   a. 6.7 amperes
   b. 0.670 amperes
   c. 67 amperes
   d. 0.067 amperes
   e. None of the above

5. 8.2 Megohms as a resistor value is written as_____ in scientific notation.
   a. 8200
   b. 8,200,000
   c. 8200×10^3
   d. 8.2×10^3
   e. 8.2×10^6

6. 3766 Microamperes is written as_____in scientific notation.
   a. 3.766×10^{-6}
   b. 376.6×10^{-3}
   c. 3.76K
   d. 3.766×10^{-2}
   e. 3.77×10^{-3}

7. 3 Millivolts is written as_____in the metric prefixed form.
   a. 0.003 volts
   b. 3 mV
   c. 3×10^{-3}
   d. 3×10^{-6} volts
   e. None of above

8. 75 pico amperes is written as_____in decimal form.
   a. 75×10^{-6}
   b. 75×10^{-12}
   c. 0.000000000075
   d. 0.075×10^{-9}
   e. 75 pA

9. 12,000,000,000,000 cycles per second are many times expressed as _____gigahertz:
   a. 12×10^{12}
   b. 12
   c. 12×10^6
   d. 12,000
   e. 1.2

10. Numbers in scientific notation can be added when:
    a. The exponent of the base 10 are the same
    b. There are three significant digits
    c. All numbers are in decimal form
    d. The decimal point is moved to the right
    e. The decimal point is moved to the left

11. Moving the decimal point one place to the left to change the power of 10 notation for a number:
    a. Increases the power of 10 by 1
    b. Decreases the power of 10 by 1
    c. Increases the power of 10 by 10
    d. Decreases the power of 10 by 10
    e. None of above

12. In a power of 10 notation for a number, to decrease the power of 10:
    a. Move the decimal point one place to the left
    b. Move the decimal point to the right
    c. Move the decimal point two places to the right
    d. Move the decimal point one place to the right
    e. All of above except a

13. Subtraction with numbers in scientific notation:
    a. Can't be done
    b. Has a completely separate set of rules from addition
    c. Follows the same rules as addition
    d. Follows the same rules as division
    e. Follows the same rules as multiplication

14. When multiplying numbers in scientific notation, after multiplying the leading numbers:
    a. Multiply the exponents
    b. Algebraically subtract the exponents
    c. Algebraically add the exponents
    d. Keep the same exponents
    e. None of the above

15. Exponents are changed in sign when _____ numbers in scientific notation.
    a. Multiplying
    b. Dividing
    c. Subtracting
    d. Adding
    e. None of the above

16. 36,000 minus 3×10^4 is equal to:
    a. 33,000
    b. -0.6×10^4
    c. $-5,000$
    d. 0.6×10^4
    e. 12×10^4

17. 9.41×10^5 plus 8.21×10^{-2} is equal to:
    a. 941,000
    b. 1.20×10^5
    c. 1.20×10^3
    d. 17.62×10^3
    e. None of above

18. 6.5×10^{-5} minus 6.5×10^5 is equal to:
    a. 6.5×10^{-5}
    b. Zero
    c. -13×10^{-5}
    d. -6.5×10^5
    e. 13

19. 2.20×10^{10} multiplied by 1.1×10^{-4} is equal to:
    a. 2.42×10^{10}
    b. 2.42×10^{-4}
    c. 2.42×10^6
    d. 2.20×10^{10}
    e. 2.20×10^6

20. 39,500 divided by 2.17×10^{-3} is equal to:
    a. 18.2×10^6
    b. 18.2×10^3
    c. 18.2
    d. 39.5×10^3
    e. 39.5×10^6

Lesson 4

Ohm's Law and Power

This lesson thoroughly discusses and illustrates *Ohm's law* and the interrelationship of voltage, current, and resistance. Included is the necessary coverage of multiplication and division with scientific notation. *Power* and *power dissipation* are defined and demonstrated. Formulas for the calculation of the power dissipated in a circuit are developed and their use is illustrated with some sample problems. The concepts of *squares* and *square roots* are introduced and explained.

LESSON 4. OHM'S LAW AND POWER

• **Objectives**

This lesson discusses in detail the most important concept in electricity: the interrelationship of voltage, current, and resistance as predicted and described in Ohm's law. At the end of this lesson you should be able to:

1. *Write* a simple statement of Ohm's law, discussing its meaning.

2. Given a simple circuit schematic of the type shown, *write* the correct Ohm's law formula, and use it to *calculate* the unknown quantity in each of three cases.

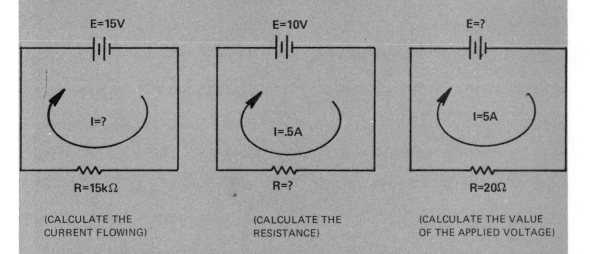

(CALCULATE THE CURRENT FLOWING)   (CALCULATE THE RESISTANCE)   (CALCULATE THE VALUE OF THE APPLIED VOLTAGE)

3. Correctly *multiply* or *divide* numbers using scientific (powers of ten) notation when making calculations involving Ohm's law.

4. *Write* a definition of the word power, and given schematic diagram and known quantities of the type shown below, *calculate* the power dissipated by the resistor in each case.

5. *Write* definitions of the term "square" and "square root." Using a calculator or square root tables as necessary, *calculate* the square or square root of any number as may be needed in the manipulation of the power formula $P = I^2R$.

- **Voltage**
- **Electron Current**

In this lesson the important concepts of Ohm's law and power, which were briefly introduced in Lesson 2, will be discussed in detail. Ohm's law is absolutely the most common and most powerful single relationship in the study of electricity or electronics today. In addition, knowledge of the concept of electrical power will be essential as you begin circuit construction in your laboratory exercises.

Before discussing the detailed objectives of this lesson, consider a brief review of what's been covered up to this point. By now you should be able to clearly define the terms voltage, current, and resistance. Since these terms need to be fresh in your mind as you study Ohm's law, the following review of these terms will be helpful.

Voltage — Volts or voltage is a measure of the push on each electron, and it is this voltage, the pressure or the push that makes the electrons move (Figure 4.1).

VOLTAGE

A MEASURE OF THE "PUSH" ON EACH ELECTRON.

THIS PRESSURE OR PUSH MAKES ELECTRONS MOVE.

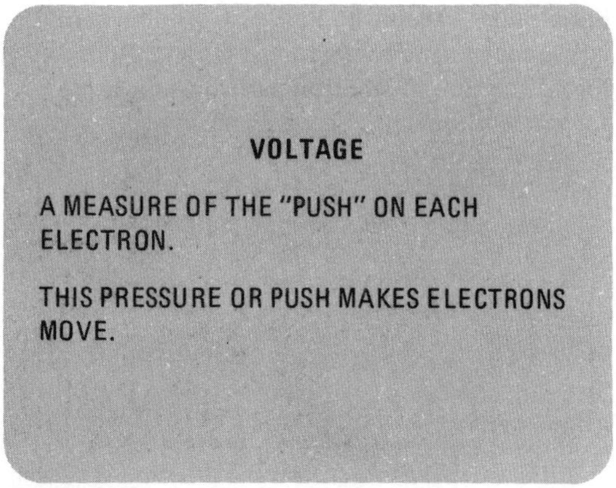

Figure 4.1

Electron Current — Electron current is the flow of free electrons in a conductor in response to an applied voltage (Figure 4.2).

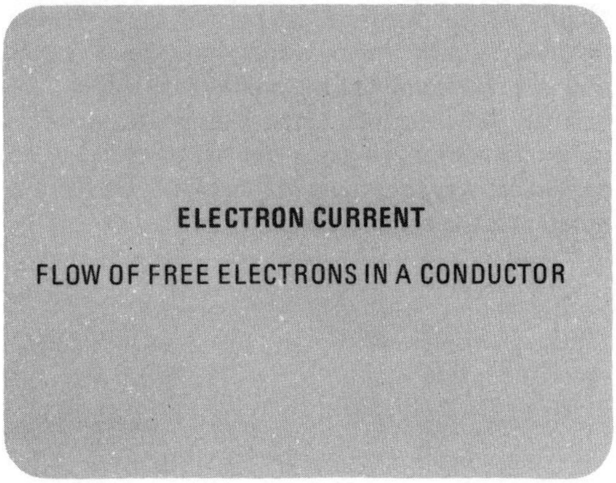

ELECTRON CURRENT

FLOW OF FREE ELECTRONS IN A CONDUCTOR

Figure 4.2

Resistance — Resistance is that property of a material which opposes the flow of free electrons in a conductor (Figure 4.3).

As has been discussed in previous lessons, in various electric circuit situations you will be dealing with voltage, current, and resistance and their interrelationship. Often the numbers involved in this interrelationship may be extremely small or large. In Lesson 3, the shorthand notation using the metric prefixes for these quantities (such as micro, milli, kilo, or mega) was introduced. These prefixes will be used throughout the remainder of the dc circuits course.

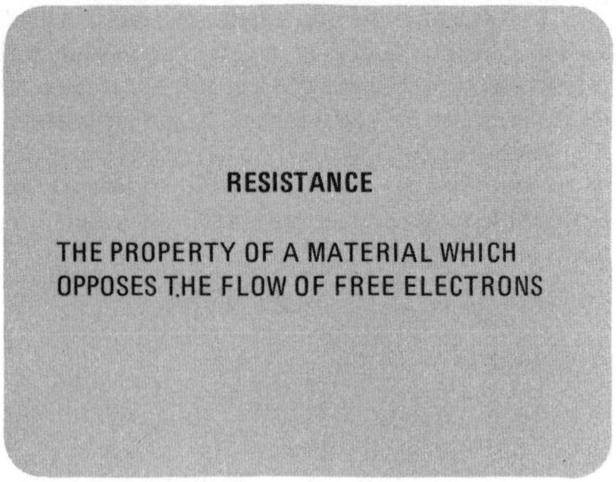

RESISTANCE

THE PROPERTY OF A MATERIAL WHICH OPPOSES THE FLOW OF FREE ELECTRONS

Figure 4.3

Basic Units — As shown in Figure 4.4, the basic units used to specify the three key quantities are the volt for potential difference, the ampere for current, and the ohm for resistance.

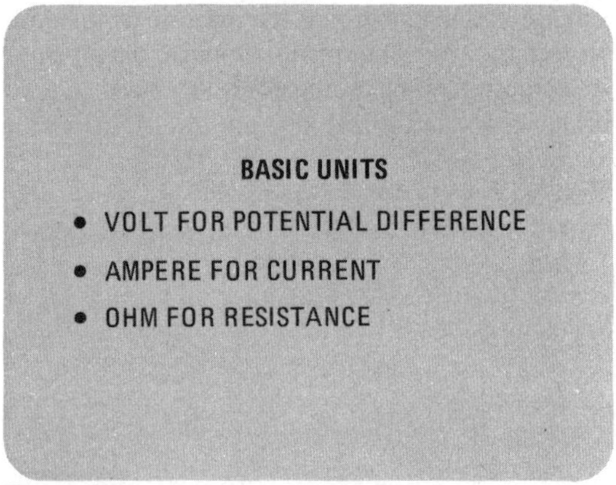

BASIC UNITS

- VOLT FOR POTENTIAL DIFFERENCE
- AMPERE FOR CURRENT
- OHM FOR RESISTANCE

Figure 4.4

Mega — As was discussed in Lesson 3, each of these units may appear with a metric prefix. For example, when you use the prefix mega in the quantity "10 megohms," it enables you to write ten million ohms in a much shorter, more convenient way, as shown in Figure 4.5. Use of this notation is the common practice.

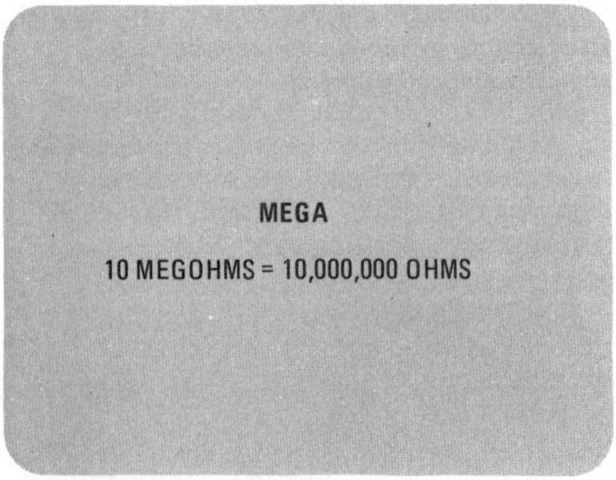

MEGA

10 MEGOHMS = 10,000,000 OHMS

Figure 4.5

Kilo — Similarly, as shown in Figure 4.6, 5 kV or 5 kilovolts is the common way to express five thousand volts in electronic shorthand.

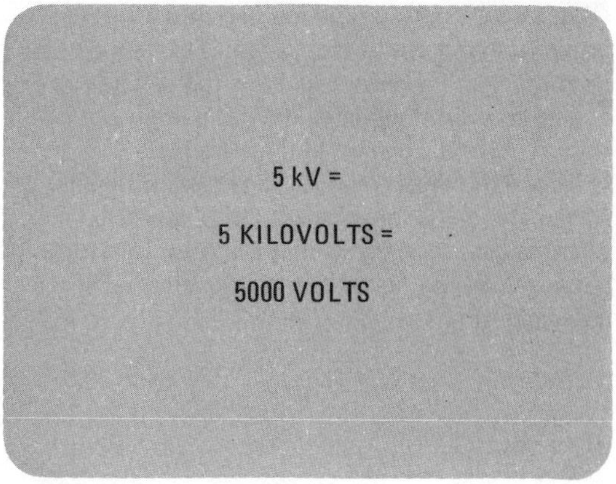

5 kV =

5 KILOVOLTS =

5000 VOLTS

Figure 4.6

Powers of Ten Notation — Powers of ten notation, sometimes called scientific notation (Figure 4.7), was also introduced and it was shown how prefixed and decimal numbers could be converted to the powers of ten format. As you'll be seeing, the key advantage of powers of ten notation is that it makes manipulations of very large or small numbers much easier than when handling them in decimal form. The simple rules for adding, subtracting, multiplying, and dividing with powers of ten were discussed in Lesson 3 and will be further expanded in this lesson.

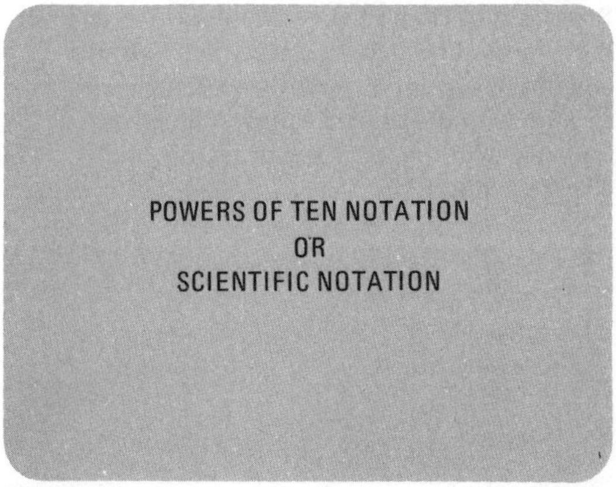

POWERS OF TEN NOTATION
OR
SCIENTIFIC NOTATION

Figure 4.7

Lesson Objectives — So, before getting down to business, look at the objectives for this lesson as summarized in Figure 4.8. When you finish this lesson, you should be able to illustrate Ohm's law graphically, and use it to predict the behavior of voltage, current, and resistance in simple circuits. You should also be able to calculate the power dissipated by a resistor in a simple circuit. This power calculation will involve some familiarity with the mathematical concepts of the *square* of a number, and the *square root* of a number.

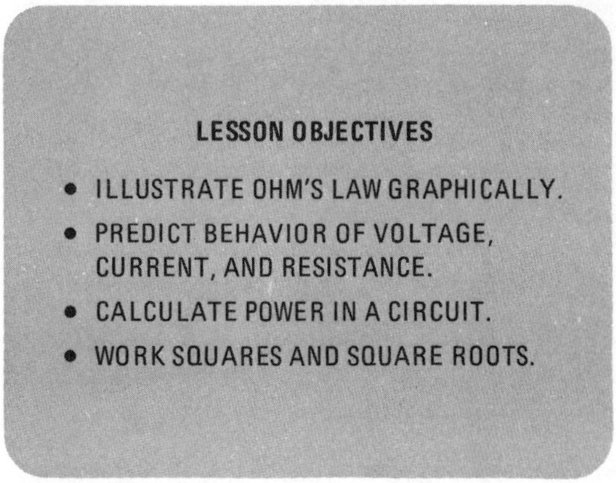

LESSON OBJECTIVES

- ILLUSTRATE OHM'S LAW GRAPHICALLY.
- PREDICT BEHAVIOR OF VOLTAGE, CURRENT, AND RESISTANCE.
- CALCULATE POWER IN A CIRCUIT.
- WORK SQUARES AND SQUARE ROOTS.

Figure 4.8

Ohm's Law — Our discussion begins with the important concept of Ohm's law. This law will be used again and again throughout the dc circuits course because it explains the basic interrelationship of voltage, current, and resistance (Figure 4.9). These relationships were originally formulated by George Simon Ohm, an early scientist who discovered that the *basic* relationship between voltage, current, and resistance is especially simple.

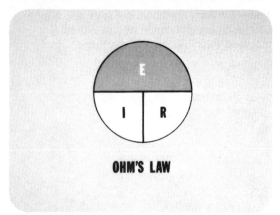

Figure 4.9

Statement of Ohm's Law — Ohm's law, in its most basic form states that: in simple materials, the amount of current through the material varies *directly* with the applied voltage, and varies *inversely* with the resistance of the material (Figure 4.10).

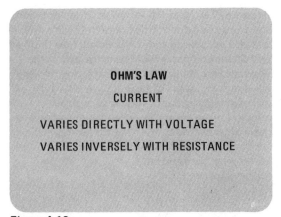

Figure 4.10

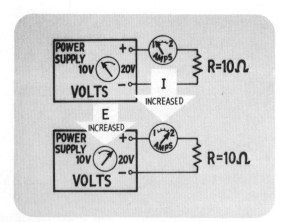

Figure 4.11

Current Varies Directly with Applied Voltage — Anyone tinkering with a simple circuit will discover changing the voltage applied to a circuit will cause the current flowing in the circuit to change. If the resistance is held constant, the current change will *follow* the pattern of the voltage change as shown in Figure 4.11. Doubling the voltage will double the current; halving the voltage will halve the current, etc. This is what is meant by "the voltage *varies directly* with the current."

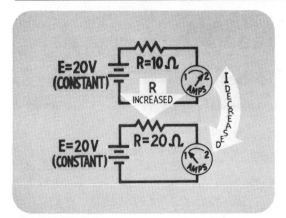

Figure 4.12

Changing the resistance in a circuit will also cause a change in current flow. If the voltage applied to a circuit is held constant, and the resistance in the circuit is *increased*, what happens? As shown in Figure 4.12, with *more opposition* to current flow in the circuit, the circuit current will *decrease*. On the other hand, if the *resistance* is decreased, the amount of current flow in the circuit *increases*. This is what is meant by "current through a material or circuit *varies inversely* with the resistance of the material or circuit."

Voltage vs Current — The direct relationship between voltage and current can be shown with a graph, which is simply an illustration of the relationship between two quantities. In Figure 4.13, voltage is marked off along the bottom line, or axis with current marked off along the vertical line or axis.

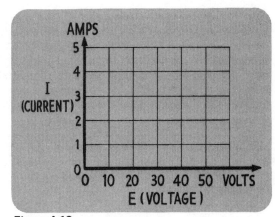

Figure 4.13

Circuit with Ammeter and Voltmeter — To examine this relationship, the voltage is increased across the single resistor in Figure 4.14. If the voltage across the resistor is measured with a voltmeter and the current through the resistor is measured with an ammeter in series as shown, with a fixed or constant resistance, a change in voltage will cause the current to change in what is called a *linear* or *straight-line* manner.

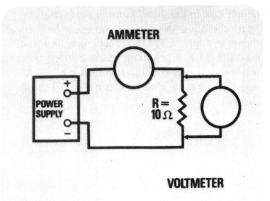

Figure 4.14

If the value of the resistor in the circuit is 10 ohms, and the voltage is set to 10 volts, the ammeter would read a current of 1 ampere or one amp. This is plotted on the graph in Figure 4.15.

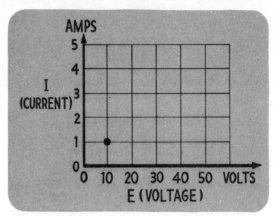

Figure 4.15

If the voltage is now changed to 20 volts, the current should rise until there are 2 amps flowing through the resistor. These values and higher voltage and current values are also plotted on the graph in Figure 4.16. As you can see, each jump in voltage would be met with a corresponding jump in current. If all the points are connected together, a *straight line* is produced, which is the reason the relationship between voltage and current, in this case, is described as *linear*. Materials and devices that behave in this manner with respect to voltage and current are called *ohmic* devices, and are said to "satisfy" Ohm's law.

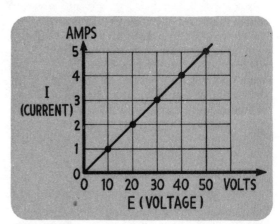

Figure 4.16

Ohm's Law in Equation Form — This linear relationship is often expressed in equation form as shown in Figure 4.17. The most common equation used is E = I X R or voltage equals current times resistance. If you know any two parts of this equation for Ohm's law, you can find the third part.

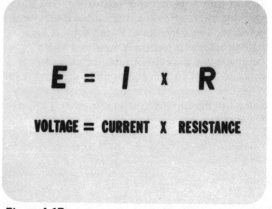

$$E = I \times R$$

VOLTAGE = CURRENT X RESISTANCE

Figure 4.17

Ohm's Law in Circle Form — Since you'll be using Ohm's law again and again as you study electricity, you will need a convenient way to remember and use the law. One simple memory key you may find useful is shown in Figure 4.18. Simply draw a circle; divide it in half with a horizontal line. Then divide the bottom half with a vertical line. Put the symbol for voltage, E, in the top half; and the letters I and R representing current and resistance on either side of the vertical line in the bottom half. The vertical line indicates multiplication of the quantities on either side of the line. The horizontal line implies division; the bottom quantity divides into the top.

As an illustration of how to use this circle, assume you are given a 100-ohm resistor, and you know 1 amp is flowing through it. You want to find the third part, the voltage across the resistor necessary to produce 1 amp of current. To use this circle, *cover the part you want to find with your thumb*, and then look at the position of the remaining letters. Their position tells you the procedure to follow. Remember, a vertical line means multiply and a horizontal line means divide the bottom into the top.

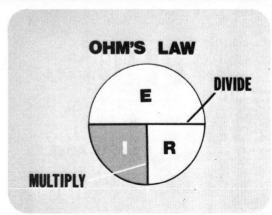

Figure 4.18

An Illustration of Ohm's Law in Circle Form — Since, in this case the unknown is voltage, cover the E with your thumb, which leaves I on one side of the vertical line and R on the other (Figure 4.19). This tells you to *multiply current times resistance* which is the original equation, $E = I \times R$. In this case, I equals 1 amp and R is 100 ohms. Multiply 1 amp times 100 ohms to get 100 volts as your answer.

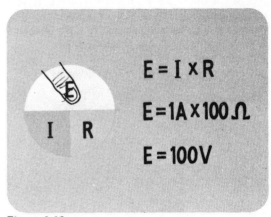

$$E = I \times R$$

$$E = 1A \times 100\,\Omega$$

$$E = 100V$$

Figure 4.19

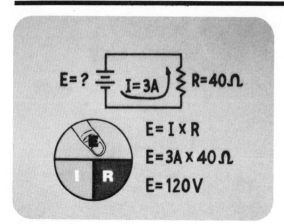

Figure 4.20

Figure 4.20 shows another problem where voltage is the unknown factor. In this case, you know that I = 3A and R = 40 Ω from the schematic diagram. You need to calculate E, so cover the E in the circle with your thumb. The remaining letters and the vertical line between them indicate I X R. So, your formula is E = I X R, substitute 3 amps and 40 ohms for I and R and your calculated result is 120 volts.

Finding Current with Ohm's Law in Circle Form — Ohm's law can also be used in another way. Suppose you want to find the current through a 2-ohm resistor when you know the voltage across the resistor is 1 volt. Again, you can use Ohm's law in circle form, as shown in Figure 4.21. This time the unknown quantity you need to calculate is the current, I. So cover the I with your thumb. The remaining letters are E and R, and the E is over the R, which tells you to divide E by R. So this gives you another key formula from the basic Ohm's law relationship: current is equal to voltage divided by resistance, or E over R. Any time you write one number over another (as in a fraction) this means to divide the top number by the bottom number. For instance, 1 volt divided by 2 ohms is 1/2 amp.

If you divide the top number by the bottom number, or divide 1 by 2, you get the decimal fraction .5. That is, the fraction 1/2 is the same as .5 in decimal form. So the answer may be expressed as either 1/2 amp or .5 amp, or in the more commonly used form: 500 milliamps.

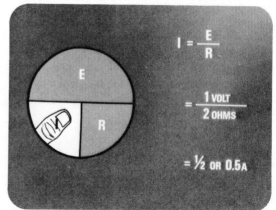

Figure 4.21

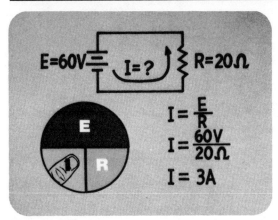

Figure 4.22

Figure 4.22 shows another problem that requires you to find the current. In this circuit you know that E = 60 volts, and R = 20 ohms, and you need to calculate the current, I. Cover the I with your thumb in the circle diagram: the remaining letters are E over R. So the formula used is I = E/R. Substituting E = 60 volts and R = 20 ohms, the calculated result for I is 3 amps.

Finding Resistance with Ohm's Law in Circle Form — A third way you can use Ohm's law is to *find resistance when you know voltage and current.* This time, since the resistance, R, is your unknown, you would cover R in the circle, as shown in Figure 4.23. The position of the remaining letters tells you that resistance equals E over I; or voltage divided by current. So, if 100 volts applied to a circuit results in a current of 2 amps flowing through the circuit, the resistance of the circuit equals E over I, or 100 volts divided by 2 amps, which equals 50 ohms.

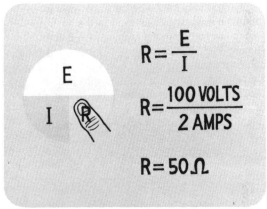

Figure 4.23

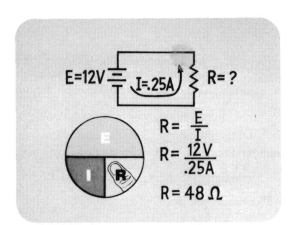

Figure 4.24

Remember, the horizontal line means to *divide* the upper number by the lower number. Figure 4.24 shows another example. In this circuit schematic, you're given E = 12 volts, and I is equal to .25 amps. You can now find R by first covering the R in the circle diagram. The remaining letters, E over I, tell you the Ohm's law formula in this case is R = E/I. Substituting for E and I and dividing, the calculated result for R is 48 ohms.

So notice — from Ohm's law come three basic and very powerful formulas to aid in solving circuit problems.

- **Simple Circuit**
- **Applying Ohm's Law**
- **Division with Powers of Ten**

Simple Circuit — To help you learn the concepts introduced in Lesson 3, this lesson shows you how to use *scientific notation* with Ohm's law to find the current in simple circuits. For example, the simple circuit in Figure 4.25 consists of a 10-volt battery wired to a 33-kilohm resistor. The one *unknown* factor is the *current flowing* in the circuit.

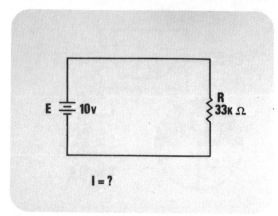

Figure 4.25

Applying Ohm's Law — Since the current is the unknown quantity, cover the I in the circle for Ohm's law to obtain the formula: I = E/R (Figure 4.26). When ten volts is substituted for E and thirty-three thousand ohms for the value of R, the resulting fraction is ten over thirty-three thousand. Now convert these quantities to powers of ten form. Ten simply is $1 \times 10^{+1}$ and thirty-three thousand in powers of ten form is $3.3 \times 10^{+4}$.

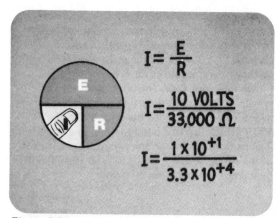

Figure 4.26

Division with Powers of Ten — Follow through the steps necessary to carry out the division with powers of ten.

First, write down the form in which you expect to write the answer: point some number times ten to the some exponent (Figure 4.27).

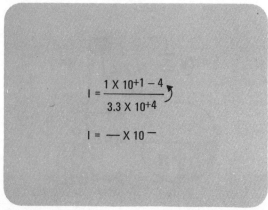

Figure 4.27

When a fraction is divided with powers of ten, the *bottom exponent is first brought up to the top*, across the division line and *its sign is changed*. As shown in Figure 4.28, the exponents on the top of the fraction are plus 1, and minus 4.

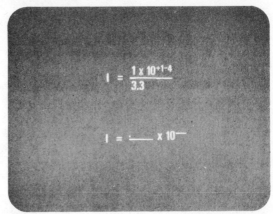

Figure 4.28

Combine the Exponents — When the exponents are combined, as shown in Figure 4.29, a plus 1 and a minus 4 equal a minus 3. This minus 3 becomes the exponent in the answer. After the division of the leading numbers (1 divided by 3.3 equals .303), the answer is .303 X 10^{-3} (Figure 4.30).

COMBINE THE EXPONENTS

$$\begin{array}{r} +1 \\ \underline{-4} \\ -3 \end{array}$$

Figure 4.29

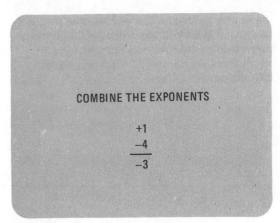

$$I = \frac{1 \times 10^{+1\ -4}}{3.3}$$

$$I = .303 \times 10^{-3}$$

Figure 4.30

Conversion to Decimal Form — To convert the answer to decimal form, remember that the exponent is minus 3, which tells you to move the decimal three places to the left. When the decimal point is moved and the spaces between it and the number are filled in with zeros, the answer in decimal form is seen to be .000303 (Figure 4.31).

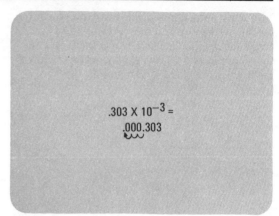

Figure 4.31

Conversion to a Metric Prefixed Number — As the final step, Figure 4.32 shows how .000303 amp is converted to metric prefixed form by using the metric *chart*. As you see, .000303 amp equals 303 microamps (303 μA).

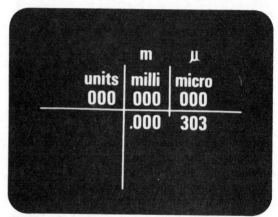

Figure 4.32

Smaller Resistance — The left-hand side of Figure 4.33 shows the first example circuit labeled with its current flow of 303 microamps. As a second example, consider the same circuit but with a *smaller* resistor. In the right-hand circuit the 33-kilohm resistor has been replaced with a smaller resistor, a 1 kilohm. To determine how much current flows in the second circuit, proceed exactly as before. You are solving for current, so cover up the I with your thumb on the circle to get E over R.

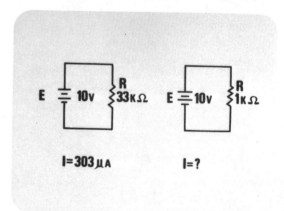

Figure 4.33

4-16

- **Current in the Second Circuit**
- **Steps in Division**
- **Conversion to a Metric Prefixed Number**

Current in the Second Circuit — If you substitute ten volts in place of E in the formula, and 1 kilohm in place of R in the formula, you then have the fraction 10 over 1000. Figure 4.34 shows these numbers converted to powers of ten form: 10 volts becomes $1 \times 10^{+1}$, 1000 ohms becomes $1 \times 10^{+3}$.

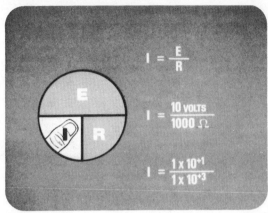

Figure 4.34

Steps in Division — Figure 4.35 shows the steps you use to divide these numbers. Again your answer will be written as some number times ten to some power. Take the lower exponent and bring it up above the division line, changing its sign. Combine the exponents; +1 and −3 gives you a −2; then divide the leading numbers. Here 1 over 1 is 1, so the answer is 1×10^{-2}, which in decimal form is .01.

$$I = \frac{1 \times 10^{+1-3}}{1 \times 10^{+3}}$$

$$I = \frac{1 \times 10^{+1-3}}{1}$$

$$I = \frac{1 \times 10^{-2}}{1} \qquad \begin{array}{r} +1 \\ -3 \\ \hline -2 \end{array}$$

$$I = 1 \times 10^{-2} = .01$$

Figure 4.35

Conversion to a Metric Prefixed Number — Again use the metric chart as shown in Figure 4.36 to convert the result to metric prefixed form. Here .01 amp becomes *ten milliamps* of current.

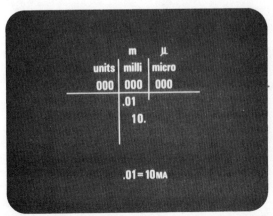

Figure 4.36

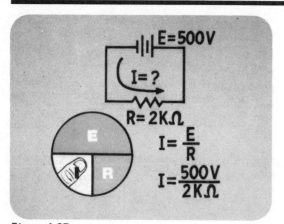

Figure 4.37

Additional Circuit Example — Figure 4.37 shows another example. In this circuit, you are given that 500 volts is applied to a 2-kilohm resistor, and have to calculate the resulting current flow. Correct use of Ohm's law in circle form tells you that the formula needed for this calculation is I = E/R. Substituting 500 volts and 2 kilohms you need only to divide 500 volts by 2000 ohms to find I.

$$I = \frac{5 \times 10^{+2-3}}{2 \times 10^{+3}}$$

$$I = \frac{5 \times 10^{+2-3}}{2} \quad \begin{array}{r} +2 \\ -3 \\ \hline -1 \end{array}$$

$$I = \frac{5 \times 10^{-1}}{2}$$

$$I = 2.5 \times 10^{-1} = .25A$$

Figure 4.38

Calculation with Powers of Ten — Written in powers of ten form, you have: $5 \times 10^{+2}$ divided by $2 \times 10^{+3}$ (Figure 4.38). As in the last example, bring the bottom exponent up above the division line, changing its sign as you perform the operation. This gives you a plus 2 and a minus 3 for the exponent above the line; combining these results gives you an exponent of minus 1. The result of the division of the leading numbers (5/2) is 2.5. So the result of this calculation may be written as 2.5×10^{-1}, which is equal to .25 amp.

UNITS	m milli	µ micro	
000	000	000	
	.25		
	250.		
	.25A = 250mA		

Figure 4.39

Metric Prefixed Form — Using the metric conversion chart as shown in Figure 4.39, you can convert the .25 amp to 250 milliamps.

Smaller Resistance — Larger Current — Now that you've had some practice with basic Ohm's law calculations, it will be helpful to go back to the two circuits discussed a moment ago to study the general circuit behavior Ohm's law describes. Focus your attention on the circuits shown in Figure 4.40. The right-hand circuit with ten volts of pressure and a 1-kilohm resistor has ten milliamps of current flowing in it. The left-hand circuit with ten volts applied and a *larger* resistance of 33 kilohms, has a much *smaller* current flowing, 303 microamps. An examination of the two circuits in Figure 4.40 shows a very important relationship. In circuits with the same voltage applied, as the resistance decreases, the current increases. When the resistance decreased from 33 kilohms to 1 kilohm the current increased from 303 microamps to 10 milliamps.

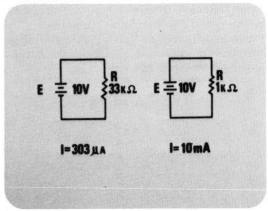

Figure 4.40

Inverse Relationship — This is an important relationship, so it is restated here. If the voltage applied to a circuit is held constant, and the value of the resistance *decreases*, the amount of current flowing will *increase* (Figure 4.41). As has been mentioned, this relationship between the current flowing and resistance present in a simple circuit is called an *inverse relationship*.

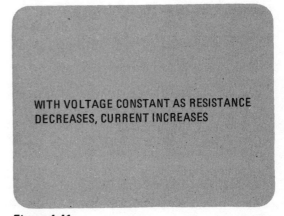

WITH VOLTAGE CONSTANT AS RESISTANCE DECREASES, CURRENT INCREASES

Figure 4.41

Larger Voltage — The circuit shown in Figure 4.42 can be used to examine another relationship expressed by Ohm's law. This is the same circuit as that shown in Figure 4.40, except that the voltage has been increased to 20 volts.

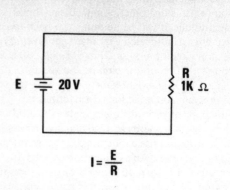

$$I = \frac{E}{R}$$

Figure 4.42

Current in the Third Circuit — The current in this third circuit can be found with the same formula as before, I = E/R (Figure 4.43). Substitute the known voltage of 20 volts and the known resistance value in its proper place, and convert each of these to powers of ten form. Twenty volts is $2 \times 10^{+1}$; and the 1-kilohm resistor becomes $1 \times 10^{+3}$. To divide these two numbers, again bring the lower exponent up above the division line and change its sign. Divide the leading numbers, and combine the exponents. Two divided by 1 is 2, and a +1 and −3 is a −2. The answer is then 2×10^{-2}.

$$I = \frac{E}{R} = \frac{20 \text{ VOLTS}}{1 \text{ K}\Omega}$$

$$I = \frac{2 \times 10^{+1-3}}{1 \times 10^{+3}}$$

$$I = \frac{2 \times 10^{+1-3}}{1} \quad \begin{matrix} +1 \\ -3 \\ \hline -2 \end{matrix}$$

$$I = 2 \times 10^{-2}$$

Figure 4.43

Conversion to a Metric Prefixed Number — As shown in Figure 4.44, 2×10^{-2} equals .02 amp or 20 milliamps of current.

$$2 \times 10^{-2} = .02 = 20 \text{ MA}$$

Figure 4.44

Larger Voltage — Larger Current — Figure 4.45 shows a comparison of the two circuits, and illustrates another of the key relationships from Ohm's law. For circuits with the *same resistance*, when voltage *increases*, current *increases*; thus, voltage and current are *directly related*.

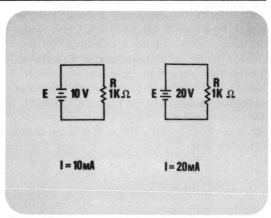

Figure 4.45

Power — Now that you've seen Ohm's law and the interrelationship of voltage current, and resistance, focus your attention on another important topic in the study of electricity: power (Figure 4.46). The word power may call to mind many things from your everyday life. In the study of electricity, however, the word power has a rather specific meaning. When you use electricity, you usually want it to do some sort of work, either to move something or often to generate heat. Power is specifically the *rate at which work is done*, or the *rate at which heat is generated*.

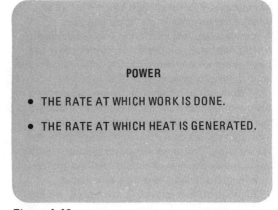

Figure 4.46

Previously you learned that voltage represents an electrical force that causes current to move in a closed circuit. *Work* is done whenever a force causes motion. If the force exists *without* causing any motion, no work is done. In an open circuit, where a voltage exists but no current flows, no work is done. Whenever a closed circuit is carrying current and voltage is causing electrons to move, work is done. The instantaneous *rate* at which this work is done is called the Electrical Power Rate.

The unit commonly used to specify electric power is the *watt*. In equations you'll find power abbreviated with the letter P, and watts, the unit of measure for power, is abbreviated in the letter W. The wattage rating of an appliance gives you an indication of how fast it does work, or the rate at which it generates heat. For example, a 60-watt electric drill can drill through a 2-inch board in about ten seconds, but a 200-watt drill can do it in much less time.

Horsepower — Large electric motors are often rated in horsepower in the same way that automobile engines are rated. One horsepower equals 746 watts (Figure 4.47). Horsepower and wattage are simply two different units that may be used to specify power.

Work is also often done by appliances through the generation of heat. As an example, a 500-watt hair dryer contains a special resistance wire to provide heat, along with a fan to blow air by the heated wire. This, then, generates enough heat to comfortably dry hair. An industrial heat gun is constructed similarly, but may be rated at 1500 watts of heat output. It can supply enough heat to char and burn a piece of paper.

1 HORSEPOWER = 746 WATTS

Figure 4.47

Resistors — Any *resistance will generate heat as current passes through it*. Resistors also have a power or wattage rating which is somewhat *different* from the wattage rating given to an appliance. The power rating of an appliance is usually the rate at which it is using electrical energy to do work for us. The power rating of a resistor is a *measure of its ability to give off unwanted heat* (Figure 4.48). A resistor gives off heat, or it is said to *dissipate* heat, by the convection of air moving around the resistor. A bigger resistor comes in contact with more air, and so it can handle more power, and thus has a higher wattage rating. The wattage rating of a resistor is

RESISTOR'S POWER OR WATTAGE RATING

MEASURE OF THE ABILITY TO GIVE OFF UNWANTED HEAT

Figure 4.48

the highest power the resistor can be operated at without overheating.

Selection of a Resistor — As you begin building actual circuits, you'll have to be able to select resistors with the correct wattage rating for any given application (Figure 4.49). In order to select a resistor with sufficient wattage rating, so that it will not be damaged in a circuit, it will first be necessary to *calculate* the power that the resistor will dissipate in the circuit. Then you'll choose one whose power rating is much higher (usually twice) this calculated value, except in some higher power situations.

TO SELECT A RESISTOR

- CALCULATE THE POWER THE RESISTOR WILL DISSIPATE
- THEN CHOOSE A RESISTOR WITH A POWER RATING WHICH IS TWICE THE CALCULATED VALUE (EXCEPT IN HIGH POWER APPLICATIONS).

Figure 4.49

Wattage of Resistors — Figure 4.50 shows several 1-kilohm resistors with wattage ratings from 1/10th of a watt to 2 watts. The 2-watt resistor is usually the largest carbon composition resistor used in electronics. The actual sizes of these resistors may vary, but usually conform to the size reference stated in Lesson 2. The resistor size reference chart is reproduced in the Appendix for your reference.

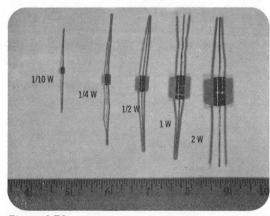

Figure 4.50

Wirewound Resistor — Figure 4.51 shows a wirewound resistor. Notice that the core is usually hollow in this type of resistor, which allows lots of air to move around and through it. This means it can dissipate a lot more heat than smaller composition resistors. Its ceramic core is made of materials that can withstand higher temperatures and consequently these resistors may operate safely at a much higher temperature than most carbon composition resistors.

Figure 4.51

Formula for Power — Back to the question of how you can calculate the power dissipated by a resistor. As first introduced in Lesson 2, Figure 4.52 shows *one* formula that may be used to calculate power: P equals I times E. Note that the letters in this formula spell PIE, and that they can be arranged in a circle similar to the circle memory aid used with Ohm's law. Using this memory aid and the P = I X E formula you can easily find how much power is dissipated in a given resistor if you know the current through, and the voltage across the resistor.

Note again that in writing formulas, the letter P is used to symbolize power while a capital W is used to abbreviate the unit of power, watts. So you may see "the power equals five watts" written more simply as "P = 5 W."

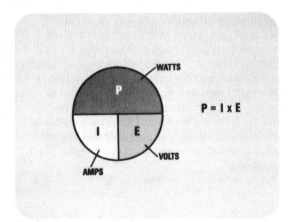

Figure 4.52

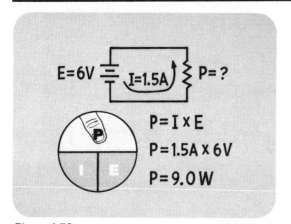

Figure 4.53

Example — In the circuit shown in Figure 4.53, you are given that the voltage applied to the resistor is 6 volts, and the current flowing through it is 1.5 amp. To calculate the power, cover the P in the new circle diagram; and you'll see that the formula you need is P = I X E. Substituting for I and E, *multiply* 1.5 amps by 6 volts to yield the calculated result of 9 watts of power dissipated.

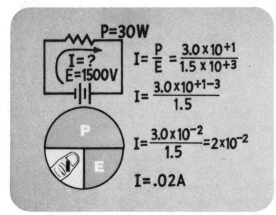

Figure 4.54

Calculating I and E, Given P — If the power is known, either of the quantities I or E may be calculated if one of these two values is known. For instance, if in a circuit you know that P = 30 W, and E = 1500 V, the current I may be found by correctly using the circle formula (Figure 4.54). Covering the I with your thumb, you see that the resulting formula needed for this calculation is I = P/E. This time the numbers will be handled in powers of ten form. Substituting for P and E, to find I you just divide $3.0 \times 10^{+1}$ by $1.5 \times 10^{+3}$. The current flowing in the circuit is 2×10^{-2} A or .02 amp or 20 milliamps.

Another Power Formula — As you may recall from Lesson 2, the first power formula can be used with Ohm's law to derive another formula for power. All that's necessary is to substitute one of the forms of Ohm's law into the first power formula; that is, one part of the power formula is replaced with its equivalent from Ohm's law, as shown in Figure 4.55. The result is that power can now be found by multiplying current times current times resistance.

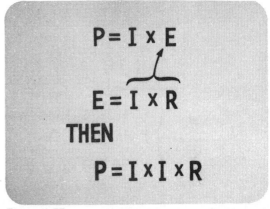

Figure 4.55

Second Circle Formula for Power — Another way to write this is shown in Figure 4.56. I X I may be written in the shorter form I^2, which is pronounced "I squared." To "square" any number means simply to multiply it by itself.

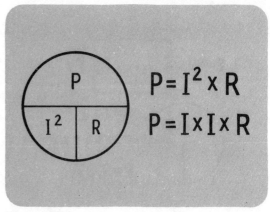

$$P = I^2 \times R$$
$$P = I \times I \times R$$

Figure 4.56

Perfect Squares — The concept of the square of a number may be illustrated by looking at what are called *perfect squares*. If whole numbers are multiplied by themselves, the results are what are called perfect squares, as shown in Figure 4.57. Thus 1 times 1 equals 1 squared (1^2) which equals 1; 2 times 2 equals 2^2 (2 squared) which equals 4, 3^2 equals 9, and so forth, Again, the *square* of a number is just that number multiplied by itself. So the formula $P = I^2 R$ actually means $P = I$ times I times R; or power equals current, times current, times resistance.

WHOLE NUMBERS		PERFECT SQUARES
1 X 1	=	1
2 X 2	=	4
3 X 3	=	9
4 X 4	=	16
5 X 5	=	25
ETC.		

Figure 4.57

$$R = 10\,\Omega$$
$$I = 3A \qquad P = ?$$

$$P = I^2 R$$
$$P = 3A \times 3A \times 10\,\Omega$$
$$P = 90 \text{ WATTS}$$

Figure 4.58

Examples — In using the circle formula illustrated in Figure 4.58, you must remember that I is *not* called for. The value of I X I, or I^2 is used in this diagram. For example in the circuit shown in the illustration, I is 3 amps and R is 10 ohms, and you need to calculate the power dissipated, P. Covering the P in this circle diagram, you find the formula needed in the usual way: $P = I^2 R$. Substituting 3 X 3 or 9 for I^2, and 10 ohms for R, the power dissipation calculated for this problem is 90 watts.

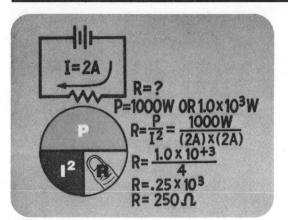

Figure 4.59

The "PI^{2}R" circle can be used to find any of the three values, as long as the other two values are known. For instance, suppose you know "P" and "I" and want to find "R." In Figure 4.59, P = 1000 watts and I = 2 amps, and you need to calculate the value of R. Covering the R in the circle diagram tells you to divide P by I^2. In scientific notation form you would have 1.0 X 10^{+3} divided by 4. One divided by four is equal to .25. Bring down the power of ten for the problem and the answer is .25 X 10^{+3}, or 250 ohms.

Square Roots — Often in your study of electricity you may have to go through the process of squaring a number in *reverse*, that is, you may know the *square* of a certain current, and need to calculate the current itself. This leads to a new idea — that of the *square root*.

Each of the numbers on the right side of Figure 4.60 are called square roots. What is meant by the expression "square root of a number"? Well, the square root of a number is another number, which must be multiplied by itself, in order to obtain the original number. The symbol used to indicate square roots is shown on the left side of the illustration. Notice that the square root of 1 is 1, the square root of 4 is 2, the square root of 9 is 3. In other words, to find the square root of 9 you have to ask yourself: What number when multiplied by itself equals 9? Well, 3 X 3 equals 9, so 3 is the square root of 9.

In mathematical shorthand notation the statement "the square root of 9 is 3" would be written $\sqrt{9} = 3$. In Figure 4.60 the square roots of numbers that are *perfect squares* are illustrated. These square roots are easy to determine, because they're whole numbers. In other situations special techniques must be used if you have to find the square root of a number. A special pencil and paper technique for determining square roots of numbers is included in the Appendix.

SQUARE ROOTS

$\sqrt{1}$	=	1
$\sqrt{4}$	=	2
$\sqrt{9}$	=	3
$\sqrt{16}$	=	4
$\sqrt{25}$	=	5

Figure 4.60

For the purposes of this course, it is suggested that you use either a calculator or square root tables to find square roots in the problems you work. Briefly consider how to use these tools.

First of all, when you use a calculator of a type similar to the TI SR-10, SR-11, or SR-16 to find the square root of a number, simply enter the number and press the square root key (Figure 4.61). The number's square root will appear in the display. On calculators of this type, the square roots of numbers that are in powers of ten form may also be determined.

TO FIND THE SQUARE ROOT OF A NUMBER WITH A CALCULATOR:
1. ENTER THE NUMBER
2. PRESS THE SQUARE ROOT KEY
3. THE NUMBER'S SQUARE ROOT WILL APPEAR IN THE DISPLAY

Figure 4.61

SCIENTIFIC CALCULATOR SQUARE ROOT SEQUENCES

ENTER	PRESS	DISPLAY
2277	$\sqrt{x}$	47.71792116 ROUND TO 47.7
4.1	EE	4.1　00
3	$\sqrt{x}$	6.40312 42 01 ROUND TO 64

Figure 4.62

For example, take the number 2277. To find the square root of this number, first enter it into the calculator, then press the square root key (Figure 4.62). The square root of 2277 is 47.71792116. This number multiplied by itself should be equal to 2277. To check it, simply press the x^2 key. In practical application, the square root of a number usually need not contain more than three significant digits. Simply round the third digit in the same manner as was discussed in Lesson 3.

Numbers written using scientific notation may also have their square roots taken. To find the square root of 4100, enter the number into the calculator in correct powers of ten form: 4.1, enter exponent key, 3, and square root. The answer, also in scientific notation when rounded off is $6.4 \times 10^{+1}$, or 64.

If no calculator is available, you may wish to use a square root table. Simply find the number for which you need to determine the square root in the table and look across to the column labeled square roots (Figure 4.63). Since in most cases you will only need three significant digits, if the exact number you want is not listed in the table, you can use the square root of a number that is closest to the one you want. If you need the square root of a number which is larger or smaller than those listed in the table, you may break the number up into a product and look up the square root of each of the terms. Then these square roots are multiplied to find the square root of the original number. A complete square root table, along with a brief recap on how to use it, is included in the Appendix.

n	n^2	$\sqrt{n}$	n^3	$\sqrt[3]{n}$
1	1	1	1	1
2	4	1.414		
1000	1,000,000	31.62278		

Figure 4.63

$$250 = 25 \times 10$$

$$\sqrt{250} = \sqrt{25} \times \sqrt{10}$$

$$15.81 = 5 \times 3.162$$

$$15.81^2 = 249.95$$

Figure 4.64

Here's how to use the table to find the square root of a simple number, say 250. Looking in the n^2 column of the square root table, you cannot see 250, so 250 must not be a perfect square. The numbers closest to 250 are 256 (which is the square of 16) and 225 (which is the square of 15). The square root of 250 must be between 15 and 16, but how can the table be used to find the exact number? Easy, just break the number, in this case 250, into two terms that equal 250 when multiplied together. In this case, you can use 25 and 10. As shown in Figure 4.64, 25 times 10 is 250, so the square roots of ten and 25 multiplied together should equal the square root of 250. The square root of 10 is 3.162 according to the table.

$$\sqrt{60} = ?$$

$$\sqrt{60} = \sqrt{6} \times \sqrt{10}$$

$$7.743 = 2.449 \times 3.162$$

$$(7.743)^2 = 59.96$$

Figure 4.65

The square root of 25 is 5. Five times 3.162 is equal to 15.81. To check the validity of this calculation, simply square 15.81. This number squared is equal to 249.95 . . . close enough!

Using this method and the square root table, find the square root of 60. Since 60 is not a perfect square and thus is not listed in the n^2 column of the square root table, you must find two numbers that *are* listed in the table and whose product is 60. How about 10 and 6? As shown in Figure 4.65, the square root of 10 is 3.162 and the square root of 6 is 2.449; 3.162 times 2.449 is equal to 7.743. Square this number, and you arrive at 59.96; again, pretty close to the original number.

Sample Circuit and Calculation — To get back to circuits, consider a few additional examples involving power calculations. Figure 4.66 is an example of how to find the power dissipated by a resistor in a simple circuit. You know there are 2 amps of current flowing through a resistor, the value of the resistor is 50 ohms, and you need to calculate the power it is dissipating. Proceed just as before. Because you know the current and resistance, you can use the formula $P = I^2R$. Next, substitute the current and resistance values in the formula. Square 2 by multiplying it times itself. Next, write the formula with the square of 2 in it, and you then have 4 times 50. Carry out the multiplication and the answer is 200 watts.

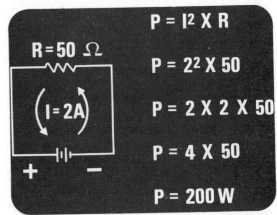

$$P = I^2 \times R$$
$$P = 2^2 \times 50$$
$$P = 2 \times 2 \times 50$$
$$P = 4 \times 50$$
$$P = 200\ W$$

Figure 4.66

Calculation with Powers of Ten — Figure 4.67 shows another example with the same formula, but this time with powers of ten notation. In this circuit there are 10 milliamps of current and 20 kilohms of resistance. Proceed just as you did before.

First, write out your formula $P = I^2R$. Next, substitute the current and resistance values into the formula. You must square 10 milliamps and multiply that result by 20 kilohms to get your answer. Ten milliamps may be expressed in powers of ten form as 1×10^{-2} amps. 1×10^{-2} times 1×10^{-2} (remember the rules for multiplying with powers of ten) equals 1×10^{-4}. Twenty kilohms equals $2 \times 10^{+4}$ in powers of ten form. So

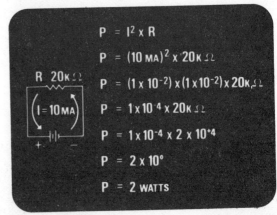

$$P = I^2 \times R$$
$$P = (10\ \text{MA})^2 \times 20\text{K}\ \Omega$$
$$P = (1 \times 10^{-2}) \times (1 \times 10^{-2}) \times 20\text{K}\ \Omega$$
$$P = 1 \times 10^{-4} \times 20\text{K}\ \Omega$$
$$P = 1 \times 10^{-4} \times 2 \times 10^{+4}$$
$$P = 2 \times 10^{0}$$
$$P = 2\ \text{WATTS}$$

Figure 4.67

your final result is $1 \times 10^{-4} \times 2 \times 10^{+4}$, which equals 2×10^{0} or *2 watts*.

Squaring Numbers Less Than One — In this last example, an interesting thing happened. When 10 milliamps was multiplied by itself, the answer was 1×10^{-4} which is 100 microamps, and this is *smaller* than the original number. At first you might think that whenever you square a number, you should get a larger number for your answer, but this is not the case. Consider what happens when you square 1/2, for example. As shown in Figure 4.68, 1/2 times 1/2 equals 1/4. What you are actually saying here is that .5 X .5 is actually 1/2 of 1/2, which is 1/4 or 250 milli. In general, any number less than one, when squared, results in a smaller number than the original.

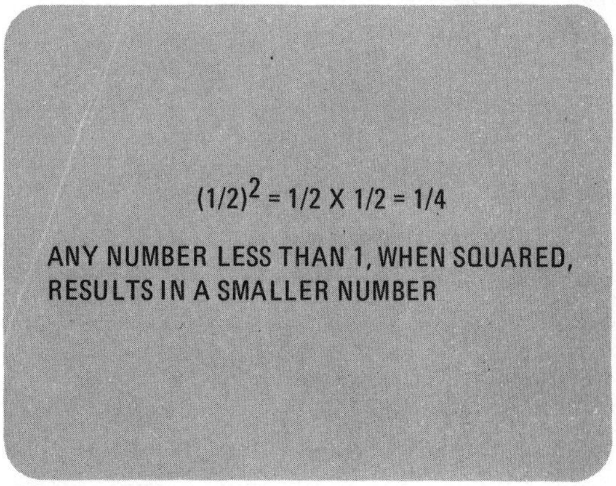

$$(1/2)^2 = 1/2 \times 1/2 = 1/4$$

ANY NUMBER LESS THAN 1, WHEN SQUARED, RESULTS IN A SMALLER NUMBER

Figure 4.68

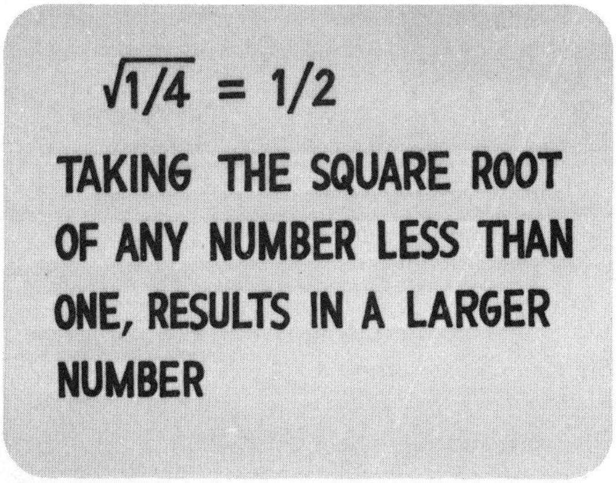

$$\sqrt{1/4} = 1/2$$

TAKING THE SQUARE ROOT OF ANY NUMBER LESS THAN ONE, RESULTS IN A LARGER NUMBER

Figure 4.69

Taking the Square Root of Numbers Less Than One — This relationship also has interesting effects when acting in reverse. That is, if you take the square root of a number that is less than one; the square root turns out *larger* than the original number. For example, $\sqrt{1/4} = 1/2$ (as seen from Figures 4.68 and 4.69).

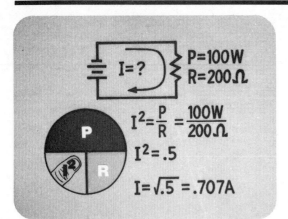

$$I^2 = \frac{P}{R} = \frac{100W}{200\,\Omega}$$

$$I^2 = .5$$

$$I = \sqrt{.5} = .707A$$

Figure 4.70

Example — To see this in an electronic example, consider the circuit in Figure 4.70. If P = 100 watts and R is 200 ohms, the current flowing, I, can be calculated in the same manner, again, using the circle. The circle instructs you to divide 100 by 200 which gives you a value for I^2 equal to .5. The square root of I^2 is I, so you must also find the square root of .5 which is .707 amp, a larger number! Remember, whenever the square root of a number less than 1 is taken, the result will be a number that is *larger* than the original number!

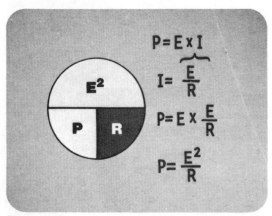

$$P = E \times I$$

$$I = \frac{E}{R}$$

$$P = E \times \frac{E}{R}$$

$$P = \frac{E^2}{R}$$

Figure 4.71

Third Formula for Power Calculations — One more formula is available that may come in handy in calculating power in circuit situations where a resistor's ohmic value and the voltage across it are known and you need to calculate the power being dissipated. Using the original formula for power, P = E X I, requires that you know both the voltage across and current flowing through the resistor. One form of Ohm's law enables you to solve for current: I = E/R. If you substitute this formula for current in the proper place in the power formula, the result is a new equation (as shown in Figure 4.71).

$$P = E \times E/R$$
or
$$P = E^2/R$$

This new formula is useful in examples of the following type.

- Example
- **Summary**

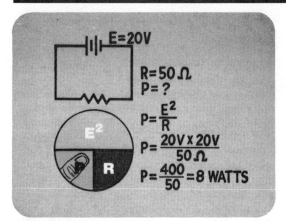

Figure 4.72

Example — If you know the voltage across a resistor to be 20 volts and its resistance to be 50 ohms, you can use this new formula to derive the answer directly. You simply cover the P in the circle formula as shown in Figure 4.72. Substituting 20 volts and 50 ohms in the formula $P = E^2/R$ and squaring yields 400 divided by 50 ohms. The power dissipated by the resistor is 8 watts.

Summary — This lesson has discussed Ohm's law and has shown a graphical meaning of the *linear relationship* between voltage and current. Several sample problems were used to conclude that *voltage and current vary directly*; that is, in a circuit if the resistance is held constant and the voltage is increased, the current flowing increases. Also, it was shown that current and resistance have an *inverse relationship*; that is, in a circuit, if the voltage applied is held constant, as the resistance is decreased, the current will increase.

The concept of power was also discussed in detail, and some everyday applications of power rating were shown. The special power rating of resistors was reviewed, and size and construction were shown to determine their wattage rating. Several formulas for power, and the concepts of squares and square roots were introduced. Finally, several sample calculations for the power dissipated by a resistor in simple circuit situations were shown.

At this point in your study, four of the most basic electrical quantities have been discussed. These are voltage (E), current (I), resistance (R), and power (P). The units for each (volts, amps, ohms, and watts) have also been explained. It is of key importance that you thoroughly understand the interrelations of these quantities. You should understand how any of these quantities either controls or are controlled by the others in a circuit situation.

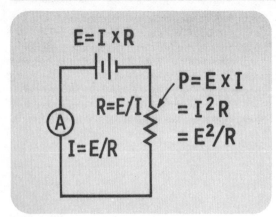

Figure 4.73

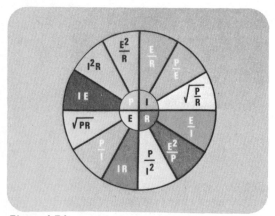

Figure 4.74

In your study you've now seen three Ohm's law formulas and where they may be used, and as shown in Figure 4.73. Also three power formulas have been covered. In practice, because of the mathematical relationship of the quantities E, I, R, and P, you should be able to express any of these four quantities in terms of any two of the others. What this means is that there are three formulas available for each quantity. Each of these three formulas involves a different combination of the other quantities, for a total of 12. A convenient way to summarize all 12 of these basic formulas is shown in Figure 4.74. The four basic quantities, E, I, R, and P, are at the center of the illustration. Next to each quantity are three segments. In each quarter of the circle there are three formulas. For example in the upper right-hand quarter, I is the basic quantity and it is equal to those expressions in the three segments.

$$I = E/R, P/E, \text{ or } \sqrt{P/R}.$$

Note that this circle is different from the other circle formulas that have been used. It is designed to easily catalog the formulas for E, I, R, and P; but does not serve as a memory aid in the same way the simple circle formulas for Ohm's law and power.

LESSON 4. OHM'S LAW AND POWER

● **Worked Through Examples**

The following examples are designed to reinforce your understanding of the use of Ohm's law and power formulas with scientific notation.

All of the examples involve the use of powers of ten with one exception. Example 5 illustrates the proper use of the P-I^2-R circle formula to find the current flowing in a simple circuit. This involves taking the square root of a number. Since the procedure for finding square roots of quantities expressed in scientific notation is covered in Lesson 10, only very simple numbers are used in the example.

1. In the simple circuit shown, a voltage is applied to a resistor and current flows through the resistor. Use Ohm's law to find the applied voltage if the current is 5 mA and the resistance is 3 kilohms.

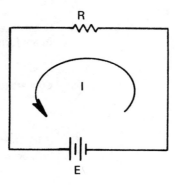

    a.   Draw the circle formula for Ohm's law.

    b.   Cover the quantity you want to find with your thumb; in this case, cover E. Remember, a vertical line tells you to multiply the quantities on either side of the line and a horizontal line tells you to divide the bottom quantity into the top.

$E = I \times R$

    c.   The resulting formula is $E = I \times R$.

$E = 5 \text{ mA} \times 3 \text{ k}\Omega$

    d.   Substitute the values of voltage and current in the formula.

$E = 5 \times 10^{-3} \times 3 \times 10^{+3}$

    e.   Convert to powers of ten.

$E = 15 \times 10^{0}$

    f.   Multiply the leading numbers and combine the exponents. In this case, +3 and −3 equal zero.

$E = 15 \text{ volts}$

    g.   Convert to metric prefixed form. Because $10^{0} = 1$, the answer can best be expressed directly in units, $E = 15$ volts.

2. Given the same circuit as in Example 1, use Ohm's law to find the current flowing when the voltage is 12.6 volts and the resistance is 820 ohms.

$$I = \frac{E}{R}$$

a. Draw the circle formula for Ohm's law.

b. Cover the quantity you want to find with your thumb; in this case, cover I.

c. The resulting formula is I = E/R.

$$I = \frac{12.6 \text{ V}}{820}$$

d. Substitute the values of voltage and resistance in the formula.

$$I = \frac{1.26 \times 10^{+1}}{8.2 \times 10^{+2}}$$

e. Convert to powers of ten.

$$I = \frac{1.26 \times 10^{+1-2}}{8.2}$$

f. Bring the bottom exponent across the division line, up to the top and change its sign.

$$I = \frac{1.26 \times 10^{-1}}{8.2}$$

g. Combine the exponents. Here a +1 and a −2 equal −1.

$$I = .154 \times 10^{-1}$$

h. Divide the leading numbers and round off.

$$I = 15.4 \text{ mA}$$

i. Convert to metric prefixed form.

3. Again considering the same circuit as before, find the power dissipated by the resistor when the applied voltage is 45 volts and the current flowing through the resistor is 16 mA.

$$P = I \times E$$

a. Draw the P-I-E circle formula for power.

b. Cover the quantity you want to find with your thumb; in this case, cover P.

c. The resulting formula is P = I X E.

$$P = 16 \text{ mA} \times 45 \text{ V}$$

d. Substitute the values of voltage and current in the formula.

$$P = 16 \times 10^{-3} \times 4.5 \times 10^{+1}$$

e. Convert to powers of ten.

$$P = 72 \times 10^{-2}$$

f. Multiply the leading numbers and combine the exponents. In this case, −3 and +1 equal −2.

P = 720 mW

g. Convert to metric prefixed form. A 1-watt resistor would be appropriate for this example.

4. In a simple circuit, find the power dissipated by a 100-ohm resistor when the current flowing through it is 50 mA.

a. Since you *know* I and R, and *need to find* P, select the P-I^2-R circle formula for power.

b. Cover the quantity you want to find with your thumb; in this case, cover P.

$P = I^2 \times R$

c. The resulting formula is $P = I^2 \times R$.

$P = (50 \text{ mA})^2 \times 100\ \Omega$

d. Substitute the values of current and resistance in the formula.

$P = (5 \times 10^{-2}) \times (5 \times 10^{-2}) \times 1 \times 10^{+2}$

e. Convert to powers of ten. Since 50 mA equal 5×10^{-2}, the square of 50 mA equals 5×10^{-2} times 5×10^{-2}.

$P = 25 \times 10^{-4} \times 1 \times 10^{+2}$

f. Multiply 5×10^{-2} by itself remembering to add the exponents. Here -2 and -2 equals -4.

$P = 25 \times 10^{-2}$

g. Multiply again and add the exponents. Here -4 and $+2$ equals -2.

$P = 250 \text{ mW}$

h. Convert to metric prefixed form. A 1/2-watt resistor would be used in this example.

5. Given a simple circuit, find the current flowing through a 4-ohm resistor when the resistor is dissipating 100 watts of power.

a. Here you know P and R, and need I, so draw the P-I^2-R circle formula.

b. Cover the quantity you want to find with your thumb; in this case, cover I^2.

$I^2 = \dfrac{P}{R}$

c. The resulting formula is $I^2 = P/R$.

$I^2 = \dfrac{100 \text{ W}}{4\ \Omega}$

d. Substitute the values of power and resistance in the formula.

$I^2 = 25$

e. Divide; note that the result is the square of current.

$I = \sqrt{25}$

f. To get the current, you must then find the square root of 25.

$I = 5$ A

g. If you have a calculator with a square root key, enter 25, press the square root key, and the answer, 5, will appear in the display. You may also use the square root tables in the Appendix. To find the square root of 25, look up 25 in the table, and look across to the column labeled square roots ($\sqrt{}$) where you should see "5."

6. If the voltage applied to a 3.3-kilohm resistor in a simple circuit is 15 volts, find the power dissipated by the resistor.

a. Here you know E and R and need to find P, so draw the E^2-R-P circle formula.

b. Cover the quantity you want to find with your thumb; in this case, cover P.

$P = \dfrac{E^2}{R}$

c. The resulting formula is $P = E^2/R$.

$P = \dfrac{(15 \text{ V})^2}{3.3 \text{ k}\Omega}$

d. Substitute the values of voltage and resistance in the formula.

$P = \dfrac{(1.5 \times 10^{+1}) \times (1.5 \times 10^{+1})}{3.3 \times 10^{+3}}$

e. Convert to powers of ten. Since 15 volts equals $1.5 \times 10^{+1}$, the square of 15 volts equals $1.5 \times 10^{+1}$ times $1.5 \times 10^{+1}$.

$P = \dfrac{2.25 \times 10^{+2}}{3.3 \times 10^{+3}}$

f. Square $1.5 \times 10^{+1}$ by multiplying it by itself, remembering to add the exponents.

$P = \dfrac{2.25 \times 10^{+2-3}}{3.3}$

g. Bring the bottom exponent across the division line, up to the top and change its sign.

$P = \dfrac{2.25 \times 10^{-1}}{3.3}$

h. Combine the exponents; here +2 and −3 equals −1.

P = .682 X 10^{-1} | i. | Divide the leading numbers and round off.

P = 68.2 mW | j. | Convert to metric prefixed form.

7. In the simple circuit shown below, the applied voltage forces current to flow through the resistor. If the voltage is increased while the resistance remains constant, the current will increase. Remember, in a circuit with a constant resistance, voltage and current vary directly.

On the chart next to the circuit, the increase in voltage is indicated by an arrow pointing up (↑), the constant resistance is indicated by a dot (·), and the resulting increase in current flow is also indicated by an arrow pointing up (↑). In a *direct* relationship when one quantity increases, the other quantity also increases. In an *inverse* relationship, when one quantity increases, the other quantity decreases. Using this information and considering the simple circuit shown, complete the chart by filling in the blank spaces with the appropriate symbol:

↑ means the quantity increases

↓ means the quantity decreases

· means the quantity remains constant

	E	I	R
Example	↑	↑	·
1	·	↑	
2	↓		·
3	·	↓	
4	·		↓
5		·	↑

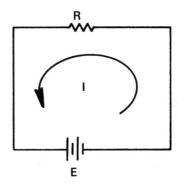

Solution:

Solution	E	I	R
1	·	↑	↓
2	↓	↓	·
3	·	↓	↑
4	·	↑	↓
5	↑	·	↑

1. In a circuit with a constant voltage, current and resistance vary inversely. In order for current to increase, resistance must decrease.

2. In a circuit with a constant resistance, voltage and current vary directly. When voltage decreases, current must decrease.

3 & In a circuit with a constant voltage, current and resistance vary inversely. In order for current to
4. decrease, resistance must increase; and in order for current to increase, resistance must decrease.

5. In a circuit with constant current, voltage and resistance vary directly. If the voltage increases, the resistance must be increased proportionally in order for the current to be held constant.

Note: This type of description of circuit behavior is a good one to use in getting a "feel" for what is happening in circuits; and is sometimes called "variational analysis." It consists of varying some quantity, holding others constant, and considering what happens in the rest of the circuit.

LESSON 4: OHM'S LAW AND POWER

● **Practice Problems**

The key objective of this lesson was to enable you to correctly use Ohm's
law and the power formulas in analyzing basic circuits of the type shown below.
You will know that you have achieved this objective when you can solve problems
of this type without referring back to the lesson for help. In each case, solve for
the quantities indicated with a question mark. Fold over the page to check your
answers.

Fold Over

1.

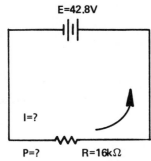

$I = \underline{2.68\ mA}$

$P = \underline{114\ mW}$

2.

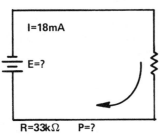

$E = \underline{\hspace{2cm}}$

$P = \underline{\hspace{2cm}}$

3.

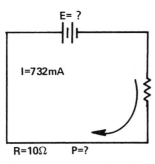

$E = \underline{\hspace{2cm}}$

$P = \underline{\hspace{2cm}}$

4.

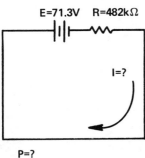

$I = \underline{\hspace{2cm}}$

$P = \underline{\hspace{2cm}}$

Solutions

1. $I = 2.68$ mA

    $P = 115$ mW

2. $E = 594$ V

    $P = 10.7$ W

3. $E = 7.32$ V

    $P = 5.36$ W

4. $I = 148 \,\mu$A

    $P = 10.5$ mW

Fold Over

5.

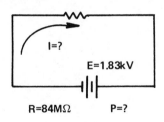

I=?

E=1.83kV

R=84MΩ   P=?

I = _____

P = _____

6.

E=782V

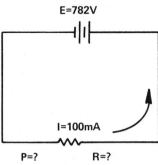

I=100mA

P=?   R=?

P = _____

R = _____

7.

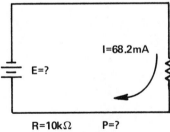

I=68.2mA

E=?

R=10kΩ   P=?

E = _____

P = _____

8.

E=14.14V

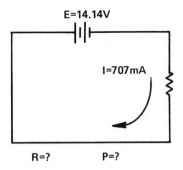

I=707mA

R=?   P=?

R = _____

P = _____

Solutions

5.   I = 21.8 μA

     P = 39.9 mW

6.   P = 78.2 W

     R = 7.82 kΩ

7.   E = 682 V

     P = 46.5 W

8.   R = 20.0 Ω

     P = 10 W

Fold Over

9.

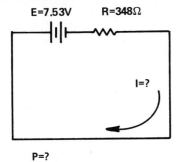

E=7.53V   R=348Ω

I=?

P=?

I = _____

P = _____

10.

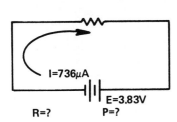

I=736μA

R=?   E=3.83V   P=?

R = _____

P = _____

11.

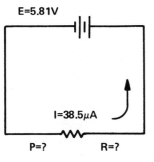

E=5.81V

I=38.5μA

P=?   R=?

P = _____

R = _____

12.

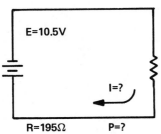

E=10.5V

I=?

R=195Ω   P=?

I = _____

P = _____

Solutions

9.   I = 21.6 mA

   P = 163 mW

10.   R = 5.2 kΩ

   P = 2.82 mW

11.   P = 224 μW

   R = 151 kΩ

12.   I = 53.8 mA

   P = 565 mW

Fold Over

13.

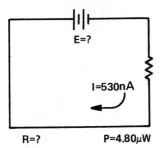

E = _____

R = _____

14.

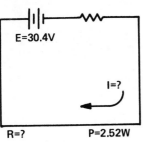

I = _____

R = _____

15.

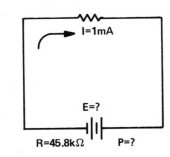

P = _____

E = _____

16.

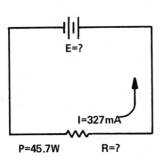

E = _____

R = _____

• **Practice Problems**

Solutions

13. E = 9.06 V

     R = 17.1 MΩ

14. I = 82.9 mA

     R = 367 Ω

15. P = 45.8 mW

     E = 45.8 V

16. E = 140 V

     R = 427 Ω

Fold Over

17.

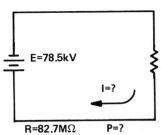

R=82.7MΩ   P=?

I = _____

P = _____

18.

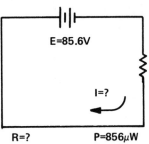

R=?   P=856μW

I = _____

R = _____

19.

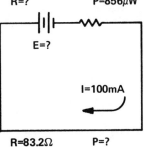

R=83.2Ω   P=?

E = _____

P = _____

20.

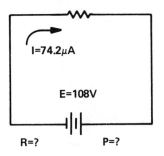

R=?   P=?

R = _____

P = _____

Solutions

17. I = 949 μA

P = 74.5 W

18. I = 10 μA

R = 8.56 MΩ

19. E = 8.32 V

P = 832 mW

20. R = 1.46 MΩ

P = 8.01 mW

Fold Over

21.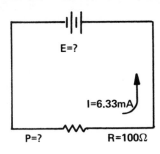

E = _____

P = _____

22.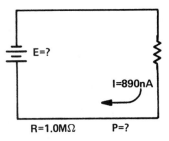

E = _____

P = _____

23.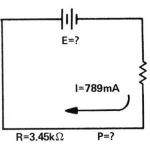

E = _____

P = _____

24.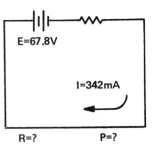

R = _____

P = _____

Solutions

21. E = 633 mV

P = 4.01 mW

22. E = 890 mV

P = 792 nW

23. E = 2.72 kV

P = 2.15 kW

24. R = 198 Ω

P = 23.2 W

Fold Over

25.

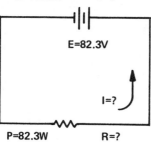

P = _____

I = _____

26.

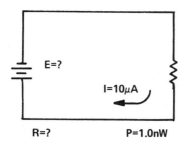

I = _____

R = _____

27.

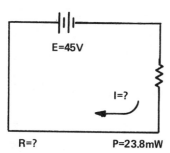

R = _____

E = _____

28.

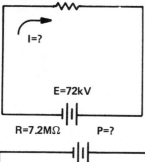

I = _____

R = _____

Solutions

25. P = 720 W

    I = 10.0 mA

26. I = 1.0 A

    R = 82.3 Ω

27. R = 10.0 Ω

    E = 100 μV

28. I = 529 μA

    R = 85.1 kΩ

Fold Over

29.

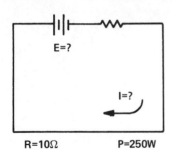

E=?

I=?

R=10Ω   P=250W

E = ___50 V___

I = ___5 A___

30.

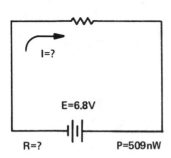

I=?

E=6.8V

R=?   P=509nW

R = _____

I = _____

Solutions

29. E = 50.0 V

    I = 5.0 A

30. R = 90.8 MΩ

    I = 74.9 nA

1. $E = IR$:
   a. Is known as Ohm's Law
   b. Shows that the current in a closed circuit containing a constant resistance is directly proportional to the voltage applied
   c. Shows that current and voltage vary in a linear manner.
   d. All of the above
   e. a only

2. The rate at which work is done in overcoming the resistance to current flow in a closed circuit is known as:
   a. The first law of electrostatics
   b. Coulomb's law
   c. Power dissipation
   d. Resistor rating
   e. None of above

3. Current flowing through a circuit resistance generates power in the form of:
   a. Electron flow
   b. Heat
   c. Coulombs
   d. Amperes
   e. Volts

4. From Ohm's law, what voltage must be applied to a circuit with 50 ohms resistance if 2 amperes of current is to flow?
   a. 100 volts
   b. 25 volts
   c. 10 volts
   d. 0.04 volts
   e. None of above

5. A circuit has a resistance of 2500 ohms and a battery of 10 volts. How much current will flow in the closed circuit?
   a. 4 amperes
   b. .004 amperes
   c. 4×10^{-3} amperes
   d. 4 milliamperes
   e. None of the above
   f. b, c, and d above

6. In a circuit that obeys Ohm's Law, the current varies _____ with the resistance in the circuit if the voltage is constant.
   a. Directly
   b. Inversely
   c. Exponentially
   d. Squarely
   e. None of the above

7. In a circuit that obeys Ohm's Law, the resistance must vary_____with the voltage in the circuit to keep the current constant.
   a. Directly
   b. Inversely
   c. Exponentially
   d. Squarely
   e. None of above

8. If a closed circuit has a 1 volt battery and 0.1 milliampere of current flows, what is the circuit resistance?
   a. 10K ohms
   b. 10 ohms
   c. 10×10^3 ohms
   d. 10×10^{-3} ohms
   e. 10,000 ohms
   f. a, c and e above
   g. All of above

9. What is the unit of measure of power?
   a. Volt
   b. Ampere
   c. Coulomb
   d. Watt
   e. Ohm

10. With the voltage across and the current through a resistor, what equation is used to determine the power dissipated in the resistor?
    a. $P = E \times 1$
    b. $P = E \times R$
    c. $P = I \times R$
    d. $P = I^2 \times E$
    e. $P = E^2 \times I$

11. The current and resistance in a circuit is known. What equation is used to determine the power dissipated in the resistance?
    a. $P = I^2 \times R$
    b. $P = I^2 \times E$
    c. $P = I \times R$
    d. $P = E^2 \times R$
    e. $P = E \times R$

12. The equation for power dissipated in a
    simple closed circuit containing a resistor
    and a battery is:
    a. $E = IR$
    b. $P = EI$
    c. $P = I^2R$
    d. $P = \dfrac{E^2}{R}$
    e. b, c and d above
    f. a above
    g. All of above

13. The heat dissipated in a resistor
    varies_____with the resistance when the
    voltage is constant.
    a. Directly
    b. Exponentially
    c. Inversely
    d. Squarely
    e. Quadratically

14. A resistor has 20 volts across it and 2
    amperes of current through it. What power is
    dissipated in the resistor?
    a. 40 watts
    b. 4 watts
    c. 4 milliwatts
    d. 10 watts
    e. 40 milliwatts

15. A circuit has 2 amperes of current and 50
    ohms of resistance. How much power is
    dissipated in the resistance?
    a. 25 watts
    b. 100 watts
    c. 200 watts
    d. 25 milliwatts
    e. 100 Kilowatts

16. A circuit has 120 volts applied to a motor
    that has 10 ohms of resistance. How much
    power is dissipated in the motor?
    a. 12 watts
    b. 1200 watts
    c. 144 watts
    d. 1.44 Kilowatts
    e. None of above

17. A circuit has 22 Kilohms of resistance across
    a battery and 10 milliamperes of current is
    flowing. What is the applied voltage and the
    power dissipated in the resistance?
    a. 220 volts, 2.2 milliwatts
    b. 220 volts, 2.2 watts
    c. 220 Kilovolts, 2.2 watts
    d. 22 volts, 2.2 watts
    e. 220 volts, 0.44 watts

18 A closed circuit with a 60 volt battery and a
    resistor has 30 milliamperes of current
    flowing. What is the value of the resistor and
    how much power is being dissipated?
    a. 2K, 1800W
    b. 2K, 1.8W
    c. 2MΩ, 1.8mW
    d. 2 ohms, 1.8 KW
    e. None of above

19. A closed circuit with a battery has a resistor
    and 1 milliampere of current flowing. 1
    milliwatt of power is dissipated in the
    resistor. What is the value of the resistor and
    the battery voltage?
    a. 1M, 1KV
    b. 100 ohms, 10V
    c. 10K, 1V
    d. 1K, 100V
    e. 1K, 1V

20. 16 watts is dissipated in a 4 ohm resistor in
    a closed circuit with a battery. What current
    is flowing and what voltage is applied?
    a. 1A, 16V
    b. 0.25A, 1V
    c. 4A, 16 V
    d. 2A, 8V
    e. None of above

Lesson 5

Series Circuits

This lesson defines a *series circuit* and discusses the *three laws* governing the *behavior* of series circuits. In addition, the *polarity* of voltage drops and the topic of *circuit grounds* are discussed.

● **Objectives**

Now that Ohm's law relating voltage, current and resistance has been introduced, its use in predicting the behavior of simple circuits will be covered. At the end of this lesson you should be able to identify and analyze *series* circuits.

1. Given several circuit schematics, *identify* which are series circuits. *Write* a definition of the term "series circuit."

2. *Write* the three laws that describe the behavior of voltage, current, and resistance in a series circuit. *Write* a formula for each.

3. Given a schematic of several resistors connected in series of the type shown, *calculate* their *total* resistance, R_T.

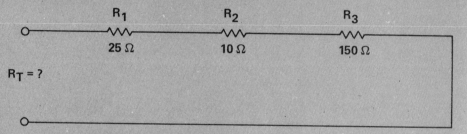

$R_T = ?$

4. Given a schematic diagram for a circuit of the type shown, *calculate* the following:

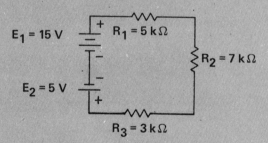

    a.   The total current flowing

    b.   The voltage across each resistor

    c.   The voltage polarity across each resistor

    d.   The total power dissipated by the circuit.

5. *Define* what is meant by the term "IR drop."

6. *Define* a reference point (ground) in a circuit. *Identify* the schematic symbol for a chassis ground and an earth ground, and *state* where each is commonly used.

7. In a circuit, given a schematic of the type shown, *calculate* the voltage and *state* the polarity that would be measured from the reference point to points A, B, and C.

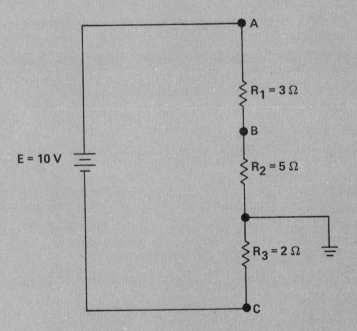

8. Given a circuit of the type shown, with known quantities labeled on the schematic; *calculate* all of the unknown quantities in the circuit including the power dissipated by each resistor.

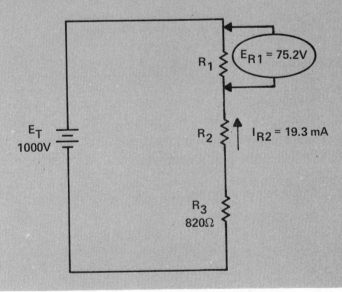

- **Ohm's Law**
- **Inverse Relationship**

At this point in the course a picture of electricity and the basics of the electrical circuit have been introduced. In addition, some of the most important basic circuit components have been discussed. In particular, the *resistor* and its use in a circuit have been described in detail.

This lesson continues to concentrate on one of the key types of circuit you'll encounter, the *series circuit*. A *series circuit* will be defined, and the three basic laws that describe the behavior of these circuits will be introduced. This discussion will introduce several new concepts — the concept of *voltage polarity* and its measurement, and the important concept of reference points for voltage or circuit grounds.

Ohm's Law — Before beginning a detailed discussion of series circuits, it will be important for you to recall the fundamental relationships of voltage, current, and resistance that were set down in the discussion of Ohm's law in Lesson 4. Recall that a circle memory aid, shown in Figure 5.1, was introduced that allowed you to easily recall and write down the three Ohm's law formulas interrelating voltage, current, and resistance.

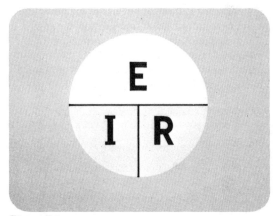

Figure 5.1

Inverse Relationship — Ohm's law and the "common sense of circuits" were used to show that when *resistance* in a circuit *increases*, the current flowing in the circuit *decreases* provided that the voltage applied to the circuit is held constant (Figure 5.2). Now any time an increase in one factor causes a decrease in another factor, it is said that the two factors vary *inversely*. If a constant voltage is providing a push on electrons in a circuit, the amount of current flowing depends on the amount of opposition to current flow (resistance) present in the circuit. If the resistance is doubled, the current flow will be halved, and vice versa. Again, this is called an *inverse* relationship.

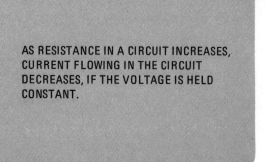

AS RESISTANCE IN A CIRCUIT INCREASES, CURRENT FLOWING IN THE CIRCUIT DECREASES, IF THE VOLTAGE IS HELD CONSTANT.

Figure 5.2

Direct Relationship — A second important relationship that's been discussed is the *direct relationship between voltage and current* in a circuit where the resistance is held constant. When voltage increases across a circuit containing a constant resistance, current through that resistance also increases (Figure 5.3). If the voltage were doubled, the current would double, and vice versa.

The study of these basic relationships moves on to an investigation of the behavior of the circuits. At the conclusion of this lesson, you should understand what a *series circuit* is, and using Ohm's law and fundamental circuit rules, you should be able to solve for the voltage, E, current, I, and resistance, R in a series circuit containing several resistors.

Series Circuit Definition — First of all, how is a series circuit defined? *A series circuit is a circuit containing only one path for current flow.* In Figure 5.4, current leaves the voltage source, and must travel through the one single path to return to the other side of the voltage source.

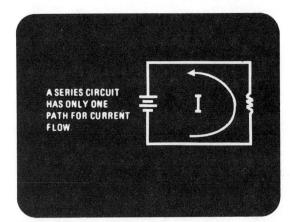

AS VOLTAGE INCREASES ACROSS A CIRCUIT CONTAINING CONSTANT RESISTANCE, CURRENT INCREASES.

Figure 5.3

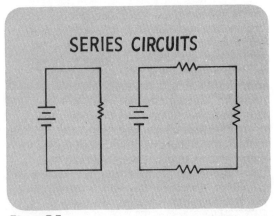

A SERIES CIRCUIT HAS ONLY ONE PATH FOR CURRENT FLOW.

Figure 5.4

Types of Circuits — As shown in Figure 5.5, a series circuit may have only one or it may have several resistors in it, but *there is still only one path for current flow.* Current can only flow in this path.

SERIES CIRCUITS

Figure 5.5

Parallel Circuit — If, however, another resistor is put in the circuit, as shown in Figure 5.6, notice that the current will have *two paths* to follow. Thus, *this is not a series circuit*. This is another type of circuit called a *parallel* circuit, which will be discussed in a following lesson. If the second path were removed, the circuit would once again become a series circuit because current would have only one path to flow through.

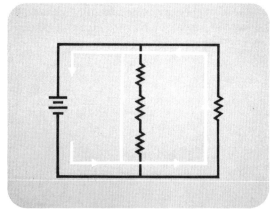

Figure 5.6

Laws of Series Circuits — There are three basic laws that describe the behavior of series circuits, and these are listed for you in Figure 5.7. The three laws will be stated briefly for you, and then each one will be examined in detail to see what each means and how each applies in series circuit situations.

The first law states that *the individual resistances in a series circuit add up to the total circuit resistance*. The second law states that *the current has the same value at any point within a series circuit*. The third law states that *the individual voltages across the resistors in a series circuit add up to the total voltage applied to that circuit*.

SERIES CIRCUIT LAWS

1. INDIVIDUAL RESISTANCES IN A SERIES CIRCUIT ADD UP TO THE TOTAL CIRCUIT RESISTANCE.
2. CURRENT HAS THE SAME VALUE AT ANY POINT WITHIN A SERIES CIRCUIT.
3. INDIVIDUAL VOLTAGES ACROSS RESISTORS IN A SERIES CIRCUIT ADD UP TO THE TOTAL APPLIED VOLTAGE.

Figure 5.7

Resistance in a Series Circuit — The first law states that in a series circuit the sum of the individual resistances in the circuit adds up to the total resistance in the circuit. This law can be written in formula form as shown in Figure 5.8. The dots at the end of the formula indicate that the formula can be expanded to include all resistors in any given series circuit.

Now focus your attention for a second on the shorthand form of writing being used here. The R sub T (R_T), the R sub one (R_1), R sub two (R_2), and so forth are written in this manner so that each of the various resistors can be identified. Writing a number below the main line like this is called *subscripting*. Subscripting is often done for all

$$R_T = R_1 + R_2 + R_3 + R_4 \ldots$$

Figure 5.8

types of components to help identify each one of many components of the same type that may be in one circuit.

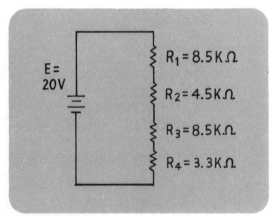

Figure 5.9

For example, consider the schematic diagram (Figure 5.9) of a series circuit containing four resistors. By using subscripted notations, it is very easy to identify, either in printed text or in formulas, any resistor that must be singled out for consideration. If a problem stated "find the current through R_3," you would know to begin examining the third resistor from the top, whose ohmic value is 8.5 kilohms. As your coursework progresses, you'll see this subscripted shorthand extended to other types of circuit components.

Simple Series Circuit — Now several series circuits will be examined to illustrate how the total resistance of series circuits is determined. Figure 5.10 shows a circuit containing only one 6-kilohm resistor. This is a basic circuit of the type that has been analyzed in our course material so far. This circuit can be identified as a series circuit since there is *only one path for current to flow through*. Also, in this circuit, since there's only one single resistor, R, the total resistance is merely the value of R, or 6 kilohms.

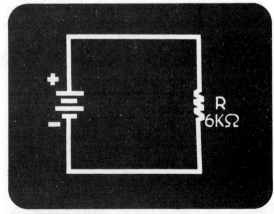

Figure 5.10

Series Circuit with Three Resistors — Figure 5.11 shows another series circuit, which contains three 2-kilohm resistors. When the values for R_1, R_2, and R_3 are substituted in the formula ($R_T = R_1 + R_2 + R_3$), the total resistance in the circuit is 2 kilohms plus 2 kilohms plus 2 kilohms. This addition is simple since all resistors have "kilohm" values. This means that they are all raised to the same power of ten and when they are added, the sum equals $2 \times 10^{+3} + 2 \times 10^{+3} + 2 \times 10^{+3}$, or $6 \times 10^{+3}$, or 6 kilohms (Figure 5.12). If all of a series of numbers have the same metric prefix, the numbers may be added and the total will have the same prefix: 2 k + 2 k + 2 k = 6 k ohms.

One way this calculation could be verified in the laboratory would be to remove the voltage from the circuit, and then connect an ohmmeter across all the resistors. The ohmmeter would indicate the circuit's total resistance and should read approximately 6 kilohms, within the tolerance limits of the resistors.

An important point to notice is that when an ohmmeter is placed across all three resistors in series, the ohmmeter measures only the *total* resistance. It does not matter that there may be three individual 2-kilohm resistors in the circuit; the meter indicates the resistance of the circuit just as if there were one 6-kilohm resistor present. So no matter how many resistors are in a series circuit, the sum of their individual ohmic values is the total resistance of the circuit.

Not only does the ohmmeter regard the three resistors as one 6-kilohm resistor, but the power supply does also. As far as the power supply is concerned, there is only one 6-kilohm resistor connected in this circuit, even though there may actually be many resistors adding up to the 6 kilohms. It is the total resistance of a series circuit that controls the amount of current which will flow in that circuit. The circuit *acts* as if a *single resistance* of 6 kilohms is opposing (or controlling) the current flow.

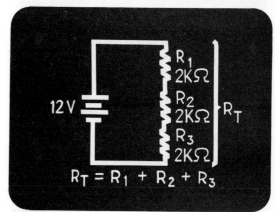

Figure 5.11

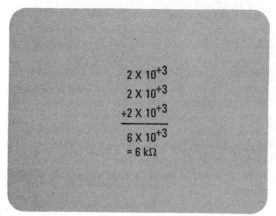

Figure 5.12

Current in a Series Circuit — The second law governing the behavior of series circuits states that *the current in a series circuit has the same value at any point in the circuit.* In series circuits the amount of current flow that reaches any point in a circuit has to equal the amount that leaves that point, because there is no other path for current to flow. To really see this, stop and examine each point in the series circuit shown in Figure 5.13 carefully. There is no point in a circuit such as this where the current could divide and take more than one path. As a result, exactly the same amount of current flows through every part of the circuit, through every resistor, device, wire, etc. So a key point to remember is that the value of the current is the same at all points in a series circuit.

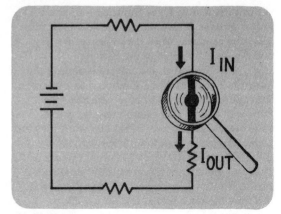

Figure 5.13

Total Resistance and Total Current — Recall from the last lesson in the discussion of Ohm's law, a circuit was described which contained a power supply and a single resistor. As shown in Figure 5.14, the single resistor controls how much current flows in the circuit. As has been said, in a series circuit containing several resistors, it appears to the power supply that there is one resistor whose value is the total of all the resistors in the circuit. Thus, if the circuit's *total resistance is known*, Ohm's law can be used to determine the *total current* that flows through all parts of a series circuit.

Note that in the following calculations the factors total resistance and total voltage will be used. Total voltage means the total voltage applied to the circuit by the battery or power supply.

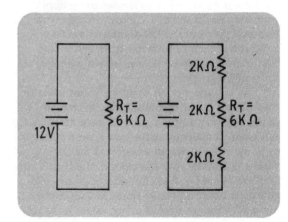

Figure 5.14

Total Current in Ohm's Law — As shown in Figure 5.15, Ohm's law can be rewritten in the form total current equals total voltage divided by the circuit's total resistance. Using Ohm's law in this form, the total current in the circuit can be determined. You can now substitute the values of the applied voltage and the total circuit resistance in the formula as shown in the illustration. When you divide, you should find that there are 2 milliamps of current flowing throughout this circuit. So in the series circuit of Figure 5.15, 2 milliamps of current are flowing through the 6-kilohm resistor, R_T. If R_T were actually made up of three 2-kilohm resistors in series, as shown in the figure, the 2 milliamps of current would be flowing through *each* of the three individual 2-kilohm resistors.

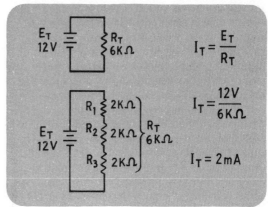

Figure 5.15

Current is the Same at All Points in a Series Circuit — It can be demonstrated that this amount of current is flowing everywhere in the circuit by actually building the circuit and connecting an ammeter to measure the current at various points throughout the circuit. No matter where the ammeter is connected, it will read the same current at each point — 2 milliamps (Figure 5.16).

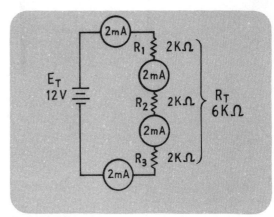

Figure 5.16

Voltage in a Series Circuit — With a picture of the current flow in series circuits in mind, turn your attention to the behavior of *voltages* in series circuits. First, examine a resistor in the circuit of Figure 5.17. Notice that since this is the same series circuit discussed previously, you know that the resistance of each resistor is 2 kilohms and the current flowing through all parts of the circuit is 2 milliamps. Whenever a current flows through a resistance, a voltage is developed across it which equals I times R, by Ohm's law. This voltage that develops across a resistor carrying a current is called an *IR drop or voltage drop*. A voltage drop across one resistor reduces the voltage available for other resistors remaining in the series circuit. Thus, it is said to "drop" available voltage.

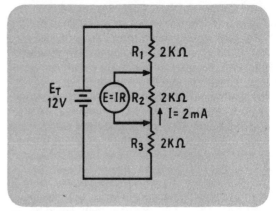

Figure 5.17

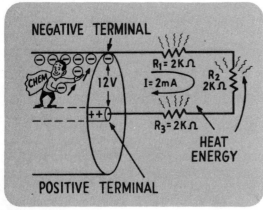

Figure 5.18

Voltage Drop — Recall in previous lessons you learned how a battery or power supply sets up a *difference in potential* between the positive and negative terminals. This is done through a chemical action causing an excess of electrons at the negative terminal and a corresponding lack of electrons, or positive charges, at the positive terminal. Figure 5.18 depicts what might look like the action inside a battery. In this sketch, a chemical action is moving electrons from the positive terminal, through a 12-volt potential difference to the negative terminal. The electrons at the negative terminal have been given energy by the battery, energy that now can be used to do work if a complete circuit is connected across the battery terminals. When such a circuit is connected, the 12-volt potential difference creates a push on all the electrons in the circuit. Electrons flow from the negative terminal of the battery, through the circuit, to the positive terminal under the pressure of the potential difference. For every electron leaving the negative terminal, the battery provides a replacement, keeping the potential difference across the battery terminals constant, even though current is flowing. Each electron has been *given* energy by the battery, and the amount of energy it has is determined by the *potential difference* of the battery. When the electrons move around the circuit, they give up this energy so that when they finally return to the positive terminal, they have lost all the energy the battery gave them.

The energy they have lost shows up as *heat* in the resistors in the circuit. The farther through the circuit the electrons move, the more energy they lose. This loss in energy is what is called a *voltage drop*, or *IR drop*.

Here's another important point. The *voltage drop* across the entire circuit will always be equal to the *voltage provided* by the battery. No matter how many resistors the circuit contained, the total voltage dropped, in this case, would be 12 volts. Therefore, the total voltage dropped in a circuit always equals the source voltage.

Calculation of E_{R1} — The voltage drop across a resistor can be determined by using Ohm's law in the form shown in Figure 5.19. This is one form of Ohm's law you should be familiar with, $E = IR$. It states the relationship between voltage, current, and resistance of individual circuit components, as well as entire circuits. In this equation Ohm's law is written in a form that allows you to calculate the voltage drop across resistor R_1 in the circuit shown in Figures 5.17 and 5.18. Since a series circuit can contain many different resistors of varying sizes, it is necessary to identify which resistor is dropping what voltage in some way. This is done by subscripting the voltage with the name of the resistor it is dropped across. This is shown in Figure 5.20: E_{R1} means the "voltage drop across resistor R_1." Similarly for current: I_{R1} means "the current flowing through resistor R_1." This is basically Ohm's law formula ($E = I \times R$), except that now the voltage drop across one resistor is being calculated using the current through R_1 times the ohmic value of R_1. Using Ohm's law you can calculate the voltage across R_1, called E_{R1}, which equals the current flowing through R_1 (called I_{R1}) times the resistance of R_1.

If you substitute the values of I_{R1} and R_1 from the circuit in Figure 5.17 into the Ohm's law formula, the voltage across R_1 is determined.

Recall that there are 2 milliamps of current flowing in this circuit. This means that 2 milliamps are flowing through *each* circuit component,

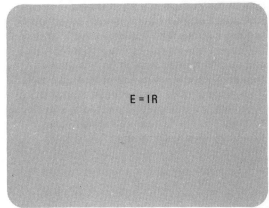

$$E = IR$$

Figure 5.19

$$E_{R1} = (I_{R1})(R_1)$$

$$= (2\text{ mA})(2\text{ k}\Omega)$$

$$= (2 \times 10^{-3})(2 \times 10^{+3})$$

$$= 4 \times 10^{0}$$

$$= 4\text{ VOLTS}$$

Figure 5.20

including R_1. So I_{R1} is 2 milliamps and R_1 is 2 kilohms as shown. This gives you 2×10^{-3} times $2 \times 10^{+3}$, which equals 4×10^0 or 4 volts. Thus, E_{R1}, the voltage across R_1, is 4 volts.

Sum of the Voltages in a Series Circuit — Notice that since the other two resistors in the circuit (R_2 and R_3) have the same 2-kilohm value, and since the current flowing through each of them is the same as that flowing through R_1, then both resistors R_2 and R_3 must also drop 4 volts each. So if you were to build a series circuit such as this in the laboratory and apply 12 volts to it, you would find that you would measure 4 volts (within tolerance) across each resistor (Figure 5.21).

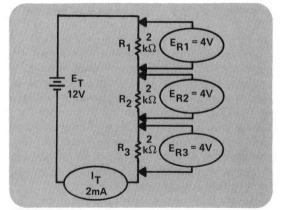

Figure 5.21

Voltage Calculation — As mentioned earlier, the third law that applies to series circuits states that *the sum of the individual voltages across the resistors in a series circuit equals the applied voltage.*

This is expressed in formula form as
$$E_T = E_{R1} + E_{R2} + E_{R3} \cdots$$

In the sample series circuit, it's been calculated that there is a 4-volt drop across each of the three resistors R_1, R_2 and R_3. Notice that in Figure 5.22 these three 4-volt drops *add up* to the total of 12 volts which is the total applied voltage provided by the battery.

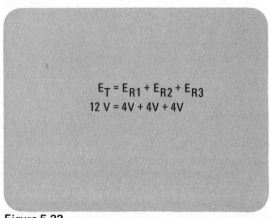

Figure 5.22

At this point, the total current flowing in this circuit and the voltage drop across each resistor have been determined. All of the basic "unknown" quantities about this circuit have been calculated, and it is common to say that this circuit has been "solved." This series circuit appears simple, but quite a bit can be learned from examining this circuit.

Series Circuit with Calculated Values — For example, an important relationship can be seen if you analyze the voltages between different pairs of points in this circuit. As shown in Figure 5.23, the voltage across the two resistors R_2 and R_3 equals the sum of E_{R2} plus E_{R3} or 8 volts. So 4 volts is dropped across R_1, the single 2-kilohm resistor, 8 volts across R_2 and R_3, and the full 12 volts across the three resistors. Notice then that this series circuit has actually *divided* the applied voltage, and at different points throughout the circuit different voltages will be available.

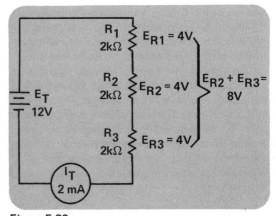

Figure 5.23

Voltage versus Resistance — This points out an important relationship. The greater the resistance between two points in a series circuit, the greater the voltage drop between those points will be, provided the current through that circuit is held constant (Figure 5.24). Thus, in a series circuit with *constant current* flowing, the *voltage* dropped between two points varies directly with the *resistance* between those points.

THE GREATER THE RESISTANCE, THE GREATER THE VOLTAGE DROP, PROVIDED THE CURRENT IS MAINTAINED CONSTANT THROUGH THE RESISTANCE.

Figure 5.24

- **Demonstration Circuit**
- **Demonstration Circuit With Applied Voltage**
- Additional Example

Demonstration Circuit — This can be illustrated more clearly with the use of a circuit containing a wirewound resistor. As shown in Figure 5.25, the wire windings of this resistor are exposed. If an ohmmeter is connected so that more of the resistor's wire is put between the probes, a greater resistance will be measured. With the ohmmeter's common lead connected at one end, as the probe is touched to the resistance wire and slid to positions A, B, and C, the ohmmeter needle will indicate progressively higher and higher resistance readings. When the probe is connected at point D, the resistor's entire value (whatever it may be) will be indicated on the ohmmeter.

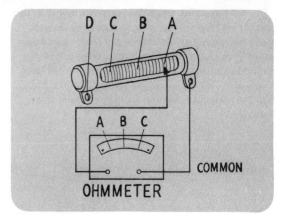

Figure 5.25

Demonstration Circuit With Applied Voltage — If the ohmmeter is disconnected and a voltage source of 12 volts is connected to the resistor, a voltmeter may be used to measure the voltage between various pairs of points on the resistor. Again, if the voltmeter is connected as shown in Figure 5.26, the voltage reading will increase as you slide the probe from point A through points B and C. At point D, the meter will indicate the total voltage drop of 12 volts across the entire resistor. Notice, the more resistance wire that is between the voltmeter's probes, the higher the measured voltage drop will be.

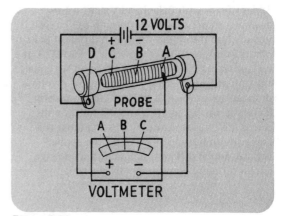

Figure 5.26

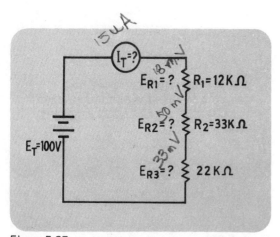

Figure 5.27

Additional Example — Before moving to another type of series circuit problem, consider the additional example circuit shown in Figure 5.27. This series circuit problem is of the same type as the one just discussed. At the outset you are given the *total applied voltage* E_T = 100 V, and *the ohmic value of each resistor*; R_1 = 12 kΩ, R_2 = 33 kΩ, and R_3 = 22 kΩ. (Notice that these are not the only quantities you may be given at the start of a series circuit problem, different problem types will be covered later.) In this circuit you need to calculate the current flowing, the total resistance, and the voltage drop across each resistor.

The best way to begin in a problem of this sort is to move toward a calculation of the total current flowing in the circuit. Since the current is the same at all points in a series circuit, you will then know the current through each resistor. Knowing the current through each resistor along with the resistance value of each, you can then calculate the voltage dropped across each resistor.

$$R_T = R_1 + R_2 + R_3$$
$$R_T = 12\,k\Omega + 33\,k\Omega + 22\,k\Omega$$
$$R_T = 67\,k\Omega$$

Figure 5.28

Calculation of R_T — How do you calculate the total current? Your first move in this case should be to calculate the circuit's total *resistance*. You will then be able to use that, along with the applied voltage and Ohm's law, to calculate the total current. The total resistance of this circuit is found by just adding the resistance of all the series connected resistors as in Figure 5.28:
$R_T = R_1 + R_2 + R_3$. Substituting the circuit values from the schematic, R_T equals 12 kilohms plus 33 kilohms plus 22 kilohms for a total of 67 kilohms.

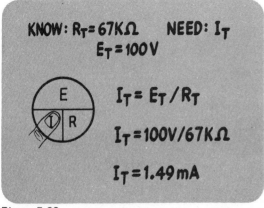

$$\text{KNOW: } R_T = 67\,K\Omega \qquad \text{NEED: } I_T$$
$$E_T = 100\,V$$

$$I_T = E_T / R_T$$
$$I_T = 100V / 67\,K\Omega$$
$$I_T = 1.49\,mA$$

Figure 5.29

Calculation of I_T — You can proceed to calculate the total current in the circuit as shown in Figure 5.29. You know R_T and E_T, and need to calculate I_T using Ohm's law. Using the circle diagram, cover what *you need to find*, I; and the formula you need is given: $I_T = E_T/R_T$. Substituting your circuit values $I_T = 100$ volts divided by 67 kilohms or 1.49 milliamps.

- Calculation of E_{R1}, E_{R2} and E_{R3}
- Series Circuit: Second Type of Sample Problem
- Calculation of E_{R1} and E_{R2}

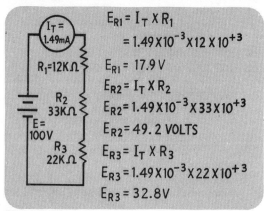

$$E_{R1} = I_T \times R_1$$
$$= 1.49 \times 10^{-3} \times 12 \times 10^{+3}$$
$$E_{R1} = 17.9 \, V$$
$$E_{R2} = I_T \times R_2$$
$$E_{R2} = 1.49 \times 10^{-3} \times 33 \times 10^{+3}$$
$$E_{R2} = 49.2 \, VOLTS$$
$$E_{R3} = I_T \times R_3$$
$$E_{R3} = 1.49 \times 10^{-3} \times 22 \times 10^{+3}$$
$$E_{R3} = 32.8V$$

Figure 5.30

Calculation of E_{R1}, E_{R2} and E_{R3} — Knowing the total circuit current, you can now calculate the voltage drop across each resistor, using Ohm's law in the form $E = I \times R$ as shown in Figure 5.30. In each case the current, I, is the total circuit current, 1.49 milliamps (1.49×10^{-3} amps). Substituting for R_1, R_2, and R_3; you can calculate that E_{R1} = 17.9 V, E_{R2} = 49.2 V; and E_{R3} = 32.8 V. These three voltages should add up to the total applied voltage according to the voltage law for series circuits. The circuit solution is complete when 17.9 plus 49.2 plus 32.8 equals 99.9 or 100 volts.

For an additional worked through example of this type, see Example 5 at the end of this lesson.

Series Circuit: Second Type of Sample Problem — Now that the key features of series circuits have been discussed, turn your attention to some other types of actual series circuit situations. A simple extension of the principles you've learned thus far should enable you to predict and control their behavior. Figure 5.31 shows a series circuit that you can analyze with the use of the three laws for series circuit behavior just described. R_1 is 2.5 kilohms and R_2 is 1.5 kilohms. Suppose that in this circuit you know in advance that there is *1 milliamp of current* flowing in the circuit, and that the applied voltage is 9 volts. The problem is to calculate the voltage drop across R_3, E_{R3}, and its resistance value, as well as the other unknown quantities in the circuit, which are E_{R1} and E_{R2}.

Calculation of E_{R1} and E_{R2} — You can determine the voltage across R_1. Think first about the quantities you *know*, then about those you need to calculate. You know the current flowing through R_1 and its resistance value and you need to *calculate* the voltage across it. Using the circle memory aid, *cover what you want to calculate* with your thumb, and the remaining letters are the formula you need. In this case (Figure 5.32), Ohm's law can be used in the form:
$E_{R1} = I_{R1} \times R_1$.

Substituting the value of the current through R_1 (1 milliamp) and its resistance (2.5 kilohms) and carrying out the multiplication, you will find the voltage across R_1 to be 2.5 volts.

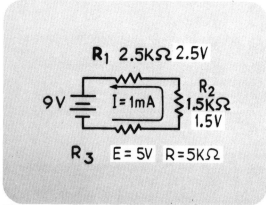

Figure 5.31

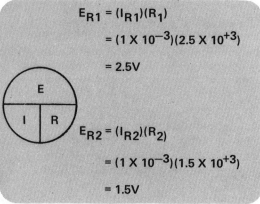

$$E_{R1} = (I_{R1})(R_1)$$
$$= (1 \times 10^{-3})(2.5 \times 10^{+3})$$
$$= 2.5V$$

$$E_{R2} = (I_{R2})(R_2)$$
$$= (1 \times 10^{-3})(1.5 \times 10^{+3})$$
$$= 1.5V$$

Figure 5.32

Similarly, if you focus your attention on R_2, you know the current through it (or any of the resistors in this series circuit) must also be 1 milliamp. Following the same procedures as above, you can determine that E_{R2} is 1.5 volts.

Calculation of E_{R3} — How do you calculate E_{R3}? As shown in Figure 5.33, the formula for *total voltage* in a series circuit is $E_T = E_{R1} + E_{R2} + E_{R3}$. You now know that E_{R1} is 2.5 volts and E_{R2} is 1.5 volts, and it was given that E_T is 9 volts. The voltage formula can be manipulated to solve for E_{R3}. As shown, $E_{R3} = E_T - (E_{R1} + E_{R2})$. What this formula means is that since the total applied voltage is known, and the sum of two of the three voltage drops in the circuit is known, the sum of the two individual voltage drops can be subtracted from the total applied voltage, and the voltage remaining will be the voltage across R_3.

$$E_T = E_{R1} + E_{R2} + E_{R3}$$

$$E_{R3} = E_T - (E_{R1} + E_{R2})$$

Figure 5.33

Substituting the values of E_T, E_{R1} and E_{R2} in the formula as shown in Figure 5.34, you should get 9 volts, minus the sum of 2.5 and 1.5, or 4 volts. Thus, you can see that the final result for E_{R3} is actually 5 volts.

$$E_{R3} = E_T - (E_{R1} + E_{R2})$$

$$= 9 - (2.5 + 1.5)$$

$$= 9 - 4$$

$$= 5V$$

Figure 5.34

- **Calculation of R_3**
- **Twice the Resistance, Twice the Voltage**

Calculation of R_3 — In the circuit shown in Figure 5.31, R_1 dropped only 2.5 volts and R_2 dropped only 1.5 volts. This left 5 volts to be dropped by R_3 which is greater than the voltage drop across other two resistors. This indicates that the value of R_3 must be larger than either of the other resistors in the circuit, because, as mentioned earlier, the more resistance there is between two points in a series circuit, the greater the voltage drop between those two points will be.

Now that the voltage across R_3 and the current through R_3 are known, Ohm's law can be used in the form $R_3 = E_{R3}/I_{R3}$ to solve for R_3 as shown in Figure 5.35. The result is 5 kilohms, which is larger than either resistor R_1 or R_2. Note also that the voltage drop across R_3 is exactly twice the voltage drop across R_1, which follows from the fact that R_3 is twice the value of R_1.

$$R_3 = \frac{E_{R3}}{I_{R3}}$$

$$= \frac{5}{1} \times 10^{-3}$$

$$= 5 \times 10^{+3}$$

$$= 5\ \mathrm{k\Omega}$$

Figure 5.35

Twice the Resistance, Twice the Voltage — This points up another interesting and useful relationship to keep in mind. That is, in a series circuit, if one resistor has twice the resistance of another one, the larger resistor will have twice the voltage drop of the smaller one (Figure 5.36).

IN A SERIES CIRCUIT IF A RESISTOR HAS TWICE THE RESISTANCE OF ANOTHER RESISTOR, IT WILL HAVE TWICE THE VOLTAGE DROP.

Figure 5.36

Series Circuit with Chart — As you progress from analyzing simple circuits, such as the one in Figure 5.37, to more complex circuits, the number of known and unknown quantities you have to keep track of may present quite a large and complex picture to you. However, a simple chart can be used to help you organize the solution of complex circuit problems.

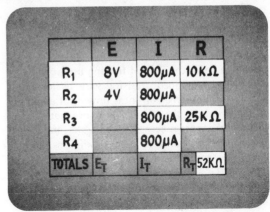

	E	I	R
R_1	8V	800μA	10KΩ
R_2	4V	800μA	
R_3		800μA	25KΩ
R_4		800μA	
TOTALS	E_T	I_T	R_T 52KΩ

Figure 5.37

Since in most circuit analysis problems voltage, current and resistance have to be calculated, the chart shown in Figure 5.37 includes a column for voltage, a column for current, and a column for resistance. The series circuit in Figure 5.38 will be used as an example in completing the chart. The chart will be of help in keeping track of all your calculations. Since there are four resistors in the circuit, there are four rows labeled R_1, R_2, R_3, and R_4. At the bottom is a row for totals. As you go through the solution of this circuit step by step, on the chart, you will see how the chart will help in keeping track of answers and in checking your work.

Notice that in most of the circuit analysis problems you will encounter, the known values (the ones you are given at the start of the problem) will be labeled on the schematic diagram. Therefore the first step in solving this problem will be to fill in the known values on the chart. From the schematic diagram you know at the beginning of the problem that R_1 is 10 kilohms and has 8 volts across it. R_2 has 4 volts across it, R_3 is 25 kilohms, and the total circuit resistance is also given, 52 kilohms.

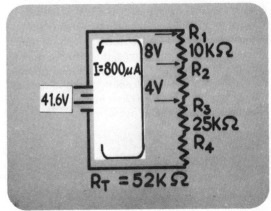

Figure 5.38

Calculation of I_{R1} — A good line of action to follow when solving series circuit problems is to solve for the current flowing through one resistor as soon as you can. Remember, in a series circuit, once you know the current flowing through any one circuit component (resistor, wire, device, etc.), you know that same current is flowing at all points of the circuit. In the circuit you know that the value of the voltage across R_1 is 8 volts, and the value of the resistance of R_1 is 10 kilohms. So, Ohm's law can be used to calculate the current through R_1. This calculation is shown in Figure 5.39, and the result for I_{R1} is 800 microamps.

$$I_{R1} = \frac{E_{R1}}{R_1}$$

$$= \frac{8}{1} \times 10^{+4}$$

$$= 8 \times 10^{-4}$$

$$I_{R1} = 800 \ \mu A$$

Figure 5.39

Circuit Current — Remember this is the current flowing through R_1, but the current is the *same anywhere within a series circuit*. So, if the current through R_1 is 800 microamps, it is also the current through all of the other resistors, and it is also the total current. Thus, at this point, all the squares in the current column of the chart in Figure 5.40 have been filled with 800 microamps.

	E	I	R
R_1	8V	800μA	10KΩ
R_2	4V	800μA	
R_3		800μA	25KΩ
R_4		800μA	
TOTALS	E_T	I_T 800μA	R_T 52KΩ

Figure 5.40

Calculation of R_2 — Moving on with the solution, focus your attention on R_2. At this point you know that R_2 has 4 volts across it and that the current through it is 800 microamps. Ohm's law can be used in the form $R_2 = E_2/I_2$ to calculate the resistance of R_2 (Figure 5.41). Thus, R_2 equals 5 kilohms, and this solution may be added to your chart.

$$R_2 = \frac{E_{R2}}{I_{R2}}$$

$$= \frac{4}{8} \times 10^{-4}$$

$$= 5 \ k\Omega$$

Figure 5.41

- Calculation of E_{R3}
- Calculation of R_4
- Formula for Calculating R_4

Calculation of E_{R3} – Moving down the chart, you will see that another unknown quantity you need to calculate is the voltage across R_3, E_{R3}. The ohmic value of R_3 is 25 kilohms. As shown in Figure 5.42, Ohm's law can be used to find the voltage across R_3 which turns out to be 20 volts.

$$E_{R3} = (I_{R3})\,(R_3)$$
$$= (8 \times 10^{-4})\,(25 \times 10^{+3})$$
$$= 20V$$

Figure 5.42

Calculation of R_4 – Another unknown quantity is the resistance value of R_4. Up to this point, all that is known about R_4 is that 800 microamps of current flows through it. However, you are given that the *total* resistance of the circuit is 52 kilohms. To calculate R_4 you can use the formula in Figure 5.43 which simply states that the sum of the individual resistances within a series circuit equals the total resistance. This formula can be manipulated to solve for R_4, since you know the value of R_T, and all the other resistor values except R_4.

$$R_T = R_1 + R_2 + R_3 + R_4$$

Figure 5.43

Formula for Calculating R_4 – The formula for total resistance can be rearranged as shown in Figure 5.44 to get $R_4 = R_T - (R_1 + R_2 + R_3)$. Then substituting the appropriate values into the formula, you will get 52 kilohms minus the sum of R_1 through R_3, which is 40 kilohms. Thus, R_4 equals 12 kilohms, and another unknown value may be filled in on the chart.

$$R_4 = R_T - (R_1 + R_2 + R_3)$$
$$= 52\,k\Omega - (10\,k\Omega + 5\,k\Omega + 25\,k\Omega)$$
$$= 52\,k\Omega - 40\,k\Omega$$
$$= 12\,k\Omega$$

Figure 5.44

- Calculation E_{R4}
- Calculation of E_T

Calculation E_{R4} – Now that the resistance value of R_4 is known, Ohm's law can be used to calculate the voltage across R_4. The calculation is shown in Figure 5.45, E_{R4} equals 9.6 volts.

$$E_{R4} = (I_{R4})(R_4)$$
$$= (8 \times 10^{-4})(12 \times 10^{+3})$$
$$= 9.6V$$

Figure 5.45

Calculation of E_T – At this point there is only one unknown left to be calculated, E_T. Notice now that there are two ways you may solve for it. As shown in Figure 5.46, E_T could be found by using the formula: $E_T = E_{R1} + E_{R2} + E_{R3} + E_{R4}$. That means to simply add up all the voltages in the voltage column of the chart.

$$E_T = E_{R1} + E_{R2} + E_{R3} + E_{R4}$$

Figure 5.46

Another way is to use Ohm's law in the form $E_T = I_T \times R_T$ as shown in Figure 5.47. These two solutions can be used as a check against each other. No matter which formula you choose for this solution, the total voltage of the circuit is 41.6 volts.

$$E_T = (I_T)(R_T)$$
$$= (8 \times 10^{-4})(5.2 \times 10^{+4})$$
$$= 41.6V$$

Figure 5.47

Completed Solution — You'll find that this simple chart (Figure 5.48) will be of great help in keeping track of the quantities involved in circuit analysis problems of this sort. You have also seen how all of the voltage, current, and resistance values are calculated in a couple of typical series circuit problems. Several additional examples are worked out for you at the end of this lesson to help reinforce the "how to" of applying Ohm's law and the basic rules for series circuits.

Be sure to work several of the problems provided for you. Notice that the *objective* has been achieved only if you can *solve* series circuit problems of this type. At this point, you should be able to approach the problem with some circuit sense, set up a correct solution, decide on the steps to follow, and carry out the solution. *Practice* is really the key to developing proficiency in handling circuits of this sort. An additional worked through example of the type discussed above is Example 8 included at the end of this lesson.

	E	I	R
R_1	8V	800µA	10KΩ
R_2	4V	800µA	5KΩ
R_3	20V	800µA	25KΩ
R_4	9.6V	800µA	12KΩ
TOTALS	E_T 41.6V	I_T 800µA	R_T 52KΩ

Figure 5.48

Voltage Source in Series — Notice that in the above discussion, it was shown how resistors can be added when connected in series to find the circuit's total resistance. Now quite similarly, *voltage sources* may also be added when connected in series to find the total voltage applied to a series circuit. If two voltage sources, such as two batteries, are connected so that *their applied voltages are in the same direction*, as shown in Circuit A of Figure 5.49, the *sum* of their voltages is applied to the circuit. That is, if the terminals of the voltage sources are connected *negative to positive*, their voltages *can simply be added together*. In a common flashlight, two 1.5-volt batteries are connected this way for a total of three

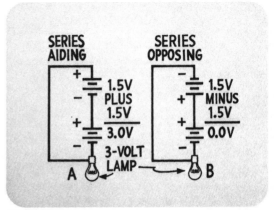

Figure 5.49

volts applied to the bulb. This type of connection for voltage sources is called *series aiding*.

Voltage sources may also be wired in what is called *series opposing* fashion as shown in Circuit B of Figure 5.49. That is, where their voltages are opposite to, or oppose, each other. When voltage sources are wired series opposing, the *voltages are subtracted* from each other. For example, the result of reversing *one* flashlight battery is 1.5 volts minus 1.5 volts which equals zero volts applied to the bulb, and as a result the bulb does not light. So voltages in series add if they are connected in the same direction or series aiding; and subtract if they are connected in opposite directions, or series opposing.

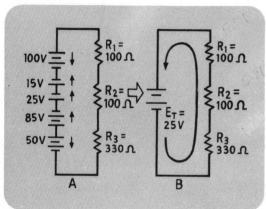

Figure 5.50

Additional Example — Consider the series circuit of Figure 5.50 which contains five voltage sources and five resistors, all of whose values are given. How do you solve for the total current flowing in this circuit, as well as the voltage across each resistor? First, examine the voltage sources and notice the *direction* in which each is wired. The 100-volt voltage source is wired to push electrons down the left-hand side of the circuit — forcing electron current downward, then counterclockwise through the current as indicated by the little arrow in Figure 5.50. The voltage source below that one (15 volts) is wired to push electrons in an opposite direction to the first (series opposing), as are the 25-volt and 85-volt sources. The 50-volt source at the bottom pushes electrons in the same direction as the 100-volt source. These voltage sources are all having a "pushing contest" with the electrons; and the strongest team wins. Two sources are pushing electrons counterclockwise, 100 volts plus 50 volts for a total of 150 volts in that direction; while three sources push clockwise 15 volts, plus 25 volts, plus 85 volts for a total of 125 volts in that direction. Subtracting these two opposing voltages yields a net voltage of 25 volts pushing electron current counterclockwise as shown in Figure 5.50(B). This circuit is equivalent to one having one single 25-volt source supplying three series resistors. The rest of the solution is similar to those you have seen before.

- Calculation of R_T and I_T
- Complete Solution
- **Power in a Series Circuit**

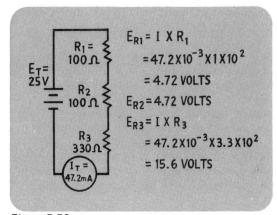

$$R_T = R_1 + R_2 + R_3$$

$$R_T = 100\,\Omega + 100\,\Omega + 330\,\Omega$$

$$R_T = 530\,\Omega$$

$$I_T = \frac{E_T}{R_T} = \frac{25V}{530\,\Omega}$$

$$I_T = 47.2\text{ mA}$$

Figure 5.51

Calculation of R_T and I_T — First calculate R_T for the circuit, as shown in Figure 5.51. R_T is just the sum of the circuit resistances, 100 ohms plus 100 ohms plus 330 ohms equals 530 ohms. Using Ohm's law in the form $I_T = E_T/R_T$, calculate the total circuit current: 25 volts divided by 530 ohms, equals 47.2 milliamps for I_T.

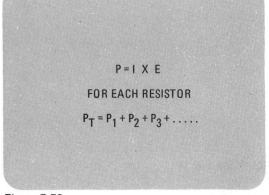

$$E_{R1} = I \times R_1$$
$$= 47.2 \times 10^{-3} \times 1 \times 10^2$$
$$= 4.72 \text{ VOLTS}$$
$$E_{R2} = 4.72 \text{ VOLTS}$$
$$E_{R3} = I \times R_3$$
$$= 47.2 \times 10^{-3} \times 3.3 \times 10^2$$
$$= 15.6 \text{ VOLTS}$$

Figure 5.52

Complete Solution — The voltage across each resistor can be calculated using Ohm's law in the form $E = I \times R$. In each case the current is the total circuit current of 47.2 milliamps (47.2×10^{-3} A) as shown in Figure 5.52. Since R_1 and R_2 each equals 100 ohms, $E_{R1} = E_{R2} = 47.2 \times 10^{-3} \times 100$ ohms or 4.72 volts. For R_3, $E_{R3} = 47.2 \times 10^{-3} \times 3.3 \times 10^{+2}$, or 15.6 volts and the circuit solution is complete. One additional problem of this type is worked for you in Example number 9 included at the end of this lesson.

Power in a Series Circuit — Now turn your attention to the power dissipated in a series circuit. Recall that the basics of power were discussed in a previous lesson. If a resistor has a voltage, E, across it and is carrying a current, I, the power it dissipates is equal to I times E (Figure 5.53). In a series circuit the power that *each resistor* dissipates may be calculated in this same way. The total power dissipated in the circuit is just equal to the sum of the power dissipated by each resistor, or $P_T = P_1 + P_2 + P_3$ and so on for as many resistors as a circuit may contain.

$$P = I \times E$$

FOR EACH RESISTOR

$$P_T = P_1 + P_2 + P_3 + \ldots$$

Figure 5.53

• Power Example
• Calculation of P_1 and P_2
• Calculation of P_3 and P_T

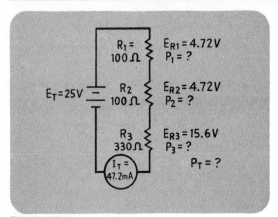

$R_1 =$ 100 Ω $E_{R1} = 4.72V$ $P_1 = ?$

$E_T = 25V$

R_2 100 Ω $E_{R2} = 4.72V$ $P_2 = ?$

R_3 330 Ω $E_{R3} = 15.6V$ $P_3 = ?$

$I_T = 47.2mA$ $P_T = ?$

Figure 5.54

Power Example — As an example, consider how you would calculate the total power dissipated in the circuit shown in Figure 5.54. (This is the circuit that was previously analyzed.) You are given the current flowing in the circuit and the voltage across each resistor. Your first step is to calculate the power dissipated by each of the three resistors P_1, P_2, and P_3. Then you add these to find the total circuit power. To calculate the power dissipated, you can use either of the two power formulas you have learned, either $P = I \times E$, or $P = I^2R$. Since resistors R_1 and R_2 have the same resistance, voltage drop, and current, their power dissipation is the same. Just for practice, both formulas can be used in determining the resistor's power.

$$P_1 = I_{R1} \times E_{R1}$$
$$P_1 = 47.2 \times 10^{-3} \times 4.72V$$
$$P_1 = 2.23 \times 10^{-1} = .223 \, W$$

$$P_2 = I_{R2}^2 \times R_2 = I_{R2} \times I_{R2} \times R_2$$
$$P_2 = (47.2 \times 10^{-3}) \times (47.2 \times 10^{-3}) \times 100$$
$$P_2 = 2.23 \times 10^{-1} = .223 \, W$$

Figure 5.55

Calculation of P_1 and P_2 — To calculate P_1, use the formula $P_1 = I_{R1} \times E_{R1}$ as shown in Figure 5.55. Substituting 47.2 milliamps for I_{R1} and 4.72 volts for E_{R1}, and carrying out the required multiplication, you get 0.223 watt for the power dissipated in R_1. The power dissipated by R_2 should be the same no matter what formula is used to calculate it. If you use $P_2 = I_{R2}^2 \times R_2$ as in Figure 5.55, substituting the total current and value of R_2 in the formula, you get $(47.2 \times 10^{-3}) \times (47.2 \times 10^{-3}) \times (100)$. (Remember that squaring a number just means to multiply it by itself.) The result for P_2 is also 0.223 watt.

$$P_3 = I_{R3} \times E_{R3}$$
$$P_3 = 47.2 \times 10^{-3} \times 15.6V$$
$$P_3 = .736 \, W$$

$$P_T = P_1 + P_2 + P_3$$
$$P_T = .223 + .223 + .736$$
$$P_T = 1.18 \, W$$

Figure 5.56

Calculation of P_3 and P_T — To finish this problem, calculate P_3, using either of two formulas, in Figure 5.56, $P_3 = I_{R3}^2 \times R_3$ is used, 47.2×10^{-3} amps $\times$ 15.6 volts equals 0.736 watt. The power dissipated by each resistor is now known. To calculate the total power dissipated by this circuit, add the power dissipation of each part:
$$P_T = P_1 + P_2 + P_3$$
$$P_T = 0.223 \, W + 0.223 \, W + 0.736 \, W$$
$$P_T = 1.18 \, W \, (total).$$
An additional example involving power considerations in series circuits is Example 9, included at the end of this lesson.

Other Features of Series Circuits — Consider some other useful aspects of series circuits, focusing your attention on some features you may discover and examine while in the laboratory. Two new concepts present themselves for discussion when you consider the operations involved in measuring the voltage drop across various pairs of points in a series circuit. One of these is the idea of a *reference point* for voltage measurement, and the second concerns the polarity of voltages in a series circuit.

Reference Point — Consider the idea of a *voltage reference* point. As you recall, a voltage measurement is always a reading across *two* points. To read the voltage drop correctly across a resistor; you put the probes on either side of it. It is common in circuit situations to select one point in the circuit and compare all other circuit points to it. This arbitrarily chosen point is commonly called a *reference point*. Your voltmeter usually has one of its leads labeled "reference," or "common," or often simply negative, or minus, —. This lead is considered as the voltage reference point by the meter, and the meter will indicate the potential difference between the reference point and the point in the circuit where the other probe is applied. Any point in a series circuit may be chosen as a reference point, and as mentioned, different voltage readings will be measured throughout the circuit, depending on the reference point chosen. In addition, the *directions* your meter's needle will swing will depend on where your reference is chosen. This is because the voltage drops across the resistors in the circuit and has a direction or *polarity* which depends on the direction of electron current flow through them.

As shown in Figure 5.57, when you measure a positive or negative voltage with respect to a reference point, you are determining the polarity of the voltage with respect to that point.

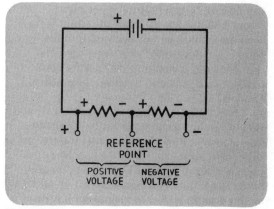

Figure 5.57

In the circuit shown, if your reference point is selected as the point between these two resistors and you place the reference probe of the meter there, a positive voltage will be measured across the resistor on the left while a negative voltage will be measured across the resistor on the right.

Polarity of Voltage — The polarity of voltage drops *across* the resistors in a series circuit can be determined with a simple rule as shown in Figure 5.58. *The side of any resistor closest to the positive terminal of the battery or power supply may be labeled with a positive polarity and that side closest to the negative terminal may be labeled with a negative polarity.*

POLARITY OF VOLTAGE DROPS

THE SIDE OF ANY RESISTOR CLOSEST TO THE POSITIVE TERMINAL OF THE BATTERY OR POWER SUPPLY MAY BE LABELED WITH A POSITIVE POLARITY; THAT SIDE CLOSEST TO THE NEGATIVE TERMINAL MAY BE LABELED WITH A NEGATIVE POLARITY

Figure 5.58

Example of Voltage Polarity — For example, in Figure 5.59, the left-hand side of resistor R_1 is connected closest to the positive side of the battery. So, it is labeled with a plus sign. The right-hand end is closest to the negative battery terminal, so it is labeled negative. R_2 is labeled similarly.

Note that the voltages labeled *across* R_1 and R_2 are now in the correct direction for electron flow clockwise through the resistors and the rest of the circuit. Notice an interesting item here: the point *between* the resistors now has a *plus and minus label*. That is because each pair of polarities on this diagram applies to the resistor it is near.

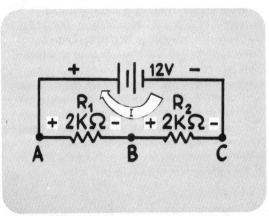

Figure 5.59

The point between the resistors, point B, is negative with respect point A at the left side of R_1 and positive with respect to point C at the right side of R_2.

Voltage Polarity Measurement — This means with the voltmeter's reference probe connected to the left side of R_1 (Figure 5.60), a *negative* voltage would be measured if you connected the other probe to point B. With reference probe at point C, a *positive* voltage would be indicated if the other probe was touched to point B. So, notice the *voltage polarity of a point depends on which point you chose as your reference point.*

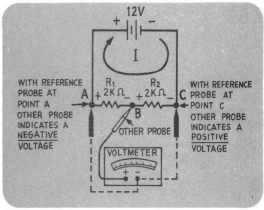

Figure 5.60

Voltage Across a Wire — If you actually tried to measure voltage across the wire *between* the resistors, it would be zero volts (Figure 5.61). This is because there is just a wire with little or no resistance between the probes. With no resistance present, there is no voltage drop since voltage drop equals I times R, and in this case R is zero.

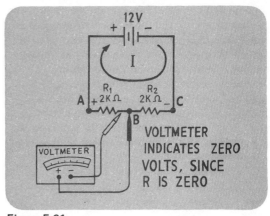

Figure 5.61

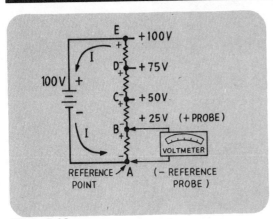

Figure 5.62

Circuit Reference Illustration — As a further illustration of the concept of a circuit reference point, consider the simple series circuit shown in Figure 5.62. This circuit's reference point is selected to be at point A. Each of the resistors in this circuit has the same value, so the applied voltage is distributed equally across each resistor. The potential at point B is 25 volts more positive than point A. A voltmeter connected as shown would read +25 volts from the reference at A to point B. If the positive probe were now moved to point C, a +50 volts would be measured, and so on, +75 volts at D, and +100 volts at E.

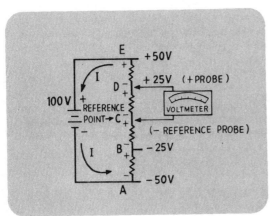

Figure 5.63

Circuit with Moved Reference Point — Consider this same circuit but with the reference point moved to point C instead of point A (Figure 5.63). The voltmeter would show that the potential at point D is 25 volts more positive than the new reference point at point C. Moving the probe would show that point E is at +50 volts with respect to point C. Point B would indicate a —25 volts with respect to point C. Point A would show a —50 volts. Again, both the size and the polarity of the voltages you would measure in a series circuit depend on the *reference point* selected.

Component Chassis — Consider another interesting point about series circuits concerning actual circuit construction. Often when large electrical circuits are constructed, a metal frame known as a chassis is used to secure and house the electrical components (Figure 5.64). With this type of construction, the chassis itself, because it is a conductor, is often used as part of the circuit. It is used for this purpose most often for economy and simplicity because such construction eliminates a lot of wires.

Figure 5.64

- Chassis Wiring
- Voltage Test Point Chart
- **Symbol for Chassis Connection**

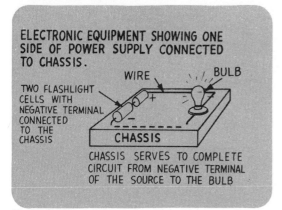

ELECTRONIC EQUIPMENT SHOWING ONE SIDE OF POWER SUPPLY CONNECTED TO CHASSIS.

TWO FLASHLIGHT CELLS WITH NEGATIVE TERMINAL CONNECTED TO THE CHASSIS

WIRE

BULB

CHASSIS

CHASSIS SERVES TO COMPLETE CIRCUIT FROM NEGATIVE TERMINAL OF THE SOURCE TO THE BULB

Figure 5.65

VOLTAGE TEST POINTS
(ALL VOLTAGES MEASURED WITH RESPECT TO CHASSIS REFERENCE)

A	+25V	G	−10V	M	−10V
B	+35V	H	−105V	N	+15V
C	+150V	I	+110V	O	+20V
D	−10V	J	+110V	P	+25V
E	−100V	K	+5V	Q	+35V
F	−35V	L	+5V	R	−10V

Figure 5.66

Symbol for Chassis Connection — Now a very important new term is introduced. A common connection to a chassis of the type discussed above is called a *chassis ground*. A point connected to a chassis (or chassis ground point) is symbolized in schematic diagrams with the symbol shown in Figure 5.67. A circuit wired with a metal chassis joining part of the circuit could then be illustrated schematically as shown in the illustration.

The term *ground* that has just been introduced is an important one. Much has been written on the topics of grounds and grounding, and you will find that often a problem exists because the term *ground* is given different

Chassis Wiring — In a typical piece of constructed equipment, you may find that one side of the power supply is connnected to the chassis as shown in Figure 5.65. Either the positive or the negative side of the battery or supply can be connected to the chassis depending upon the needs of the circuit.

When one side of the power supply is connected to the chassis, wire leads need only be connected between the other side of the power supply and one side of a circuit component for components which are electrically isolated from the chassis. The other side of the component is connected directly to the chassis to complete the circuit.

Voltage Test Point Chart — Often in a constructed circuit, the manufacturer will provide a schematic or table showing what the measured voltages should be at different points in the circuit (with the chassis used as the reference point) during normal operation (Figure 5.66). A chart such as this is an invaluable aid in locating circuit problems.

If you know that the voltage between the chassis and a given point in the circuit should be +25 volts, and the reading you get is +150 volts, chances are a problem exists in the circuitry connected to that point. Problems can often be traced through circuits in this fashion through *careful* voltage measurements.

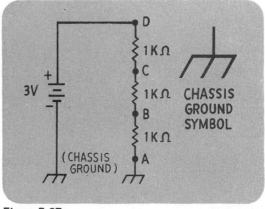

D

1KΩ

C

1KΩ

CHASSIS GROUND SYMBOL

3V

B

1KΩ

(CHASSIS GROUND)

A

Figure 5.67

meanings in different situations. Consider now some of the different uses of the term "ground."

Chassis Ground and Earth Ground — The term chassis ground refers to a chassis that is used as a reference point and as part of the circuits on it. An *earth* ground refers to a point that is at the potential of the earth or something in direct electrical connection to the earth such as water pipes. The schematic symbol for a point connected to an *earth ground* is shown in Figure 5.68.

Often a chassis may be connected to an *earth ground*, but often it may not, depending on the application of the circuit wired on the chassis. For this reason the earth ground symbol may be used for a chassis ground that is connected to an earth ground.

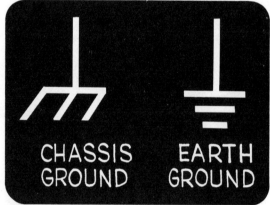

Figure 5.68

TROUBLE SHOOTING SERIES CIRCUITS

- SHORT CIRCUITS
- OPEN CIRCUITS

Figure 5.69

Common Circuit Problems: Troubleshooting in Series Circuits — Now that you are familiar with some of the basics of series circuit operation, consider the symptoms you can expect if a problem develops somewhere in an actual series circuit you are testing. By far the most common types of problems you will encounter in circuitry are *short circuits* or *open circuits* (Figure 5.69). "Troubleshooting" a circuit in general usage means to locate and correct a circuit fault or malfunction. What are the effects of these two common malfunctions in series circuits and how would you locate and correct them?

As was mentioned earlier, a *short circuit* is a circuit path with very *low* resistance. In common usage the term short circuit or "short" refers to a path for current which should not exist. A short is usually created by some mishap or malfunction in the circuit by pieces of wire or other conducting material dropped into the circuit, or by wire insulation worn through by heat or vibration. When a short is created in a series circuit, the total circuit *resistance is always reduced* so the total current flowing will increase. In some cases the current may increase to the point where circuit components become damaged.

Series Circuit-Normal Operation — Consider the series circuit shown in Figure 5.70 which consists of a 100-volt supply connected to three series resistors. In normal operation this circuit will carry a current, I_T which you can easily determine from Ohm's law. $I_T = E_T/R_T$; R_T in this case is 100 ohms; E_T is 100 volts; so $I_T = 1$ amp.

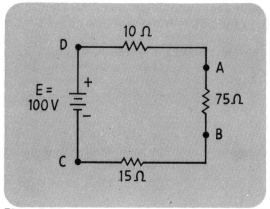

Figure 5.70

Series Circuit with "Short" — If a short developed somehow across the 75-ohm resistor, as shown in Figure 5.71, notice what happens. Since current will always follow the path of least resistance, the 75-ohm resistor is bypassed or "shorted out" reducing the effective total resistance of this circuit. This is where the term short circuit originated. Part of the normal circuit load has been bypassed and the circuit is now "electrically shorter." With the short present, the total resistance of the circuit between points A and B is reduced from 75 ohms without the short present down to a value that is essentially zero ohms, which is the resistance of whatever is causing the short such as a piece of wire or exposed metal, etc.

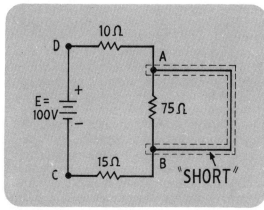

Figure 5.71

The total resistance of the circuit is then reduced from 100 ohms to 25 ohms, and the current flowing increases to 4 amps. In this case part of the load was shorted and so this situation is called a *partial short*. The higher current now flowing may be enough to damage other parts of the circuit.

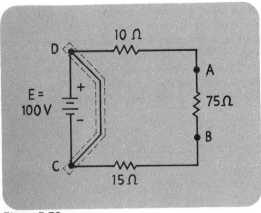

Figure 5.72

Series Circuit with Direct Short — Another potentially more damaging situation occurs when a short is created directly across the power supply as shown in Figure 5.72. In this case the total resistance between points C and D has dropped from 100 ohms to a very small value, and, hence, the power supply current will become very high. This type of circuit situation is called a *direct short*, and the resulting current, which is very high, will usually damage the supply unless some sort of protective device is included in the power supply circuit.

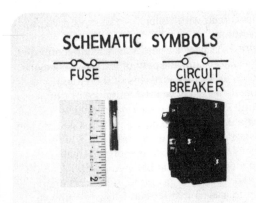

Figure 5.73

Protective Devices — Common protective devices used in power supply circuits are *fuses* or *circuit breakers* (Figure 5.73). A fuse is a special component designed to burn out or "blow" when the current through it exceeds the ampere value for which it is rated. A wide variety of fuse types is available but they all essentially employ some material with a low melting point which melts when heated by excessive current and *opens* the circuit stopping the flow of current.

A *circuit breaker* is a device designed to perform the same function as a fuse with the added advantage that once it opens the circuit or "trips," it can be reset. Once a fuse has blown, it must be replaced. Before a fuse is replaced or a breaker is reset, the short (or other problem) causing it to open the circuit must be located and corrected.

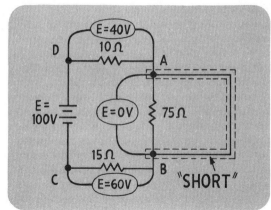

Figure 5.74

Short Circuits: Voltage Test — Shorts can be located in circuits with either a voltmeter, if the circuit can still be operated, or an ohmmeter, if the circuit is fully disconnected from the power source. Any abnormal voltage reading indicates some sort of fault in the circuit. A *short circuit* usually will read zero volts right across the short and higher than normal voltage reading elsewhere. In the circuit of Figure 5.74, if a short exists between points A and B, a voltmeter would indicate the readings shown. Notice that a reading of zero volts across the 75-ohm resistor would indicate that it is shorted out of the circuit.

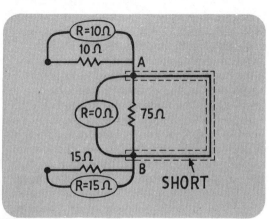

Figure 5.75

Short Circuits: Resistance Test — This same circuit fault could be located using an ohmmeter. First, make sure *no power* is applied to the circuit and then test each resistor. The readings would appear as shown in Figure 5.75. A reading of zero ohms across the 75-ohm resistor would confirm the presence of the short.

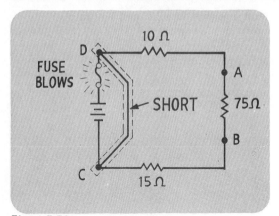

Figure 5.76

Open Circuits — Another type of circuit fault commonly encountered is the *open circuit* or *open*. In a series circuit, if any part of the circuit is broken, all current flow ceases. There is no longer a complete path for electrons. A circuit may be opened due to a blown fuse or tripped breaker as shown in Figure 5.76. This too indicates trouble, because something had to cause that to happen. A short is usually the cause, and it must be corrected before the fuse is replaced.

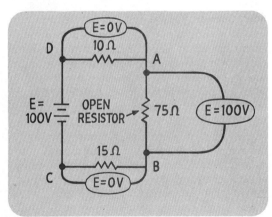

Figure 5.77

Full Voltage Appears Across Open — Other components in the circuit may open due to overheating, excessive wear, or maybe actually being shaken apart or loose in severe situations. As was mentioned, if a component opens in a series circuit, no current flows. Because of this, a voltmeter can be used to detect the open part. All *normal* parts will show *no voltage drop* across them since no current is flowing through them. Here's the trick. The open component is what causes the current flow to stop. An open component presents an infinitely high resistance to the circuit and the source voltage cannot push any electrons through it. As a result, the *total source voltage* will appear across an open component as shown in Figure 5.77.

Open Circuit Resistance Check — An ohmmeter also can be used to locate circuit opens once the circuit is fully disconnected from the power source. A resistor lableld 75 ohms that reads infinity on the ohmmeter is clearly a source of trouble and must be replaced (Figure 5.78).

Often as you troubleshoot circuits, a component's appearance may easily give away the fact that it is burned out, or a short circuit may be clearly visible. On occasion, however, the sneakiest, toughest situation to see may be causing a circuit short or open. In this case the meter will be your guide in locating the problem. Then your patience, careful ingenuity, and perseverence will be needed in correcting it. Proceed *carefully* as you examine

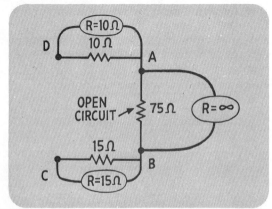

Figure 5.78

all trouble areas and replace components or "disconnect" shorts. Remember to disconnect all power and to observe caution when working on circuits.

Summary — In this lesson you have been presented with many important concepts that you will continue to need throughout your study of dc electricity. Series circuits were defined and introduced along with the three rules governing their behavior: The sum of the individual resistances in a series circuit equals the total circuit resistance; the current is the same at any point in a series circuit; and the sum of the individual voltages across the resistors in a series circuit equals the total voltage applied to that circuit.

Also key topics *related* to series circuits were discussed including series voltage sources and the polarity of voltages in a series circuit. The idea of a *reference* point for voltage measurement was introduced, and this led to an investigation of the important topics of the chassis and earth grounding.

At the end of this lesson there are several series circuit problems for you to practice on and work out. It is suggested that you work them all to really enable you to analyze, predict, and control the behavior of series circuits.

LESSON 5. SERIES CIRCUITS

● **Worked Through Examples**

1. Find the total resistance, R_T, of this circuit.

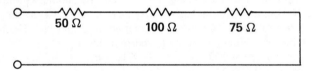

Since there is only one path for current flow in this circuit, each resistor in the circuit stands in the path of the current. Any current flowing in the circuit must flow through a 50-ohm resistor, a 100-ohm resistor, and a 75-ohm resistor for a total of 225 ohms. The formula used to find total resistance is $R_1 + R_2 + R_3 = R_T$.

2. How does current behave in a series circuit? Explain.

A series circuit consists of only one path through which current can flow. Because of this, all of the current flowing must flow through *all* of the circuit. Since current flow is considered to be the simultaneous shift of electrons through a circuit, the current is the same at all points in the circuit. This may be explained mathematically by stating that the total current or I_T is equal to I_1, the current through resistor number 1, which is equal to I_2, which is equal to I_3, and so on, depending on the total number of resistors in the circuit.

3. How does voltage behave in a series circuit? Explain.

As current flows through a series circuit, and a resistance is encountered, something called a voltage drop or IR drop takes place across the resistor. The source voltage is applying pressure to the circuit causing current to flow. All of the energy given to the electrons flowing in the circuit is dropped across the resistors in the circuit. The sum of the voltage drops in the circuit must equal the applied voltage.

4. What is the voltage drop across R_1?

$$R_1 = 50\ \Omega \quad R_2 = 100\ \Omega$$

$$E_T = 10\ V$$

There are two ways to solve this problem. You may use the ratio method or Ohm's law. The ratio method consists of looking at the circuit and determining the resistance ratio between the two resistors. To find the ratio, divide the circuit resistances by the value of the smallest resistor, in this case, 50 ohms. Fifty divided by 50 is equal to 1; 100 divided by 50 is equal to 2. The resistance ratio is then 1:2. This means that the 100-ohm resistor will have a voltage drop *twice* that of the 50-ohm resistor. The applied or source voltage will be split into two voltages. One resistor, the 50-ohm resistor, will drop 1/3 the total voltage, while the other resistor, the 100-ohm resistor, will drop 2/3 of the total voltage. One third of ten is 3.33 volts; 2/3 of ten is equal to 6.67 volts.

The second method, Ohm's law, is the more valuable of the two methods because the circuit current can be found. First, find the total resistance in the circuit by adding R_1 and R_2. Fifty ohms plus 100 ohms is equal to 150 ohms. The total current may now be found by dividing the total voltage, 10 volts, by the total resistance, 150 ohms, as specified by Ohm's law. The total current is equal to 10/150 amp or 0.066 amp. Now you know the current flowing through the entire circuit. The voltage across R_1 may be found by using Ohm's law. $R_1 \times I_T = E_{R1}$. Substituting the values into this equation, you have 50 X 0.066 which is equal to 3.3 volts.

5. In the circuit shown below, calculate:

    a.   The total resistance, R_T 1950 Ω

    b.   The total current flowing, I_T 51 mA

    c.   The voltage drop across each resistor.

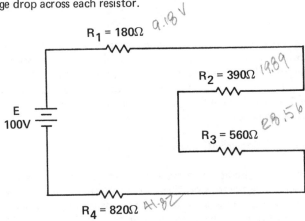

Solution: First of all, recognize that this is a series circuit. Even though the circuit pathway is not a straight line, there is only one path for current to follow.

To find the total resistance, just use the rule for series resistances: the total resistance of a series circuit equals the sum of the individual series resistances.
In formula form:

$$R_T = R_1 + R_2 + R_3 + R_4$$

Substituting the resistance values from the circuit schematic and adding:

$$R_T = 180\ \Omega + 390\ \Omega + 560\ \Omega + 820\ \Omega$$
$$R_T = 1950\ \Omega$$

Now you know the total voltage applied to this circuit is 100 volts, and that its total resistance is 1950 ohms. You next need to calculate the current, using Ohm's law. In the circle diagram, cover what you need to find, and the formula appears, $I = E/R$.

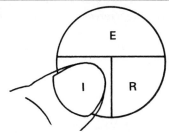

In this case, use this formula in the form $I_T = E_T/R_T$ or total current equals total voltage divided by total resistance, and substitute as shown in the following equation.

$$I_T = \frac{E_T}{R_T} = \frac{100 \text{ V}}{1950 \text{ }\Omega}$$

$$I_T = 51.3 \text{ mA}$$

This 51.3 milliamps is the total current flowing in this circuit, and because this is a series circuit, you know that this current is flowing through each resistor in it. Knowing the current through each resistor as well as the resistance value of each, you can now use Ohm's law in the form $E = I \times R$ to calculate the voltage drop across each resistor.

$E_{R1} = I_T \times R_1$
$E_{R1} = 51.3 \text{ mA} \times 180 \text{ }\Omega$
$E_{R1} = 51.3 \times 10^{-3} \times 1.8 \times 10^{+2}$
$E_{R1} = 9.23 \text{ V}$

$E_{R2} = I_T \times R_2$
$E_{R2} = 51.3 \text{ mA} \times 390 \text{ }\Omega$
$E_{R2} = 51.3 \times 10^{-3} \times 3.9 \times 10^{+2}$
$E_{R2} = 20 \text{ V}$

$E_{R3} = I_T \times R_3$
$E_{R3} = 51.3 \text{ mA} \times 560 \text{ }\Omega$
$E_{R3} = 51.3 \times 10^{-3} \times 5.6 \times 10^{+2}$
$E_{R3} = 28.7 \text{ V}$

$E_{R4} = I_T \times R_4$
$E_{R4} = 51.3 \text{ mA} \times 820 \text{ }\Omega$
$E_{R4} = 51.3 \times 10^{-3} \times 8.2 \times 10^{+2}$
$E_{R4} = 42.1 \text{ V}$

To check these results, recall the rule for voltages in a series circuit, the sum of the voltage drops equals the applied voltage. In this case, check to see that it does: $E_T = 100$ V. Does $E_{R1} + E_{R2} + E_{R3} + E_{R4}$ also equal 100 volts? Adding 9.23 + 20 + 28.7 + 42.1 gives 100.0 volts. The voltages are correct and the solution is complete.

6. Define the term "ground."

"Ground" is simply the voltage reference point in a circuit. This point in a circuit may be connected to the earth, in which case it is referred to as an *earth ground*.

7. What is the polarity of point A as compared to a reference at point B? Why?

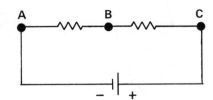

Negative. As current flows from the power supply, all resistors are polarized negative on the end closest to the negative terminal of the source and positive on the end closest to the positive potential. So, point A is negative with respect to point B.

8. In the circuit shown to the right, several known quantities are labeled. Calculate all of the unknown quantities in this circuit.

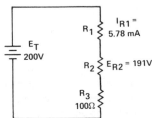

Solution: In a circuit situation such as this with "mixed" known and unknown quantities, it is a good idea to use a chart such as that shown to keep track of what you know and what you need to find.

When beginning the solution of a problem like this, first focus your attention carefully on the quantities you are given. Specifically, in a series circuit, see if you are given the current at any point. Once you know the current at one point, you know it everywhere. In this circuit, you are given that the current through R_1 is 5.78 milliamps, so you can fill in all the spaces in the current column of the chart with the same value.

	E	I	R
R_1		5.78 mA	
R_2	191V		
R_3			100Ω
TOTALS	200V		

	E	I	R
R_1		5.78 mA	
R_2	191V	5.78 mA	
R_3		5.78 mA	100Ω
TOTALS	200V	5.78 mA	

Once the current column on the chart is filled in, focus your attention on the rows where two out of three of the spaces are filled in. In these rows you can use Ohm's law to find the third quantity. For example, for R_2 you know E and I and need to calculate R. In the circle diagram, cover the R with your thumb, and the position of the remaining letters tells you the formula you need, R = E/I.

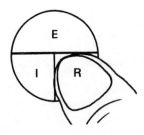

$$R_2 = \frac{E_2}{I_2} = \frac{191\ V}{5.78\ mA}$$

$$R_2 = \frac{191}{5.78 \times 10^{-3}} = 3.3 \times 10^4$$

$$R_2 = 33\ k\Omega$$

Substituting the known values from the chart, you can calculate that $R_2 = 33\ k\Omega$.

Similarly for R_3 you know I and R_3, therefore E_{R3} can be calculated using Ohm's law in the form $E = I \times R$.

$E_{R3} = 5.78\ mA \times 100\ \Omega$

$E_{R3} = 5.78 \times 10^{-3} \times 100 = 0.578$

$E_{R3} = 0.58\ V$

In the last column on the chart, you know the total voltage and current flowing. Using Ohm's law in the form $R = E/I$, you can calculate the total circuit resistance as follows:

$R_T = E_T/I_T$

$R_T = \dfrac{200}{5.78\ mA} = \dfrac{200}{5.78 \times 10^{-3}} = 3.46 \times 10^4$

$R_T = 34.6\ k\Omega$

At this point the chart will be filled in up to the point shown:

	E	I	R
R_1		5.78 mA	
R_2	191V	5.78 mA	33kΩ
R_3	.58V	5.78 mA	100Ω
TOTALS	200V	5.78 mA	34.6 kΩ

There are two ways you can now proceed. Either R_1 or E_{R1} may be found by remembering that sum of the resistances in this circuit equals the total resistance, and the sum of the voltage drops equals the applied voltage. In the right-hand column you will see that the total resistance of the circuit is 34.6 kilohms; R_2 and R_3 only add up to 33,000 plus 100 or 33,100 ohms of this total. Subtracting 33,100 from the 34,600 yields 1500 or 1.5 kilohms for R_1. Once R_1 is known, Ohm's law may be used to solve for E_{R1} as follows:

$E_{R1} = I_T \times R_1$

$E_{R1} = 5.78\ mA \times 1.5\ k\Omega$

$E_{R1} = 5.78 \times 10^{-3} \times 1.5 \times 10^3$

$E_{R1} = 8.67\ V$

and the solution is complete. To check your calculations, see if the sum of the voltage drops is equal to the applied voltage.

$E_{R1} = \quad 8.67$ V

$E_{R2} = 191 \quad$ V

$+E_{R3} = \quad 0.58$ V

200.25 V = 200 volts when rounded off. The answer checks.

9. In the circuit of Example 8, calculate the total power dissipated by the circuit. The complete schematic for the circuit, with values calculated is shown below.

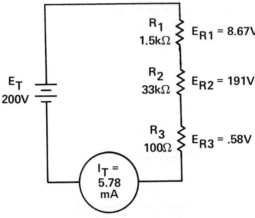

Solution: Recall that the total power in a series circuit is equal to the sum of the power dissipated by each resistor. Therefore, first calculate the power each resistor dissipates using the formula P = I X E as shown below.

R_1: $P_1 = I_T \times E_{R1}$

$P_1 = 5.78$ mA X 8.67 V

$P_1 = 0.05$ W

$P_1 = 50$ mW

R_2: $P_2 = I_T \times E_{R2}$

$P_2 = 5.78$ mA X 191

$P_2 = 1.10$ W

$P_2 = 1,100$ mW

R_3: $P_3 = I_T \times E_{R3}$

$P_3 = 5.78$ mA X 0.58

$P_3 = 3.35 \times 10^{-3}$ W

$P_3 = 3.35$ mW

Then, to find the total power, just add these values together:

$P_T = P_1 + P_2 + P_3$

$P_T = 50$ mW + 1,100 mW + 3.35 mW

$P_T = 1153.35$ mW

$P_T = 1.15$ W

10. Find I_T in the following circuit.

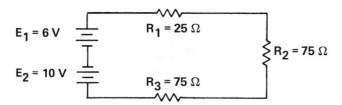

Before I_T can be found, E_T must be found. The 6-volt battery and the 10-volt battery are connected in series opposing so their voltages partially cancel. The resultant voltage is 4 volts. The net effect achieved here is that of a 4-volt battery pushing current around the circuit in a clockwise direction.

In order to use Ohm's law to find I_T, you must divide the total voltage by the total resistance. Four volts divided by $R_1 + R_2 + R_3$ is equal to I_T. The total resistance in the circuit is equal to 175 ohms. Four volts divided by 175 ohms is equal to 0.0229 amp or 22.9 milliamps.

11. One morning you try to start your car and it won't crank over. Checking the battery, you find it is good, even applying a jumper cable with a good battery doesn't help. Describe how you would troubleshoot this situation.

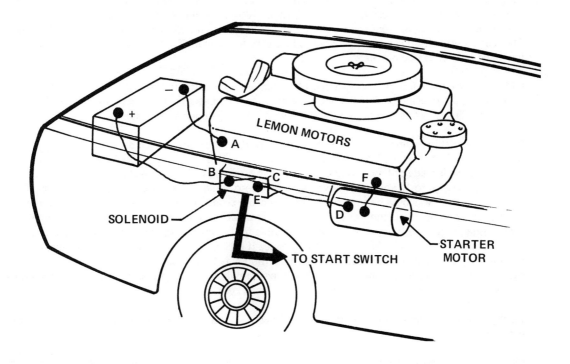

This boils down to troubleshooting a series circuit which you can do by using a voltmeter and techniques discussed in this lesson. The schematic of this circuit in simplified form is shown below.

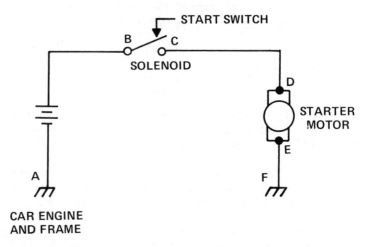

This circuit includes two new components which can be treated fairly simply in this troubleshooting situation. The solenoid is a special switch that is activated, turned on and off by your ignition switch. Your starter motor draws very heavy current, about 100 to 150 amps, and hence a special heavy-duty switch, tougher than your ignition switch, is needed to run the motor. This is why the solenoid is used. The starter motor itself can be considered to be a very heavy load. That means you can think of it as a low-value resistor as you proceed.

While someone turns the ignition switch on for you, use your voltmeter to test for an open in the circuit. How do you find one? If the battery voltage appears between any two points of this circuit other than directly across the battery, you know the circuit is open at that point.

Check the cables including the grounding cables by putting the voltmeter across them.

Since the cable is made of low-resistance wire, the voltage across this low resistance should be zero. If the voltage is equal to 12 volts, or any voltage above zero, a high resistance is indicated. Typical high resistance in this case could be caused by a broken cable producing infinite resistance or by corrosion on the cable clamps which connect to the battery terminals, or the starter motor or frame. Here's where visual inspection is important. Check the cable for breaks or corrosion. If all cables show zero volts dropped across them, check the solenoid to see if it is open or closed. It may be sticking, need replacement, or the circuit connecting it to the ignition may have problems. (This circuit can be examined for trouble in the same way.) Remember if a switch is open, the full battery voltage appears across it. If it is closed, the voltage across it should read zero.

Finally, check the starter motor itself. If it has 12 volts across it and isn't running, that's probably the problem ($$$).

(Hint — In most cases, corrosion on the battery terminals is the problem. Removing them and cleaning them thoroughly is usually the solution.)

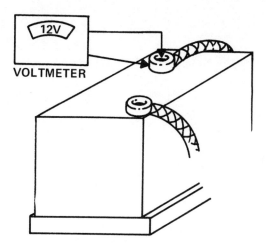

VOLTMETER

If you measure several volts between the center battery post and cable clamp, there is a high resistance between them. Clean them.

LESSON 5: SERIES CIRCUITS

- **Practice Problems**

Fold over the page to check your answers.   **Fold Over**

1. Find the following values for this circuit.

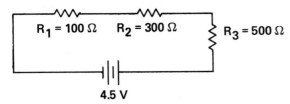

$R_1 = 100\ \Omega$ $R_2 = 300\ \Omega$ $R_3 = 500\ \Omega$

4.5 V

a. $E_{R1} =$ _____

b. $E_{R2} =$ _____

c. $E_{R3} =$ _____

d. $I_T \ \ =$ _____

e. $R_T \ \ =$ _____

2. Find the following values for this circuit.

$E_T = 100\ V$ $R_1 = 2.2\ k\Omega$ $R_2 = 3.3k\Omega$ $R_3 = 4.7k\Omega$ P_T = total power dissipated

a. $E_{R1} =$ _____

b. $E_{R2} =$ _____

c. $E_{R3} =$ _____

d. $I_T \ \ =$ _____

e. $R_T \ \ =$ _____

f. $P_T \ \ =$ _____

Answers

1. a.   0.5 V

    b.   1.5 V

    c.   2.5 V

    d.   5 mA

    e.   900 Ω

2. a.   21.6 V

    b.   32.3 V

    c.   46.1 V

    d.   9.8 mA

    e.   10.2 kΩ

    f.   980 mW

Fold Over

3. Find the following values for this circuit.

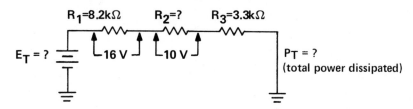

a. E_T = _____

b. R_T = _____

c. I_T = _____

d. R_2 = _____

e. E_{R3} = _____

f. P_T = _____

4. Find the following values for this circuit.

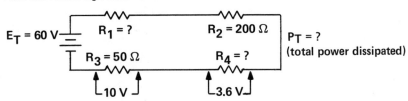

a. I_T = _____

b. R_T = _____

c. R_1 = _____

d. R_4 = _____

e. E_{R1} = _____

f. E_{R2} = _____

g. P_T = _____

Answers

3.   a.   32.4 V

      b.   16.6 kΩ

      c.   1.95 mA

      d.   5.13 kΩ

      e.   6.44 V

      f.   63.2 mW

4.   a.   0.2 A

      b.   300 Ω

      c.   32 Ω

      d.   18 Ω

      e.   6.4 V

      f.   40 V

      g.   12 W

Fold Over

5. Using "ground" as the reference point, what is the amount of voltage and polarity measured at point:

A = _____

B = _____

C = _____

D = _____

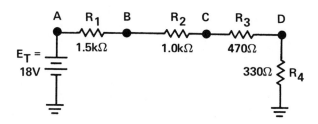

6. Using point "B" as a reference point, what is the amount of voltage and polarity measured at point:

A = _____

B = _____

C = _____

D = _____

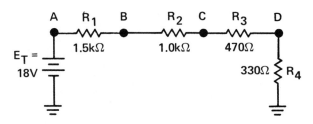

Answers

5. A = +18 V

    B = +9.82 V

    C = +4.36 V

    D = +1.8 V

6. A = +8.18 V

    B = 0 V

    C = −5.45 V

    D = −8.01 V

Fold Over

7. Fill in the chart.

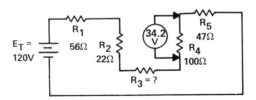

P$_1$		R$_1$		E$_{R1}$		I$_{R1}$	
P$_2$		R$_2$		E$_{R2}$		I$_{R2}$	
P$_3$		R$_3$		E$_{R3}$		I$_{R3}$	
P$_4$		R$_4$		E$_{R4}$		I$_{R4}$	
P$_5$		R$_5$		E$_{R5}$		I$_{R5}$	
P$_T$		R$_T$		E$_T$		I$_T$	
	P		R		E		I

(Note: In problems 7, 8, and 9 your answers may vary from the answers given on the back of the page. This is due to rounding off between steps in the solutions and to the different ways in which any one answer can be found. In any case, the first two significant digits in your answer should agree with the answers given here but the third significant digit may vary widely. If your answers do not agree, recheck your work.)

8. Fill in the chart.

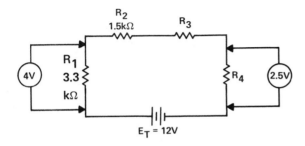

R$_1$		E$_{R1}$		I$_{R1}$	
R$_2$		E$_{R2}$		I$_{R2}$	
R$_3$		E$_{R3}$		I$_{R3}$	
R$_4$		E$_{R4}$		I$_{R4}$	
R$_T$		E$_T$		I$_T$	
	R		E		I

Answers

7.

6.58W	56Ω	19.2V	.342A
2.58W	22Ω	7.52V	.342A
14.7W	126Ω	43.0V	.342A
11.7W	100Ω	34.2V	.342A
5.52W	47Ω	16.1V	.342A
41.1W	350Ω	120V	.342A
P	R	E	I

8.

3.3kΩ	4V	1.21mA
1.5kΩ	1.8V	1.21mA
3.06kΩ	3.7V	1.21mA
2.06kΩ	2.5V	1.21mA
9.92kΩ	12V	1.21mA
R	E	I

Fold Over

9. Fill in the chart.

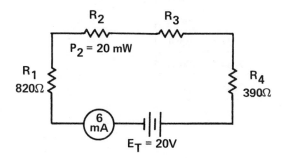

P₁		R₁		E_R1		I_R1	
P₂		R₂		E_R2		I_R2	
P₃		R₃		E_R3		I_R3	
P₄		R₄		E_R4		I_R4	
P_T		R_T		E_T		I_T	
P		**R**		**E**		**I**	

Answers

9.

29.5mW	820Ω	4.92V	6mA
20mW	555Ω	3.33V	6mA
56.5mW	1.53kΩ	9.21V	6mA
14mW	390Ω	2.34V	6mA
120mW	3.3kΩ	20V	6mA
P	R	E	I

LESSON 5 — QUIZ

1. A series circuit contains:
   a. 1 path for voltage flow
   b. 3 paths for current flow
   c. 2 paths for current flow
   d. 1 path for current flow

2. The sum of the individual resistances in a series circuit is equal to:
   a. The total voltage
   b. 100 ohms
   c. The total resistance
   d. The total current

3. A series circuit has a constant voltage applied. If the circuit resistance is decreased, the circuit current will:
   a. Remain the same
   b. Increase
   c. Decrease
   d. Rise to maximum

4. In any given series circuit containing two or more resistors, if one of the resistors is increased in value, the voltage drop across it will:
   a. Increase
   b. Decrease
   c. Stay the same
   d. Fall to zero

5. The sum of the individual voltage drops in a series circuit is equal to:
   a. Total resistance
   b. Total current
   c. Average current
   d. Total applied voltage

6. In a series-wired light set, one lamp burns out. The result is:
   a. All of the other lamps get brighter
   b. An increase in the total circuit current
   c. All of the other lamps go out
   d. A decrease in the total circuit resistance

7. The following symbol stands for:

   a. Earth ground
   b. Positive potential
   c. Multicell battery
   d. Chassis ground

8. A "chassis" ground is:
   a. Always negative
   b. Usually high potential
   c. A common reference point
   d. Always safe to touch

9. 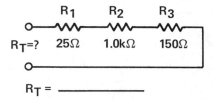 is the symbol for:
   a. Chassis ground
   b. Air ground
   c. Earth ground
   d. Positive potential

10. Calculate the total resistance

$R_T = $ _____

11. Calculate the total resistance.

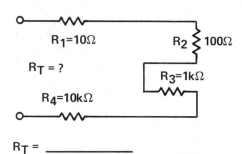

$R_T = $ _____

12. Calculate the total resistance

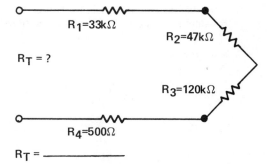

$R_T = $ _____

13. Calculate the total resistance

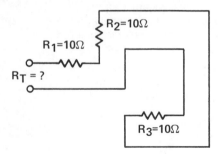

$R_T =$ _____

14. Calculate the total resistance

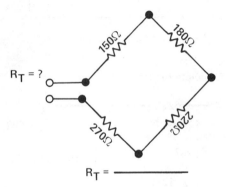

$R_T =$ _____

15. Current in a series circuit has_____value at any point in the circuit.

a. A different
b. The same
c. An unequal
d. An opposite

16. In any series circuit containing two or more resistors and a battery, when any one of the resistors is increased the current:

a. Increases
b. Decreases
c. Remains the same
d. Can't be determined

17. In any series circuit containing two or more resistors, if any one of the resistors is increased and the current is to remain constant, the voltage applied must:

a. Increase
b. Decrease
c. Stay the same
d. Be in series

18. If two batteries are connected in a series circuit, they will aid in pushing electrons through the circuit if their terminals are connected:

a. − + + −
b. + − − +
c. + − + −
d. − + − +
e. c and d above

19. When a 100 ohm resistor has 2 amperes of current through it, the voltage drop across it is:

a. 2 volts
b. 20 volts
c. 50 volts
d. 200 volts

20. When a series circuit has two resistors, each 1 K ohms resistance, and a circuit is 25 milliamperes, the voltage drop across both resistors is:

a. 25 volts
b. 25 Kilovolts
c. 50 volts
d. 2 Kilovolts

Lesson 6

Introduction to Parallel Circuits

This lesson introduces and discusses a new type of basic circuit, the *Parallel circuit*. The key features differentiating the parallel circuit from the series circuit are outlined, and then the behavior of resistances in parallel is examined in detail. Essential concepts including conductance, shunt connections, and reciprocals are introduced as they are required to explain and predict circuit behavior.

LESSON 6. INTRODUCTION TO PARALLEL CIRCUITS

● Objectives

In this lesson the *parallel* circuit is introduced and the behavior of resistances in parallel is covered in depth. The student successfully completing this lesson should be able to:

1. *Identify* parallel circuits in several given schematics.

2. *Write* a definition of "parallel circuit" describing the key differences between series and parallel circuits; including a definition of the term "branch."

3. *Write* the three laws describing the behavior of the voltage, current and resistance in parallel circuits.

4. *Define* the terms conductance, shunt, and reciprocal.

5. Given a circuit schematic diagram of the type shown below:

   a. *Write* how a quick estimate of the total resistance R_T could be obtained and estimate it.

   b. *Calculate* the circuit's total resistance and conductance using the *reciprocal formula*.

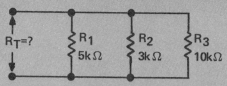

6. Given a schematic diagram of the type shown below:

   Calculate R_T, the total resistance, using the Product-over-Sum formula.

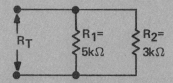

7. Given a schematic diagram of the type shown below:

   Write the rule by which the total resistance R_T could be most quickly calculated, and *calculate* it.

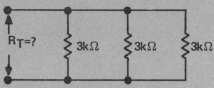

8. *Perform any of the above calculations with the aid of a calculator*, having scientific notation capability and a scratch-pad memory, such as TI-30, TI-50 or TI-55.

LESSON 6. INTRODUCTION TO PARALLEL CIRCUITS

- **Parallel Circuit**
- **Laws of Series Circuits**

This lesson introduces the topic of *parallel circuits*. As you will see, there are several distinct features that make these circuits and their operation different from the series circuit discussed in Lesson 5. As was discussed, a series circuit has *only one path for current to flow through* and as a result, the current must have the same value in all parts of the circuit, as shown in Figure 6.1.

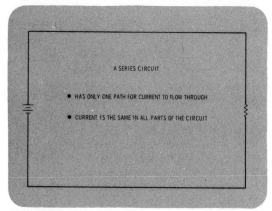

Figure 6.1

Parallel Circuit — A *parallel circuit*, such as the one shown in Figure 6.2, is defined as a circuit that has *two or more paths for current to flow through*. As you will be seeing in this lesson, the fact that parallel circuits have more than one path for current flow has several interesting effects on the behavior of the circuit current, voltage, and resistance. Before actually delving into these effects, a brief review of the laws governing series circuit behavior will be helpful.

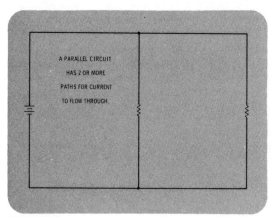

Figure 6.2

Laws of Series Circuits — In Lesson 5 you studied the behavior of series circuits and learned the three basic laws describing their behavior, namely:

1. The sum of the individual voltage drops around a series circuit equals the applied voltage.
2. Current has the same value at any point within a series circuit.
3. The sum of the individual resistances within a series circuit equals the total resistance of that circuit. Voltage polarity was also discussed as well as the topic of circuit grounds.

You also saw how each of these laws may be expressed as a formula, as shown in Figure 6.3.

This lesson introduces *parallel circuits*. First, a parallel circuit will be defined for you, and then the three laws that describe the behavior of voltages, currents, and resistances in parallel circuits will be discussed. Subsequently, the behavior of resistance in parallel circuits will be closely examined. At the end of this lesson you should be able to *compute* the equivalent resistance of several resistors that are wired in parallel. The *concept of conductance* will be introduced and discussed. The use of a calculator to handle the mathematics involved in finding the total resistance of a parallel circuit will also be shown.

Parallel Circuit — First of all, *a parallel circuit is defined as a circuit where there is more than one path for current to flow through* (Figure 6.4). In this lesson simple parallel circuits will be discussed. These circuits will be comprised of several paths for current flow, each containing one resistor.

Schematic of a Parallel Circuit — As an illustration of a parallel circuit, consider the schematic diagram of two resistors connected in parallel shown in Figure 6.5. The arrow labeled A is one path for current, and the arrow labeled B is another path for current. It is more technically correct to call these paths *branches*. Therefore the circuit in Figure 6.5 has two resistive branches that will carry current, each containing a single resistor. (Although branches may contain more than one resistor, in this lesson only simple parallel circuits with one resistor per branch will be discussed.)

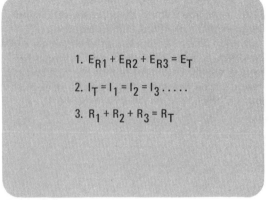

$$1. \ E_{R1} + E_{R2} + E_{R3} = E_T$$

$$2. \ I_T = I_1 = I_2 = I_3 \ldots \ldots$$

$$3. \ R_1 + R_2 + R_3 = R_T$$

Figure 6.3

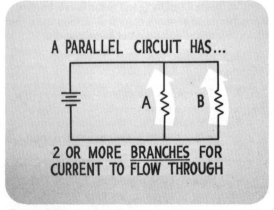

A PARALLEL CIRCUIT

A CIRCUIT WHERE THERE IS MORE THAN ONE PATH FOR CURRENT TO FLOW THROUGH.

Figure 6.4

A PARALLEL CIRCUIT HAS...

A   B

2 OR MORE <u>BRANCHES</u> FOR CURRENT TO FLOW THROUGH

Figure 6.5

Laws of Parallel Circuits — Three rules, similar to those stated for series circuits, may be applied to parallel circuit operation. In this study the three laws will first be briefly stated, and then to really gain insight into how resistances behave when wired in parallel, the *third law* will be examined in detail. The three laws are stated briefly in Figure 6.6.

The total voltage of a parallel circuit is the same across each branch of that circuit.

The total current in a parallel circuit is equal to the sum of the individual branch currents.

The total resistance in a parallel circuit is always less than or approximately equal to the value of the smallest resistive branch.

1. THE TOTAL VOLTAGE OF A PARALLEL CIRCUIT IS THE SAME ACROSS EACH BRANCH OF THAT CIRCUIT.

2. THE TOTAL CURRENT IN A PARALLEL CIRCUIT IS EQUAL TO THE SUM OF THE INDIVIDUAL BRANCH CURRENTS.

3. THE TOTAL RESISTANCE IN A PARALLEL CIRCUIT IS ALWAYS LESS OR APPROXIMATELY EQUAL TO THE VALUE OF THE SMALLEST RESISTIVE BRANCH.

Figure 6.6

Formula for Parallel Resistance — Focus your attention on the third of these stated laws which will be discussed in detail in this lesson. The law *states that the total resistance in a parallel circuit is always less than or approximately equal to the resistance of the smallest resistive branch.*

This resistance behavior is characteristic of all parallel circuits. What this statement really means to you as you study circuitry will be the key topic for discussion as the lesson proceeds.

As a first step, consider how to calculate mathematically the equivalent resistance for several resistances wired in parallel.

The formula for calculating the total resistance of a parallel circuit is shown in Figure 6.7. Stated verbally this formula is: the total resistance (of a parallel circuit) equals the reciprocal of R_1, plus the reciprocal of R_2, plus the reciprocal of R_3, all divided into one. (The dots at the end of this formula indicate that the formula continues for as many resistors as there are in the circuit.)

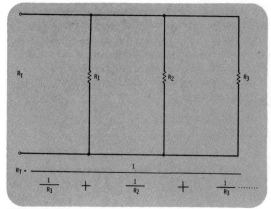

Figure 6.7

- **Reciprocal**
- **Conductance – The Reciprocal of Resistance**
- **Definition of Conductance**

Reciprocal – In this formula you will notice that a new term has been introduced. *Reciprocal*, what does that mean? As shown in Figure 6.8, the reciprocal of any number is simply that number divided into one. So in this case, when you say the reciprocal of R_1, this actually means 1 divided by R_1. (As shown in Figure 6.8, the reciprocal of 2 is 1/2; or 0.5.)

A "RECIPROCAL" OF ANY NUMBER IS SIMPLY THAT NUMBER DIVIDED INTO 1.

THE RECIPROCAL OF $R_1 = 1/R_1$

THE RECIPROCAL OF 2 = 1/2 = .5

Figure 6.8

Conductance - The Reciprocal of Resistance – The word reciprocal is sometimes used to mean "the opposite of." The *opposite or reciprocal of resistance* is a special and interesting quantity which has been given the name *conductance* (Figure 6.9). The reciprocal of any given resistance is its *conductance*.

THE *OPPOSITE* OF RESISTANCE IS CONDUCTANCE

THE *RECIPROCAL* OF RESISTANCE IS CONDUCTANCE

Figure 6.9

Definition of Conductance – You will recall that the definition of resistance is the *opposition to current flow*. Since resistance and conductance are opposites, conductance can therefore be defined as the *ability* to conduct current (Figure 6.10). When a wire, a resistor, or a branch in a parallel circuit has a large conductance, this means that it can carry or conduct a large amount of current. (A device with a *large conductance* has a *small resistance* and vice versa.)

CONDUCTANCE IS THE ABILITY TO CONDUCT CURRENT

LARGE CONDUCTANCE ALLOWS HIGH CURRENT FLOW

LOW CONDUCTANCE ALLOWS LESS CURRENT FLOW

Figure 6.10

- **The Unit Measure for Conductance**
- **Symbol for Mhos**
- **Reciprocal of Resistance**

The Unit Measure for Conductance — The unit of measure for conductance is the *mho* (Figure 6.11). Notice that the word mho is simply ohm spelled backwards, again indicating the opposite of ohms.

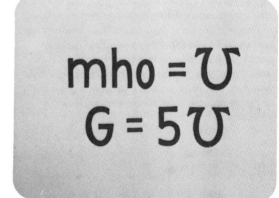

THE UNIT OF MEASURE FOR CONDUCTANCE IS THE "MHO"

Figure 6.11

Symbol for Mhos — The symbol for mhos is an upside-down omega ($\mho$). When writing formulas, a capital G is used to indicate conductance. So the statement "The conductance equals 5 mhos" may be written as G = 5 $\mho$ as shown in Figure 6.12.

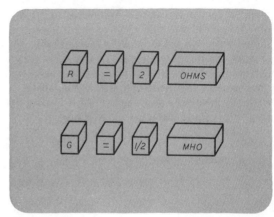

$$mho = \mho$$
$$G = 5\,\mho$$

Figure 6.12

Reciprocal of Resistance — If a resistor has a resistance, R of 2 ohms, what will its conductance be? Again, conductance is just the reciprocal (or one over) the resistance. Therefore a resistor that has 2 ohms of resistance has a conductance of 1 divided by 2 or *1/2 a mho*. So if you have a device with R = 2 Ω, then G = 1/2 $\mho$. (Figure 6.13).

R = 2 OHMS

G = 1/2 MHO

Figure 6.13

- **Formula for Total Resistance in Terms of Conductance**
- **Simple Parallel Circuit**
- **Shunt or Parallel Resistors**

Formula for Total Resistance in Terms of Conductance — The formula for total resistance in a parallel circuit may be written in terms of *conductance* as shown in Figure 6.14. This simplifies the expression and the formula now states that the total resistance, R_T, equals the sum of the individual branch conductances. This is because $1/R_1 = G_1$; $1/R_2 = G_2$; and so on.

$$\text{(a)} \quad R_T = \cfrac{1}{\cfrac{1}{R_1} + \cfrac{1}{R_2} + \cfrac{1}{R_3} \cdots}$$

$$\text{(b)} \quad R_T = \cfrac{1}{G_1 + G_2 + G_3 \cdots}$$

Figure 6.14

Simple Parallel Circuit — Up to this point you have seen how to write the formula for total resistance in a parallel circuit in two ways. Now you will see how to apply the formula to a typical parallel circuit, such as the one shown in Figure 6.15. Notice that this parallel circuit contains only two branches. The resistance of one branch is 1 kilohm while the resistance of the other branch is 5 kilohms.

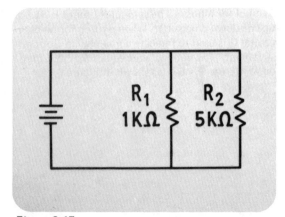

Figure 6.15

Shunt or Parallel Resistors — Another word often used to describe a parallel connection is the word *shunt*. When one resistor is wired in *shunt* with another, this means the two are connected in parallel. Also, you will sometimes see two parallel lines used as a shorthand to represent a parallel circuit connection. For instance, $R_1 \parallel R_2$ would be read as "R_1 in parallel with R_2" as shown in Figure 6.16.

Figure 6.16

Calculation of Total Resistance in a Parallel Circuit — To continue with the circuit problem to calculate the total resistance, first write the formula for total resistance in a parallel circuit as shown in Figure 6.17(a).

The second step is to substitute the values of the circuit resistors into the formula — writing them in powers of ten form. In powers of ten form, R_1 is $1 \times 10^{+3}$ ohms and R_2 is $5 \times 10^{+3}$ ohms, as shown in Figure 6.17(b).

The third step in this solution is to calculate the reciprocals of the resistors.

If you calculate 1 divided by $1 \times 10^{+3}$, you will find that it equals 1×10^{-3}. Similarly 1 over $5 \times 10^{+3}$ equals 0.2×10^{-3}, as shown in Figure 6.17(c). Next these reciprocals are added together, and note that since their powers of ten are the same, there will be no need to change the powers before adding. Adding these gives you 1.2×10^{-3} as seen in Figure 6.17(d), which now must be divided into one to get the final answer for the total resistance. Performing this final step, the total circuit resistance is calculated as $0.833 \times 10^{+3}$ ohms. Using the metric chart, this answer may be converted to its final form, *833 ohms*, as shown in Figure 6.17(d).

Recall for a moment that the law for total resistance in a parallel circuit that was stated for you at the beginning of this lesson. This law states that the *total resistance of a parallel circuit is always less than, or approximately equal to, the resistance of the smallest resistive branch*. Notice here that the total circuit resistance of 833 ohms is indeed *less* than the resistance of the smallest resistive branch, 1 kilohm. So, the calculated result follows the statement of the law.

This result can be physically proved in the laboratory by connecting an ohmmeter across the circuit with the voltage removed. If an ohmmeter were connected across the circuit, it would measure the total resistance, R_T or 833 ohms.

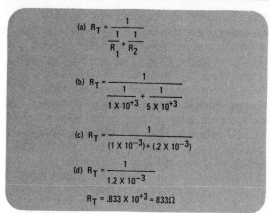

(a) $R_T = \dfrac{1}{\dfrac{1}{R_1} + \dfrac{1}{R_2}}$

(b) $R_T = \dfrac{1}{\dfrac{1}{1 \times 10^{+3}} + \dfrac{1}{5 \times 10^{+3}}}$

(c) $R_T = \dfrac{1}{(1 \times 10^{-3}) + (.2 \times 10^{-3})}$

(d) $R_T = \dfrac{1}{1.2 \times 10^{-3}}$

$R_T = .833 \times 10^{+3} = 833\,\Omega$

Figure 6.17

Circuit with Meter — Consider *why* this total resistance is less than that of the smallest resistive branch. The overall conductance of the circuit shown in Figure 6.18 will change if another resistor is added to it (as indicated by the dotted lines). Since conductance is the ability to conduct current, if the total conductance of a circuit *increases*, the total current that will flow in it must *increase*, which means that the total resistance must *decrease*. Before the second resistor is added, the circuit shown in Figure 6.18 has 10 volts applied to a 1-kilohm resistor. The current that will flow in this circuit can be *calculated by the use of Ohm's law in the form I = E/R.*

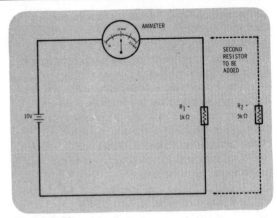

Figure 6.18

Calculation of Current Flow — As shown in Figure 6.19, the values of voltage and current have been substituted into the formula: $1 \times 10^{+1}$ for the 10 volts, divided by $1 \times 10^{+3}$ for the 1-kilohm resistor. When these are divided, the result is 1×10^{-2} which equals 0.01 amp or 10 milliamps of current. So, an ammeter connected in series as shown would indicate 10 milliamps of current flow before the second resistor is tacked on.

It was mentioned that if the conductance of a circuit is increased, total current flowing in the circuit will increase. Also, the fact that current and resistance are inversely related was discussed. If a constant voltage is applied to a circuit, and the total current increases, the total resistance *must* have decreased. Keep this fact in mind for a moment.

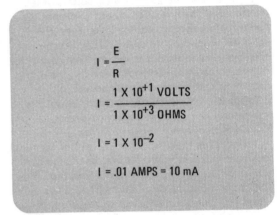

Figure 6.19

Parallel Circuit with Meter — In Figure 6.20, the second resistor has been connected in parallel with the 1-kilohm resistor. Notice how this adds a *second path* for current to flow through. This second path *increases the circuit's conductance* or the ability of the circuit to conduct current. The ammeter in the circuit will show a swing upward, indicating an *increase* in the circuit current. The circuit's total resistance had to *decrease* in order for this to happen.

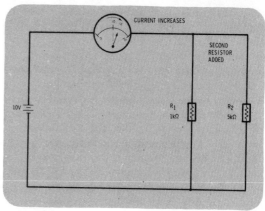

Figure 6.20

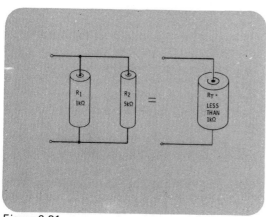

Figure 6.21

Parallel Resistances and Cross-Sectional Area — Another way of looking at the fact that the total equivalent resistance of two resistors in parallel is *less than the resistance of either one* may be seen by examining Figure 6.21.

If the resistors in a circuit were made up of cylinders of identical *length* and consisted of the same carbon composition, the picture would appear as shown. The 1-kilohm resistor would have a greater cross section than the 5-kilohm resistor; and when the two are connected in parallel, their cross-sectional areas are just combined, as shown for R_T. The total cross section available to conduct current is therefore *increased*; as the 5-kilohm resistor is added, the total circuit resistance is decreased. (Remember, as cross-sectional area increases, resistance decreases.) As a result, the addition of the 5-kilohm resistor in parallel allows more current to flow in the circuit than does the 1-kilohm resistor alone, and, therefore, the total circuit resistance is *less* than 1 kilohm. You can now make some calculations on the current flowing in this circuit.

- **Calculation of Current with R_2 Added**
- **Calculation of Total Resistance with Ohm's Law**

Calculation of Current with R_2 Added – Recall that earlier in the lesson the first law governing the behavior of parallel circuits was introduced. It will be discussed in depth in the next lesson, but it briefly states that the *voltage across all branches of a parallel circuit is the same*. Since the added 5-kilohm resistor was connected in parallel with the 1-kilohm resistor, both resistors will have the same 10 volts applied to them. Once you know there are 10 volts across this resistor, you can use Ohm's law to calculate the current flowing through this 5-kilohm resistor. In Figure 6.22, the values of voltage and resistance have been substituted into Ohm's law formula for current, $1 \times 10^{+1}$ for the voltage, divided by $5 \times 10^{+3}$ for the resistance. When the division is carried out, the result is 0.2×10^{-2} amps, which equals 0.002 amp or 2 milliamps of current.

If in an actual circuit you were to connect an additional 5-kilohm resistor in parallel with the 1-kilohm resistor, the total current should increase by 2 milliamps. In an actual circuit, the ammeter's reading would then increase from its previous reading of 10 milliamps to 12 milliamps. Again since the circuit current has *increased*, the total circuit resistance must have *decreased*. Since the total circuit voltage and total current flowing are known, this circuit's total resistance can be calculated using Ohm's law.

$$I = \frac{E}{R}$$

$$I = \frac{1 \times 10^{+1} \text{ VOLTS}}{5 \times 10^{+3} \text{ OHMS}}$$

$$I = .2 \times 10^{-2}$$

$$I = .002 \text{ AMPS} = 2 \text{ mA}$$

Figure 6.22

Calculation of Total Resistance with Ohm's Law – Figure 6.23 shows the calculation for the total resistance using the total current of 12 milliamps and the applied voltage of 10 volts. To calculate R_T, total resistance, R_T, equals total voltage applied to the circuit divided by total circuit current. These values have been substituted in the formula to get $1 \times 10^{+1}$ divided by 1.2×10^{-2} which equals $0.833 \times 10^{+3}$ or *833 ohms*, which is *less* than 1 kilohm. Again, this proves that *the total resistance of a parallel circuit is less than or approximately equal to that of the smallest resistive branch*.

$$R = \frac{E}{I}$$

$$R = \frac{1 \times 10^{+1} \text{ VOLTS}}{1.2 \times 10^{-2} \text{ AMPS}}$$

$$R = .833 \times 10^{+3}$$

$$R = 833 \text{ OHMS}$$

Figure 6.23

- **Parallel Connection of a Voltmeter**
- **Total Resistance**
- **Resistance Calculation**

Parallel Connection of a Voltmeter — Figure 6.24 is another circuit very similar electrically to the connection of a voltmeter across a resistor. Whenever the voltage across a resistor is measured with a voltmeter, the meter is connected in *parallel* with the voltage to be measured. Voltmeters in general are deliberately designed to have a very high resistance so that very little current flows in the meter branch of the circuit.

    The circuit shown consists of a 1-kilohm resistor and a 1-megohm resistor connected in parallel. Here the 1-megohm resistor simulates the very high resistance a voltmeter would have in a circuit situation such as this.

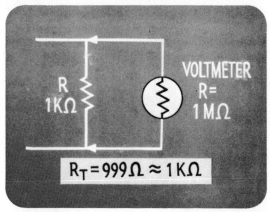

Figure 6.24

Total Resistance — In order to find the total resistance of this parallel circuit, the reciprocal formula can be used as shown previously. In Figure 6.25 the values of resistance have been substituted in the formula: $1 \times 10^{+3}$ for the 1-kilohm resistor and $1 \times 10^{+6}$ for the 1-megohm resistance of the meter. The reciprocals of these resistances are 1×10^{-3} and 1×10^{-6}. The easiest way to add two numbers such as this with quite a large difference between their values is to convert them to their decimal form.

$$R_T = \cfrac{1}{\cfrac{1}{R_1} + \cfrac{1}{R_2}}$$

$$R_T = \cfrac{1}{\cfrac{1}{1 \times 10^{+3}} + \cfrac{1}{1 \times 10^{+6}}}$$

$$R_T = \cfrac{1}{(1 \times 10^{-3}) + (1 \times 10^{-6})}$$

Figure 6.25

Resistance Calculation — As shown in Figure 6.26, 1×10^{-3} is 0.001 and 1×10^{-6} is 0.000,001. Adding these together yields the number 0.001001. This number converted back to powers of ten form equals 1.001×10^{-3}. All that remains to obtain the final answer for R_T is to take the reciprocal of 1.001×10^{-3}. When this is done, the final answer is seen to be 999 ohms. Notice the curved lines ($\approx$) in Figure 6.26. These are a shorthand notation used to indicate that 999 ohms is *approximately equal* to 1 kilohm. This problem illustrates what was meant when it was stated that the total resistance of a parallel circuit is less than or *approximately equal* to the smallest resistive

$$1 \times 10^{-3} = .001$$
$$\underline{+1 \times 10^{-6} = +.000001}$$
$$.001001$$

$$R_T = \frac{1}{.001001} = \frac{1}{1.001 \times 10^{-3}}$$

$$R_T = 999 \text{ OHMS} \approx 1 \text{ k}\Omega$$

Figure 6.26

branch. In this case, 999 ohms is approximately equal to 1 kilohm which is the resistance of the smallest resistive branch.

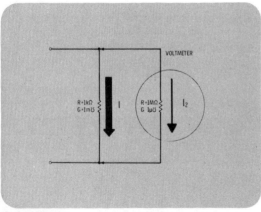

Figure 6.27

Conductance in Voltmeter Circuit — Notice what has happened here in terms of *conductance*. In Figure 6.27 if the original circuit contains a 1-kilohm resistor only, this provides 1/1 kilohm or 1 millimho (1 m℧) of conductance. Depending on the voltage applied to the circuit, a certain current, call it I_1 would flow in this circuit.

If a second circuit path (such as a voltmeter) is added that has a very high resistance, it adds very little to the circuit in terms of overall conductance. Essentially another very low conductance path is created where a little trickle of current (labeled I_2) will flow. The overall current ($I_1 + I_2$) flowing in this circuit will be only slightly different from I_1 above since I_1 is so large and I_2 is so small. *Adding branches* with very high resistance to a parallel circuit will affect its behavior very little.

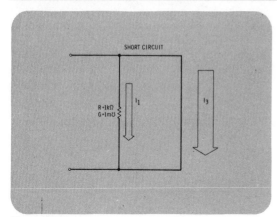

Figure 6.28

"Short" Branch Added in Parallel — Consider what would happen if a branch were added to this circuit that had a very *low* resistance (or high conductance) as shown in Figure 6.28. A plain piece of wire would do nicely. In this case, the piece of wire provides a very low resistance path which would conduct a very high current. The situation amounts to a direct short across the power supply. The total circuit resistance will be *less* than that of the wire above. The wire creates a path of very large conductance permitting a huge current (I_3) to flow through it. Adding branches with very low resistance to a parallel circuit *reduces the circuit resistance drastically* and tends to "short out" all resistances in the circuit.

Parallel Circuits with More than Two Branches — Consider how the total resistance of *several* parallel branches, each having an equal resistance, may be calculated. One straightforward method is to use the reciprocal formula shown in Figure 6.29. In this case, you would just substitute 3 kilohms for each resistance in the formula. The only difference in this formula compared to those worked previously is that there is now a third resistor, R_3.

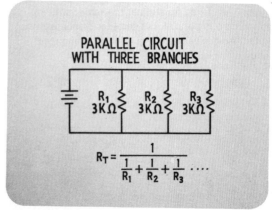

Figure 6.29

Calculating R_T — The reciprocal of 3 kilohms is $1/3 \times 10^{+3}$ or 0.333×10^{-3} as shown in Figure 6.30. Adding all the reciprocals in the bottom of the equation in the lower left corner of Figure 6.30, you see that R_T equals the reciprocal of 0.999×10^{-3}. When the reciprocal of that sum is found, the answer is 1001 ohms. Here is an interesting point. Notice that the total resistance of this circuit is very close to one-third the value of one of the resistive branches; one-third of 3 kilohms is 1 kilohm. It turns out that in cases of many equal value parallel branches, a rule exists that simplifies the calculation of R_T.

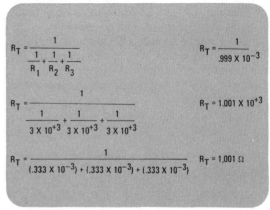

Figure 6.30

Parallel Resistors of Equal Value — Figure 6.31 illustrates this rule about resistors of equal value connected in parallel. When *two* resistors are connected in parallel and their values are the same, their total resistance is *one-half* the value of *one* resistive branch.

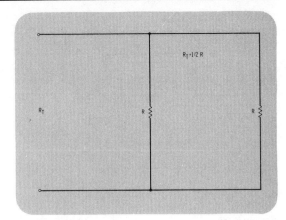

Figure 6.31

When *three* resistors *of the same value* are connected in parallel, the total resistance is *one-third* the value of one resistive branch, as shown in Figure 6.32.

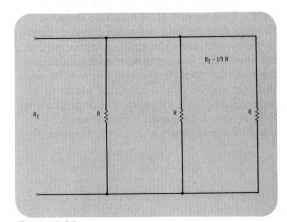

Figure 6.32

Similarly, Figure 6.33 shows that when four resistors of the same value are connected in parallel, the total resistance is *one-fourth* the value of one resistive branch. This rule continues for any number of branches that a parallel circuit might have, *as long as the resistance of all of the branches is the same value.*

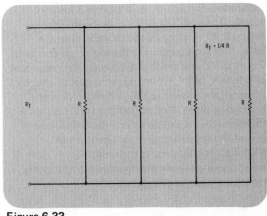

Figure 6.33

Equation for Equal Resistors in Parallel — The rule for finding the total resistance of a parallel circuit containing resistors of equal value can be expressed in equation form as shown in Figure 6.34.

Here R$_{eq}$ is the *total or equivalent resistance* of this circuit, R$_S$ stands for the resistance of *one* of the branch resistors, and N stands for the number of resistors connected in parallel.

Remember, this formula may be applied only when all of the branch resistors are of the same value.

$$R_{eq} = \frac{R_S}{N}$$

R$_{eq}$ = EQUIVALENT RESISTANCE

R$_S$ = SAME VALUE RESISTORS

N = NUMBER OF RESISTORS

Figure 6.34

Calculation of R$_{eq}$ — Using this simpler formula, you can go back and recalculate the total or equivalent resistance of the three 3-kilohm resistors in parallel. When the values are substituted in the formula, R$_S$ equals 3 kilohms and N equals 3. It is then pretty simple to calculate that the equivalent resistance; R$_{eq}$ does equal 1 kilohm. A new term, equivalent resistance, has been introduced. Equivalent resistance may be new in this discussion but the idea is not. In previous discussion on series circuits, it was said that often the terms equivalent resistance and total resistance would be used interchangeably. In general both these terms are used to mean the value of one single resistor that could be used to replace a more complex connection of several resistors.

For example in Figure 6.35, it has been calculated that all the resistors connected in parallel have an equivalent resistance of 1 kilohm. So as far as a power supply would be concerned, there is only 1 kilohm of resistance connected to it. It would make no difference whether there were actually three parallel branches with 3 kilohms per branch or a single 1-kilohm resistor connected to the circuit. Each of the circuits' total resistance and total current will be the same, if the same voltage were applied to each of these two circuits.

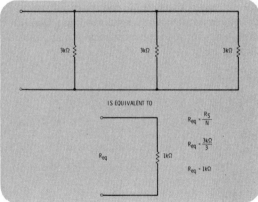

Figure 6.35

Parallel Circuit with Three Unequal Resistances —
Figure 6.36 shows a parallel circuit with three
branches, each having a different resistance. The
first branch contains a 25-kilohm resistor; the
second branch, a 24-kilohm resistor; and the third
branch, a 30-kilohm resistor. The values of these
resistors are all relatively close to one another. As
previously discussed, when three resistors of equal
value are connected in parallel, the total resistance
is one-third the value of one resistive branch. In
this case, the total resistance can be estimated by
finding one-third of a value close to 25 kilohms.
An easier number to work with is 24. One-third of
24 kilohms is 8 kilohms, so the estimate of the
total resistance is about 8 kilohms. In fact, if the
actual total resistance of this circuit were found
with the reciprocal formula, it would be
8.7 kilohms. Thus, keeping this short-cut concept
of handling equal value resistors in mind, you can
easily *estimate* the total resistance of many parallel
circuits.

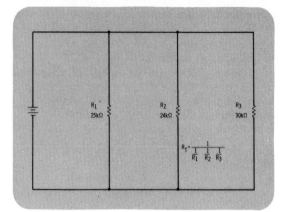

Figure 6.36

TI-50 Calculator — The total or equivalent
resistance of the circuit in Figure 6.36 can also be
found easily with the use of an electronic
calculator, such as the TI-50 shown in Figure 6.37.
There are other calculators that could be used
for the calculation. With some of these it
may be necessary to write down some of the
intermediate results before obtaining the final
answer.

    Keep in mind, however, that the procedures
used will be similar on any calculator of this type.
The circuit has three parallel resistors, a 25-kilohm,
a 24-kilohm, and a 30-kilohm resistor. The total
resistance of this circuit can be found easily with

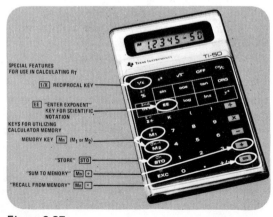

Figure 6.37

the use of the calculator, and the reciprocal formula for total resistance of a parallel circuit. The procedure used will be to find the reciprocal of each branch resistance, and then add these reciprocals together. For the final answer, take the reciprocal of the sum.

To perform this calculation you will be using special features of the calculator:

|EE| key — To enter exponents or numbers expressed in scientific notation

|1/x| key — (Reciprocal key) To take reciprocals of resistances where needed

|Mn||STO| — (Store key sequence) To store intermediate results in one of the calculator's memorys

|Mn|| + | — ("Sum memory" key sequence) To add intermediate results into one of the calculator memorys

|Mn|| = | — (Recall key sequence) To recall results stored in one of the calculator's memorys.

If your calculator doesn't have all these special features, you will have to keep track of some steps in this calculation manually. Note, however, if your calculator does not have an |EE| key (enter exponent key or some other scientific notation entry system), all numbers must be entered into the calculator in decimal form. This could prove a severe limitation in solving some problems in electricity and electronics. When purchasing a calculator for use in electrical applications, it is strongly suggested that you purchase one that allows you to handle numbers in scientific notation, even if other special features are not included.

To begin solving this problem, substitute the resistor values into the formula as shown in Figure 6.38. Notice that these resistance values are already converted into powers of ten form. Start solving the problem by finding the reciprocal of $25 \times 10^{+3}$. In the figures that follow, each operation will be tabulated for you. *Numbers* that are keyed in will be listed under the *Enter* column; *operation keys* to be pressed are listed under the *Press* column. Displayed results are included in the *Display* column. Before beginning, be sure to *clear* the calculator and its memory.

$$R_T = \frac{1}{1/R_1 + 1/R_2 + 1/R_3 \dots}$$

$$R_T = \frac{1}{\dfrac{1}{25 \times 10^{+3}} + \dfrac{1}{24 \times 10^{+3}} + \dfrac{1}{30 \times 10^{+3}}}$$

Figure 6.38

Calculation of Reciprocal — As shown in Figure 6.39, enter 25. To use the scientific notation feature of the calculator, press the $\boxed{EE}$ or "enter exponent" key. Then enter 3, which is displayed on the right as the power of ten. The display now represents $25 \times 10^{+3}$. Next, press the reciprocal key, $\boxed{1/x}$. This yields 4×10^{-5}, which is the reciprocal of R_1 expressed in scientific notation.

This number must somehow be kept so that when the other reciprocals are found, they can be added to it. As mentioned earlier, at this point, you could write down this reciprocal as an intermediate answer and add the other reciprocals

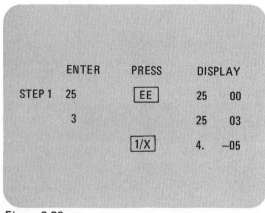

	ENTER	PRESS	DISPLAY	
STEP 1	25	$\boxed{EE}$	25	00
	3		25	03
		$\boxed{1/X}$	4.	−05

Figure 6.39

to it by hand. However, calculators such as the TI-50 have a feature called the "memory" which can store intermediate answers for you.

Store Result in Memory — On calculators with a memory feature, the key labeled $\boxed{\text{STO}}$, is the *"store"* key which is highlighted in Figure 6.40. When the desired memory key ($\boxed{\text{Mn}}$) and this key are pressed, the calculator electronically "writes down" the displayed number in the proper memory. This operation is called "storing a number in the memory." The $\boxed{\text{Mn}}$ $\boxed{=}$ or recall key sequence is used to recall the number that is stored in that particular memory. When these keys are pressed, the number in the memory appears in the display. A number in the display can be *added* to the contents of memory $\boxed{\text{Mn}}$ by pressing $\boxed{\text{Mn}}$ $\boxed{+}$. The key sequence $\boxed{\text{Mn}}$ $\boxed{-}$ is used to *subtract* the number in the display from the contents of memory $\boxed{\text{Mn}}$

Figure 6.40

To proceed with the problem, first *store* the reciprocal of R_1, which is still in the display. To do this, press [M1] [STO] as indicated in Figure 6.41 and you are ready to find the reciprocal of R_2.

Enter 24, press the [EE] key, and enter 3 for the exponent. The display now represents $24 \times 10^{+3}$. Next press the reciprocal key. The calculator should now display the reciprocal of R_2 expressed in scientific notation. Next press [M1] [+] to add the number to the contents of Memory 1. If [M1] [=] is pressed at this point, you can see that the calculator has indeed added the reciprocal of R_1 to the reciprocal of R_2.

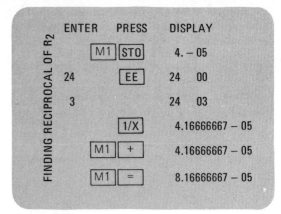

	ENTER	PRESS	DISPLAY
		[M1] [STO]	4. – 05
	24	[EE]	24 00
	3		24 03
		[1/X]	4.16666667 – 05
		[M1] [+]	4.16666667 – 05
		[M1] [=]	8.16666667 – 05

FINDING RECIPROCAL OF R_2

Figure 6.41

To continue, find the reciprocal of R_3 as shown in Figure 6.42. Enter 30, press the [EE] key, enter 3, and the calculator displays $30 \times 10^{+3}$. Press the reciprocal key and the calculator now displays the reciprocal of R_3 in scientific notation.

To add this reciprocal to the others, simply press [M1] [+]. Next press [M1] [=], and the sum of the three reciprocals is 1.15×10^{-4}. Remember, in order to get the total resistance, you must find the *reciprocal* of this sum. Press the reciprocal key once more to get 8.695, etc., times 10^{+3}. This can be rounded off and expressed in abbreviated form as 8.7 kilohms. If you recall, the original estimate of 8 kilohms for the total resistance of this circuit was fairly close.

A word of caution, many calculators are available with an extremely wide variety of special features. On some calculators, intermediate results will have to be written down while others, such as the TI-50, can allow you, if you are careful, to perform calculations such as this with even fewer steps than were outlined. Refer to your Owner's Manual to learn all you can about the use of your calculator.

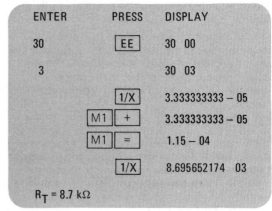

ENTER	PRESS	DISPLAY
30	[EE]	30 00
3		30 03
	[1/X]	3.333333333 – 05
	[M1] [+]	3.333333333 – 05
	[M1] [=]	1.15 – 04
	[1/X]	8.695652174 03

$R_T = 8.7 \text{ k}\Omega$

Figure 6.42

- **Product Over Sum Technique**
- **Incorrect Use of a Formula**
- **Two-Resistor Parallel Circuit Problem**

Product Over Sum Technique — One easier method that may be used to calculate the total resistance of a parallel circuit *containing only two resistors* is the use of the formula shown in Figure 6.43. Recall that any time two quantities are multiplied together, the result is called a "product," and any time two quantities are added together, the result is called a "sum." Thus, "the product-over-the-sum" formula for total resistance in a parallel circuit in Figure 6.43 is a product over a sum.

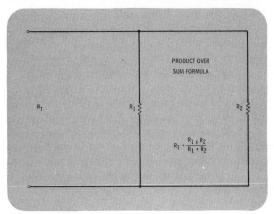

Figure 6.43

Incorrect Use of a Formula — It is important to remember that this formula can be used most effectively with only *two* resistances in parallel. If you try to extend this formula to cover three or more resistors in parallel, as shown in Figure 6.44, the result will be incorrect. You *cannot* extend this formula as shown in Figure 6.44 to cover the product of three resistances in parallel divided by the sum of three resistances in parallel.

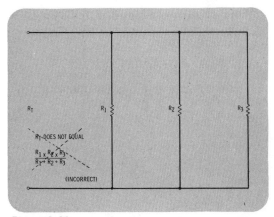

Figure 6.44

Two-Resistor Parallel Circuit Problem — This formula can be most easily used in solving for the total resistance of a parallel circuit containing two resistive branches, such as the circuit shown in Figure 6.45. R_1 is 7.5 kilohms, and R_2 is 16 kilohms

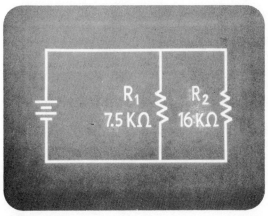

Figure 6.45

6-25

Product Over Sum Calculation — First, write down the formula as shown in Figure 6.46. (Remember the *product* term, the multiplication, is on the top of this formula.)

Next, substitute the appropriate resistance values into the formula as shown. This gives you $7.5 \times 10^{+3}$ *times* $1.6 \times 10^{+4}$ over $7.5 \times 10^{+3}$ *plus* $1.6 \times 10^{+4}$.

Next, carry out the required arithmetic. Begin by adding the values on the bottom of the fraction. Recall that in order to do this you must make sure that the powers of ten of the two terms involved are the same.

$$R_T = \frac{R_1 \times R_2}{R_1 + R_2}$$

$$R_T = \frac{(7.5 \times 10^{+3}) \times (1.6 \times 10^{+4})}{(7.5 \times 10^{+3}) + (1.6 \times 10^{+4})}$$

Figure 6.46

In the bottom of Figure 6.47, the $7.5 \times 10^{+3}$ has been left alone, but the $1.6 \times 10^{+4}$ has been changed to $16 \times 10^{+3}$.

These two numbers may be added to give $23.5 \times 10^{+3}$ in the bottom of this equation.

Putting this number in correct powers of ten form, you have $2.35 \times 10^{+4}$, as shown in Figure 6.47.

$$R_T = \frac{(7.5 \times 10^{+3}) \times (1.6 \times 10^{+4})}{(7.5 \times 10^{+3}) + (16 \times 10^{+3})}$$

$$R_T = \frac{(.5 \times 10^{+3}) \times (1.6 \times 10^{+4})}{23.5 \times 10^{+3}}$$

$$R_T = \frac{(7.5 \times 10^{+3}) \times (1.6 \times 10^{+4})}{2.35 \times 10^{+4}}$$

Figure 6.47

Moving on in the calculation, multiply the two numbers on the top to get $1.2 \times 10^{+8}$ as shown in Figure 6.48.

At this point, the two numbers can be divided, recalling the rules for powers of ten, to obtain the final result, $5.106 \times 10^{+3}$.

When this number is correctly rounded off and put in metric prefixed form, using the metric chart if necessary, you arrive at the final answer, 5.11 kilohms.

$$R_T = \frac{1.2 \times 10^{+8}}{2.35 \times 10^{+4}}$$

$$R_T = 5.106 \times 10^{+3}$$

$$R_T = 5.11 \text{ k}\Omega$$

Figure 6.48

Three Ways to Find Total Resistance of a Parallel Circuit — At this point in the course you should have acquired the knowledge and skills necessary to calculate the total resistance of any parallel circuit. As shown in Figure 6.49, three methods of performing this calculation have been discussed. One method is the *reciprocal technique*, which may be used in calculations involving any number of resistances in parallel. The second method listed is the *sum-over-sum technique*, which may be used only when considering two parallel resistances at a time. The third method shown is the special shortcut formula that may be used only if all the resistive branches being considered have the same ohmic value.

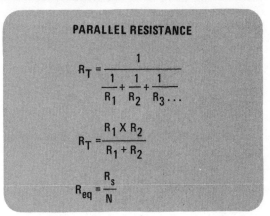

PARALLEL RESISTANCE

$$R_T = \cfrac{1}{\cfrac{1}{R_1} + \cfrac{1}{R_2} + \cfrac{1}{R_3} \ldots}$$

$$R_T = \cfrac{R_1 \times R_2}{R_1 + R_2}$$

$$R_{eq} = \cfrac{R_s}{N}$$

Figure 6.49

Notice that of the three methods listed, the reciprocal method is the most powerful. You can *always* use it to calculate the total resistance in any parallel circuit. As mentioned, the other two formulas are short cuts that can be used in special cases. For an additional look at how these formulas work, three additional examples are included here.

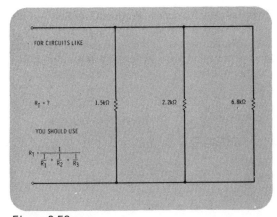

Figure 6.50

Three Resistive Branches - Reciprocal Formula — For a circuit such as that shown in Figure 6.50, with three unequal value resistors in parallel, the reciprocal formula should be used. The calculation would proceed as follows:

$$R_T = \frac{1}{1/R_1 + 1/R_2 + 1/R_3}$$

$$R_1 = 1.5 \text{ k}\Omega \quad R_2 = 2.2 \text{ k}\Omega \quad R_3 = 6.8 \text{ k}\Omega$$

Substitute

$$R_T = \frac{1}{1/1.5 \text{ k}\Omega + 1/2.2 \text{ k}\Omega + 1/6.8 \text{ k}\Omega}$$

Find the reciprocal

$$1/1.5 \text{ k}\Omega = 1/1.5 \times 10^{+3} = 6.67 \times 10^{-4}$$
$$1/2.2 \text{ k}\Omega = 1/2.2 \times 10^{+3} = 4.55 \times 10^{-4}$$
$$1/6.8 \text{ k}\Omega = 1/6.8 \times 10^{+3} = 1.47 \times 10^{-4}$$

Substituting reciprocals

$$R_T = \frac{1}{6.67 \times 10^{-4} + 4.55 \times 10^{-4} + 1.47 \times 10^{-4}}$$

Add up reciprocals in bottom of equation

$$R_T = \frac{1}{12.69 \times 10^{-4}}$$

Take reciprocal

$$R_T = 7.88 \times 10^{2}$$
$$R_T = 788 \ \Omega$$

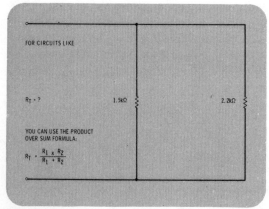

Figure 6.51

Two Resistive Branches - Product Over Sum — For a circuit such as that shown in Figure 6.51, with two unequal resistive branches in parallel, you can use the product-over-sum formula as a shortcut in calculating R_T. The calculation would proceed as follows:

$$R_T = \frac{R_1 \times R_2}{R_1 + R_2}$$

$$R_1 = 1.5 \text{ k}\Omega \quad R_2 = 2.2 \text{ k}\Omega$$

Substitute

$$R_T = \frac{1.5 \text{ k}\Omega \times 2.2 \text{ k}\Omega}{1.5 \text{ k}\Omega + 2.2 \text{ k}\Omega}$$

Convert to powers of ten

$$R_T = \frac{1.5 \times 10^{+3} \times 2.2 \times 10^{+3}}{1.5 \times 10^{+3} + 2.2 \times 10^{+3}}$$

Multiply top
Add Bottom

$$R_T = \frac{3.3 \times 10^6}{3.7 + 10^3}$$

Divide

$$R_T = \frac{3.3 \times 10^{+6 -3}}{3.7}$$

$$R_T = 0.892 \times 10^{+3}$$

$$R_T = 892 \ \Omega$$

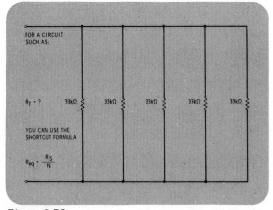

Figure 6.52

Several Equal Value Resistive Branches — For a circuit such as that shown in Figure 6.52 with several equal value resistive branches, you can use the shortcut formula $R_{eq} = R_S/N$ to calculate the total or equivalent resistance.

$$R_{eq} = \frac{R_S}{N}$$

All resistor values equal 33 kilohms. There are *five* resistors altogether.

Substitute

$$R_{eq} = \frac{33 \text{ k}\Omega}{5}$$

$$R_{eq} = \frac{33 \times 10^{+3}}{5}$$

Divide

$$R_{eq} = 6.6 \times 10^{+3}$$
$$R_{eq} = 6.6 \text{ k}\Omega$$

Power in Series and Parallel Circuits — Before completing this lesson, *power* in parallel circuits is discussed. Remember from a previous discussion about power in a series circuit, that *the powers dissipated by each individual resistance are simply added* to find the total power dissipated by the series circuit. As it turns out, this same procedure also applies to parallel circuits. If there were five resistive branches in a parallel circuit and each was dissipating 1 watt of power, the *total power* the circuit was dissipating would be 5 watts as shown in Figure 6.53. *The individual power dissipations of all resistors are added to find the total power dissipated.*

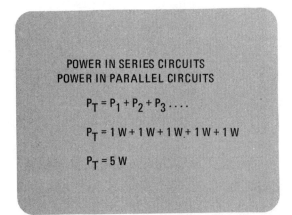

POWER IN SERIES CIRCUITS
POWER IN PARALLEL CIRCUITS

$$P_T = P_1 + P_2 + P_3 \ldots$$

$$P_T = 1 \text{ W} + 1 \text{ W} + 1 \text{ W} + 1 \text{ W} + 1 \text{ W}$$

$$P_T = 5 \text{ W}$$

Figure 6.53

A Typical Problem — Consider a typical related problem that technicians run across from time to time as shown in Figure 6.54. Suppose that you need a 500-ohm resistor and it must have a power dissipation rating of 2 watts. If the only resistors you have on hand are 1-kilohm resistors rated at 1 watt, how do you arrange these to obtain 500 ohms of resistance rated at 2 watts?

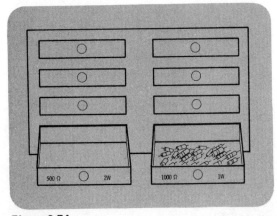

Figure 6.54

The answer is quite simple, all you have to do is connect two 1-kilohm resistors in parallel with one another and that gives you 500 ohms of resistance as shown in Figure 6.55. Remember that anytime two resistors of the same value are connected in parallel with one another, the total resistance is *half* the value of one of the resistors. Half of 1 kilohm is 500 ohms. Since the power rating of each resistor is 1 watt, the total power rating now is 2 watts. (Note that this simple analysis is true only when two *equal value* resistors are being used in parallel to replace a third resistor.)

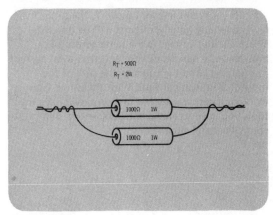

Figure 6.55

Summary — This lesson has introduced parallel circuits and the three laws governing their behavior (Figure 6.56). However, only one of these laws has been discussed at length in this lesson, that the total resistance of a parallel circuit is always less than or approximately equal to the smallest resistive branch. From this law, three formulas allowing you to calculate the total or equivalent resistance of parallel circuits in a variety of situations were developed. Since knowledge of how these formulas may be used is of such importance, most of this lesson was devoted to them.

The idea of power in a parallel circuit has also been discussed. Recall that power in a parallel circuit is handled the same as in a series circuit. Simply add all the individual powers together to find the total power the circuit is dissipating.

It is important for you to really understand the concept of total resistance in a parallel circuit and to be able to calculate it. Thus, several practice problems or circuits at the end of this lesson will allow you to use your new skills. Please try them all.

The worked out problems included will serve as a guide as you begin working with the different circuit situations. At the end of the lesson (as for all lessons) a short quiz is included to let you know how well you are doing.

CONCEPTS COVERED IN THIS LESSON

● RESISTANCE IN PARALLEL CIRCUITS
  – 3 FORMULAS
  – HOW TO USE EACH
● POWER IN PARALLEL CIRCUITS

Figure 6.56

In your next lesson, Lesson 7, the other two laws governing parallel circuits will be discussed. These are that voltage is the same across all branches of a parallel circuit; and that total current in a parallel circuit is equal to the sum of the individual branch currents. With these two laws working together with the first, you will have the tools you need to analyze, troubleshoot, and control parallel circuits.

LESSON 6. INTRODUCTION TO PARALLEL CIRCUITS

• **Worked Through Examples**

1. Find the total resistance of the following circuit.

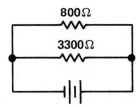

There are two options that may be taken to find R_T. The product-over-sum formula or the sum of the reciprocal formula. This first example will use the product-over-sum formula:

$$R_T = \frac{R_1 \times R_2}{R_1 + R_2}$$

First, substitute the circuit values in correct powers of ten form.

$$R_T = \frac{8.0 \times 10^2 \times 3.3 \times 10^3}{8.0 \times 10^2 + 3.3 \times 10^3}$$

To *add*, the exponents of the numbers in the denominator (or bottom) of this equation must be the same. Changing the $3.3 \times 10^{+3}$ to $33.0 \times 10^{+2}$, you have

$$R_T = \frac{8.0 \times 10^2 \times 3.3 \times 10^3}{8.0 \times 10^2 + 33.0 \times 10^2}$$

Add

$$R_T = \frac{8.0 \times 10^2 \times 3.3 \times 10^3}{41 \times 10^2}$$

Multiply the numbers on top. (Remember to *add* the exponents when multiplying.)

$$R_T = \frac{26.4 \times 10^5}{41 \times 10^2}$$

Now you may divide 26.4×10^5 by 41×10^2. (Remember to do this you bring the bottom exponent up above the division line and change its sign.)

$$R_T = \frac{26.4 \times 10^{5-2}}{41}$$

Then combine these top exponents

$$R_T = \frac{26.4 \times 10^3}{41}$$

$$R_T = 6.44 \times 10^2 = 644 \ \Omega$$

2. Find the total resistance of the circuit shown below.

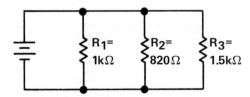

This time the reciprocal formula will be used to solve this problem.

First, substitute the circuit values into the formula:

$$R_T = \frac{1}{1/R_1 + 1/R_2 + 1/R_3}$$

$$R_T = \frac{1}{\dfrac{1}{1.0 \times 10^3} + \dfrac{1}{8.2 \times 10^2} + \dfrac{1}{1.5 \times 10^3}}$$

Find the *reciprocals* of the resistance values. (Divide the resistance value into 1.) This gives you the individual conductances which go into the bottom of this equation.

$$R_T = \frac{1}{1 \times 10^{-3} + 1.22 \times 10^{-3} + 6.67 \times 10^{-4}}$$

Add all individual conductances in the bottom of this equation. (Remember to change all exponents to the same number; here, 10^{-3}.)

$$R_T = \frac{1}{2.89 \times 10^{-3}}$$

Now divide 2.89×10^{-3} into 1 to find the total resistance.

$$R_T = 3.46 \times 10^2 = 346 \ \Omega$$

3. Find the approximate resistance of the circuit shown below. (Use the quickest method.)

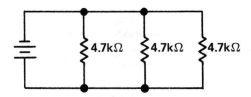

Since the three resistors are equally sized, the "shortcut" formula may be used.

$$R_{eq} = \frac{R_s}{N}$$

R_s = Same size resistor resistance (4.7 kΩ).

N = Number of resistors (3).

Substituting

$$R_{eq} = \frac{4.7 \ k\Omega}{3}$$

Change 4.7 kΩ to proper powers of ten notation

$$R_{eq} = \frac{4.7 \times 10^3}{3}$$

Divide

$$R_{eq} = 1.57 \times 10^3 = 1570 \ \Omega$$

4. Define the term "Branch."

A branch in an electrical circuit is simply a separate path through which electrical current can flow. In other words, a series circuit has only one branch. A parallel circuit has two or more branches.

5. Find the R_{eq} of the following circuit.

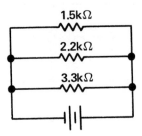

This problem will be worked using an SR-50 type calculator. The reciprocal formula will be used to solve the problem.

Enter the first number in correct powers of ten form.

1.5 $\boxed{\text{EE}}$ 3

Press the reciprocal key and store that number in the calculator's memory.

$\boxed{\text{1/X}}$ $\boxed{\text{STO}}$

Enter the other two numbers using the same procedure as outlined above except rather than pressing the "STO" key, press the $\boxed{\Sigma}$ key which adds the displayed number to the number held in memory.

2.2 $\boxed{\text{EE}}$ 3 $\boxed{\text{1/X}}$ $\boxed{\Sigma}$

3.3 $\boxed{\text{EE}}$ 3 $\boxed{\text{1/X}}$ $\boxed{\Sigma}$

The reciprocals of all three numbers have been found and added together.

This number may be recalled by pressing the "RCL" key.

$\boxed{\text{RCL}}$

Now, this number must be divided into 1, so press the reciprocal key.

$\boxed{\text{1/X}}$

Your answer appears on the display.

7.021276596 02

This number is rounded to

7.02 X 10^2 or
702 Ω

LESSON 6. INTRODUCTION TO PARALLEL CIRCUITS

● **Practice Problems**

The key objective of this lesson has been achieved if you can calculate the total resistance of any basic parallel circuit. To gain some practice in this area, the problems below are provided. Fold over the page to check your answers.

Depending upon the approach you use to solve these problems and how you round off intermediate results, your answers may vary slightly from those given here. However, any differences you encounter should only occur in the third significant digit of your answer. If the first two significant digits of your answers do not agree with those given here, recheck your calculations.

Find R_T for each of the following circuits.   **Fold Over**

1.

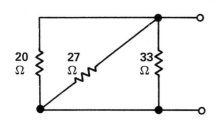

$R_T =$ _____

2.

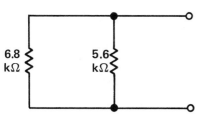

$R_T =$ _3.07 KΩ_

3.

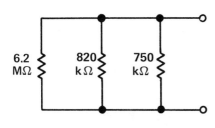

$R_T =$ _____

4.

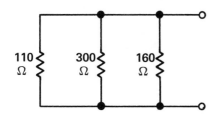

$R_T =$ _____

Answers

1.   8.52 Ω

2.   3.07 kΩ

3.   368 kΩ

4.   53.5 Ω

5.

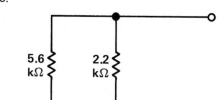

Fold Over

R_T = _____

6.

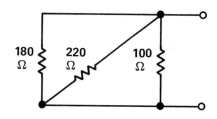

R_T = _____

7.

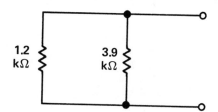

R_T = _____

8.

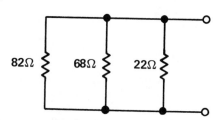

R_T = _____

9.

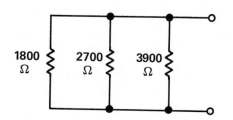

R_T = _845 Ω_

Answers

5.　1.58 kΩ

6.　49.7 Ω

7.　918 Ω

8.　13.8 Ω

9.　846 Ω

Fold Over

10.

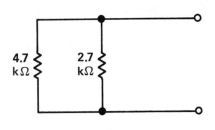

$R_T =$ _____

11.

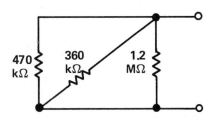

$R_T =$ _____

12.

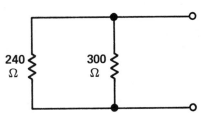

$R_T =$ _____

13.

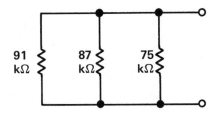

$R_T =$ _____

14.

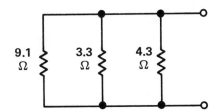

$R_T =$ 1.55 Ω

Answers

10. 1.71 kΩ

11. 174 kΩ

12. 133 Ω

13. 27.9 kΩ

14. 1.55 Ω

15.

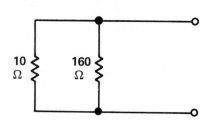

R_T = _____

Fold Over

16.

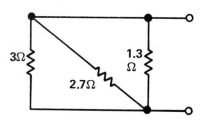

R_T = _____

17.

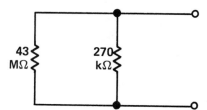

R_T = _____

18.

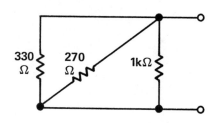

R_T = _____

19.

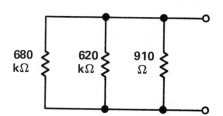

R_T = _____

Answers

15. 9.41 Ω

16. 679 mΩ

17. 268 kΩ

18. 129 Ω

19. 907 Ω

20.

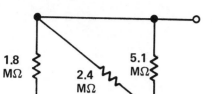

Fold Over

$R_T =$ _____

21.

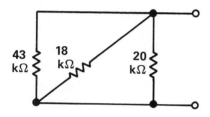

$R_T =$ _____

22.

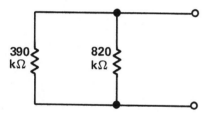

$R_T =$ _____

23.

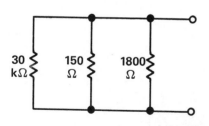

$R_T =$ _____

24.

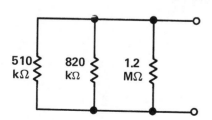

$R_T =$ _____

Answers

20. 856 kΩ

21. 7.76 kΩ

22. 264 kΩ

23. 138 Ω

24. 249 kΩ

25.

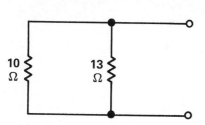

$R_T =$ _____

26.

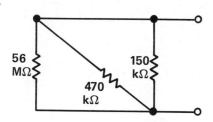

$R_T =$ _____

27.

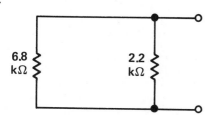

$R_T =$ _____

28.

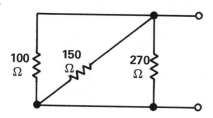

$R_T =$ _____

29.

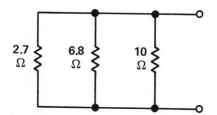

$R_T =$ _____

Fold Over

Answers

25. 5.65 Ω

26. 113 kΩ

27. 1.66 kΩ

28. 49.1 Ω

29. 1.62 Ω

30.   **Fold Over**

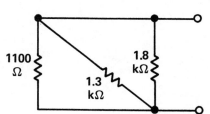

$R_T =$ _____

• **Practice Problems**

Answers

30.   448 Ω

1. A circuit which has two or more current paths is a_____circuit.
   a. Linear
   b. Series
   c. Open
   d. Parallel

2. The total voltage of a parallel circuit is_____across each branch of that circuit.
   a. Unlike
   b. Totally different
   c. Equal to zero
   d. The same

3. The total current in a parallel circuit is equal to_____of the individual branch currents.
   a. Each
   b. The difference
   c. The sum
   d. Opposite

4. The total resistance in a parallel circuit is_____or approximately_____the value of the smallest resistive branch.
   a. Always less, equal to
   b. Always greater, equal to
   c. Always equal, less than
   d. Sometimes greater, equal to

5. _____is the ability to conduct current.
   a. Resistance
   b. EMF
   c. Conductance
   d. Reliability

6. $R_T = \dfrac{1}{\dfrac{1}{R_1} + \dfrac{1}{R_2} + \dfrac{1}{R_3} \cdots}$ is the equation used to calculate the total resistance in:
   a. A parallel circuit
   b. A parallel circuit with more than two branches
   c. A series circuit
   d. A circuit with unequal values of resistors
   e. a and b above

7. Conductance can be defined as:
   a. The opposite of resistance
   b. The reciprocal of resistance
   c. Resistance in series
   d. a and b above
   e. None of above

8. $R_T = \dfrac{R_S}{N}$ . . . can be used to calculate the total resistance in a parallel circuit when:
   a. The current is flowing backwards
   b. All resistors are in series in each branch
   c. R_S is the same in all N branches
   d. N is a large number

9 Conductance is indicated by a capital G and has the unit:
   a. ampere
   b. mho
   c. ohm
   d. volt
   e. coulomb

10. The product over sum equation for total resistance in a parallel circuit is:
    a. $R_T = R1 \times R2$
    b. $R_T = R1 + R2$
    c. $R_T = \dfrac{R1R2}{R1 + R2}$
    d. $R_T = \dfrac{R1 + R2}{R1}$

11. The equation $R_T = \dfrac{R1R2}{R1 + R2}$ is used to calculate total resistance in a parallel circuit with:
    a. Two branches
    b. All resistors in series
    c. All branches open
    d. No voltage in the circuit

12. The total power dissipated in a circuit with at least one battery is equal to the sum of the power dissipated in each resistor in:
    a. A parallel circuit
    b. A series circuit
    c. One branch of the circuit
    d. A closed circuit
    e. a, b and d above
    f. a above only

13. Solve for R_T in this circuit:

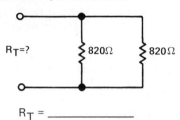

R_T = _____

14. Solve for R_T in this circuit:

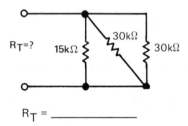

R_T = _____

15. Solve for R_T in this circuit:

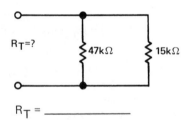

R_T = _____

16. Solve for R_T in this circuit:

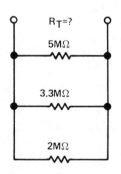

R_T = _____

17. Solve for R_T in this circuit:

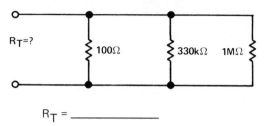

R_T = _____

18. Solve for R_T in this circuit:

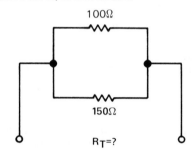

R_T = _____

19. Solve for R_T in this circuit:

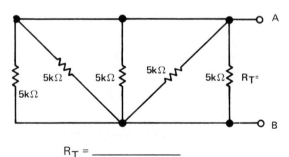

R_T = _____

20. If a 100 volt battery is connected to points A and B above, what current will flow and what power is dissipated?
   a. 10A, 1KW
   b. 0.1A, 10W
   c. 100A, 10KW
   d. 40A, 4KW

Lesson 7

Parallel Circuit Analysis

This lesson continues the discussion of parallel circuits with an examination of the laws governing the behavior of voltage and current in parallel circuits. The differences between series circuits and parallel circuits are well illustrated with a demonstration of each type of circuit operation. The lesson concludes by showing the calculations necessary to find voltage, current, and resistance values in several sample parallel circuits.

LESSON 7. PARALLEL CIRCUIT ANALYSIS

● Objectives

1. *Write* an explanation of the essential difference between series and parallel circuits as far as voltage behavior is concerned.

    Write the three laws describing the behavior of parallel circuits and also the formula for each.

2. Given two 12-volt batteries or power supplies, *sketch* a schematic of how they would be wired to produce a single 12-volt supply with greater current output than either single supply.

3. Given a schematic diagram of the type shown:

    a. *Write* down the voltage across points A-B, C-D, E-F.

    b. *Calculate* the current flowing in each branch of this circuit (A-B, C-D, E-F).

    c. *Calculate* the total resistance of this circuit, R_T.

    d. *Calculate* the total current of this circuit, I_T.

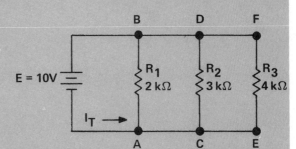

4. Given a schematic diagram of the type shown:

    a. *Calculate* the applied voltage and voltage across each branch.

    b. *Calculate* R_1.

    c. *Calculate* the current in the branch containing the 3-kilohm resistor and the total current I_T.

    d. *Calculate* the total resistance of the circuit.

    e. *Calculate* the total power dissipated by this circuit.

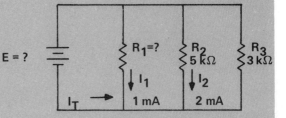

5. *Perform* any of the calculations described above using a calculator with memory such as the TI-30, TI-50 or TI-55.

- **Parallel Circuit Laws**
- **Series Circuit with One Bulb**
- **Series Circuit with Three Bulbs**

Parallel Circuit Laws — This lesson continues the discussion of parallel circuits. As you remember from the last lesson, *total resistance* within a parallel circuit was discussed. It was shown that *the total resistance of a parallel circuit is always less than, or approximately equal to, that of the smallest resistive branch* (Figure 7.1). This is just one of the three laws governing the behavior of parallel circuits, and it was discussed in depth in Lesson 6.

This lesson now goes on to examine the other two laws and how they are used in analyzing parallel circuits. *First, voltage is the same across all branches of a parallel circuit, and second, the total or "main line" current in a parallel circuit is equal to the sum of the individual branch currents.*

Series Circuit with One Bulb — In order to actually see the voltage behavior of series and parallel circuits in action and to allow you to compare the two, consider the operation of several basic demonstration circuits. The first one (Figure 7.2) is a *series circuit* containing one light bulb, a switch, and a 6-volt power supply. Switches are simply devices which are used to easily complete (close or "make") or open (break) a circuit. The details and mechanics of switches and their use are covered in a laboratory exercise in this series. When the switch is closed, the circuit is completed and the light bulb burns brightly.

Series Circuit with Three Bulbs — However, if three more light bulbs are added in series (Figure 7.3) and power is applied, the light bulbs will burn very dimly. Why? Because each of the bulbs requires 6 volts across it to burn brightly, but the voltage across each bulb is now actually much less than 6 volts, 1.5 volts to be exact. As was mentioned in an earlier lesson, in a *series circuit* the sum of the individual voltage drops in the circuit equals the total applied voltage. In this series circuit when the three bulbs are added, the original applied voltage of 6 volts has not changed, but it is now divided among four bulbs so the bulbs burn very dimly. (Each bulb now receives only 25% of the voltage it needs to operate normally.) If the lights in your

PARALLEL CIRCUIT LAWS

1. THE TOTAL RESISTANCE OF A PARALLEL CIRCUIT IS ALWAYS LESS THAN OR APPROXIMATELY EQUAL TO THE SMALLEST RESISTIVE BRANCH.

2. THE VOLTAGE IS THE SAME ACROSS ALL BRANCHES OF A PARALLEL CIRCUIT.

3. THE TOTAL OR "MAIN LINE" CURRENT IN A PARALLEL CIRCUIT IS EQUAL TO THE SUM OF THE INDIVIDUAL BRANCH CURRENTS.

Figure 7.1

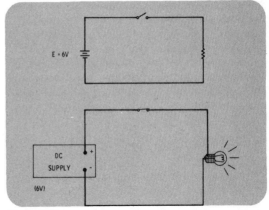

Figure 7.2

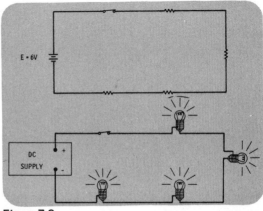

Figure 7.3

house were wired in series, as you turned on more
and more lights, they would burn dimmer and
dimmer.

Series String of Lights — Another effect seen in a
series-wired circuit is that if the circuit is open at
any point, no current can flow anywhere in the
circuit. An example of this is the old-style
Christmas tree lights that were all wired in series.
When one light bulb burned out, they all went out.
The burned out bulb created an open circuit, and
current cannot pass through an open circuit.

Figure 7.4

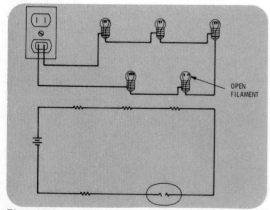

Figure 7.5

In a series circuit connection (Figure 7.5),
each load (bulb, resistor, etc.) forms a part of the
whole path for current. This same current flows
through every load. If any of these loads should
fail in such a way as to become *open*, the entire
operation of the circuit ceases. ("If one burns out,
they all go out.")

Parallel Lights — However, most modern Christmas tree lights are wired in parallel so that one light can burn out without the whole string of lights going out (Figure 7.6).

Figure 7.6

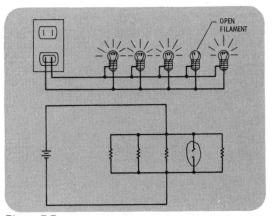

Figure 7.7

In a parallel circuit (Figure 7.7), the full supply voltage is applied to each load and current from the power source is divided between the loads. No one single load carries the entire circuit current. If one of several parallel loads should open, current would still flow in the rest of the circuit since there are still one or more complete paths for current flow through the other loads. ("If one burns out, the rest will stay lit.")

Parallel Circuit — The behavior of voltages and currents in parallel circuits can be illustrated with another simple demonstration circuit. Figure 7.8 shows the same four light bulbs used before, but this time they are connected in parallel across the 6-volt supply. A switch has also been placed in each branch. If each switch is closed, the bulb in the branch with that switch will burn brightly. In a circuit such as this, more light bulbs can be turned on or off without affecting the brightness of the others.

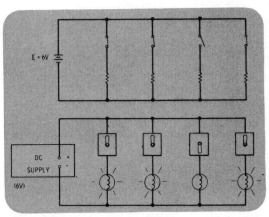

Figure 7.8

Voltage Measurement in Parallel Circuit — If a voltmeter were actually used to measure the voltage across each branch, the voltage measured would be very nearly the *same* across all branches of the circuit (Figure 7.9). Allowing for slight variations in voltage due to resistance in wires and connections, the voltage across the first branch would measure 6 volts, the second branch 6 volts, the third branch 6 volts, and the fourth branch 6 volts. The reason for this is that all the lights are essentially connected directly across the 6-volt source. *This is why the voltage is said to be the same across all branches of a parallel circuit.*

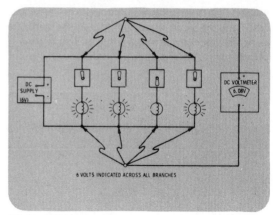

Figure 7.9

Parallel Wiring for Home Use — The fact that the voltage is the same across all branches of a parallel circuit is an advantage of parallel wiring over series wiring for home use (Figure 7.10). The appliances are wired to your house supply in parallel, many appliances may be plugged in and turned on or off, and they will all receive the same voltage and will not affect the operation of other appliances. Disconnecting any one appliance will not break the circuit supplying current to any other. For example, in your kitchen you can plug in a toaster, a coffee pot, and a mixer, and since they are all connected in parallel, it makes no difference whether one is on or all three are on. They all receive the voltage they require to operate properly.

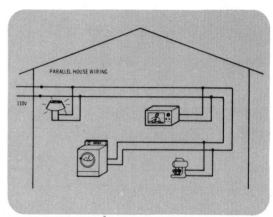

Figure 7.10

Voltage Drop Across Parallel Loads — Another way of considering the voltage situation in a parallel circuit is to consider that the wires from the battery to the loads just transmit the battery's potential difference to the load. The wires in any circuit generally represent *lines of equal potential.* Since it is usually considered that a wire has no resistance, there is no IR drop across a simple piece of wire. So all along the length of any unbroken conductor there should be no potential difference. (If there were a potential difference across a conducting wire, a "short circuit" situation would exist.) So in Figure 7.11 wire 1 is at the potential of the positive battery terminal all along its length right up to the loads. The difference of potential of

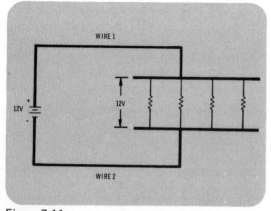

Figure 7.11

the battery terminal is 12 volts as shown. So the potential difference across each load is 12 volts.

Formula for Voltage in a Parallel Circuit — The voltage behavior of a parallel circuit can be simply summarized in formula form as shown in Figure 7.12. The total applied voltage, E_T, equals the voltage across each branch.

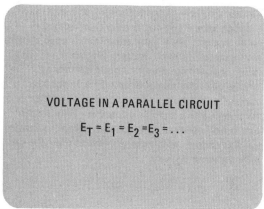

VOLTAGE IN A PARALLEL CIRCUIT

$$E_T = E_1 = E_2 = E_3 = \ldots$$

Figure 7.12

Parallel Voltage Sources — Thus far, only circuits with a single voltage source have been discussed. Consider now what effect two parallel voltage sources would have on a circuit.

As shown in Figure 7.13, when two equal batteries are connected in parallel with one another, the voltage is still the same across each branch, but the current that *can be supplied* to the circuit has increased. Notice that each battery has the capability of producing 1 amp of current, but that the total current that can be produced is 2 amps. In other words, this circuit with two identical batteries in parallel has *twice the current capability* of a similar circuit with only one battery. Thus, the current capability has been

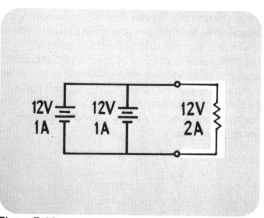

Figure 7.13

increased to 2 amps. This means that the batteries in parallel have the capability to safely deliver 2 amps if the resistance of the circuit is low enough to allow 2 amps of current flow. Note, that this does *not* mean that 2 amps flow in all cases. As discussed previously, the current flowing in a circuit and the total circuit resistance are *inversely* related. In this case, if the resistance across the batteries is made low enough, the current will increase to 2 amps. If the resistance is reduced further and more than 2 amps is drawn, the batteries may be damaged if each battery or power supply is listed as being able to safely deliver a maximum of 1 amp.

Jumper Cables — A good example of a common situation where two batteries are connected in parallel occurs when jumper cables are used to help start a car with a weak battery (Figure 7.14). The cables are used to connect a good battery in parallel with the weak battery. This is done by connecting the two positive terminals together and the two negative terminals together. This parallel connection does not increase the voltage, it simply supplies enough additional current to start the car with the weak battery.

Figure 7.14

Total Current in a Parallel Circuit — Now turn your attention to the final law governing parallel circuits which says that *the total current in a parallel circuit is equal to the sum of the individual branch currents*. When written mathematically as shown in Figure 7.15, this law states that I_T equals I_1 plus I_2 plus I_3. Remember the dots mean that the formula continues for as many branches as there are in the circuit.

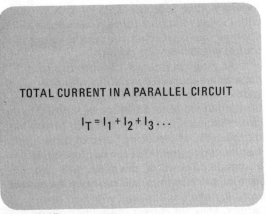

TOTAL CURRENT IN A PARALLEL CIRCUIT

$$I_T = I_1 + I_2 + I_3 \cdots$$

Figure 7.15

- Total Current = "Main Line" Current
- **Sample Circuit**
- **Calculation of I$_1$**

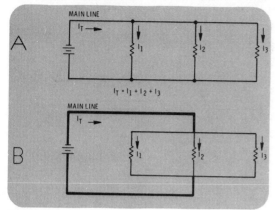

Figure 7.16

Total Current = "Main Line" Current — Another name often given to the total current in a parallel circuit is the *main line* current. The main line of the parallel circuit refers to those wires that carry the *total circuit current* out to the loads. In circuit A of Figure 7.16, the wires directly from the battery conduct the total current I$_T$ (equal to I$_1$ + I$_2$ + I$_3$) to the loads. The circuit may be redrawn as shown in circuit B to clarify the main line and the law governing current flow in parallel circuits.

Sample Circuit — To illustrate the mathematics of this law in action, consider the parallel circuit shown in Figure 7.17. The voltage, 10 volts, is the same across all three resistive branches. R$_1$ is 2 ohms, R$_2$ is 5 ohms, and R$_3$ is 10 ohms. Now follow through the steps necessary to calculate the *current flowing in each branch.*

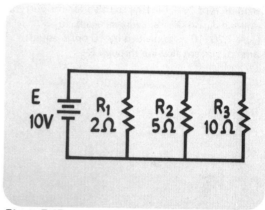

Figure 7.17

Calculation of I$_1$ — To solve for the amount of current in branch 1, use Ohm's law in the form I = E/R. As shown in Figure 7.18, E/R equals 10 volts (since there are 10 volts across each branch) divided by 2 ohms because R$_1$ = 2 Ω. Thus, there are 5 amps of current flowing through R$_1$.

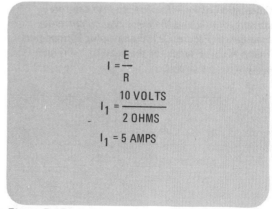

Figure 7.18

Calculation of I_2 — To solve for the current through R_2, use the same procedure. In Figure 7.19, using $I = E/R$ for this branch gives you 10 volts divided by 5 ohms or 2 amps of current flowing through R_2.

$$I = \frac{E}{R}$$

$$I_2 = \frac{10 \text{ VOLTS}}{5 \text{ OHMS}}$$

$$I_2 = 2 \text{ AMPS}$$

Figure 7.19

Calculation of I_3 — Finally, to solve for the current in branch 3, use Ohm's law once again. In Figure 7.20, 10 volts divided by 10 ohms equals 1 amp of current flowing through R_3.

$$I = \frac{E}{R}$$

$$I_3 = \frac{10 \text{ VOLTS}}{10 \text{ OHMS}}$$

$$I_3 = 1 \text{ AMP}$$

Figure 7.20

Calculation of I_T — *In order to find the total current, simply add the three branch currents together.* In Figure 7.21, 5 amps plus 2 amps plus 1 amp equals 8 amps for the total or main line current in this circuit.

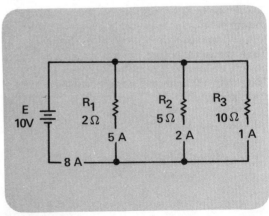

Figure 7.21

Circuit with More Common Resistance Values —
The simple example just covered was primarily
intended to quickly run through the procedures
involved in calculating the currents in a parallel
circuit. Consider a more realistic example showing
these calculations. Figure 7.22 shows a parallel
circuit with some more common resistance values.
In this example the total applied voltage and
resistance of each branch are known. To complete
the analysis of this circuit, first, the individual
branch currents will be calculated, then these will
be added together to get the total current, and
finally, the total circuit resistance will be
calculated. In this example you will be shown how
a calculator with memory, such as the TI SR-50 or
a similar model, can be used to make the
calculations easier. Remember, in the previous
lesson the memory functions were used to store,
add, and recall the reciprocals of the branch
resistances in a parallel circuit. In this example, the
memory will be used again, but this time to store,
add, and recall the branch currents. Again, note
that if your calculator is not equipped with a
memory, the branch currents calculated as
intermediate results can be written down until
needed.

As you analyze this circuit, keep in mind the
logic of your approach. What quantities do you
know? Which do you need? In a circuit situation
such as this, it's a good idea to *begin by calculating
the branch currents*, then these may be added
together to find the total current. Finally, the total
current can be divided into the total voltage to
determine the circuit's total resistance.

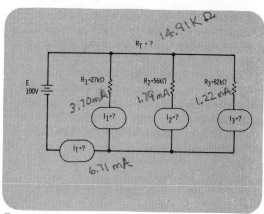

Figure 7.22

- **Calculator Operations**
- **Calculation of I_1**

Calculator Operations – Begin the problem by finding the current flowing in the first branch. The steps you will need on the calculator will be illustrated as the calculation proceeds. The calculator's steps will be listed for you in table form as they were in the previous lesson. Numbers to be entered will fall under the "Enter" column, operation keys to be pressed under the "Press" column, and displayed results under "Display."

CALCULATOR OPERATIONS

ENTER	PRESS	DISPLAY
LIST ACTUAL NUMBERS THAT MUST BE KEYED IN	LISTS OPERATION KEYS (+, –, X, ÷, 1/X, STO, RCL, Σ, ETC) TO BE PRESSED	LISTS CONTENTS OF DISPLAY AS CALCULATION PROCEEDS.

Figure 7.23

Calculation of I_1 – In order to find the first branch current, use Ohm's law in the form current equals voltage divided by resistance or I = E/R.

To begin the problem (Figure 7.24), first enter 1 for the 100 volts. Remember, to use scientific notation, press the [EE] ("enter exponent") key and then enter 2 for the exponent. The calculator should now display the number 100 in scientific notation. Next press the [÷] key. For the branch resistance of 27 kilohms, enter 27, press [EE], and enter 3. The calculator's display (27 03) now represents 27 X 10^{+3}. When the [=] key is pressed, the calculator performs the division, and displays 3.70, etc., times 10^{-3}. This can be rounded off and put in abbreviated form to get 3.7 milliamps of current through R_1.

To store this value of current in the memory so that it may be added to the other branch currents later, press the [STO] store key.

$I_{R1} = E/R_1 = 100\ V/27\ k\Omega$

ENTER	PRESS	DISPLAY
1	[EE]	1 00
2		1 02
	[÷]	1. 02
27	[EE]	27 00
3		27 03
	[=]	3.703703704 – 03
	IR_1 = 3.70 mA	
	[STO]	3.703703704 – 03

Figure 7.24

- Calculation of I_{R2}
- Calculation of I_{R3}

Calculation of I_{R2} — You are now ready to find the current through R_2 which has a resistance value of 56 kilohms (Figure 7.25). For the 100 volts, enter 1, press ⌷EE⌷ and enter 2. Then press the ⌷÷⌷ key, enter 56, press the ⌷EE⌷ key again and enter 3. The calculator is now ready to divide 100 volts by 56 kilohms. When the ⌷=⌷ key is pressed, the calculator performs the division and displays 1.78, etc., times 10^{-3}. This is the current through R_2 expressed in scientific notation. When this is rounded off and put in abbreviated form, it equals 1.79 milliamps of current through R_2.

To add this to the current through R_1, press the sum-to-memory key. (The labeling of this key will vary on different calculators. On calculators of the SR-50 type, the sum-to-memory key is labeled with a Greek letter sigma, Σ.) The sum of the two currents should now be stored in the memory.

$$I_{R2} = \frac{E}{R_2} = \frac{100}{56\,k\Omega}$$

ENTER	PRESS	DISPLAY
1	⌷EE⌷	1 00
2		1 02
	⌷÷⌷	1. 02
56	⌷EE⌷	56 00
3		56 03
	⌷=⌷	1.785714286 – 03
$I_{R2} = 1.79\ mA$		
	⌷Σ⌷	1.785714286 – 03

Figure 7.25

Calculation of I_{R3} — To find the current through R_3 (Figure 7.26), enter 1, press ⌷EE⌷, and enter 2 for the voltage across R_3. Then press the ⌷÷⌷ key, enter 82, press ⌷EE⌷ again and enter 3. The calculator is now ready to divide 100 volts by 82 kilohms. When the ⌷=⌷ key is pressed, the calculator performs the division and displays 1.21, etc., times 10^{-3}, which is the current through R_3 expressed in scientific notation. When this is rounded off and put in abbreviated form, it equals 1.22 milliamps of current.

$$I_{R3} = \frac{E}{R_3} = \frac{100\ V}{82\,k\Omega}$$

ENTER	PRESS	DISPLAY
1	⌷EE⌷	1 00
2		1 02
	⌷÷⌷	1. 02
82	⌷EE⌷	82 00
3		82 03
	⌷=⌷	1.219512195 – 03
$I_{R3} = 1.22\ mA$		

Figure 7.26

Calculation of Total Current — To find the total current (Figure 7.27), press the sum-to-memory key (Σ) once more, then press the $\boxed{\text{RCL}}$ key. Expressed in scientific notation, the total current is 6.70, etc., times 10^{-3}. Thus, the sum of the individual branch currents in the circuit is the total current of 6.71 milliamps. Note, do not turn off your calculator yet or you will lose what has been stored in the memory, if it has a memory. Proceed to the next calculation.

$$I_T = \underbrace{I_{R1} + I_{R2} + I_{R3}}_{\text{STORED}}$$

ENTER	PRESS	DISPLAY
	$\boxed{\Sigma}$	1.219512195 −03
	$\boxed{\text{RCL}}$	6.708930185 −03

$\boxed{I_T = 6.71 \text{ mA}}$

Figure 7.27

Calculation of R_T — To calculate the total resistance of the circuit, use Ohm's law in the form $R_T = E/I$ (Figure 7.28). If the total voltage of 100 volts is divided by the total current, the result will be the total resistance.

Again using the calculator as you proceed through the calculation, first enter 1, press $\boxed{\text{EE}}$, enter 2, and press the $\boxed{\div}$ key. Now remember, the value of total current is still stored in the memory, so press the $\boxed{\text{RCL}}$ key and then press the $\boxed{=}$ key. The calculator should display the total resistance expressed in scientific notation. When this is rounded off and put in abbreviated form, it equals 14.9 kilohms.

$$R_T = \frac{E_T}{I_T} = \frac{100 \text{ V}}{6.71 \text{ mA}}$$

ENTER	PRESS	DISPLAY
1	$\boxed{\text{EE}}$	1 00
2		1 02
	$\boxed{\div}$	1. 02
	$\boxed{\text{RCL}}$	6.708930185 −03
	$\boxed{=}$	1.490550613 04

$\boxed{R_T = 14.9 \text{ k}\Omega}$

Figure 7.28

Complete Circuit Solution — (At this point your calculation is complete, so be sure to turn off your calculator.) Figure 7.29 shows the circuit with all calculated values drawn in.

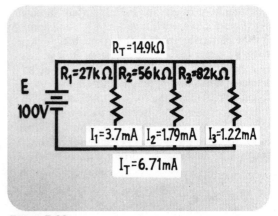

Figure 7.29

Circuit with Several Unknown Quantities — By now you should be fairly familiar with the parallel circuit laws, and you should be getting a "feel" for how to analyze a parallel circuit. Turn your attention to an even more typical circuit situation where several unknown quantities are involved. Figure 7.30 shows the schematic diagram of the circuit with the known and unknown quantities labeled. You need to calculate voltage applied to the circuit, as well as several unknown voltages and currents, as labeled.

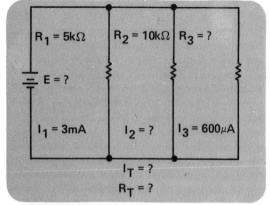

Figure 7.30

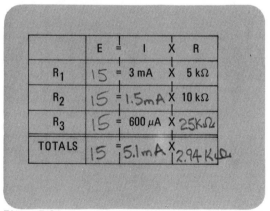

Figure 7.31

Chart for Knowns and Unknowns — As an aid in keeping track of the known and unknown quantities in circuit problems, you can use a chart of the type shown in Figure 7.31. Recall a similar chart was used in analyzing series circuits earlier. Included here are columns for voltages, currents, and resistances in all parts of the circuit and a row for totals. All of the known values have been filled in as you would do for your first step in analyzing this circuit. Note that in the case of a parallel circuit, all of the entries in your *voltage* column will be the same since voltage is the same across all branches of a parallel circuit. In a series circuit problem, all the *current* entries will be the same since current is the same everywhere in a series circuit. Notice also that the first three rows across this chart follow Ohm's law in the form E = I X R for each branch of the circuit. The final row gives you $E_T = I_T \times R_T$ for the total circuit.

7-17

Calculation of E_T — Keeping the parallel circuit laws in mind, it would be a good idea to first solve for the total applied voltage in this circuit. Then this information can be used to solve for the other unknowns. Notice in Figure 7.32 that in branch 1 you know that 3 milliamps of current is flowing through a resistance of 5 kilohms. (Also notice that since this is the only branch in this circuit where two quantities are known, this is a good place to begin your analysis.) These two values may be used with Ohm's law in the form $E = I \times R$ to solve for the voltage across branch 1. Then, since the voltage is the same across all branches of a parallel circuit, the voltage calculated for branch 1 will be equal to the total applied voltage as well as all the other branch voltages.

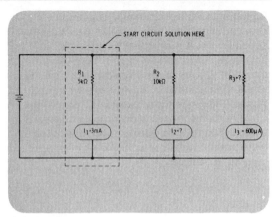

Figure 7.32

Calculation of E_{R1} — In Figure 7.33, the values of current and resistance have been substituted in the formula to get 3×10^{-3} for the 3 milliamps of current, and $5 \times 10^{+3}$ for the 5 kilohms. When these are multiplied together, the voltage across R_1 is found to be 15 volts. So now, you know that the voltage of the battery is actually 15 volts, which is also the voltage across branch 2 and branch 3.

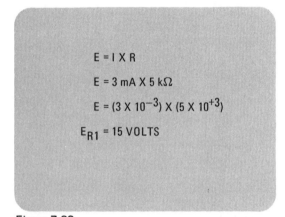

$$E = I \times R$$
$$E = 3\,mA \times 5\,k\Omega$$
$$E = (3 \times 10^{-3}) \times (5 \times 10^{+3})$$
$$E_{R1} = 15\ VOLTS$$

Figure 7.33

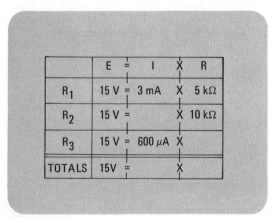

	E	=	I	X	R
R_1	15 V	=	3 mA	X	5 kΩ
R_2	15 V	=		X	10 kΩ
R_3	15 V	=	600 μA	X	
TOTALS	15V	=		X	

Figure 7.34

Partial Chart of Values — Now you can fill in the voltage values on your chart. Remember the voltage across any one branch of a parallel circuit is equal to the voltage across all other branches of the circuit. Therefore, the "voltage" column in Figure 7.34 can be filled in with 15 volts.

Calculation of I_2 — Using that new knowledge, your next step is to solve for the *current* in branch 2 by using Ohm's law in the form $I = E/R$. In Figure 7.35, the values of voltage and resistance have been substituted in the formula to get $1.5 \times 10^{+1}$ for the 15 volts, divided by $1 \times 10^{+4}$ for the 10 kilohms of resistance. Then divide to get 1.5×10^{-3}, which equals 1.5 milliamps. Thus, you now know that there are 1.5 milliamps of current flowing through R_2.

$$I = \frac{E}{R}$$

$$I_2 = \frac{15 \text{ VOLTS}}{10 \text{ k}\Omega}$$

$$I_2 = \frac{1.5 \times 10^{+1}}{1 \times 10^{+4}}$$

$$I_2 = 1.5 \times 10^{-3} = 1.5 \text{ mA}$$

Figure 7.35

Calculation of R_3 — In branch 3 of the circuit, the value of R_3 is still unknown, although you now know the voltage across it and the current through it. Use Ohm's law in the form $R = E/I$ to calculate the resistance in this branch. Substitute the values in the formula as shown in Figure 7.36 to get $1.5 \times 10^{+1}$ for the 15 volts divided by 6×10^{-4} for the 600 microamps of current. Then divide to get $0.25 \times 10^{+5}$ which equals 25 kilohms. Thus, R_3 equals 25 kilohms. All there is to know about *each individual branch* has been calculated.

$$R = \frac{E}{I}$$

$$R_3 = \frac{15 \text{ VOLTS}}{600 \text{ } \mu A}$$

$$R_3 = \frac{1.5 \times 10^{+1}}{6 \times 10^{-4}}$$

$$R_3 = .25 \times 10^{+5} = 25 \text{ k}\Omega$$

Figure 7.36

Parallel Circuit with Calculated Values — You might want to notice an interesting feature of the circuit you have been analyzing. As in any parallel circuit, the same voltage is applied across all branches of the circuit. However, note that in branch 2 of Figure 7.37, there is exactly one-half as much current flowing as in branch 1. Why do you suppose that is happening here? Look carefully and you will see that R_2 is 10 kilohms which is twice the value of R_1.

Recall from earlier discussions about Ohm's law that current and resistance are inversely related. In other words, if the resistance increases in a circuit that has a constant voltage applied, the current must decrease. The same voltage is applied

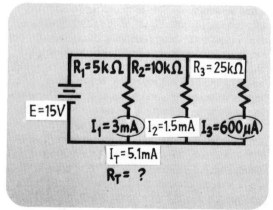

Figure 7.37

to R_1 and R_2. In this case, the resistance of R_2 is exactly double the resistance of R_1. So, the current flowing through R_2 must be just half of that flowing through R_1. One-half of 3 milliamps is 1.5 milliamps. It is always a good idea to look for this type of relationship in a circuit since it may save you some calculations.

Total Current — To solve for the *total current* in the circuit, remember that the total current in a parallel circuit is equal to the sum of the individual branch currents. The formula is shown in Figure 7.38. Remember before these branch currents can be added, all the exponents in the numbers must be the same. So change 6×10^{-4} to 0.6×10^{-3}. Now add 3×10^{-3} plus 1.5×10^{-3} plus 0.6×10^{-3}, to get a total current of 5.1×10^{-3}, which in metric prefixed form is 5.1 milliamps. So the total current flowing in this circuit is 5.1 milliamps. That is the amount of current being drawn out of the battery.

$$I_T = I_1 + I_2 + I_3$$

$$I_1 = 3mA \quad = 3 \times 10^{-3} \quad = 3 \times 10^{-3}$$
$$I_2 = 1.5\ mA \quad = 1.5 \times 10^{-3} \quad = 1.5 \times 10^{-3}$$
$$I_3 = 600\ \mu A \quad = 6 \times 10^{-4} \quad = \underline{.6 \times 10^{-3}}$$

$$I_T = 5.1 \times 10^{-3} = 5.1\ mA$$

Figure 7.38

Total Resistance — The only unknown quantity left to tackle in this circuit is the total resistance, R_T. To solve for total resistance of this circuit, you could use the reciprocal technique. However, at this point you know the total applied voltage, E_T, and total main line current, I_T, for the circuit. So, in this situation, it is much easier to use Ohm's law in the form $R_T = E_T/I_T$ as shown in Figure 7.39. To find the total resistance, simply substitute the total applied voltage value and total current value in this formula, 15 volts for the total voltage divided by 5.1 milliamps for the total current. If you correctly convert this to powers of ten form, you get $1.5 \times 10^{+1}$ divided by 5.1×10^{-3}. This

$$R_T = \frac{E_T}{I_T}$$

$$R_T = \frac{15\ VOLTS}{5.1\ mA}$$

$$R_T = \frac{1.5 \times 10^{+1}}{5.1 \times 10^{-3}}$$

$$R_T = .294 \times 10^{+4} = 2.94\ k\Omega$$

Figure 7.39

equals $0.294 \times 10^{+4}$ which equals 2.94 kilohms. This is the total or equivalent resistance of the circuit.

Parallel Circuit with Total Resistance — Does this equivalent resistance agree with the law for total resistance in a parallel circuit? The law states that the total resistance in a parallel circuit is always less than or approximately equal to the smallest resistive branch. As shown in Figure 7.40, the total resistance of 2.94 kilohms is less than the smallest resistance of 5 kilohms. Thus, the equivalent resistance does agree with the total resistance law.

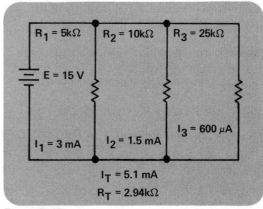

Figure 7.40

Completed Chart — At this point your chart (Figure 7.41) should be completely filled in. If you examine it, you will note that the total circuit current is just the sum of the three branch currents listed above it. Notice now that the total resistance, which was determined using $R_T = E_T/I_T$ is *not* just equal to the sum of the resistances listed above it, but is *less than the smallest branch resistance of 5 kilohms.*

	E	=	I	X	R
R_1	15 V	=	3 mA	X	5 kΩ
R_2	15 V	=	1.5 mA	X	10 kΩ
R_3	15 V	=	600 μA	X	25 kΩ
TOTALS	15 V	=	5.1 mA	X	2.94 kΩ

Figure 7.41

More Complex Circuit — The more practice examples you work through, the more your "circuit sense" will develop. You will begin sensing where voltages are applied, where currents divide in various circuits, etc. So, before completing this lesson, one more parallel circuit problem is analyzed in detail. Follow along and focus your attention on the steps involved in its solution. First, examine the circuit schematic shown in Figure 7.42. Note that this circuit problem is somewhat more complex than that covered in the previous example.

In the circuit, the known quantities are the total applied voltage of 33 volts, the total or main line current of 6.33 milliamps, the current through R_1 of 1 milliamp, and the resistance of R_2 is 100 kilohms. The unknown quantities which you must find are the resistance of R_1, the current through branch 2, and in branch 3, both the current and the resistance. Notice that although neither the resistance nor current is known in branch 3, the voltage applied to that branch is 33 volts since voltage is the same across all branches of a parallel circuit.

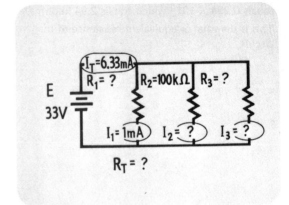

Figure 7.42

	E	=	I	X	R
R_1	33 V	=	1 mA	X	33 KΩ
R_2	33 V	=	330 uA	X	100 kΩ
R_3	33 V	=	5 mA	X	6.6 KΩ
TOTALS	33 V	=	6.33 mA	X	5.21 KΩ

Figure 7.43

Circuit Chart with Known Values — As a first step in solving this problem, you might want to again set up a chart to keep track of all the circuit values you know, all those you will be calculating, and their relationship. Figure 7.43 shows the chart with all of the quantities known about this circuit filled in. Notice that all of the voltage values are inserted since once you know the voltage applied to a parallel circuit, you automatically know the voltage across each branch. Examine this chart and remember that across each row it follows Ohm's law, $E = I \times R$. So, at this point you can solve for any unknown quantity that has two known entries in the same row. This means that you could immediately proceed to solve for R_1, I_{R2}, or R_T.

- **Calculation of R_1**
- **Calculation of I_2**

The only unknowns that *cannot* be calculated at this point are I_{R3} and R_3.

Calculation of R_1 — First use Ohm's law to solve for the resistance value of R_1 as shown in Figure 7.44. You know that there are 33 volts across R_1 and there is 1 milliamp of current flowing through it. So you can solve for R_1 by using Ohm's law in the form R = E/I. One milliamp divided into 33 volts equals $3.3 \times 10^{+1}$ divided by 1×10^{-3}. Thus, the resistance of branch 1 is 33 kilohms.

$$R = \frac{E}{I}$$

$$R_1 = \frac{33\ \text{VOLTS}}{1\ \text{mA}}$$

$$R_1 = \frac{3.3 \times 10^{+1}}{1 \times 10^{-3}}$$

$$R_1 = 3.3 \times 10^{+4} = 33\ \text{k}\Omega$$

Figure 7.44

Calculation of I_2 — A logical "next step" in this calculation is to solve for the current through branch two as shown in Figure 7.45. At this point, you know that the voltage across R_2 is 33 volts and that its resistance value is 100 kilohms. You can solve for the current through R_2 by using Ohm's law again, but this time in the form I = E/R. Substituting the appropriate values, 33 volts divided by 100 kilohms equals $3.3 \times 10^{+1}$ divided by $1 \times 10^{+5}$. This equals 3.3×10^{-4} or 330 microamps of current flowing through R_2.

$$I = \frac{E}{R}$$

$$I_2 = \frac{33\ \text{VOLTS}}{100\ \text{k}\Omega}$$

$$I_2 = \frac{3.3 \times 10^{+1}}{1 \times 10^{+5}}$$

$$I_2 = 3.3 \times 10^{-4} = 330\ \mu\text{A}$$

Figure 7.45

- **Parallel Circuit Showing R₁ and I₂**
- **Partially Completed Chart**
- **Calculation of I₃**

Parallel Circuit Showing R_1 and I_2 — Look at the circuit of Figure 7.46 carefully. At this point in your analysis you know how much current is flowing through two branches, and you know the *total* main line current flowing out of the power supply.

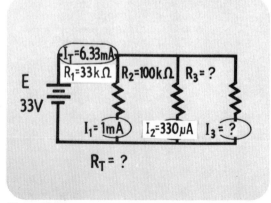

Figure 7.46

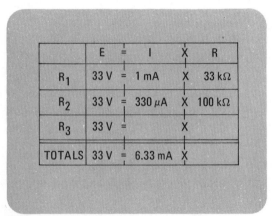

Figure 7.47

Partially Completed Chart — Examining the chart after inserting the values you've just calculated leads you on to the next step in this circuit solution. The only unknown listing in the column for current at this point is the current flowing through R_3, I_{R3} (Figure 7.47).

Calculation of I_3 — Since the total current flowing must equal the sum of the individual branch currents, if two of the branch currents are known, then the remainder of the total current must flow in the third branch. This idea can be expressed as a formula as shown in Figure 7.48. Substitute the current values that you know into the formula and carry out the required mathematics to get 6.33 milliamps minus 1.33 milliamps. Thus, the value of I_3 is 5 milliamps. Notice that two values for branch 3 are now known: the amount of current flowing through the branch, and the voltage applied to it. It should be a simple matter for you to calculate R_3 at this point using Ohm's law.

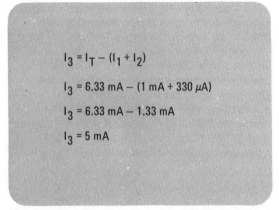

Figure 7.48

- An Important Relationship
- Calculation of R_3
- Calculation of R_T

An Important Relationship — Before you finish this calculation, take time to study this circuit's behavior for a moment. Recall an important relationship discussed in your lesson on Ohm's law: if the voltage in a circuit is held constant, the current and resistance vary inversely. In branch 1 (Figure 7.49), the current is 1 milliamp, and the resistance is 33 kilohms. Branch 2 has 100 kilohms of resistance, and the current is much less than 1 milliamp, 330 microamps. Branch 3 has 5 milliamps of current. Right away you should be able to infer something about the way this circuit operates. Since the current in branch 3 is by far the largest of the three branch currents, branch 3 must have the smallest resistance.

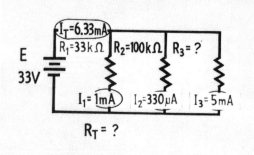

Figure 7.49

Calculation of R_3 — To calculate the resistance of R_3, use Ohm's law as shown in Figure 7.50. Again the applied voltage of 33 volts is divided by 5 milliamps which equals $3.3 \times 10^{+1}$ divided by 5×10^{-3}. The result is $0.66 \times 10^{+4}$. In abbreviated form this equals 6.6 kilohms, which is the smallest resistance in the circuit.

$$R = \frac{E}{I}$$

$$R_3 = \frac{33 \text{ VOLTS}}{5 \text{ mA}}$$

$$R_3 = \frac{3.3 \times 10^{+1}}{5 \times 10^{-3}}$$

$$R_3 = .66 \times 10^{+4} = 6.6 \text{ k}\Omega$$

Figure 7.50

Calculation of R_T — The only quantity that has not been found at this point is the total resistance of the circuit. To find this, use Ohm's law as shown in Figure 7.51. Substitute the values in the formula to get 33 volts divided by 6.33 milliamps. Converting to powers of ten form, this equals $3.3 \times 10^{+1}$ divided by 6.33×10^{-3}, which gives a total resistance of $0.5213 \times 10^{+4}$ ohms. Written in abbreviated form, this is 5.21 kilohms.

$$R_T = \frac{E_T}{I_T}$$

$$R_T = \frac{33 \text{ VOLTS}}{6.33 \text{ mA}}$$

$$R_T = \frac{3.3 \times 10^{+1}}{6.33 \times 10^{-3}}$$

$$R_T = .5213 \times 10^{+4} = 5.21 \text{ k}\Omega$$

Figure 7.51

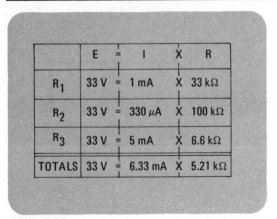

Figure 7.52

Completed Chart — With the final entries filled in on your chart (Figure 7.52), you have a complete picture of the circuit's operation. The total voltage across each branch is 33 volts. The branch currents 1 milliamp, 330 microamps, and 5 milliamps add up to the main line current of 6.33 milliamps. The circuit's total resistance of 5.21 kilohms is less than the lowest branch resistance ($R_3 = 6.6$ kΩ), and the circuit analysis is complete.

Connection of a Voltmeter — A review of the operation of another common parallel circuit may be helpful, especially as you continue in your laboratory work. You may recall that Lesson 6 discussed a simple circuit consisting of a very large value resistor connected in parallel with one of much lower value. At this point, if you were asked to compute the total resistance of a circuit such as this, you should immediately be thinking "a large value resistor in parallel with a much lower valued one — mmm — the total resistance should be close to that of the *small* resistor, probably just a little less." You would be right in thinking that. In fact, as has been discussed, this is just the situation that occurs when a voltmeter is used to measure the voltage drop across a resistor as shown in Figure 7.53.

As an example, consider the circuit in Figure 7.54 which has 25 volts applied to a 2.5-kilohm resistor. A quick Ohm's law calculation will tell you that R_1 is drawing 10 milliamps of current. If a voltmeter which has 500 kilohms of internal resistance is connected as shown, a very small amount of current will flow through the voltmeter. Recall that a voltmeter is always connected in parallel or to use an equivalent term, in *shunt*, with a voltage being measured. The connection of the 500 kilohms of resistance that the meter represents will cause an increase in the total current flowing or the total current drain on the power supply. This is because the meter has created an additional path for current to flow

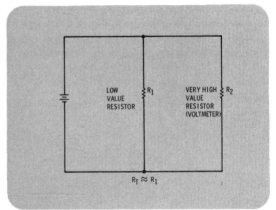

Figure 7.53

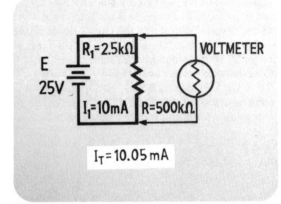

Figure 7.54

through this circuit. Any time the amount of current from the power supply is increased, it can be said that the *load* on the power supply is increased.

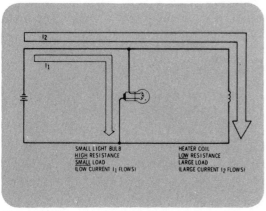

Figure 7.55

Circuit Loading — An important point to remember is that the word *load* refers in general to two things: (1) a *device* that draws power from a source, and (2) the actual *power* the source must deliver. When someone says "the load on this circuit has increased," it means that some device has been added or some change has taken place to cause the supply to deliver more power. In a parallel circuit, each time a branch is added another path for current is created and the overall *load* on the supply increases. Note the *lower the resistance* of any added branch, the more current it draws, and the *greater the load* is on the supply. On the other hand, the *greater the resistance* of an added branch, the less current it draws, and the *smaller the load* is on the supply. These relationships are illustrated in Figure 7.55.

- Current Through the Meter
- Total Current

Current Through the Meter — You can calculate the amount of current flowing through the meter branch added to this circuit by simply using Ohm's law in the form $I = E/R$, as shown in Figure 7.56. Substitute in the formula, $2.5 \times 10^{+1}$ for the 25 volts applied, divided by $5 \times 10^{+5}$ for the 500 kilohms representing the internal meter resistance. This equals 0.5×10^{-4} or 50 microamps of current flowing through the voltmeter

$$I = \frac{E}{R}$$
$$I = \frac{25 \text{ VOLTS}}{500 \text{ k}\Omega}$$
$$I = \frac{2.5 \times 10^{+1}}{5 \times 10^{5}}$$
$$I = .5 \times 10^{-4} = 50 \ \mu A$$

Figure 7.56

Total Current — The two branch currents can be added to calculate the total current now flowing in this circuit as shown in Figure 7.57. Before adding you must make sure that the powers of ten of the numbers you are adding are the same. Change the exponents to get 1×10^{-2} plus 0.005×10^{-2}. The sum is 1.005×10^{-2} which equals 10.05 milliamps of current flowing in the total circuit.

So, what actually happened when the voltmeter was connected in parallel with R_1 (Figure 7.54) is that the amount of total current drain from the power supply increased from 10 milliamps to 10.05 milliamps which is an increase of only 0.05 milliamps. If you were to calculate the percent of increase that occurred when the meter was added (0.05 milliamps/10 milliamps), you would find that the current increase was only one-half of one percent. This is not very much, and so this voltmeter is said to have a *low* loading effect on the circuit. Remember that a greater load draws more current.

A voltmeter's loading effect refers to the fact that when the meter is connected to a circuit, another path for current flow is created, and hence the power supply must deliver more current to the circuit. When a meter is used to make measurements in a circuit, it is desirable that the meter doesn't change or disrupt the operation of the circuit in any way. For this reason, voltmeters are designed to have as *high* an internal resistance as possible so that when they are connected in

$$I_T = I_1 + I_2$$
$$I_T = 10 \text{ mA} + 50 \ \mu A$$
$$I_T = (1 \times 10^{-2}) + (5 \times 10^{-5})$$
$$I_T = (1 \times 10^{-2}) + (.005 \times 10^{-2})$$
$$I_T = 1.005 \times 10^{-2} = 10.05 \text{ mA}$$

Figure 7.57

parallel with circuit components, they create as small a load as possible.

Summary — This basically concludes the discussion of parallel circuits, and the three laws governing their operation. So, when analyzing or working with parallel circuits, remember that:

1. The total resistance in a parallel circuit is always less than or approximately equal to the smallest resistive branch.
2. The total current in a parallel circuit equals the sum of the individual branch currents.
3. Voltage is the same across all branches of a parallel circuit.

You will find several practice problems at the end of this lesson. To really gain an understanding of the operation of parallel circuits, review these problems carefully. If you encounter any difficulties, you may want to review material in this and the previous lesson before you move on.

Several important new techniques for solving complex parallel circuits have been presented in this lesson. With practice, you will be able to select the simplest technique and apply it correctly to parallel circuit analysis. Before closing this lesson and moving on to study other aspects of dc electricity, several other topics concerning actual parallel circuits are covered.

The first concerns techniques used in redrawing circuit schematics so that you can see more clearly how to analyze them, the second is the important topic of troubleshooting in parallel circuits, the third is a brief review of power calculations in parallel circuits.

REDRAWING CIRCUITS FOR CLARITY

TRACE CURRENT PATHS

LABLE JUNCTIONS

RECOGNIZE POINTS AT THE SAME POTENTIAL

VISUALIZE "STRETCH" OF WIRES

RESKETCH CIRCUIT (THROUGH STAGES IF NECESSARY) INTO SIMPLER FORM.

Figure 7.58

Redrawing Circuit Schematics for Clarity — You will notice that most of the schematic diagrams you have been working with have shown parallel circuits drawn as neat square figures, with each branch easily identified. Actual wired circuits and more complex schematics, however, rarely are laid out in this simple form. For this reason, it is important to recognize that circuits can be drawn in a variety of ways, and to learn some of the simple techniques for redrawing them in simplified form.

When redrawing a circuit to clarify it, several simple ideas can be of help (Figure 7.58). Start at the source and trace the path of current flow through the circuit. At points where the current *divides*, called junctions or nodes, parallel branches begin. These junctions are key points of reference in any circuit, so it is a good idea to label them as you find them. Note also that wires in circuit schematics are assumed to have no resistance. This means that there is *no voltage drop* along any wire, so any unbroken piece of wire is at the same voltage all along its length, until it is interrupted by a resistor, battery, etc. In any redrawing problem, a wire in a schematic can be "stretched" or shrunk as much as you like without changing its role in the circuit.

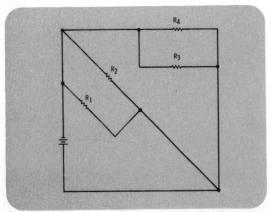

Figure 7.59

Circuit Redrawing Problem — Consider the circuit in Figure 7.59. To simplify it, start at the power source and trace the paths for current that exist in the circuit back to the source. As you redraw the circuit, draw it in simple, box-like form. Each time you reach a junction, a new branch is created.

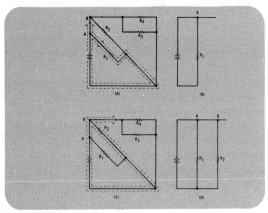

Figure 7.60

Redrawing Circuit — Starting at the negative terminal of the power source and tracing the current flow, you come to a point where the current divides. Label this junction A, as shown in Figure 7.60 and notice that one complete path for current flow exists through R_1. This path can be drawn as one branch of a parallel circuit in your redrawn schematic as shown in Figure 7.60(B). Moving on from point A, you come almost immediately to another point where current divides. Label this junction point B, as shown in Figure 7.60(C), and notice that another complete path for current flow back to the source exists through R_2. This path will form a second branch in your redrawn circuit as shown in Figure 6.60(D).

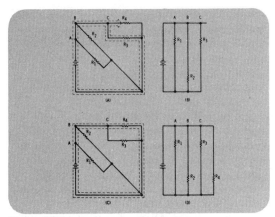

Figure 7.61

Now moving on from point B, as shown in Figure 7.61(A), you will reach a third point where current divides. Label this junction point C and note that another complete path for current flow through R_3 exists, which forms an additional branch in your redrawn circuit [Figure 7.61(B)]. One last complete path for current flow through R_4 can be identified in your original circuit [Figure 7.61(C)], and this may be drawn as a final branch in your redrawn schematic [Figure 7.61(D)].

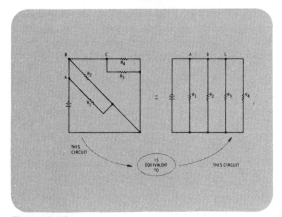

Figure 7.62

Therefore a circuit which at first appears complex, is actually just four resistors in parallel (Figure 7.62). The redrawn circuit may be handled easily using the techniques you have learned in this lesson and the preceding one. The original schematic might have looked tricky at first, but as you learn to redraw circuits for clarity, even very complex looking circuits may often be redrawn in simpler form and then tackled in a straightforward way.

Problems in Parallel Circuits: Troubleshooting —
The two most common sources of trouble in any
circuit, as mentioned when trouble in series circuits
was discussed, are *short* circuits and *open* circuits.
Because of the differences in operation between
series and parallel circuits, symptoms and problem
tracing techniques will be different when
examining parallel circuits for possible trouble.

First of all, a short circuit across any parallel
branch causes drastic problems for a parallel
circuit. By now the reason why should be clear to
you. The total resistance of a parallel circuit is less
than that of the smallest resistive branch. If one
branch becomes shorted, its resistance is zero, and
so the total circuit becomes one big short directly
across the power supply.

Effect of Short in Parallel Circuit — Figure 7.63(A)
is a schematic of a simple parallel circuit working
normally. By now you should easily see that its
total resistance is just ($R_T = R_S/N = 30$ kΩ/3) 10
kilohms, and it will therefore draw a total of
1 milliamp of current. If a short should develop
across any branch, such as is shown in
Figure 7.63(B), the effective resistance of this
branch goes to zero and a huge current flow
through this branch results. If this circuit is lucky
enough to have a fuse in its main line, the line
carrying the total current, the fuse will blow
immediately creating an open in the main line and
all activity will stop. If no fuse is present,
something will give. Maybe the power supply will
fail, maybe a wire will burn up; but the huge
current flow that the short creates usually cannot
go on for very long. So a short in a parallel circuit
usually creates an open before long. The open may
be in the main line, in the branch containing the
short, or anywhere along the path to the short.

Figure 7.63

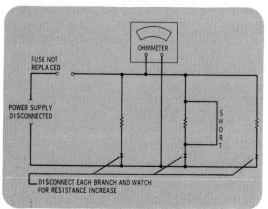

FUSE NOT
REPLACED

OHMMETER

POWER SUPPLY
DISCONNECTED

SHORT

DISCONNECT EACH BRANCH AND WATCH
FOR RESISTANCE INCREASE

Figure 7.64

Locating Shorts in Parallel Circuits — To find a short in a parallel circuit, look at the circuit. Visible smoke or soot may be present, showing you where the short occurred. If there is no visible evidence, other methods may be used. First disconnect the power supply, and *do not* replace any fuses or reset any circuit breakers until the problem is located and corrected. Connect an ohmmeter across the circuit, as shown in Figure 7.64, and disconnect each branch in turn. If the resistance stays close to zero, the short is in another branch. If you disconnect the branch, and the resistance abruptly rises, that's your problem area. Dig in and correct the problem. If disconnecting all the branches still results in a zero ohm reading on the meter, the short has developed directly across the main line and not along any branch.

Opens Often Caused by Shorts — As has been mentioned, usually if a short develops in a parallel circuit and is allowed to continue for very long, an open circuit will result. A wire, resistor, or other component will usually burn open, and a visible sign of this will often be present. If one branch of a parallel circuit has opened, the rest of the circuit will continue to function normally. The only difference will be that the circuit will be operating with reduced *total* current. In troubleshooting, the safest way to proceed is to disconnect the power to the circuit and use an ohmmeter to test each branch. Remember, if a *short* has caused one branch to open up, the *short* must be corrected

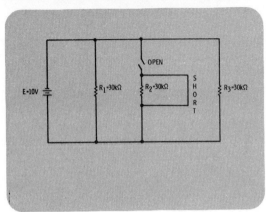

Figure 7.65

before the open can be repaired as indicated in Figure 7.65.

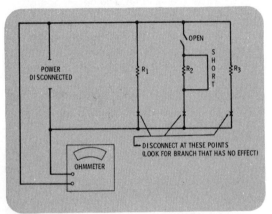

Figure 7.66

Locating Opens Caused by Shorts — If this is the situation, again *disconnect all power* and connect an ohmmeter across the main line as shown in Figure 7.66. Disconnect each branch, one at a time. As each branch is disconnected or opened, the total circuit resistance should *increase*. If disconnecting a branch has no effect on the total circuit resistance you are measuring, that branch is *already* open somewhere.

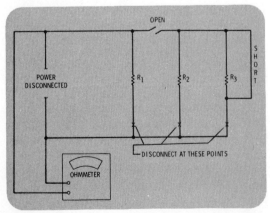

Figure 7.67

Continue Checking — Be sure at this point to continue checking all the other circuit branches. You may find that *more* than one branch may be disconnected with no effect on the ohmmeter reading. If this is the case the open is probably somewhere *between* branches as shown in Figure 7.67. Using a little logic, you can fairly easily narrow down where the open is. Again, be certain to check for *shorts* in each branch that gave you no change in the total resistance when removed. A short may have caused the circuit to open in the first place. If the open has occurred in the circuit's main line, the entire circuit will stop operating. At this point, check the whole circuit for shorts as outlined previously.

The troubleshooting methods outlined here are not the only possible ones to use. With *caution*, common sense, and some experience with circuits, you will develop a knack for sniffing out circuit problems.

Power Calculations in Parallel Circuits — As a final topic, consider a brief review of how to calculate the total power dissipated in a parallel circuit. As mentioned in Lesson 6, to calculate the total power dissipated by a parallel circuit all you have to do is calculate the power dissipated by each resistor, and then add these together for the total. The total power dissipated in a parallel circuit equals the sum of the powers dissipated by each resistor in the circuit. To calculate the power each resistor dissipates, you can use one of the three formulas you have been given for this calculation;

$$P = I \times E$$
$$P = I^2 R$$
or $\qquad P = E^2/R$ (for each resistor).

As an alternative method, if you have already calculated the total or main line current flowing in the circuit, the total circuit voltage, and/or the total circuit resistance, you can calculate the total power the circuit dissipates by using the three formulas listed above on the total circuit quantities, as shown in Figure 7.68.

Total Power Dissipation Equals:

$$P_T = I_T \times E_T$$
or $\qquad P_T = I_T^2 \times R_T$
or $\qquad P_T = E_T^2/R_T$

POWER DISSIPATED IN A PARALLEL CIRCUIT EQUALS THE SUM OF THE INDIVIDUAL POWERS DISSIPATED BY EACH RESISTOR.

USE

$P = I \times E$ | ON EACH $\qquad P_T = I_T \times E_T$

$P = I^2 R$ | RESISTOR OR $P_T = I_T^2 R_T$

$P = E^2/R$ | & ADD $\qquad P_T = E_T^2/R_T$

ON TOTAL CIRCUIT QUANTITIES.

Figure 7.68

- **Example**
- **Alternate Methods**
- Comparison Chart

Example — Consider the circuit of Figure 7.69. (This is a circuit that has already been analyzed completely in this lesson, see Figure 7.29.) To calculate the total power this circuit dissipates, you could apply the formula $P = I \times E$ to each resistor in the circuit and then add the results. The result would be 0.67 watts, as shown. Alternatively, you could apply $P_T = I_T \times E_T$, and substitute the total circuit current and voltage to get the same result.

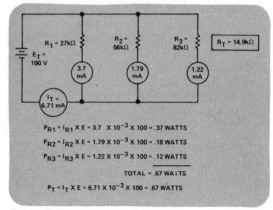

Figure 7.69

Alternate Methods — There are several other alternative methods you can use in this calculation, as shown in Figure 7.70. You could use the calculated total circuit values E_T, I_T, and R_T and the two other power formulas ($P_T = I_T^2 R_T$, or $P_T = E_T^2/R_T$) and as shown in the figure, you would get the same result.

$$P_T = I_T^2 R_T$$

$$P_T = (6.71 \times 10^{-3})^2 (14.9 \times 10^{+3})$$

$$P_T = (4.50 \times 10^{-5}) (14.9 \times 10^{+3})$$

$$P_T = 67.05 \times 10^{-2}$$

$$P_T = .67 \text{ W}$$

$$P_T = E_T^2/R_T = \frac{(100)^2}{14.9 \times 10^3} = \frac{10,000}{14.9 \times 10^3}$$

$$P_T = .67 \text{ W}$$

Figure 7.70

Comparison Chart — The following chart compares the laws and formulas for series and parallel circuits. A study of this chart will certainly be helpful in analyzing any series or parallel circuit.

COMPARISON CHART FOR SERIES AND PARALLEL CIRCUITS

	LAWS	FORMULAS
SERIES CIRCUITS	1. Individual resistances in a series circuit add up to the total circuit resistance.	$R_T = R_1 + R_2 + R_3 + \ldots$
	2. Individual voltages across resistors in a series circuit add up to the total applied voltage.	$E_T = E_{R1} + E_{R2} + E_{R3} + \ldots$
	3. Current has the same value at any point within a series circuit.	$I_T = I_{R1} = I_{R2} = I_{R3} = \ldots$
PARALLEL CIRCUITS	1. The total resistance of a parallel circuit is always less than or approximately equal to the smallest resistive branch.	$R_T = \dfrac{1}{\dfrac{1}{R_1} + \dfrac{1}{R_2} + \dfrac{1}{R_3}} \ldots$
	2. The voltage is the same across all branches of a parallel circuit.	$E_T = E_1 = E_2 = E_3 \ldots$
	3. The total or "main line" current in a parallel circuit is equal to the sum of the individual branch currents.	$I_T = I_1 + I_2 + I_3 + \ldots$

● **Worked Through Examples**

1. Find I_T in the circuit shown below.

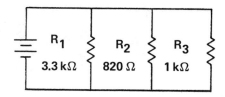

There are two ways I_T may be found. You can find the individual branch currents (by dividing E_T/R_1, etc.) and add them together to arrive at the total current or you could calculate R_T, then use its value to divide into E_T, producing I_T. This problem will be worked so that the branch currents are found first, then added to produce the total current. First of all, the voltage across all the resistors is the same because they are connected in parallel with the source voltage. The source is equal to 12 V, so E_{R1}, E_{R2} and E_{R3} are all equal to 12 V. You now know two things about each resistor. Using the resistance value and the voltage drop, Ohm's law may be used to determine the current flowing through each resistor. E_{R1} is 12 V. R_1 is 3.3 kilohms. 12 V divided by 3.3 kilohms is equal to 3.64 milliamps.

$$E_{R1}/R_1 = I_{R1}$$
$$12\ V/3.3\ k\Omega = 3.64\ mA$$

The same procedure may be used to determine I_{R2}. Twelve volts divided by 820 ohms is equal to 14.6 milliamps.

$$E_{R2}/R_2 = I_{R2}$$
$$12\ V/820\ \Omega = 14.6\ mA$$

Finally, I_{R3} may be found by dividing 12 volts by 1 kilohm. Twelve volts divided by 1 kilohm is equal to 12 milliamps.

$$E_{R3}/R_3 = I_{R3}$$
$$12\ V/1\ k\Omega = 12\ mA$$

The branch currents are:

$$I_{R1} = \ \ 3.64\ mA$$
$$I_{R2} = 14.6\ mA$$
$$\underline{I_{R3} = 12\ mA}$$
$$I_T \ \ = 30.24\ mA$$

In a parallel circuit, the total current flowing is equal to the *sum* of the individual branch currents. The total current flowing is then 30.24 milliamps.

2. Find the current flowing through R_2.

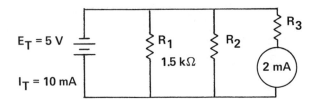

As the problem now stands, nothing is known about resistor 2. A knowledge of the parallel circuit laws is needed to solve this problem. The voltage across all branches of a parallel circuit is the same, so each resistor in the circuit has 5 volts across it. In order to keep up with the numbers in this problem, the values will be charted.

Branch	E	I	R
1	5V		1.5kΩ
2	5V		
3	5V	2mA	
Total	5V	10mA	

Looking at the chart, you notice that all known values have been entered. The voltage value of 5 volts has been inserted in each row because the voltage is the same across each branch of a parallel circuit. Ohm's law can be used to find I_{R1} because E_{R1} and R_1 are both known. Five volts divided by 1.5 kilohms is equal to 3.33 milliamps.

$$E_{R1}/R_1 = I_{R1}$$
$$5 \text{ V}/1.5 \text{ k}\Omega = 3.33 \text{ mA}$$

This value is now added to the chart.

Branch	E	=	I	x	R
1	5V		3.33mA		1.5kΩ
2	5V				
3	5V		2mA		
Total	5V		10mA		

Looking back to the schematic you will notice that the total current is 10 milliamps. The total current flowing through branches 1 and 3 is equal to 2 milliamps plus 3.33 milliamps or 5.33 milliamps. This leaves the balance of the current flowing through branch 2. This value can be found by subtracting 5.33 milliamps from 10 milliamps, and you will find that 4.67 milliamps is flowing through resistor 2.

$$\begin{array}{r} 10.00 \\ -5.33 \\ \hline 4.67 \text{ mA} \end{array}$$

Since this value was the only one called for in the problem, you can stop here. However, all of the other circuit values can be found by using the chart and Ohm's law.

3. An automobile has a run-down battery. Draw a schematic showing how to connect another battery in conjunction with the original battery to start the car.

Solution:

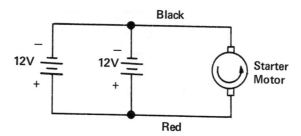

In most modern automobile electrical systems, a "negative ground" is used. This means that the negative pole of the battery is connected to the automobile chassis. When connecting the jumper cables, be sure to connect the *negative poles together, and the positive poles together*.

Batteries connected as shown below can explode due to the high surge of current flow that results. Severe damage to the car's alternator may result also.

INCORRECT

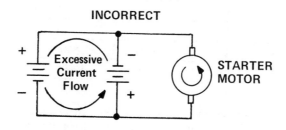

4. Find the unknown circuit values for the circuit shown below.

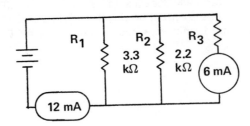

First, list the values you know and the values you want to find. A chart will again be of help. In the chart below the known values from the circuit schematic have been inserted.

Branch	E	=	I	x	R
1					
2					3.3kΩ
3			6mA		2.2kΩ
Total			12mA		

Notice that E_{R3} may be found because both R_3 and I_{R3} are known.

$$E_{R3} = I_{R3} \times R_3$$
$$E_{R3} = 6 \text{ mA} \times 2.2 \text{ k}\Omega$$
$$E_{R3} = 13.2 \text{ V}$$

Place this value into the chart, and note that since this is a parallel circuit, the voltage across all branches must then be 13.2 volts. Place this information into the chart also.

Branch	E	=	I	x	R
1	13.2V				
2	13.2V				3.3kΩ
3	13.2V		6mA		2.2kΩ
Total	13.2V		12mA		

Looking at the chart, you see that two values are known about resistor 2. Its voltage is 13.2 volts and its resistance value is 3.3 kΩ. The current flowing through the resistor may be found by using Ohm's law. $I_{R2} = E_{R2}/R_2$.

$$I_{R2} = 13.2/3.3 \text{ k}\Omega$$
$$I_{R2} = 4 \text{ mA}$$

This value is now placed into the chart in the shaded area.

Branch	E	=	I	x	R
1	13.2V				
2	13.2V		4mA		3.3kΩ
3	13.2V		6mA		2.2kΩ
Total	13.2V		12mA		

Analyzing the chart, you can see that the current through two branches ($I_{R2} + I_{R3}$) is known. Using the current law for a series circuit, you know that $I_{R1} + I_{R2} + I_{R3} = I_T$. Substituting the known numbers into this equation you have I_{R1} + 4 mA + 6 mA = 12 mA. This equation may be rewritten to give the value of I_{R1}:

$$I_{R1} = 12 \text{ mA} - 4 \text{ mA} - 6 \text{ mA}$$
$$I_{R1} = 12 \text{ mA} - 10 \text{ mA}$$
$$I_{R1} = 2 \text{ mA}$$

This value is now added to the chart.

Branch	E	=	I	x	R
1	13.2V		2mA		
2	13.2V		4mA		3.3kΩ
3	13.2V		6mA		2.2kΩ
Total	13.2V		12mA		

The last two unknowns listed on the chart may now be found by using Ohm's law.

$$R_1 = E_{R1}/I_{R1}$$
$$R_1 = 13.2 \text{ V}/2 \text{ mA}$$
$$R_1 = 6.6 \text{ k}\Omega$$

$$R_T = E_T/I_T$$
$$R_T = 13.2 \text{ V}/12 \text{ mA}$$
$$R_T = 1.1 \text{ k}\Omega$$

These values should now be added to the chart.

Branch	E	=	I	x	R
1	13.2V		2mA		6.6kΩ
2	13.2V		4mA		3.3kΩ
3	13.2V		6mA		2.2kΩ
Total	13.2V		12mA		1.1kΩ

Notice how the chart facilitates the solution of circuit problems. All known values are charted, and the unknown values can be spotted easily. It is suggested that you adopt the "chart method" of solving circuit problems where the knowns and unknowns are "mixed" in nature.

LESSON 7. PARALLEL CIRCUIT ANALYSIS

- **Practice Problems**

The key objective of this lesson has been achieved if you can solve for any unknown quantity in any basic parallel circuit situation. The following problems will give you practice in this area and by folding over the page you can check your accuracy and progress.

Depending upon the approach you use in solving these problems and how you round off intermediate results, your answers may vary slightly from those given here. However, any differences you encounter should only occur in the third significant digit of your answer. If the first two significant digits of your answers do not agree with those given here, recheck your calculations.

1. In each of the following circuits the applied voltage and values of all resistors are given. Calculate each branch current, the total current, the total circuit resistance, and total power dissipated in each case.

Fold Over

a.

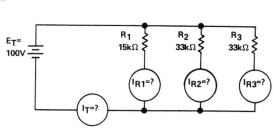

$I_{R1} = \underline{6.6\,mA}$

$I_{R2} = \underline{3\,mA}$

$I_{R3} = \underline{3\,mA}$

$I_T = \underline{12.6\,mA}$

$R_T = \underline{7.86\,K\Omega}$

$P_T = \underline{1.26\,W}$

b.

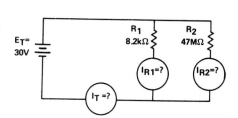

$I_{R1} = \underline{63.8\,\mu A}$

$I_{R2} = \underline{3.66\,mA}$

$I_T = \underline{}$

$R_T = \underline{8.2\,K\Omega}$

$P_T = \underline{}$

c.

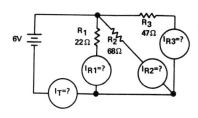

$I_{R1} = \underline{}$

$I_{R2} = \underline{}$

$I_{R3} = \underline{}$

$I_T = \underline{}$

$R_T = \underline{}$

$P_T = \underline{}$

Answers

1.

a. I_{R1} = 6.67 mA

I_{R2} = 3.03 mA

I_{R3} = 3.03 mA

I_T = 12.73 mA

R_T = 7.86 kΩ

P_T = 1.27 W

b. I_{R1} = 3.66 mA

I_{R2} = 638 nA

I_T = 3.66 mA

R_T = 8.2 kΩ

P_T = 110 mW

c. I_{R1} = 273 mA

I_{R2} = 88.2 mA

I_{R3} = 128 mA

I_T = 489 mA

R_T = 12.28 Ω

P_T = 2.94 W

Fold Over

d.

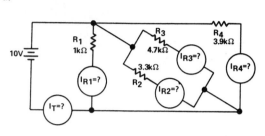

I_{R1} = __10 mA__

I_{R2} = __3.03 mA__

I_{R3} = __2.13 mA__

I_{R4} = __2.56 mA__

I_T = __17.72 mA__

R_T = __564.27 Ω__

P_T = __177 mW__

e.

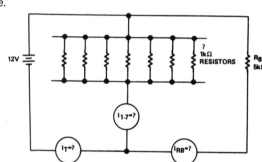

I_{1-7} _____
(Total current through all seven 1-kΩ resistors)

I_{R8} = _____

I_T = _____

R_T = _____

P_T = _____

2. In each of the following problems, several known quantities are given, and the unknowns labeled with a question mark. Calculate the unknowns in each case.

a.

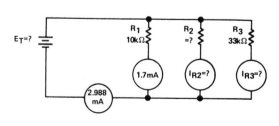

E_T = _____

R_2 = _____

R_T = _____

I_{R2} = _____

I_{R3} = _____

P_T = _____

Answers

d. $I_{R1} = 10$ mA

$I_{R2} = 3.03$ mA

$I_{R3} = 2.13$ mA

$I_{R4} = 2.56$ mA

$I_T = 17.72$ mA

$R_T = 564 \, \Omega$

$P_T = 177$ mW

e. $I_{1-7} = 84$ mA

$I_{R8} = 2.4$ mA

$I_T = 86.4$ mA

$R_T = 139 \, \Omega$

$P_T = 1.04$ W

2.

a. $E_T = 17$ V

$R_2 = 22.0$ kΩ

$R_T = 5.68$ kΩ

$I_{R2} = 773 \, \mu A$

$I_{R3} = 515 \, \mu A$

$P_T = 51$ mW

Fold Over

b.

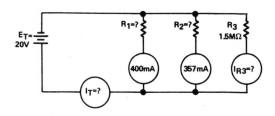

$I_T =$ _____

$R_T =$ _____

$R_1 =$ _____

$R_2 =$ _____

$I_{R3} =$ 13.3 µA

$P_T =$ _____

c.

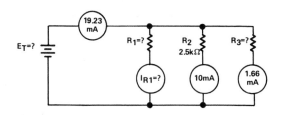

$R_T =$ _____

$R_1 =$ _____

$E_T =$ _____

$I_{R1} =$ _____

$R_3 =$ _____

$P_T =$ _____

d.

$E_T =$ _____

$I_{R2} =$ _____

$R_3 =$ _____

$I_T =$ _____

$R_T =$ _____

$P_T =$ _____

Answers

b. I_T = 757 mA

R_T = 26.4 Ω

R_1 = 50 Ω

R_2 = 56 Ω

I_{R3} = 13.3 μA

P_T = 15.1 W

c. R_T = 1.3 kΩ

R_1 = 3.3 kΩ

E_T = 25 V

I_{R1} = 7.57 mA

R_3 = 15.1 kΩ

P_T = 480 mW

d. E_T = 5 V

I_{R2} = 714 mA

R_3 = 15 Ω

I_T = 1.55 A

R_T = 3.23 Ω

P_T = 7.74 W

Fold Over

e.

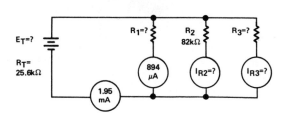

$E_T =$ _____

$R_1 =$ _____

$R_3 =$ _____

$I_{R3} =$ _____

$I_{R2} =$ _____

$P_T =$ _____

f.

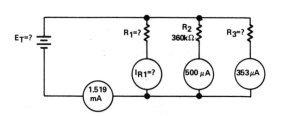

$E_T =$ _____

$R_T =$ _____

$R_1 =$ _____

$R_3 =$ _____

$I_{R1} =$ _666μA_

$P_T =$ _____

g.

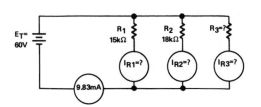

$R_T =$ _____

$R_3 =$ _____

$I_{R1} =$ _____

$I_{R2} =$ _____

$I_{R3} =$ _____

$P_T =$ _____

Answers

e. $E_T = 49.92$ V

$R_1 = 55.8$ kΩ

$R_3 = 111.4$ kΩ

$I_{R3} = 448$ μA

$I_{R2} = 608$ μA

$P_T = 97.3$ mW

f. $E_T = 180$ V

$R_T = 118$ kΩ

$R_1 = 270$ kΩ

$R_3 = 509$ kΩ

$I_{R1} = 666$ μA

$P_T = 273$ mW

g. $R_T = 6.1$ kΩ

$R_3 = 24$ kΩ

$I_{R1} = 4$ mA

$I_{R2} = 3.33$ mA

$I_{R3} = 2.5$ mA

$P_T = 590$ mW

Fold Over

h.

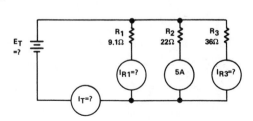

E_T = _____

I_T = _____

R_T = _____

I_{R1} = _____

I_{R3} = _____

P_T = _____

i.

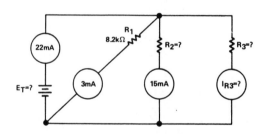

E_T = _____

R_T = _____

R_2 = _____

R_3 = _____

I_{R3} = _____

P_T = _____

j.

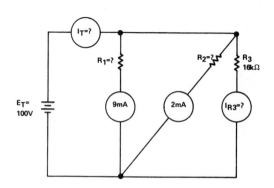

I_T = _____

R_T = _____

R_1 = _____

R_2 = _____

I_{R3} = _____

P_T = _____

Answers

h. $E_T = 110$ V

$I_T = 20.2$ A

$R_T = 5.45$ Ω

$I_{R1} = 12.1$ A

$I_{R3} = 3.06$ A

$P_T = 2.22$ kW

i. $E_T = 24.6$ V

$R_T = 1.12$ kΩ

$R_2 = 1.64$ kΩ

$R_3 = 6.15$ kΩ

$I_{R3} = 4$ mA

$P_T = 540$ mW

j. $I_T = 17.25$ mA

$R_T = 5.79$ kΩ

$R_1 = 11.1$ kΩ

$R_2 = 50$ kΩ

$I_{R3} = 6.25$ mA

$P_T = 1.73$ W

1. A circuit which has two or more paths for current flow is called a_____circuit.

   a. Series
   b. Parallel
   c. Open
   d. Voltage

2. The voltage across any branch of a parallel circuit is equal to_____voltage.

   a. The applied
   b. The opposite branch
   c. A similar
   d. A negative

3. The sum of the individual branch currents in a parallel circuit is equal to_____current.

   a. Each branch
   b. A negative
   c. An opposite
   d. The total

4. The equation for total voltage in a parallel circuit is:

   a. $E = E1 + E2 + E3 \ldots$
   b. $E_T = E1 = E2 = E3 \ldots$
   c. $E_T = E1 + E2 - E3 \ldots$
   d. $E_T = \dfrac{E1\,E2}{E1 + E2}$

5. The equation for total current in a parallel circuit is:

   a. $I_T = I1 + I2 + I3 \ldots$
   b. $I_T = I1 = I2 = I3 \ldots$
   c. $I_T = I1 + I2 - I3 \ldots$
   d. $I_T = \dfrac{I1\ I2}{I2 + I2} \ldots$

6. When the current output of a power supply increases, it is caused by_____in the load resistance.

   a. An open
   b. An increase
   c. A decrease
   d. A large tolerance

7. The total power dissipation of a parallel circuit is equal to the_____of the individual powers dissipated in the circuit.

   a. Subtraction
   b. Difference
   c. Sum
   d. Negative

8. If current increases in a circuit which has a constant voltage applied, its resistance has:

   a. A constant value
   b. Decreased
   c. Gone negative
   d. Opened

9. When a voltmeter is connected across a resistor in a circuit, one of the following values will always increase:

   a. R_S
   b. R_T
   c. E_T
   d. I_T

10. A circuit consists of five $1\,M\Omega$ resistors connected in parallel across a battery. If a sixth $1\text{-}M\Omega$ resistor is connected across the original circuit, the total_____will decrease.

    a. Voltage
    b. Resistance
    c. Current
    d. Power

11. The equation for the total applied voltage in a series circuit is:

    a. $E_T = E1 = E2 = E3 \ldots$
    b. $E_T = E_{R1} + E_{R2} + E_{R3} \ldots$
    c. $E_T = E1 + E2 - E3$
    d. $E_T = \dfrac{E1\,E2}{E1 + E2}$

12. The equation for the total resistance of a series circuit is:

    a. $R_T = R1 + R2 + R3 \ldots$
    b. $R_T = R1 = R2 = R3 \ldots$
    c. $R_T = R1 - R2 + R3 \ldots$
    d. $R_T = \dfrac{R_1 R_2}{R_1 + R_2}$

13. The equation for the total current in a series circuit is:

    a. $I_T = I_1 = I_2 = I_3 = \ldots$
    b. $I_T = I_1 + I_2 + I3 \ldots$
    c. $I_T = I_1 - I_2 + I_3 \ldots$
    d. $I_T = \dfrac{I1\,I2}{I1 + I2}$

The following circuit is used for questions 14 through 20

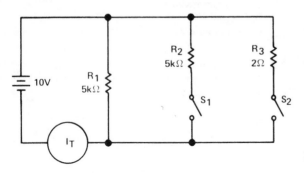

14. With switches S1 and S2 open, what does the ammeter read?

    a. 2 amperes
    b. 1 milliampere
    c. 2 mA
    d. 50 mA

15. With switch S1 closed and S2 open, the total resistance of the circuit is:

    a. 10 K ohms
    b. $10 \times 10^{+3}$ ohms
    c. 2500 ohms
    d. 2.5K ohms
    e. c and d above

16. With switch S1 closed and S2 open, the ammeter reads:
    a. 4 amperes
    b. 1 milliampere
    c. 4 mA
    d. 10×10^{-3} amperes

17. With switch S1 and S2 closed, the ammeter reads:
    a. over 5 amperes
    b. 4 mA
    c. 4 amperes
    d. 10×10^{-3} amperes

18. With switch S1 and S2 closed, the power dissipation in the circuit will be:
    a. 90 MW
    b. Above 50 watts
    c. 40 MW
    d. None of above

19. With switch S1 open and S2 closed, the total resistance will be:

    a. Approximately 2 ohms
    b. Much greater than with S1 closed
    c. Much less than with S1 closed
    d. None of above

20. With switch S1 closed and S2 open, the power dissipation is:

    a. 1 watt
    b. 4 watts
    c. 40 watts
    d. 40 mW

Lesson 8

Parallel-Series Circuits

Now that series and parallel circuit analyses have been introduced, this lesson begins discussing how these methods may be combined to analyze circuits that are a simple combination of the two: PARALLEL-SERIES circuits. The techniques discussed are essentially an extension of the principles covered in Lessons 5, 6, and 7. The circuit examples discussed find many applications in everyday life, including the basic layout of the automotive wiring system.

LESSON 8. PARALLEL-SERIES CIRCUITS

●   Objectives

This lesson begins to put together the techniques of parallel and series circuit analysis and to apply them to a more complex circuit, the parallel-series circuit. The basic principles covered are essentially a logical extension of those you have learned thus far. The circuit examples you will work with in this lesson include a basic automotive wiring system. At the end of this lesson you should find yourself able to:

1.   *Define* parallel-series circuit and given a series of schematics, *distinguish* which represent series, parallel, series-parallel, and parallel-series circuits.

2.   Given a schematic diagram of the type shown below, *use circuit reduction techniques* to *calculate* the total equivalent resistance of the circuit.

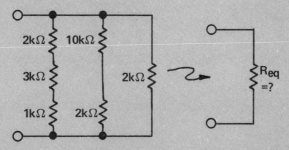

3.   Given a schematic diagram of the type shown, you should be able to use *circuit reduction techniques* and Ohm's law to: *calculate* the voltage across and current through each resistor in the circuit, *find* the total current in the circuit, and *calculate* the circuit's equivalent resistance.

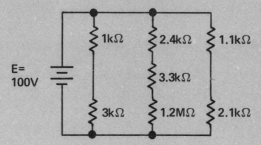

4.   Given a schematic diagram of the type shown and using *current reduction techniques* and Ohm's law, you should be able to: *calculate* all unknown currents, voltages, and resistances.

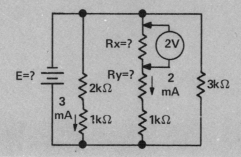

- **Series-Parallel Circuit**
- **Parallel-Series Circuit**

In Lessons 5, 6 and 7, series and parallel circuits were discussed. The essential features differentiating these two basic types of circuit were outlined along with the specific rules that describe the operation of each type of circuit.

In this lesson you will see how several of the key features of parallel circuits may be combined with those of series circuits in forming what is called a *parallel-series circuit*. Later on in Lesson 9, another circuit type, the *series-parallel circuit*, will be covered. These new circuit names, series-parallel and parallel-series, sound pretty much alike, but there are some key distinctions between these two types of circuits that will affect the methods used in analyzing them. In presenting their analysis it is important that you distinguish the two types in your mind.

Series-Parallel Circuit — Figure 8.1 shows the *series-parallel* circuit which will be covered in detail in Lesson 9. This circuit is called a series-parallel circuit because at least one circuit component, in this case R_1, lies in *series* with the total current. Notice that in this schematic, resistor R_1 is connected so that the total circuit current flows through it. It is wired in series with other circuit components which may include parallel wired combinations of resistors such as R_2 and R_3.

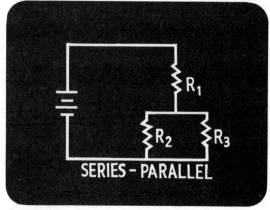

Figure 8.1

Parallel-Series Circuit — Figure 8.2 shows a *parallel-series* circuit. The key difference between this type of circuit and the series-parallel circuit is that there is *no* single component that lies in the path of the total circuit. This circuit consists of several branches wired in parallel, but notice that each of the parallel branches may contain one or more resistors connected in series with one another. The circuit contains several parallel branches which may each contain several resistors wired in series; hence, this is called a parallel-series circuit.

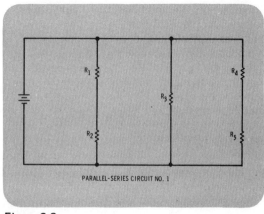

Figure 8.2

- **Series Voltage Law and Formula**
- **Series Current Law and Formula**

In order to analyze circuits of this type, you will be applying both the series and parallel circuit laws. To help keep these laws in mind, review them now briefly before examining your first parallel-series circuit.

Series Voltage Law and Formula — Series circuit behavior is summarized in three rules. Recall that as far as voltage behavior is concerned the rule to use is, "The sum of the individual voltage drops around a series circuit is equal to the total applied voltage." Putting this law in formula form, the law appears as shown in Figure 8.3:
$E_T = E_{R1} + E_{R2} + E_{R3}$.

LAWS OF SERIES CIRCUITS

1. THE SUM OF THE INDIVIDUAL VOLTAGE DROPS AROUND A SERIES CIRCUIT EQUALS THE APPLIED VOLTAGE

$$E_T = E_{R1} + E_{R2} \, E_{R3} + \cdots$$

Figure 8.3

Series Current Law and Formula — The current behavior of series circuits is summarized in a law which states "The current has the same value at any point within a series circuit," as shown in Figure 8.4. The current law for series circuits written as a formula indicates that the total circuit current is the same as that current through each of the circuit resistors: $I_T = I_{R1} = I_{R2} = I_{R3}$.

2. CURRENT HAS THE SAME VALUE AT ANY POINT WITHIN A SERIES CIRCUIT

$$I_T = I_{R1} = I_{R2} = I_{R3} = \cdots$$

Figure 8.4

- Series Resistance Law and Formula
- Parallel Voltage Law and Formula
- Parallel Current Law and Formula

Series Resistance Law and Formula — The law governing the behavior of resistance in series circuits states: "The sum of the individual resistances within a series circuit equals the total resistance of the circuit." To find the total (or equivalent) resistance of a string of resistors wired in series, simply add the resistance of each series wired resistor, as shown in Figure 8.5.

As you recall, three similar laws governing the operation of parallel circuits have also been covered. The laws point out the key differences between series and parallel circuit behavior.

3. THE SUM OF THE INDIVIDUAL RESISTANCES WITHIN A SERIES CIRCUIT EQUALS THE TOTAL RESISTANCE OF THE CIRCUIT

$$R_T = R_1 + R_2 + R_3 + \ldots.$$

Figure 8.5

Parallel Voltage Law and Formula — The law describing voltage behavior in parallel circuits states: "The total applied voltage in a parallel circuit is the same across each branch of that circuit." This law may be written in formula form as shown in Figure 8.6. Note that in the formula the subscripts have been changed from E_{R1} to E_{B1}. The "B" subscript is used to indicate "Branch" in this formula, and it will be of help to focus your attention on the behavior of parallel *branches* in this lesson. Notice that the only real difference between "regular" parallel circuits and parallel-series circuits is that *parallel-series circuits may contain several resistors in any branch.*

LAWS FOR PARALLEL CIRCUITS

1. THE TOTAL VOLTAGE OF A PARALLEL CIRCUIT IS THE SAME ACROSS EACH BRANCH OF THAT CIRCUIT

$$E_T = E_{B1} = E_{B2} = E_{B3}$$

Figure 8.6

Parallel Current Law and Formula — The law governing current behavior in a parallel circuit states, "The total (or main line) current in a parallel circuit is equal to the sum of the individual branch currents." Figure 8.7 shows how this law is written in formula form. Notice once again a "B" subscript is used to indicate branch.

2. THE TOTAL CURRENT IN A PARALLEL CIRCUIT IS EQUAL TO THE SUM OF THE INDIVIDUAL BRANCH CURRENTS.

$$I_T = I_{B1} + I_{B2} + I_{B3} + \ldots.$$

Figure 8.7

Parallel Resistance Law — Finally, the law governing resistance in parallel circuits states, "The total resistance of a parallel circuit is always less than, or approximately equal to, that of the smallest resistive branch (Figure 8.8).

> 3. THE TOTAL RESISTANCE IN A PARALLEL CIRCUIT IS ALWAYS LESS THAN OR APPROXIMATELY EQUAL TO THE SMALLEST RESISTIVE BRANCH.

Figure 8.8

Parallel Resistance Formula — The most general formula used in determining parallel resistance is "Sum of the Reciprocals Formula," shown in Figure 8.9. This formula can be used in calculating equivalent resistance of any parallel resistor combination.

$$R_T = \cfrac{1}{\cfrac{1}{R_1} + \cfrac{1}{R_2} + \cfrac{1}{R_3} \cdots}$$

Figure 8.9

Circuit Laws — The parallel and series circuit rules will be the basic tools you will need as you proceed to analyze any parallel-series circuit. The only other formulas you will need will be those given you by Ohm's law. The real key to analyzing more complex parallel-series circuits boils down to the question, "How do I know which circuit laws to apply and where to apply them?" The answer to this question is fairly straightforward, as you might have guessed. Simply apply the series circuit laws to those portions of the circuits that are wired in series (Figure 8.10). Those parts of the circuits you encounter that are connected in parallel are subject to the parallel circuit laws.

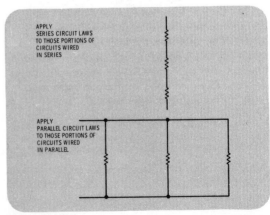

APPLY SERIES CIRCUIT LAWS TO THOSE PORTIONS OF CIRCUITS WIRED IN SERIES

APPLY PARALLEL CIRCUIT LAWS TO THOSE PORTIONS OF CIRCUITS WIRED IN PARALLEL

Figure 8.10

All of these problems depend on your ability to recognize various parts of circuits as being in either series or parallel with other circuit components. Hopefully, by now you can fairly easily guess at the one sure way to gain expertise in this area. Practice! Your "circuit sense," that is, your ability to just examine a circuit and sense what is going on right away, develops only after you have examined many different types of circuits.

Basic Parallel-Series Circuit — Figure 8.11 shows a basic parallel-series circuit. First, focus your attention on the fact that R_1 and R_2 are connected in series with each other, and that R_3 and R_4 are also connected in series with each other. Now, focus on the whole circuit. Notice that branch 1, which contains R_1 and R_2, is in *parallel* with branch 2, which contains R_3 and R_4.

To solve for voltage, current, and resistance in this circuit, you must simply apply the series circuit rules to the series-connected *resistors*, and then the parallel-circuit rules to the parallel-connected *branches*. Remember, your general aim is to solve for all the voltage drops, currents, and total resistance of the circuit. Also, recall that once the circuit's total or equivalent resistance is known, this value can be used to help you find the total or main line current it draws.

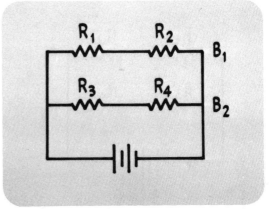

Figure 8.11

Ohm's Law — The procedure for finding the total current simply involves using Ohm's law as shown in Figure 8.12. Total current I_T equals the applied voltage, E_A, divided by the circuit's total equivalent resistance, R_{eq}.

The process by which the total equivalent resistance, R_{eq}, is found is called *circuit reduction*. The term "circuit reduction" refers to the processes by which a complex circuit is reduced to one that is simpler. A circuit is reduced by alternately applying the series and parallel circuit laws.

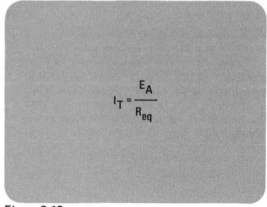

$$I_T = \frac{E_A}{R_{eq}}$$

Figure 8.12

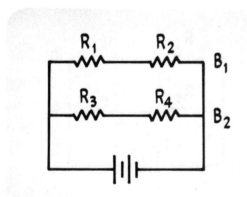

Figure 8.13

Basic Parallel-Series Circuit (Circuit Reduction) — How do you begin the process of circuit reduction? A good idea is to begin with the simplest parts of the circuit focusing your attention on how they may be further simplified. For example, in the basic parallel-series circuit shown in Figure 8.13 the simplest thing you can do immediately is to add the series resistors in each branch using the series circuit rule. You could then replace the two resistors in each branch with one equivalent resistor whose value is the sum of the two resistances. The sum of the values of R_1 and R_2 is labeled R_{1-2}. The sum of the resistances R_3 and R_4 is R_{3-4}.

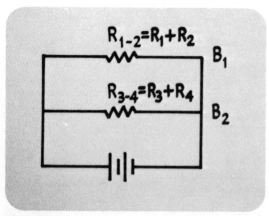

Figure 8-14

Basic Parallel-Series Circuit (First Reduction) — You can redraw the circuit as shown in Figure 8.14 using the reduced values, R_{1-2} and R_{3-4}. Notice that you now have a simple parallel circuit with a single resistor in each branch. You can proceed to further reduce this circuit to a circuit containing one single resistance by combining R_{1-2} and R_{3-4} using the law for adding resistances in parallel. The final single resistance R_{eq} can be obtained using the formula:

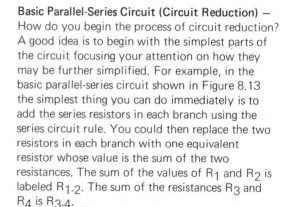

$$R_{eq} = \frac{1}{\dfrac{1}{R_{1-2}} + \dfrac{1}{R_{3-4}}}$$

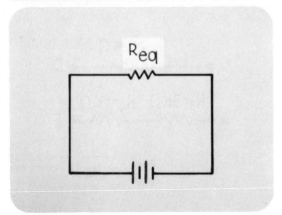

Figure 8.15

Parallel-Series Circuit No. 2 (Third Reduction) —
This finally reduces the circuit to the simple form
shown in Figure 8.15. Now you may easily solve
for the total or main line current that would flow
in this circuit. This current value will then aid in
solving for the *rest* of the voltages and currents in
this circuit.

Reexamine this same circuit but this time
consider the circuit reduction using realistic ohmic
values for the four resistors.

Parallel-Series Circuit No. 2 — Figure 8.16 shows
the same basic parallel-series circuit with resistance
values inserted. Branch 1 consists of resistor R_1
with 1 kilohm of resistance in series with resistor
R_2, which is a 2-kilohm resistor. Resistors R_3 and
R_4 in the second branch have resistance values of
3 kilohms and 1 kilohm, respectively. First focus
on the upper branch of this circuit, between
points C and D. Resistors R_1 and R_2 may be
combined using the law for addition of series
resistances. This enables you to replace R_1 and R_2
with *one* equivalent resistance, $R_{1\text{-}2}$.

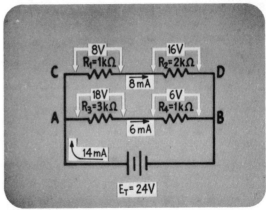

Figure 8.16

Series Resistance Formula — Using the resistance
law to find $R_{1\text{-}2}$ (as shown in Figure 8.17), $R_{1\text{-}2} =$
$R_1 + R_2$ or $R_{1\text{-}2} = 1$ kilohm + 2 kilohms for a
total of 3 kilohms.

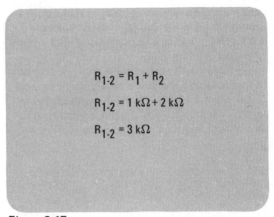

$$R_{1\text{-}2} = R_1 + R_2$$

$$R_{1\text{-}2} = 1\ k\Omega + 2\ k\Omega$$

$$R_{1\text{-}2} = 3\ k\Omega$$

Figure 8.17

- **Basic Parallel-Series Circuit (First Reduction)**
- **Series Resistance Formula**
- **Basic Parallel-Series Circuit (Second Reduction)**

Basic Parallel-Series Circuit (First Reduction) — The two resistors in the upper branch may be replaced with one single resistor, R_{1-2}, of 3 kilohms. Similarly, the second branch may be simplified by the combination of R_3 and R_4 resulting in an equivalent resistance R_{3-4} as you can see in Figure 8.18.

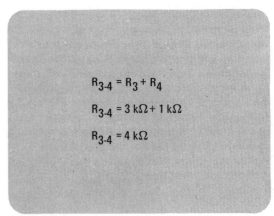

Figure 8.18

Series Resistance Formula — Again applying the series circuit resistance formula, R_{3-4} equals 3 kilohms plus 1 kilohm or 4 kilohms (Figure 8.19).

$$R_{3-4} = R_3 + R_4$$
$$R_{3-4} = 3\ k\Omega + 1\ k\Omega$$
$$R_{3-4} = 4\ k\Omega$$

Figure 8.19

Basic Parallel-Series Circuit (Second Reduction) — Resistors R_3 and R_4 may be replaced by one resistor, R_{3-4}, having a resistance of 4 kilohms and illustrated in Figure 8.20.

At this point, notice that this circuit may be further reduced because it is now a simple parallel circuit. Since this circuit has only *two* resistors in parallel, the product-over-the-sum formula can be used to calculate the total equivalent resistance.

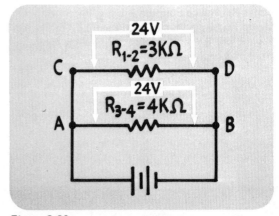

Figure 8.20

Parallel Resistance Formula — All necessary calculations are shown in Figure 8.21. Begin with the product-over-the-sum-formula and substitute the correct circuit values to get 3 kilohms *times* 4 kilohms *over* 3 kilohms *plus* 4 kilohms. Having done the indicated multiplication and addition, you will get 12 X 10^6, divided by 7 X 10^3. Completing the calculation, the total equivalent resistance is found to be equal to 1.714 X 10^{+3}, or 1.71 kilohms (Figure 8.21).

$$R_{eq} = \frac{R_{1-2} \times R_{3-4}}{R_{1-2} + R_{3-4}}$$

$$R_{eq} = \frac{3k \times 4k}{3k + 4k}$$

$$R_{eq} = \frac{12 \times 10^6}{7 \times 10^3}$$

$$R_{eq} = 1.714 \times 10^3 = 1.71 \text{ k}\Omega$$

Figure 8.21

Basic Parallel-Series Circuit (Final Reduction) — This 1.71 kilohms of equivalent resistance can be placed into a new circuit that, as far as total resistance is concerned, is equivalent to the original circuit. The new resistor has been labeled R_{eq} for equivalent. Notice that by using the process of circuit reduction or circuit simplification, the complex parallel-series circuit you began with has now been reduced to a circuit that contains only one resistor (Figure 8.22).

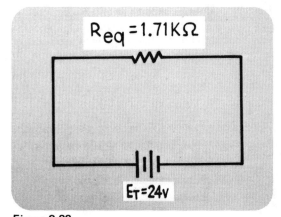

Figure 8.22

Ohm's Law (I_T) — Using this equivalent resistance, you can find the total or main line current that will flow in this circuit. The voltage applied to the circuit is 24 volts. The circuit's total resistance is also known, 1.71 kilohms. These values can be substituted into the Ohm's law formula $I_T = E_T/R_{eq}$ to find the total current (Figure 8.23). Divide 24 volts by 1.71 kilohms and you will find that 14 milliamps of main line current is flowing in the circuit.

To proceed to the next part of the circuit analysis, look at the reduced circuit carefully. It can be seen that the entire 24 volts is dropped across the equivalent resistance, R_{eq}. Now, here's a

$$I_T = \frac{E_T}{R_{eq}}$$

$$I_T = \frac{24 \text{ V}}{1.71 \text{ k}\Omega}$$

$$I_T = \frac{24 \text{ V}}{1.71 \times 10^3}$$

$$I_T = 14 \times 10^{-3} = 14 \text{ mA}$$

Figure 8.23

key method of procedure in circuit analysis: back
up one step at this point.

Basic Parallel-Series Circuit (Second Reduction) —
Figure 8.24 shows the same circuit just before it
was reduced to one resistance showing two resistive
branches, each with a single resistor. Recall that
the parallel circuit rules state that the voltage is the
same across all branches of a parallel circuit. In this
case both branches each have 24 volts applied
across them. This bit of information is very
significant because the voltage across each branch
may simply be divided by the branch resistance to
find the current flowing through each branch.

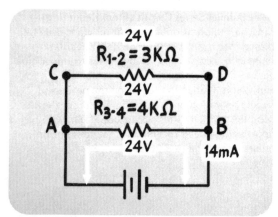

Figure 8.24

Ohm's Law (I_{1-2}) — Consider branch 1. At this
point you know that E_{R1-2} is equal to 24 volts
and that R_{1-2} is 3 kilohms. If these values are used
in an appropriate Ohm's law formula, as shown in
Figure 8.25, a current value of 8 milliamps through
R_{1-2} can be calculated. (Notice this is the current
flowing through resistor R_{1-2}, and, hence, this is
the current that flows *throughout* branch 1.)

$$E_{R1-2} = 24 \text{ V}$$

$$R_{1-2} = 3 \text{ k}\Omega$$

$$I_{1-2} = \frac{E_{R1-2}}{R_{1-2}}$$

$$I_{1-2} = \frac{24 \text{ V}}{3 \text{ k}\Omega}$$

$$I_{1-2} = 8 \text{ mA}$$

Figure 8.25

Ohm's Law (I_{3-4}) — The current flowing through the second branch containing R_{3-4} may be calculated in the same manner. The voltage across this branch is also 24 volts, and the branch resistance in this case is 4 kilohms. Using Ohm's law again, as shown in Figure 8.26, 24 volts is divided by 4 kilohms to equal 6 milliamps of current through R_{3-4}, and, *hence, all of branch 2.*

$$I_{3-4} = \frac{E_{R3-4}}{R_{3-4}}$$

$$I_{3-4} = \frac{24\ V}{4\ k\Omega}$$

$$I_{3-4} = 6\ mA$$

Figure 8.26

Basic Parallel-Series Circuit — Now that the current flow through both branches is known, the original circuit diagram can be used to find the voltages dropped across each individual resistor (Figure 8.27). For example, in branch 1, you know that 8 milliamps of current is flowing through *both* R_1 and R_2. Why? Notice that both these resistors are connected in *series*. Throughout any series-connected branch, whatever current is flowing in one resistor, flows in all of them. So, whatever current flows in the equivalent resistor R_{eq}, is flowing through all the series resistors that make it up.

You can proceed to calculate the voltage drop across R_1 and R_2 now that you know the current flowing through the top branch.

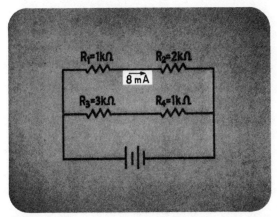

Figure 8.27

Ohm's Law (E_{R1}) — Using Ohm's law, E_{R1} can be found by simply multiplying I_{R1} by R_1 (Figure 8.28). Substituting the circuit values into the equation and completing the mathematics yields 8 milliamps times 1 kilohm, or 8 volts for the voltage across R_1.

Notice that once E_{R1} is known to be 8 volts, the series circuit voltage law may be applied to find E_{R2}. Recall that in a series circuit the sum of the individual voltage drops is equal to the applied voltage. Again recall that the total voltage applied to this branch is 24 volts.

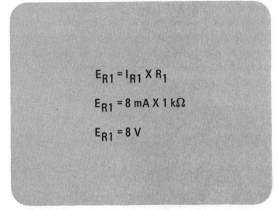

$$E_{R1} = I_{R1} \times R_1$$

$$E_{R1} = 8\ mA \times 1\ k\Omega$$

$$E_{R1} = 8\ V$$

Figure 8.28

Series Voltage Formula (E$_{R2}$) — Consequently, E$_{R1}$, which is 8 volts, plus E$_{R2}$ must equal 24 volts. Subtracting 8 volts from 24 volts yields 16 volts, which is the voltage dropped across R$_2$ (Figure 8.29).

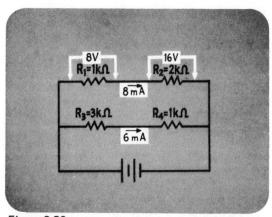

$$E_{R1} + E_{R2} = E_T$$
$$8V + E_{R2} = 24V$$
$$E_{R2} = 24V - 8V$$
$$E_{R2} = 16V$$

Figure 8.29

Basic Parallel-Series Circuit — The same procedure may now be used to calculate the voltage drops across R$_3$ and R$_4$. Previously, it was calculated that 6 milliamps flows through the second branch, and thus flows through R$_3$ and R$_4$ because of their series connection (Figure 8.30).

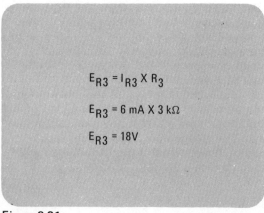

Figure 8.30

Ohm's Law (E$_{R3}$) — Using Ohm's law once again, this time in the form E$_{R3}$ = I$_{R3}$ X R$_3$ (Figure 8.31), you may now calculate the voltage drop across R$_3$. Substituting the resistance and current values into the formula, E$_{R3}$ is found to equal 6 milliamps times 3 kilohms or 18 volts.

$$E_{R3} = I_{R3} \times R_3$$
$$E_{R3} = 6 \text{ mA} \times 3 \text{ k}\Omega$$
$$E_{R3} = 18V$$

Figure 8.31

Series Voltage Formula (R$_4$) — Now that you know 18 of the 24 volts applied to this branch are dropped across R$_3$, by subtracting E$_{R3}$ (which is 18 volts) from 24 volts, the voltage dropped across R$_4$ may be found. As shown in Figure 8.32, E$_{R4}$ is 6 volts.

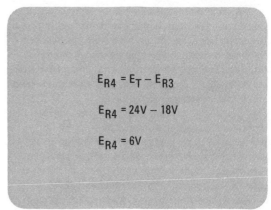

$$E_{R4} = E_T - E_{R3}$$

$$E_{R4} = 24V - 18V$$

$$E_{R4} = 6V$$

Figure 8.32

Basic Parallel-Series Circuit — Through this logical step-by-step procedure, you have seen how the voltage across and current through every component in this circuit can be calculated (Figure 8.33). The technique of circuit reduction, which you have seen illustrated, will become an important part of your "bag of tricks" as you analyze circuits. Why? One reason to learn circuit reduction as applied to parallel-series circuits is because many common electronic circuits you will encounter in your everyday life will have this general configuration. A particularly important example can be found in the typical automobile wiring system.

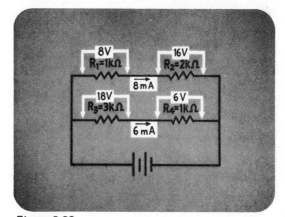

Figure 8.33

Automotive Electrical System — Figure 8.34 shows a schematic diagram of several key parts of an automobile. Notice that in this automotive situation all of the circuit branches are powered by the car's standard 12-volt battery. The branch closest to the battery contains a headlight and a switch. The headlight is designed to operate at 12 volts and will receive the total battery voltage when the switch is closed.

   The second branch is a series circuit which contains a spark coil and spark plug, a resistor called a "dropping resistor," and a momentary contact switch which simulates the switching action of the points in an automobile distributor.

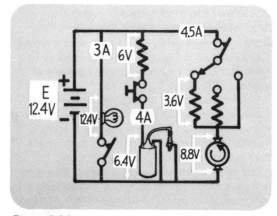

Figure 8.34

Branch 3 consists of a heater fan motor, a fan speed control switch which can be turned to one of three positions, and an on-off switch. Notice that there are resistors connected to two of the three fan speed control switch positions. These resistors are also called "dropping resistors" as is the resistor connected in series with the coil.

The term "dropping resistor" has cropped up quite a bit. Consider the function of these dropping resistors in the circuit. In many automotive applications, devices are designed to operate at a *lower voltage* than the usual 12 volts supplied by the car battery. To enable these devices to operate properly, some other device is needed which will "drop" some of the circuit voltage. This is how the "dropping resistor" got its name. In automotive applications, the dropping resistor value is selected so that when the 12 volts from the battery is applied to the branch, enough voltage is dropped across the resistor to allow a device such as the spark coil to receive the correct voltage.

Automotive Electrical System (Equivalent Schematic) — To simplify this discussion of the automotive electrical system, the circuit pictorial diagram shown in Figure 8.34 can be reduced to an equivalent circuit schematic consisting solely of resistors, as shown in Figure 8.35.

In this schematic the headlight resistance is drawn as a simple resistor in the first branch, and the spark coil is shown as a resistance in series with its dropping resistor in the second branch. In the third branch, the motor resistance is shown connected in series with a speed control dropping resistor. Examining this circuit as a whole, you can recognize this as a parallel-series circuit.

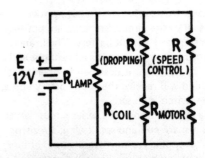

Figure 8.35

Automotive Wiring System — Figure 8.36 shows a drawing of the actual circuit. The 12-volt battery supplies power to branch 1, consisting of the headlamp and its on-off switch, branch 2 which contains the dropping resistor, on-off switch, spark coil, and spark plug. Branch 3 consists of an on-off switch, a three-pole switch, two resistors, and an automobile heater fan motor.

Imagine at this point that you need to analyze this automotive electrical system in detail. You need to calculate the total current it demands, I_T, as well as the individual branch currents, I_1, I_2, and I_3, and the voltage drop across each individual component including the dropping resistors.

When you begin your analysis of the circuit, you know absolutely nothing about it except for the total applied voltage. You will need something more than that, so how do you proceed? In this example, it will be assumed that some of the circuit voltages and currents can be *measured*, while other circuit values will have to be *calculated* using circuit laws. In analyzing or troubleshooting real circuits, you may often have to proceed in this manner. In some cases, you may have a schematic in the auto's service manual showing you generally how various components are wired throughout the car. Some of these actual components may be easily accessible allowing you convenient access for voltage or current measurements. Some components, however, may be quite inaccessible, enclosed in cabinets or in operating positions deep inside your car. As you will see, it is possible to utilize *measurements*, giving you some known values, *combined with* calculations to give information about inaccessible circuit components.

To measure voltages, remember that an appropriate voltmeter must be inserted *across* the circuit component being tested (in parallel with it). For any current values that are needed, the circuit must be broken and an ammeter inserted in *series*, so that the current being measured passes through it. For example, the voltage across the first branch may be easily measured. Notice that the voltage across this branch will be equal to the source voltage since the headlamp is connected in parallel or directly across the battery.

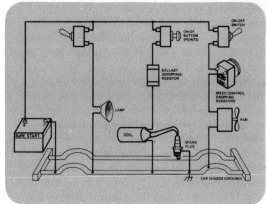

Figure 8.36

Automotive Wiring System — Suppose that the voltage measured across the branch containing lamp, E_L, was 12.4 volts and suppose you needed to find the current flowing through the lamp when it is operating. To do this you must connect an ammeter in series with the lamp. A convenient means of measuring the current would be to connect the ammeter directly across the lamp on-off switch terminals. If the switch is initially open, connecting the meter across its terminals will complete the circuit and allow the full branch current to flow through the ammeter as required for the measurement.

If you performed this measurement in a typical branch of this type containing a single headlamp, as shown in Figure 8.37, you would find that the current flowing through the lamp was about 3 amperes.

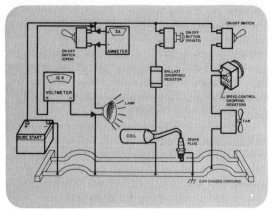

Figure 8.37

Ohm's Law (R_L) — From the two values, 12.4 volts and 3 amps, you may calculate the resistance of the lamp when in operation. Using Ohm's law in the form R equals E over I and substituting your measured values, you have 12.4 volts divided by 3 amps which is equal to 4.13 ohms (Figure 8.38). Incidentally, if you were to measure the resistance of this headlamp directly with an ohmmeter, you would read a significantly lower value for R_L. This is because as the filaments of the bulb heat up, their resistance also increases. So, in order to be accurate, the resistance of the lamp should be calculated while the lamp is operating, rather than measured while it is off.

$$R_L = \frac{E}{I}$$

$$R_L = \frac{12.4\,V}{3\,A}$$

$$R_L = 4.13\,\Omega$$

Figure 8.38

Automotive Wiring System — By applying the rules for a parallel circuit (Figure 8.39), you know that the voltage across branch 2 should be the same as the source voltage, 12.4 volts. This total voltage applied to branch 2 will be divided between the dropping resistor and the coil. To get a picture of this voltage behavior, the voltage across the dropping resistor can be measured while the coil is in operation. Then once the voltage across the resistor is known, this value may be subtracted from the total voltage applied to the branch, and the remainder is the voltage that is dropped across the coil.

If you were to connect your voltmeter across a typical automotive ballast or dropping resistor while the switch or points were closed, you would typically find that 6 volts are dropped across the resistor.

Knowing that 6 volts are dropped across the ballast resistor, you then know that the remainder of the total voltage applied to the branch is dropped across the ignition coil.

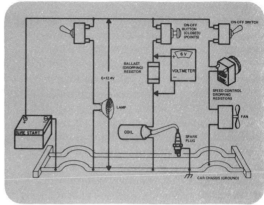

Figure 8.39

Series Voltage Formula — Subtracting the 6 volts dropped across the resistor, E_R from the 12.4 volts applied to the branch, E_T, you would find that a voltage of 6.4 volts is dropped across the coil as shown in the equation in Figure 8.40. Since any branch in this circuit is actually a series circuit, the sum of the individual voltage drops in the branch is equal to the total applied branch voltage.

$$E_T - E_R = E_C$$

$$12.4V - 6V = 6.4V$$

Figure 8.40

Automotive Wiring System — Moving on in your measurement and analysis of this circuit, pictured in Figure 8.41, you know that by connecting an ammeter in series with this branch, the branch current can be found. One way to perform this measurement would be to place the ammeter across the open switch, completing the circuit and allowing the total current to flow through the ammeter. Typically, this branch current would read about 4 amps. You would also notice that each time the circuit containing the coil was activated and then deactivated, the spark plug would "spark." This important electronic phenomenon involving sparks and coils is a key effect used in automotive ignition systems, and will be covered in detail in a later lesson.

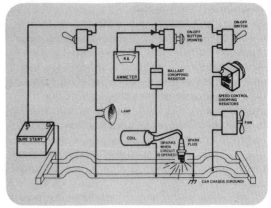

Figure 8.41

Automotive Wiring System — You could employ the same procedures to find the individual voltage drops and branch current in branch 3. Imagine for a moment, however that the heater fan was inside an enclosure that was difficult to get to. A direct voltage measurement in a case like this would require lots of disassembling. In this case, instead of measuring the voltage drop across the motor, you could calculate it after making some other measurements in the circuit. Assume that the fan motor speed control is set in the medium speed position as you begin your work (Figure 8.42).

First of all, the total branch current may be found by connecting your ammeter across the open on-off switch, thus completing the circuit. With one type of typical fan motor and motor speed control (on medium setting), you would read a branch current in the neighborhood of 4.5 amps.

While the ammeter is still connected, you could vary the setting of the motor speed control and observe its effect on current flow in this branch. You should note that when a larger value of dropping resistance is switched in, the motor slows down and the ammeter reads a lower value. When a smaller value of resistance is selected, the motor speeds up and the ammeter indicates that more current is flowing.

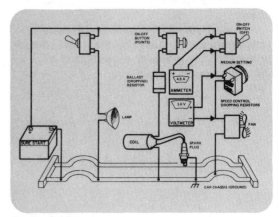

Figure 8.42

With the motor speed switch set at its medium speed position, the smaller of the two dropping resistors is in operation. In typical circuit situations, you may measure 3.6 volts dropped across this resistor. Units of this type may be mounted right behind your dashboard and be fairly accessible for measurement.

Series Voltage Formula — Since the total branch voltage is known, the voltage dropped across the speed control resistor can be subtracted from the total voltage, and the remainder will be the voltage dropped across the motor. As shown in Figure 8.43, 12.4 volts minus 3.6 volts gives a voltage of 8.8 volts across the motor.

$$E_m = E_T - E_R$$

$$E_m = 12.4V - 3.6V$$

$$E_m = 8.8V$$

Figure 8.43

Automotive Wiring System — At this point and shown in Figure 8.44, the voltage drop across each component in this automotive circuit is known along with all of the branch currents. The total or main line current the battery is providing to this circuit may be found by simply adding the individual branch currents.

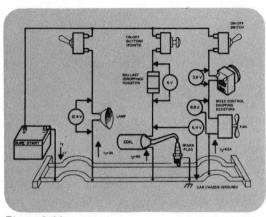

Figure 8.44

Parallel Current Formula — Using the formula for finding main line currents shown in Figure 8.45, add 3 amps of current flowing in branch 1, 4 amps flowing in branch 2, and 4.5 amps flowing through branch 3. The total current flowing in the circuit is 11.5 amps.

Your analysis of this automotive circuit example is now complete. All the voltage drops have been found, the branch currents have been measured and added together to find the total current. This example illustrates a common technique used in analyzing actual circuits. Some circuit values were first measured and then by applying the three laws governing series circuits, the three laws governing parallel circuits, and Ohm's law, the other needed values were calculated.

If you keep the laws governing series and parallel circuits firmly in your mind and practice, you will encounter little difficulty as you move on the more complex circuit.

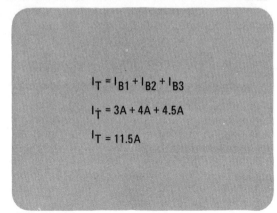

$$I_T = I_{B1} + I_{B2} + I_{B3}$$

$$I_T = 3A + 4A + 4.5A$$

$$I_T = 11.5A$$

Figure 8.45

As you have seen, your automobile is a good place to learn about dc electricity and to get some practice in handling and studying dc circuits. In a way, your car is like your own private dc circuits lab, and many laboratory and circuit analysis techniques can be learned and practiced right in your car. Be sure, as always, to use *extreme caution* when working with automotive circuitry. Also, please realize that the circuit studied in this lesson presents only part of what is happening in the average automobile electrical system, and the measured values shown in this example were only *typical* values. Actual circuit voltages and currents will vary from car to car as you might expect.

Details on automotive wiring are discussed in further detail in automotive shop manuals and repair texts that are available. You will find that most of the electrical systems in your car, gauges, charging, starting, ignition, etc., involve the basic principles of parallel-series circuit analysis that have been discussed in this lesson. Also, keep in mind that on most cars the frame or body of the car forms part of all electrical circuits completing the circuit path to the battery. This "chassis ground" is usually connected to the *negative* battery terminal in most American cars.

Before completing this lesson, one more circuit will be analyzed for you in detail. This circuit will put you through some different types of analyses and manipulation than before, but you will still be using the old standby, Ohm's law and the laws governing series and parallel circuits. This circuit, unlike the first you analyzed, will have a variety of unknown quantities and will be presented in a slightly different schematic form.

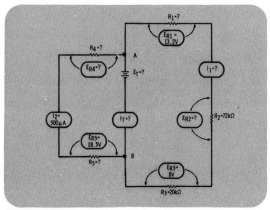

Figure 8.46

Parallel-Series Circuit — The circuit in Figure 8.46 is clearly a little different looking than those you have handled previously. The voltage source is drawn in the center of the circuit, and the branches are drawn around to either side of it. By now you should be able to recognize this as a parallel-series circuit with two branches, and you should be able to redraw this circuit if you desire in a more familiar form. Notice that the total potential difference is applied to points A and B and from there to each branch. If you can imagine for a moment that the wires in this circuit were as flexible as rubber bands, you could take the left-hand branch and stretch it back up and over the voltage source to the right of the right-hand branch without changing any circuit connections.

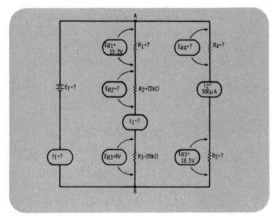

Figure 8.47

Equivalent Parallel-Series Circuit — The circuit drawn in Figure 8.46 is entirely identical to the more recognizable one shown in Figure 8.47. The knack of redrawing circuits such as this comes easily with practice and can greatly simplify complicated looking circuits into simpler ones. Doing the practice problems included at the end of this lesson will help you learn this.

Examine the known and unknown quantities in this circuit. In branch 1, the voltage across R_1 is known, but its value in ohms is not. The value of R_2 is known, but you do not know the voltage across R_2. Both the voltage across and the resistance of R_3 are known.

The branch current in the second branch is known to be 500 microamps, and the voltage across R_5 is also known. Nothing else is given. Where do you go from here? Focus your attention on the left-hand branch of this circuit and look for a component about which two values are known so that you can apply Ohm's law. R_3 fits this description. The voltage across it and its resistance value are known and thus you can use Ohm's law to find the current through it.

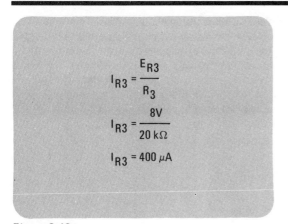

$$I_{R3} = \frac{E_{R3}}{R_3}$$

$$I_{R3} = \frac{8V}{20\ k\Omega}$$

$$I_{R3} = 400\ \mu A$$

Figure 8.48

Ohm's Law — Using Ohm's law in the form I equals E divided by R and substituting the known values for R_3, 8 volts divided by 20 kilohms yields a calculated current flowing through R_3 of 400 microamps as you can see in Figure 8.48.

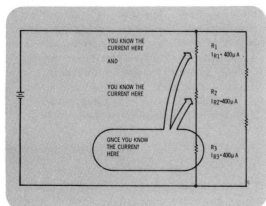

Figure 8.49

Series Current Analysis — The law for current behavior in series circuits says that current is the *same* at all points in a series circuit. Notice in Figure 8.49 that R_1, and R_2 and R_3 are all in series. Once you know the current flowing through any of these resistors, you know that the same current is flowing throughout that branch. You know 400 microamps are flowing in R_3, so you also know that the current flowing throughout branch 1 is 400 microamps. This now enables you to find the voltage across R_2, because you know that 400 microamps is flowing through R_2, and R_2 has an ohmic value of 72 kilohms.

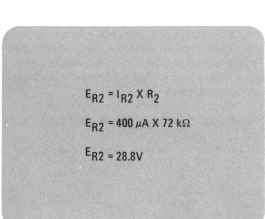

$$E_{R2} = I_{R2} \times R_2$$

$$E_{R2} = 400\ \mu A \times 72\ k\Omega$$

$$E_{R2} = 28.8V$$

Figure 8.50

Ohm's Law (E_{R2}) — The voltage across R_2 can be calculated using Ohm's law in the form E_{R2} equals the current in branch 1 times the value of R_2, which is 400 microamps times 72 kilohms which yields a voltage value of 28.8 volts as shown in Figure 8.50. With this voltage calculated, you now know the voltage drop for each resistor in branch 1.

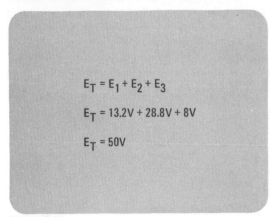

$$E_T = E_1 + E_2 + E_3$$

$$E_T = 13.2V + 28.8V + 8V$$

$$E_T = 50V$$

Figure 8.51

Series Circuit Voltage Law — Recall that the voltage law for series circuits says that the sum of the individual voltage drops in any series circuit equals the total applied voltage. Branch 1 is itself a series circuit. By adding the individual voltage drops, as shown in Figure 8.51, the total voltage across branch 1 is calculated to be 50 volts.

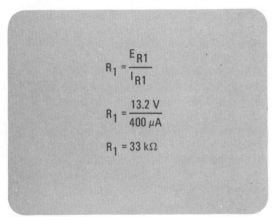

$$R_1 = \frac{E_{R1}}{I_{R1}}$$

$$R_1 = \frac{13.2 \text{ V}}{400 \text{ }\mu A}$$

$$R_1 = 33 \text{ k}\Omega$$

Figure 8.52

Ohm's Law (R_1) — The only unknown quantity remaining in the first branch is R_1. Using Ohm's law in the form R =E/I, R_1 may be calculated since you know that 13.2 volts is dropped across R_1, and 400 microamps are flowing through it. Working this problem out as shown in Figure 8.52, 13.2 volts divided by 400 microamps gives 33 kilohms of resistance for the value of R_1.

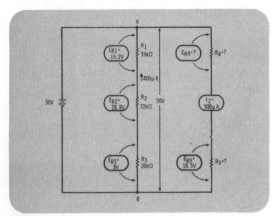

Figure 8.53

Parallel-Series Circuit: Branch 2 — In moving from one parallel branch to another, keep in mind the voltage law for parallel circuits which states that the voltage across all of the branches in any parallel circuit is the same.

Since you know there are 50 volts across branch 1, as illustrated in Figure 8.53, there must also be a 50-volt drop across branch 2. In fact, 50 volts must be the total voltage applied to this circuit by the battery or power supply.

- Calculation of E_{R4}
- Ohm's Law (R_4)
- Ohm's Law (R_5)

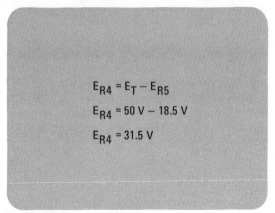

$$E_{R4} = E_T - E_{R5}$$

$$E_{R4} = 50\,V - 18.5\,V$$

$$E_{R4} = 31.5\,V$$

Figure 8.54

Calculation of E_{R4} — Since you know that there are 50 volts across branch 2 and 18.5 volts across R_5, you can readily solve for the voltage across R_4 (Figure 8.54). For a series branch such as this one, you know that the total applied voltage, E_T = 50 volts, equals the sum of individual voltage drops. One of the voltage drops, E_{R5}, is 18.5 volts, so the other, E_{R4}, must equal $E_T - E_{R5}$, which is 50 minus 18.5 or 31.5 volts.

$$R_4 = \frac{E_{R4}}{I_{R4}}$$

$$R_4 = \frac{31.5\,V}{500\,\mu A}$$

$$R_4 = 63\,k\Omega$$

Figure 8.55

Ohm's Law (R_4) — You can proceed to solve for the resistance value of R_4, as shown in Figure 8.55, because the voltage across it and the current through it are known. This time you will use Ohm's law in the form R equals E over I. Substituting the known voltage and current values into the formula yields 31.5 volts divided by 500 microamps which gives a calculated resistance of 63 kilohms for R_4.

$$R_5 = \frac{E_{R5}}{I_2}$$

$$R_5 = \frac{18.5\,V}{500\,\mu A}$$

$$R_5 = 37\,k\Omega$$

Figure 8.56

Ohm's Law (R_5) — R_5 may now be calculated quite easily. Using R = E/I and substituting the known values E_{R5} = 18.5 volts and I_2 = 500 microamps, as shown in Figure 8.56, R_5 is calculated as 37 kilohms.

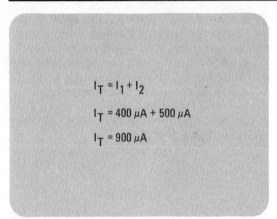

$$I_T = I_1 + I_2$$

$$I_T = 400\ \mu A + 500\ \mu A$$

$$I_T = 900\ \mu A$$

Figure 8.57

Parallel Current Formula — Now you know all there is to know about each one of the individual branches, so it will be easy to calculate the total amount of current flowing in the circuit. Use the law for parallel current which says "the sum of the individual branch currents equals the total current" in formula form, I_T equals I_1 plus I_2 (Figure 8.57). Substituting the branch currents into the formula and adding these together, 900 microamps of total current is found to be flowing in this circuit.

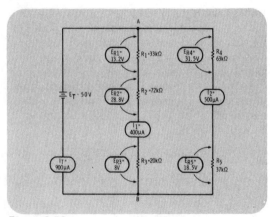

Figure 8.58

Analyzed Parallel-Series Circuit — The fully analyzed circuit with all of the known values inserted is shown in Figure 8.58 for your reference. With several of these circuit analyses behind you and the practice provided by the problems at the end of this lesson, you should be able to carry out a detailed analysis on any circuit of this type.

As you see, there are several alternative ways to proceed through each calculation and the best way to really learn how to use them is to practice these techniques.

As has been mentioned, at the end of this lesson there are several practice problems designed to enable you to really learn and practice the methods.

If you should get bogged down, there are also several additional worked through examples showing how to solve these circuits.

Procedural Steps for Parallel-Series Circuits — This lesson has shown you how to use the laws governing series and parallel circuits along with Ohm's law to analyze *parallel-series* circuits (Figure 8.59). The technique of circuit reduction was also introduced, and you have a bit of practice behind you in that area.

Basically, when you first encounter a parallel-series circuit problem, carefully examine the known or given values on the schematic and then clearly mark those circuit values that you need to calculate. Focus your attention on each circuit component and what is *known* about it. From this you may be able to draw some immediate conclusions.

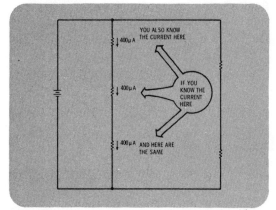

Figure 8.59

Current Through Series Resistors — As an example, see Figure 8.60. If a resistor is in a series branch and the current through it is known, then the same current flows through all resistors in that branch because of the law for currents in a series circuit.

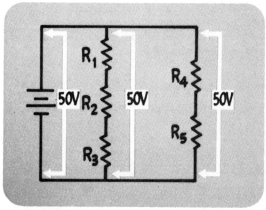

Figure 8.60

Voltage in Parallel — Remember also that once the voltage across any one complete branch is known, the parallel circuit voltage laws say that the same voltage must also be dropped across all of the circuit branches. This is shown in Figure 8.61.

Figure 8.61

- **Total Current Methods**
- **Parallel-Series Circuit**
- **Parallel-Series Circuit (First Reduction)**

Total Current Methods — Also recall that the total or main line current in a parallel-series circuit is the sum of all of the branch currents and is also equal to the applied circuit voltage divided by the circuit's equivalent resistance. This equation is shown in Figure 8.62.

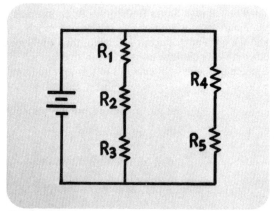

$$I_T = I_{B1} + I_{B2} + \cdots$$

$$= \frac{E_a}{R_{eq}}$$

Figure 8.62

Parallel-Series Circuit — The equivalent resistance may be found for a circuit such as the one in Figure 8.63 by systematically applying the method of circuit reduction.

Figure 8.63

Parallel-Series Circuit (First Reduction) — First, reduce *each series branch* to a single resistance value using the rules for adding resistances in series as shown in Figure 8.64.

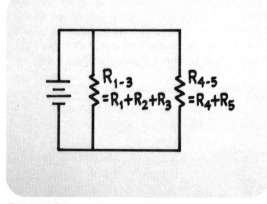

Figure 8.64

- **Parallel-Series Circuit (Second Reduction)**
- Power in Parallel-Series Circuits
- Example

Parallel-Series Circuit (Second Reduction) — Then *combine the parallel branch resistances* into a single equivalent resistance using the reciprocal formula, product-over-sum formula, or other appropriate technique as shown in Figure 8.65.

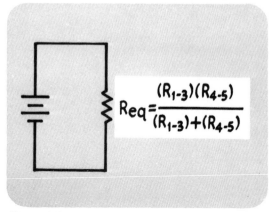

Figure 8.65

Power in Parallel-Series Circuits — Consider how to calculate the *total power dissipated* in the parallel-series circuit shown in Figure 8.66. You will find that in all dc resistive circuits power is an additive quantity. Each resistor in any circuit dissipates a certain amount of power, and to find the total power dissipated, you simply add all these individual power dissipations.

Alternatively, once a circuit has been fully analyzed, such as the circuit in the illustration, you can use the total circuit values E_T, I_T, and R_T in one of the three power formulas you have been given, and you will obtain the same result for the power dissipated.

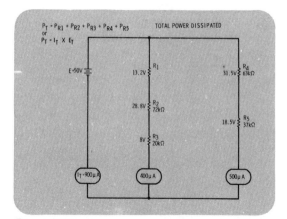

Figure 8.66

$$P = I \times E$$
$$P_{R1} = I_{R1} \times E_{R1} = (400 \times 10^{-6})(13.2)$$
$$= 5.28 \text{ MILLIWATTS}$$
$$P_{R2} = 11.52 \text{ mW}$$
$$P_{R3} = 3.2 \text{ mW}$$
$$P_{R4} = 15.8 \text{ mW}$$
$$P_{R5} = \underline{9.25 \text{ mW}}$$
$$P_T = 45.1 \text{ mW}$$
OR
$$P_T = I_T \times E_T = (900 \times 10^{-6})(50) = 45 \text{ mW}$$

Figure 8.67

Example — In the circuit of Figure 8.66 you could apply the formula $P = I \times E$ to each resistor as shown in Figure 8.67. Adding all the powers dissipated by each resistor yields a total dissipated power of 45.1 milliwatts. Alternatively, applying the formula $P_T = E_T \times I_T$ and substituting the total circuit values of 900 microamps and 50 volts, you obtain approximately 45 milliwatts.

● **Worked Through Examples**

1. Solve for all voltages and currents in the circuit shown below.

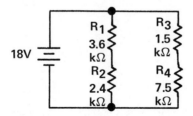

First of all, you need to simplify or reduce the circuit in order to find its equivalent resistance. The values of resistors R_1 and R_2 may be added since they are connected in series, 3.6 kilohms plus 2.4 kilohms is equal to 6.0 kilohms. This means that resistors R_1 and R_2 can be replaced by an equivalent resistance, R_{1-2} equal to 6 kilohms. The same procedure may be followed with resistors R_3 and R_4. When 1.5 kilohms and 7.5 kilohms are added, you get an equivalent resistance, R_{3-4} equal to 9.0 kilohms. The circuit redrawn to include these two equivalent resistances appears as shown below.

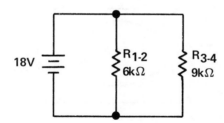

The circuit now appears as a simple parallel circuit. R_{1-2} and R_{3-4} may be combined using one of the parallel resistance formulas. This time the produce-over-the-sum formula will be used.

$$R_T = \frac{R_{1-2} \times R_{3-4}}{R_{1-2} + R_{3-4}}$$

Substitute the circuit values into the formula in powers of ten form:

$$R_T = \frac{6 \times 10^3 \times 9 \times 10^3}{(6 \times 10^3) + (9 \times 10^3)}$$

Multiply and add:

$$R_T = \frac{54 \times 10^6}{15 + 10^3}$$

Now divide:

$$R_T = \frac{54 \times 10^{6-3}}{15}$$

$$R_T = \frac{54 \times 10^3}{15}$$

$$R_T = 3.6 \times 10^3 = 3{,}600 \ \Omega$$

The total equivalent resistance of this circuit is 3,600 ohms. Now that you know the total equivalent resistance and the total applied voltage, Ohm's law may be used to find the total current flowing in this circuit.

$$I_T = E_T / R_{eq}$$

$$I_T = 18 \ V / 3{,}600 \ \Omega$$

$$I_T = 0.005 \ A = 5 \ mA$$

The equivalent circuit now looks like this:

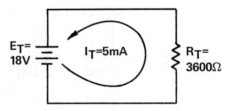

Now, an important point, back up a step. Draw the circuit as it appeared just before the final equivalent resistance was found. The circuit will look like this:

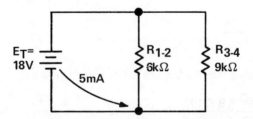

The total circuit current flows to the junction of the two resistances and splits. Some amount of current flows through the 9-kilohm resistance, and the rest flows through the other resistance. How can the exact amounts be found? This question is answered simply by applying the parallel circuit rule that says, "The voltage is the same across all branches of a parallel circuit." Since resistances R_{1-2} and R_{3-4} are connected in parallel with the source voltage, the voltage across them will be the same as the source voltage, 18 volts. Ohm's law may now be used to calculate the current flowing through each branch of the circuit.

For branch 1:

$$I_{1\text{-}2} = E_{1\text{-}2}/R_{1\text{-}2}$$

$$I_{1\text{-}2} = 18 \text{ V}/6 \text{ k}\Omega$$

$$I_{1\text{-}2} = 3 \text{ mA}$$

At this point, one of two approaches may be taken to arrive at the value of current flowing through $R_{3\text{-}4}$. Ohm's law may be used again, or a circuit law may be employed. This time, the circuit law that says "The total current in a parallel circuit is equal to the sum of the individual branch currents," will be used. The total current is equal to 5 milliamps. If 3 milliamps flow through branch 1 ($R_{1\text{-}2}$), then the rest of the current, 2 milliamps, must be flowing through the second branch ($R_{3\text{-}4}$). In formula form, the law appears like this:

$$I_T = I_{1\text{-}2} + I_{3\text{-}4}$$

To find $I_{3\text{-}4}$, $I_{1\text{-}2}$ must be subtracted from both sides of the equation to produce:

$$I_T - I_{1\text{-}2} = I_{1\text{-}2} + I_{3\text{-}4} - I_{1\text{-}2}$$

$$I_T - I_{1\text{-}2} = I_{3\text{-}4}$$

Now plug the correct values into the formula:

$$5 \text{ mA} - 3 \text{ mA} = I_{3\text{-}4}$$

$$2 \text{ mA} = I_{3\text{-}4}$$

Therefore, 2 milliamps of current flow through the resistance $R_{3\text{-}4}$. Label these current values on the original schematic.

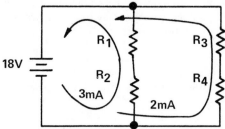

Since the current flowing through and the resistance of each resistor is known, Ohm's law may be used to find the voltage dropped across each individual resistor. R_1 is equal to 3.6 kilohms. I_1 is 3 milli-amps. Ohm's law in the form $E = I \times R$ may be used to solve this problem.

$$E_{R1} = I_{R1} \times R_1$$

$$E_{R1} = 3 \text{ mA} \times 3.6 \text{ k}\Omega$$

$$E_{R1} = 3 \times 10^{-3} \times 3.6 \times 10^{+3}$$

$$E_{R1} = 10.8 \text{ V}$$

The same procedure may be used to find E_{R2}, or a circuit law may be employed. You know that in a series circuit ($R_1 + R_2$ are in series with each other) the sum of the individual voltage drops is equal to the source voltage. Written in formula form:

$$E_{R1} + E_{R2} = E_T$$

To find E_{R2}, E_{R1} must be subtracted from both sides of the equation producing:

$$E_{R1} + E_{R2} - E_{R1} = E_T - E_{R1}$$
$$E_{R2} = E_T - E_{R1}$$

Substitute in the circuit values:

$$E_{R2} = 18 \text{ V} - 10.8 \text{ V}$$
$$E_{R2} = 7.2 \text{ V}$$

The same procedure will be followed in solving for E_{R3} and E_{R4}. First E_{R3} may be found by using Ohm's law. R_3 is known to be 1.5 kilohms. The current flowing through R_3 is 2 milliamps. Substitute these values into an Ohm's law formula and solve for E_{R3}.

$$E_{R3} = I_{R3} \times R_3$$
$$E_{R3} = 2 \text{ mA} \times 1.5 \text{ k}\Omega$$
$$E_{R3} = 2 \times 10^{-3} \times 1.5 \times 10^{+3}$$
$$E_{R3} = 3.0 \text{ V}$$

You know that 3 volts are dropped across resistor R_3. The rest of the 18 volts applied to this branch must be dropped across R_4. The voltage equation may be rearranged to solve this problem by subtracting the known value of E_{R3} from both sides:

$$E_T = E_{R3} + E_{R4}$$
$$-E_{R3} + E_T = E_{R3} + E_{R4} - E_{R3}$$
$$-E_{R3} + E_T = E_{R4}$$

Substitute in the circuit values and subtract:

$$-3 \text{ V} + 18 \text{ V} = E_{R4}$$
$$15 \text{ V} = E_{R4}$$

● **Worked Through Examples**

Place all known values on the schematic, and the circuit analysis is complete.

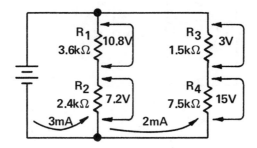

2. Solve for all of the voltages and currents in the following circuit.

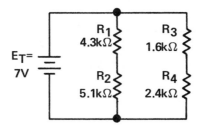

Since this circuit has the same configuration as the last example the same analysis methods will be employed in abbreviated form.

First, reduce or simplify the circuit to its equivalent resistance.

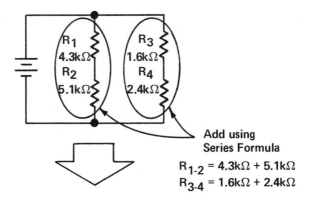

Add using
Series Formula

$R_{1-2} = 4.3k\Omega + 5.1k\Omega$
$R_{3-4} = 1.6k\Omega + 2.4k\Omega$

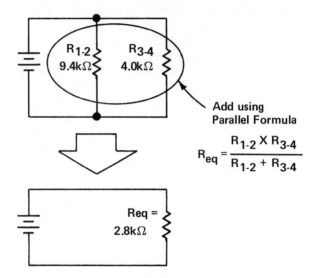

Find I_T by using Ohm's law:

$$I_T = E_T/R_{eq}$$

$$I_T = 7 \text{ V}/2.8 \text{ k}\Omega$$

$$I_T = \frac{7.0 \times 10^0}{2.8 \times 10^{+3}}$$

$$I_T = \frac{7.0 \times 10^{-3}}{2.8}$$

$$I_T = 2.5 \times 10^{-3} = 2.5 \text{ mA}$$

The branch currents may be found by using Ohm's law.

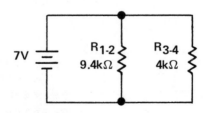

$$I_{R1\text{-}2} = \frac{E_{R1\text{-}2}}{R_{1\text{-}2}}$$

$$I_{R1\text{-}2} = \frac{7\text{ V}}{9.4\text{ k}\Omega}$$

$$I_{R1\text{-}2} = \frac{7 \times 10^0}{9.4 \times 10^{+3}}$$

$$I_{R1\text{-}2} = \frac{7 \times 10^{-3}}{9.4 \times 10}$$

$$I_{R1\text{-}2} = 0.745 \times 10^{-3} = 0.745\text{ mA}$$

$$I_{R3\text{-}4} = \frac{E_{R3\text{-}4}}{R_{3\text{-}4}}$$

$$I_{R3\text{-}4} = \frac{7\text{ V}}{4\text{ k}\Omega} = \frac{7 \times 10^0}{4 \times 10^3}$$

$$I_{R3\text{-}4} = \frac{7 \times 10^{0\text{-}3}}{4}$$

$$I_{R3\text{-}4} = 1.75 \times 10^{-3}$$

$$I_{R3\text{-}4} = 1.75\text{ mA}$$

Now that the branch currents are known, the individual voltage drops may be calculated using Ohm's law.

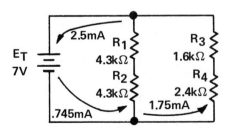

Ohm's law:

$$E_{R1} = I_{R1} \times R_1$$

$$E_{R1} = 0.745 \text{ mA} \times 4.3 \text{ k}\Omega$$

$$E_{R1} = 0.745 \times 10^{-3} \times 4.3 \times 10^{+3}$$

$$E_{R1} = 3.20 \text{ V}$$

$$E_{R2} = I_{R2} \times R_2$$

$$E_{R2} = 0.745 \text{ mA} \times 5.1 \text{ k}\Omega$$

$$E_{R2} = 0.745 \times 10^{-3} \times 5.1 \times 10^{+3}$$

$$E_{R2} = 4.0 \text{ V}$$

Moving to the right-hand branch, and again using Ohm's law:

$$E_{R3} = I_{R3} \times R_3$$

$$E_{R3} = 1.75 \text{ mA} \times 1.6 \text{ k}\Omega$$

$$E_{R3} = 1.75 \times 10^{-3} \times 1.6 \times 10^{+3}$$

$$E_{R3} = 2.8 \text{ V}$$

$$E_{R4} = I_{R4} \times R_4$$

$$E_{R4} = 1.75 \text{ mA} \times 2.4 \text{ k}\Omega$$

$$E_{R4} = 1.75 \times 10^{-3} \times 2.4 \times 10^{+3}$$

$$E_{R4} = 4.2 \text{ V}$$

Placing these values on the circuit schematic, the analysis is complete.

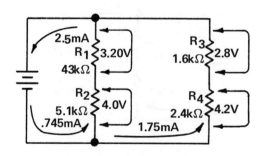

LESSON 8. PARALLEL-SERIES CIRCUITS

● **Practice Problems**

The key objective of this lesson has been achieved if you can solve for the total resistance of any parallel-series circuit using the techniques of circuit reduction, and if given any parallel-series circuit schematic with known values labeled, you can solve for any unknowns required. The following problems are designed to give you practice in both of these areas. Check your progress and accuracy by folding over the page as indicated.

Depending upon the approach you use in solving these problems and how you round off intermediate results, your answers may vary slightly from those given here. However, any differences you encounter should only occur in the third significant digit of your answer. If the first two significant digits of your answers do not agree with those given here, recheck your calculations.

1. Find the total equivalent resistance, R_{eq}, for each of these circuits. **Fold Over**

a.

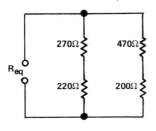

R_{eq} = _____

b.

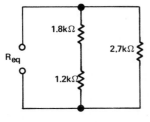

R_{eq} = _____

c.

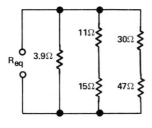

R_{eq} = _____

d.

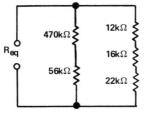

R_{eq} = _____

Answers

1.a. 283 Ω

1.b. 1.42 kΩ

1.c. 3.2 Ω

1.d. 45.7 kΩ

1. (Continued)   **Fold Over**

e.

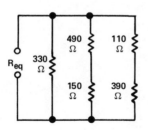

R_{eq} = _____

f.

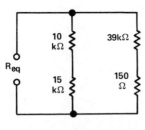

R_{eq} = _____

g.

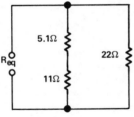

R_{eq} = _____

h.

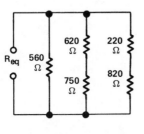

R_{eq} = _____

i.

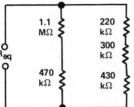

R_{eq} = _____

Answers

1.e. 152 Ω

1.f. 15.3 kΩ

1.g. 9.29 Ω

1.h. 287 Ω

1.i. 591 kΩ

1. (Continued)   **Fold Over**

j.

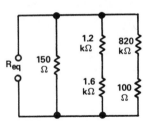

R_{eq} = _____

k.

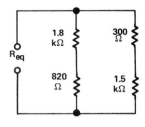

R_{eq} = _____

l.

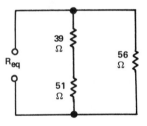

R_{eq} = _____

m.

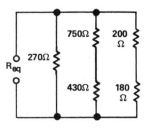

R_{eq} = _____

n.

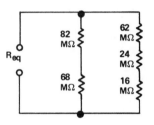

R_{eq} = _____

Answers

1.j.   142 Ω

1.k.   1.06 kΩ

1.l.   34.5 Ω

1.m. 139 Ω

1.n. 60.7 MΩ

1. (Continued)   **Fold Over**

o.

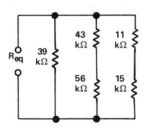

$R_{eq} = $ _____

2. In each of the following circuits, calculate the voltage across and current through each resistor as well as the circuit's total current, total resistance, and total power dissipation.

a.

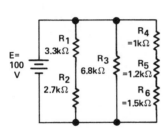

$E_{R1} = $ _____ $I_{R1} = $ _____

$E_{R2} = $ _____ $I_{R2} = $ _____

$E_{R3} = $ _____ $I_{R3} = $ _____

$E_{R4} = $ _____ $I_{R4} = $ _____

$E_{R5} = $ _____ $I_{R5} = $ _____

$E_{R6} = $ _____ $I_{R6} = $ _____

$I_T = $ _____

$R_T = $ _____

$P_T = $ _____

b.

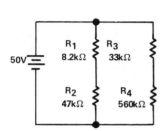

$E_{R1} = $ _____ $I_{R1} = $ _____

$E_{R2} = $ _____ $I_{R2} = $ _____

$E_{R3} = $ _____ $I_{R3} = $ _____

$E_{R4} = $ _____ $I_{R4} = $ _____

$I_T = $ _____

$R_T = $ _____

$P_T = $ _____

Answers

1.o. 13.5 kΩ

2.a. E_{R1} = 54.9 V I_{R1} = 16.7 mA

E_{R2} = 45.1 V I_{R2} = 16.7 mA

E_{R3} = 100 V I_{R3} = 14.7 mA

E_{R4} = 27 V I_{R4} = 27.0 mA

E_{R5} = 32.4 V I_{R5} = 27.0 mA

E_{R6} = 40.5 V I_{R6} = 27.0 mA

I_T = 58.4 mA

R_T = 1.71 kΩ

P_T = 5.84 W

2.b. E_{R1} = 7.43 V I_{R1} = 906 μA

E_{R2} = 42.6 V I_{R2} = 906 μA

E_{R3} = 2.78 V I_{R3} = 84.3 μA

E_{R4} = 47.2 V I_{R4} = 84.3 μA

I_T = 990 μA

R_T = 50.5 kΩ

P_T = 49.5 mW

2. (Continued)   Fold Over

c.

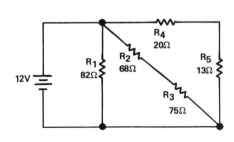

$E_{R1} =$ _____ $I_{R1} =$ _____

$E_{R2} =$ _____ $I_{R2} =$ _____

$E_{R3} =$ _____ $I_{R3} =$ _____

$E_{R4} =$ _____ $I_{R4} =$ _____

$E_{R5} =$ _____ $I_{R5} =$ _____

$I_T =$ _____

$R_T =$ _____

$P_T =$ _____

d.

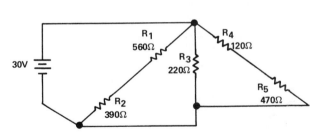

$E_{R1} =$ _____ $I_{R1} =$ _____

$E_{R2} =$ _____ $I_{R2} =$ _____

$E_{R3} =$ _____ $I_{R3} =$ _____

$E_{R4} =$ _____ $I_{R4} =$ _____

$E_{R5} =$ _____ $I_{R5} =$ _____

$I_T =$ _____

$R_T =$ _____

$P_T =$ _____

3.   In each of the circuits shown below, solve for the values indicated.

a.

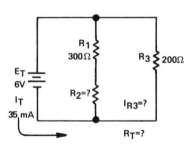

$R_T =$ _____

$R_2 =$ _____

$I_{R3} =$ _____

Answers

2.c. $E_{R1} = +12$ V $I_{R1} = 146$ mA

   $E_{R2} = 5.71$ V $I_{R2} = 83.9$ mA

   $E_{R3} = 6.29$ V $I_{R3} = 83.9$ mA

   $E_{R4} = 7.28$ V $I_{R4} = 364$ mA

   $E_{R5} = 4.73$ V $I_{R5} = 364$ mA

   $I_T = 594$ mA

   $R_T = 20.2$ Ω

   $P_T = 7.13$ W

2.d. $E_{R1} = 17.7$ V $I_{R1} = 31.6$ mA

   $E_{R2} = 12.3$ V $I_{R2} = 31.6$ mA

   $E_{R3} = 30$ V $I_{R3} = 136$ mA

   $E_{R4} = 6.1$ V $I_{R4} = 50.8$ mA

   $E_{R5} = 23.9$ V $I_{R5} = 50.8$ mA

   $I_T = 218.8$ mA

   $R_T = 137$ Ω

   $P_T = 6.55$ W

3.a. $R_T = 171$ Ω

   $R_2 = 900$ Ω

   $I_{R3} = 30$ mA

3. (Continued)   **Fold Over**

b.

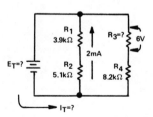

E_T = _____

I_T = _____

R_3 = _____

c.

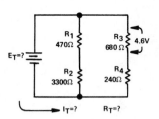

E_T = _____

I_T = _____

R_T = _____

d.

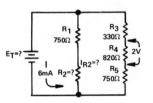

E_T = _____

R_2 = _____

I_{R2} = _____

e.

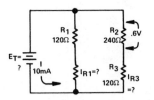

E_T = _____

I_{R1} = _____

I_{R3} = _____

f.

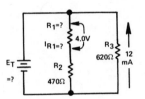

E_T = _____

I_{R1} = _____

R_1 = _____

Answers

3.b. E_T = 18 V

I_T = 3.46 mA

R_3 = 4.1 kΩ

3.c. E_T = 6.22 V

I_T = 8.42 mA

R_T = 739.5 Ω

3.d. E_T = 4.64 V

R_2 = 550 Ω

I_{R2} = 3.56 mA

3.e. E_T = 900 mV

I_{R1} = 7.5 mA

I_{R3} = 2.5 mA

3.f. E_T = 7.44 V

I_{R1} = 7.32 mA

R_1 = 546 Ω

3. (Continued)   **Fold Over**

g.

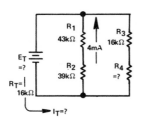

E_T = _____

I_T = _____

R_4 = _____

h.

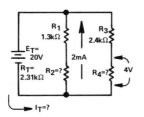

R_2 = _____

R_4 = _____

I_T = _____

i.

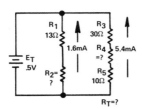

R_2 = _____

R_4 = _____

R_T = _____

j.

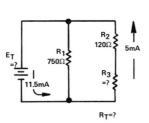

E_T = _____

R_T = _____

R_3 = _____

Answers

3.g. $E_T = 328$ V

$I_T = 20.5$ mA

$R_4 = 3.87$ kΩ

3.h. $R_2 = 8.7$ kΩ

$R_4 = 600$ Ω

$I_T = 8.66$ mA

3.i. $R_2 = 300$ Ω

$R_4 = 52.6$ Ω

$R_T = 71.4$ Ω

3.j. $E_T = 4.87$ V

$R_T = 424$ Ω

$R_3 = 854$ Ω

Identify the following circuits as being series (S), parallel (P), series-parallel (SP), or parallel-series (PS).

1.

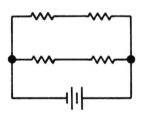

_____

2.

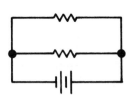

_____

3.

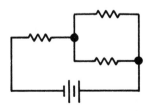

_____

4.

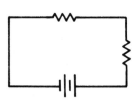

_____

5.

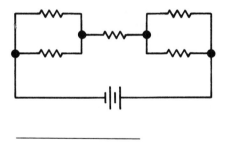

_____

6.

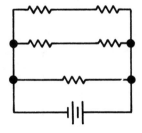

_____

7.

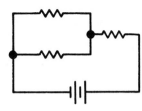

_____

8.

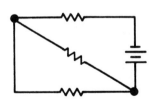

_____

9.

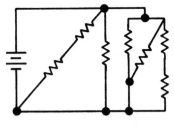

_____

10.

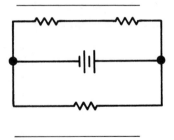

_____

Calculate R_{eq} for the following circuits:

11.

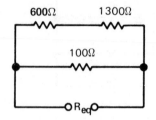

_____

12.

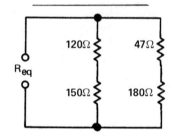

_____

13.

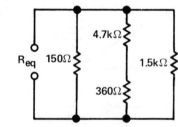

_____

14.

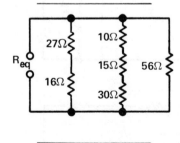

_____

15.

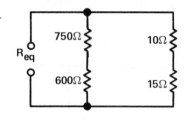

_____

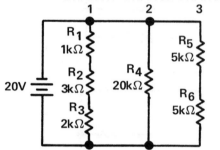

The following questions refer to the circuits above.

16. Which branch will contain the greatest amount of current?_____

17. Which resistor will have the largest voltage drop?_____

18. What is the total equivalent resistance of the circuit?_____

19. What is the value of the voltage drop across R_6_____

20. What is the value of the current flowing through R_4?_____

Lesson 9

Series-Parallel Circuits

This lesson will deal with the recognition and description of series-parallel circuits. The student will also learn to recognize series and parallel circuit configurations when they are part of larger, more complex circuits. The laws governing series and parallel circuits will be used to solve more complex circuits by applying them to one circuit segment at a time.

LESSON 9. SERIES-PARALLEL CIRCUITS

● **Objectives**

1. Given a series of schematics, *select* which represents series, parallel, series-parallel, and parallel-series circuits.

2. *Write* an explanation, using sketches, of the distinguishing features of parallel-series and series-parallel circuits.

3. Given a schematic of the type shown, *calculate* the total equivalent resistance R_{eq} for the network, using the techniques of circuit reduction.

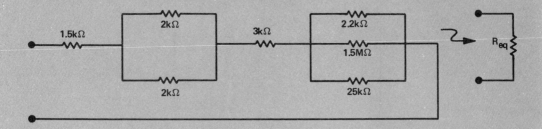

4. Using Ohm's law and circuit reduction techniques, *calculate* all of the unknown voltages and currents in any series-parallel circuit of the type shown below, given the applied voltage and resistance values as illustrated on the schematic.

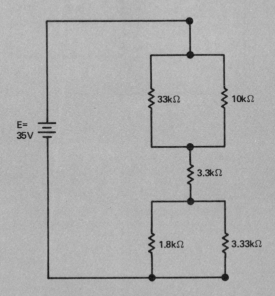

5. Using Ohm's law and circuit reduction techniques, *calculate* all unknown voltages, currents, and resistances in any series and parallel circuit of the type shown below, given a combination of known circuit values as illustrated in the schematic below.

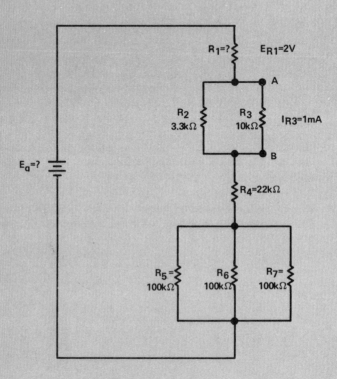

- **Parallel-Series Circuit**
- **Series-Parallel Circuit**

In Lesson 8, two new circuits were introduced, parallel-series and series-parallel circuits. The essential differences between these circuits were explained with an emphasis on how different approaches are used in solving for the voltages, currents, and resistances in these two different circuit configurations. Lesson 8 focused in deep detail on the procedures to follow in analyzing parallel-series circuits. Several example circuits were worked through in detail for you, including several circuits taken from automotive applications. In this lesson your attention will be focused on the series-parallel circuit. You will see several circuits of this type and be shown how they may be identified. You will also learn the steps to follow in analyzing this type of circuit.

An important objective of this lesson, as well as this entire course, is to enable you to recognize simple series and parallel circuit configurations when they are part of larger and more complex circuits. At the end of this lesson you should be able to use the laws governing series and parallel circuits to solve the more complex series-parallel circuits by applying the basic laws correctly to one circuit segment at a time.

To begin, the basic difference between the parallel-series circuit discussed in Lesson 8 and the series-parallel circuit discussed in this lesson is that in *series-parallel circuits* there is *either one resistor or equivalent resistance connected in series with the total current*. In the parallel-series circuit, no single element contains the total current flowing in the circuit.

Parallel-Series Circuit — To further understand the features of these two types of circuits, examine Figure 9.1 which shows one of the parallel-series circuits of the type discussed in Lesson 8. Notice that in this circuit there is no single circuit element that must carry the total or mainline current. This circuit essentially consists of several parallel branches, each of which may contain one resistor or several resistors in series.

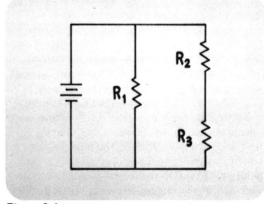

Figure 9.1

Series-Parallel Circuit — Series-parallel circuits which will be the topic for discussion in this lesson contain one circuit element (or an equivalent resistance) that lies in the path of the total current, as shown in Figure 9.2. In this series-parallel circuit, the series component is resistor R_1, and the entire circuit current will flow through it.

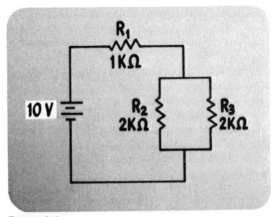

Figure 9.2

Two Parallel Circuits in Series — Notice that the definition of a series-parallel circuit specifies that the element in series with the total circuit current may be an equivalent resistance. This equivalent resistance may be made up of several resistors in parallel. For this reason, a series-parallel circuit may appear as two or more parallel circuits wired in series, as shown in Figure 9.3. The analysis of circuits with this configuration will also be discussed for you in this lesson.

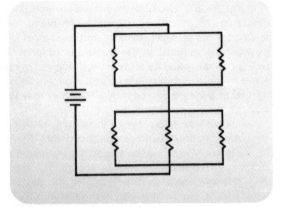

Figure 9.3

Solving Parallel-Series Circuits Review — As far as the techniques you will learn in this lesson are concerned, you will find that many of the same techniques used to solve parallel-series circuits will also be employed to solve series-parallel circuits. The only real difference is that these basic various techniques will be applied in a different order.

You may recall that when voltage, current, and resistance values in parallel-series circuits were calculated, each of the branches was examined first, as shown in Figure 9.4. The three series circuit laws were initially applied to each branch to reduce the circuit to the fewest possible elements. Then the parallel circuit laws were used to further reduce the number of elements in the circuit.

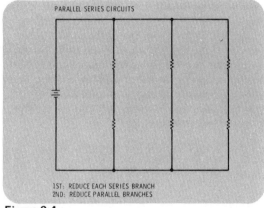

Figure 9.4

Solving Series-Parallel Circuits — When working with series-parallel circuits, you will find that the analysis will in general proceed in the reverse direction from that of parallel-series circuits (Figure 9.5). The parallel circuit "clusters" will first be reduced using the parallel circuit laws. The series circuit laws are then used to reduce the remaining series resistors and parallel equivalent resistances to one final equivalent resistance. Once the equivalent resistance is known, the total circuit current can be found and used to find the voltage across each component.

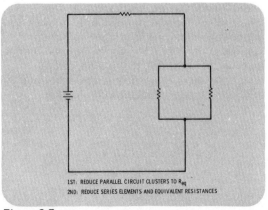

Figure 9.5

Basic Series-Parallel Circuit — Before actually going through the mechanics of a detailed circuit solution, consider in general the path of current flow in a simple series-parallel circuit. The circuit in Figure 9.6 contains one resistor connected in series with two other resistors which are connected in parallel. (This is actually the most simple form of series-parallel circuit.)

Electron current flows from the negative side of the power supply to point B where it divides, then flows through the two resistors, R_3 and R_4, in parallel. At point A, the current recombines and comes back to a single path and flows through the single resistor, R_1, and back to the other side of the power supply. Notice again that all the current in this circuit must pass through the single resistor, R_1.

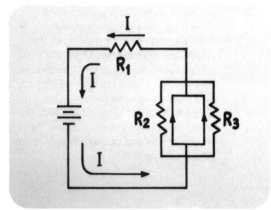

Figure 9.6

Steps in Analyzing Series-Parallel Circuits — How would you go about analyzing an actual circuit of this type? The procedural steps are listed in Figure 9.7.

1. Find the equivalent resistance of the circuit. This may be done by simplifying or reducing the circuit down to a single component. As has been mentioned, the best way to do this is to first identify those portions of the circuit that are connected in parallel, and using the rules for combining parallel resistances, reduce those parallel "clusters" to single equivalent resistors. Then using series circuit rules, combine all the series

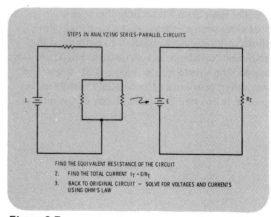

Figure 9.7

elements, including the equivalent
resistances, to a single resistance.

2. Find the total current flowing in the
circuit using Ohm's law. Once the total
circuit resistance is known, Ohm's law
may be used in the form $I_T = E/R_T$ to
find the total current.

3. Go back to the original circuit and, once
again, using Ohm's law, calculate the
voltage across and current through each
circuit component.

Basic Series-Parallel Circuit Example — With these
steps in mind, consider the circuit shown in
Figure 9.8. In this circuit the value of resistor R_1 is
1 kilohm and R_2 and R_3 are each 2 kilohms.
Following the steps in the analysis outlined above,
you will begin by reducing this circuit to a single
equivalent resistance, beginning with the parallel
resistors R_2 and R_3.

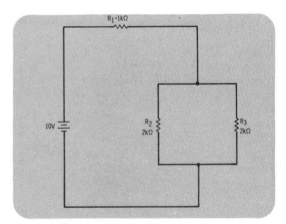

Figure 9.8

Parallel Resistance Reduction Shortcut — Note
that since both of these resistors are 2 kilohms, a
shortcut formula may be used to determine the
equivalent resistance as was discussed in a previous
lesson. The shortcut rule (Figure 9.9) states: The
equivalent resistance of two equal resistors in
parallel is one-half the value of one of the resistors.

THE RESISTANCE OF TWO RESISTORS IN
PARALLEL, WHICH HAVE THE SAME VALUE,
IS ONE-HALF THE VALUE OF ONE OF THE
RESISTORS

Figure 9.9

- **Equal Resistance Formula**
- **Simple Series-Parallel Circuit Reduced to Two Resistances**
- **Series Resistance Rule and Formula**

Equal Resistance Formula — In formula form as shown in Figure 9.10, this rule translates to, $R_{eq} = R_S/N$. When the resistance values for R_2 and R_3 are substituted into the formula, the result is 2 kilohms divided by 2, because these two equal value resistors are connected in parallel. The result is 1 kilohm, that is, resistors R_2 and R_3 are equivalent to a single 1-kilohm resistor.

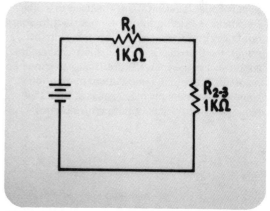

Figure 9.10

Simple Series-Parallel Circuit Reduced to Two Resistances — This is the first reduction of this circuit, and it appears as shown in Figure 9.11. The circuit consists of resistance R_1 which is 1 kilohm, in series with the equivalent resistance $R_{2\text{-}3}$, which is also 1 kilohm.

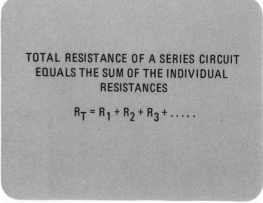

Figure 9.11

Series Resistance Rule and Formula — This circuit may be reduced even further using the law for series resistance (Figure 9.12), which states that the total equivalent resistance of a series circuit equals the sum of the individual resistances.

TOTAL RESISTANCE OF A SERIES CIRCUIT
EQUALS THE SUM OF THE INDIVIDUAL
RESISTANCES

$$R_T = R_1 + R_2 + R_3 + \ldots$$

Figure 9.12

- R_T = 2 kilohms
- **Simple Series-Parallel Circuit Reduced to a Single Resistance**
- Calculating Total Current

R_T = **2 kilohms** — In the circuit of Figure 9.13, R_T, the total equivalent resistance, simply equals R_1 plus $R_{2\text{-}3}$. One kilohm plus 1 kilohm equals 2 kilohms for the total circuit resistance.

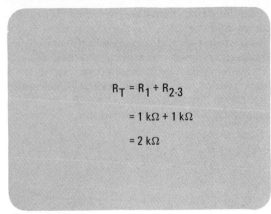

$$R_T = R_1 + R_{2\text{-}3}$$
$$= 1\ k\Omega + 1\ k\Omega$$
$$= 2\ k\Omega$$

Figure 9.13

Simple Series-Parallel Circuit Reduced to a Single Resistance — At this point it is important to point out once again that in solving for these circuit values, the first step is most usually to reduce the circuit to a single equivalent resistance as shown in Figure 9.14. Keep in mind that no matter how complex these circuits may appear at first, they reduce down to a single equivalent resistance.

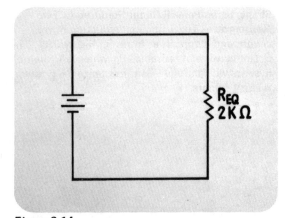

Figure 9.14

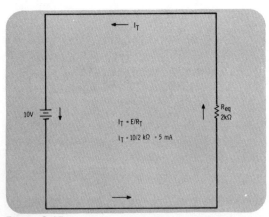

Figure 9.15

Calculating Total Current — Now that the equivalent resistance of this circuit is known, it can be used to find the total circuit current using Ohm's law in the form $I_T = E_T/R_T$ (as shown in Figure 9.15). Substituting 10 volts for the applied voltage and 2 kilohms for R_{eq}, I_T is calculated to be 5 milliamps. This 5 milliamps flows through R_{eq}. Keeping this in mind, take the circuit back to the second reduction where it was seen as two resistors in series.

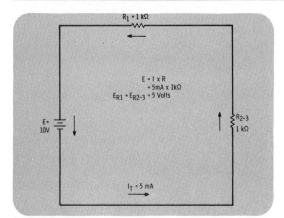

Figure 9.16

Determining Voltage Drops — In Figure 9.16 you can see that the total current, I_T, flows through both R_1 and $R_{2\text{-}3}$. Since both these resistors' values are known, the voltage drop across each can be calculated using Ohm's law in the form $E = I \times R$. Since each resistor's value is the same, the result for the voltage across each is easily seen to be 5 volts.

Notice the voltage across $R_{2\text{-}3}$ is 5 volts, *and $R_{2\text{-}3}$ is made up of two parallel resistors, R_2 and R_3.* This means that you now know the voltage across R_2 and R_3 to be 5 volts because in any parallel circuit the voltage across each branch is the same.

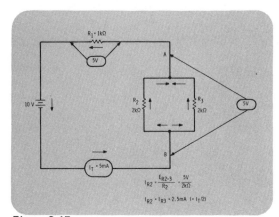

Figure 9.17

Determining the Branch Currents — At this point you are ready to put what you know about the circuit on the original unreduced schematic as shown in Figure 9.17. The currents through R_2 and R_3 are the only values remaining to be calculated, and there are two ways this can be done. Using Ohm's law in the form $I_{R2} = E_{R2\text{-}3}/R_2$, you can find the current through R_2 and also R_3 since R_2 and R_3 have the same value. Dividing 5 volts by 2 kilohms yields 2.5 milliamps for I_{R2} and I_{R3}. Consider the fact that the total current must divide at point B, to flow through two branches of equal resistance. Because of this equal resistance, exactly 1/2 of I_T must flow in each branch, and $I_T/2$ is 2.5 milliamps.

Two Parallel Circuits Connected in Series — The circuit analyzed above is actually the simplest form of a series-parallel circuit. If the single resistor R_1 were actually an equivalent resistance made up of several resistors in parallel, the circuit would appear initially more complex and resemble the circuit of Figure 9.18. As you proceed through the analysis of this circuit, however, you will see that using the same steps to do the job, the analysis is not much more complex than before. This circuit consists of a 35-volt power supply, resistors R_1 and R_2 in parallel, and resistors R_3, R_4, and R_5 which are also in parallel with each other. These two parallel circuits are then wired in series with each other.

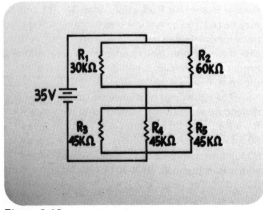

Figure 9.18

Before beginning the actual circuit analysis, it will be helpful to generally block out the steps you will be taking. The first step in solving this problem is to reduce the circuit to its equivalent resistance. Then you can use the equivalent resistance to calculate the total current flowing in the circuit. Using the value of the total current, the rest of the voltage drops, and currents in the circuit may be calculated.

In reducing this circuit to a single equivalent resistance, you must begin by first reducing the two parallel circuit clusters. The parallel combination of R_1 and R_2 may first be reduced to an equivalent resistance, and also the combination of R_3, R_4, and R_5 can be reduced to their equivalent resistance. Then, these two equivalent resistances, which are connected in series, may be added to arrive at the total equivalent resistance for the circuit.

Parallel Resistance Reduction: Step 1 — R_1 and R_2 may be reduced to a single resistance by using the rules for combining parallel resistors. Since in this case there are only two resistances in parallel, the product-over-the-sum technique will be used, as shown in Figure 9.19.

The equivalent resistance equals R_1 times R_2 divided by R_1 plus R_2. Substituting the value for $R_1 + R_2$ yields 30 kilohms (or $3 \times 10^{+4}$) times 60 kilohms (or $6 \times 10^{+4}$) over 30 kilohms plus 60 kilohms. Carrying out the multiplication and addition, this equals $18 \times 10^{+8}$ divided by $9 \times 10^{+4}$. The result is $2 \times 10^{+4}$, or 20 kilohms.

$$R_{T1\text{-}2} = \frac{R_1 \times R_2}{R_1 + R_2}$$

$$= \frac{(3 \times 10^{+4})(6 \times 10^{+4})}{(3 \times 10^{+4}) + (6 \times 10^{+4})}$$

$$= \frac{18 \times 10^{+8}}{9 \times 10^{+4}}$$

$$= 2 \times 10^{+4}$$

$$= 20 \text{ k}\Omega$$

Figure 9.19

Parallel Resistors 1 and 2 Reduced to a Single Resistance — The equivalent resistance of resistors R_1 and R_2 is 20 kilohms and is labeled R_{1-2} as shown in Figure 9.20.

The parallel combination of resistors R_3, R_4, and R_5 may be reduced, and this equivalent resistance will be labeled R_{3-5}.

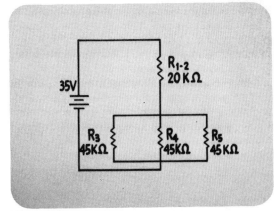

Figure 9.20

Equal Value Parallel Resistance Reduction — Since resistors R_3, R_4 and R_5 are equal in value, this parallel combination can be reduced using the shortcut formula as shown in Figure 9.21. R_{3-5} equals R_s, which is the value of one of these equal size resistors, divided by N, the total number of resistors. Substituting the values from the schematic diagram into the formula yields 45 kilohms divided by 3, which equals 15 kilohms for the equivalent resistance R_{3-5}.

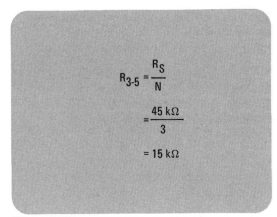

$$R_{3-5} = \frac{R_S}{N}$$
$$= \frac{45\ k\Omega}{3}$$
$$= 15\ k\Omega$$

Figure 9.21

Circuit with R_{1-2} and R_{3-5} — As seen in Figure 9.22, the two parallel resistor combinations have been reduced to their equivalent resistances, and the equivalent circuit simply consists of two resistors connected in series. At this point, another reduction can be performed using the law for combining series resistances.

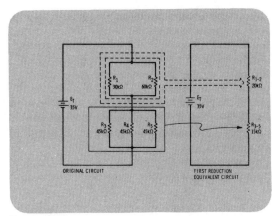

Figure 9.22

Series Resistance Formula (Final Reduction) —
The total equivalent resistance of this circuit equals
the sum of these two equivalent resistances. R$_{1-2}$
(20 kilohms) plus R$_{3-5}$ (15 kilohms) yields a total
circuit equivalent resistance of 35 kilohms as
shown in Figure 9.23. Now that the total resistance
and the total voltage in the circuit are known, you
can proceed to the second step in analyzing this
circuit, using Ohm's law to determine the circuit's
total current.

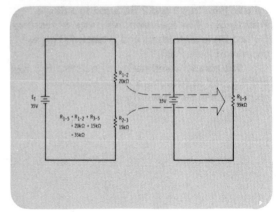

Figure 9.23

Ohm's Law to Find Total Current: Step 2 — As
you should recall, the total current in this
situation, I$_T$, equals the total applied voltage, E$_T$,
divided by the circuit's total equivalent resistance,
R$_{1-5}$. In this circuit you know that there are
35 volts applied, and this divided by the
35 kilohms just calculated for R$_{1-5}$ gives you a
total current of 1 milliamp, as shown in
Figure 9.24.

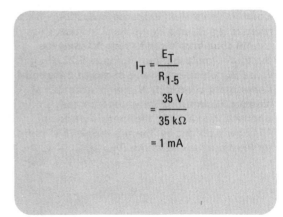

$$I_T = \frac{E_T}{R_{1-5}}$$

$$= \frac{35 \text{ V}}{35 \text{ k}\Omega}$$

$$= 1 \text{ mA}$$

Figure 9.24

Circuit with Equivalent Resistor R$_{1-5}$ — At this
point it is time to recall that an equivalent circuit
has the same essential characteristics of the larger,
more complex circuit from which it originated. An
equivalent circuit has the same amount of applied
voltage, the same total equivalent resistance, and
the same total current flowing. The only difference
is that the many resistive components have been
reduced to a single resistive component as shown in
Figure 9.25.

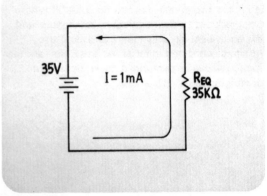

Figure 9.25

Two Parallel Circuits Connected in Series: Step 3 —
The circuit in Figure 9.25 is electrically equivalent
to the original circuit shown again in Figure 9.26.
Now that the voltage, current, and resistance of the
equivalent circuit are known, the third step in
completing the analysis of this circuit is to work
your way back to this original circuit, following
the same steps as in circuit reduction, but this time
in reverse. As you proceed you will determine all
the unknown voltage drops in the circuit and the
branch currents for each branch.

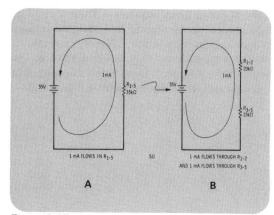

Figure 9.26

Circuit with R$_{1-2}$ and R$_{3-5}$ — Consider the circuit
shown in Figure 9.27(B). Recall that this circuit
was the next to last step in the circuit reduction,
just before the circuit was reduced to a single
equivalent resistor, Figure 9.27(A). You now know
that the total current flowing in this circuit is
1 milliamp. The current law governing series
circuits states that the current must have the same
value at any point in a series circuit. Thus,
Figure 9.27 shows that 1 milliamp of current is
flowing through both the equivalent resistance
R$_{1-2}$ and through the equivalent resistance R$_{3-5}$.
You know the current flowing through these
resistors and their value in ohms; thus, using Ohm's
law the voltage drops across R$_{1-2}$ and R$_{3-5}$ can be
found.

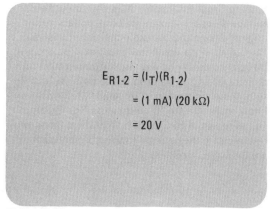

Figure 9.27

Ohm's Law (E$_{R1-2}$) — The voltage drop across the
entire equivalent resistance R$_{1-2}$ is found by using
the Ohm's law formula as shown in Figure 9.28,
E$_{R1-2}$ equals I$_T$ times R$_{1-2}$. Substituting the
circuit figures into this formula, 1 milliamp times
20 kilohms equals 20 volts dropped across R$_{1-2}$.

$$E_{R1-2} = (I_T)(R_{1-2})$$
$$= (1\ mA)\ (20\ k\Omega)$$
$$= 20\ V$$

Figure 9.28

Ohm's Law (E_{R3-5}) — Using the same formula and substituting the circuit values for R_{3-5}, you find that 15 volts is dropped across R_{3-5}, the equivalent resistance as shown in Figure 9.29.

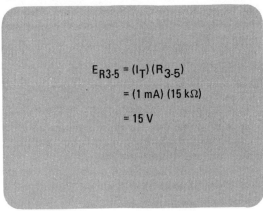

$$E_{R3-5} = (I_T)(R_{3-5})$$
$$= (1\ mA)(15\ k\Omega)$$
$$= 15\ V$$

Figure 9.29

Voltage Across Series Resistors — These calculations may be checked by adding the voltage dropped across R_{1-2}, which is 20 volts, and the voltage dropped across R_{3-5}, which is 15 volts. The sum of these voltages is 35 volts and this is equal to the total applied voltage as shown in Figure 9.30.

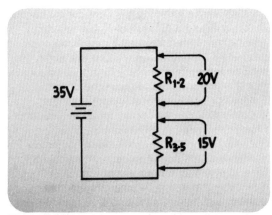

Figure 9.30

Back to Original Circuit — Now that these two voltage drops are known, you can go back to the original circuit and complete the circuit analysis. First of all, since you know that there is a drop of 20 volts across R_{1-2}, you also automatically know that there is a 20-volt drop across both R_1 and R_2 since these are connected in parallel. Likewise you know that there are 15 volts across R_3, R_4, and R_5, as shown in Figure 9.31. With these facts, you can calculate the current flowing through each of the individual resistors.

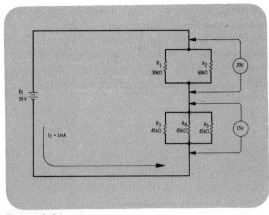

Figure 9.31

- Ohm's Law (I_{R1})
- Parallel Circuit Current Law
- Ohm's Law (I_{R2})

Ohm's Law (I_{R1}) — The current flowing through R_1 may be found by using Ohm's law as shown in Figure 9.32. Dividing 20 volts by 30 kilohms yields 666 microamps of calculated current flowing through R_1.

$$I_{R1} = \frac{E_{R1}}{R_1}$$

$$= \frac{20 \text{ V}}{30 \text{ k}\Omega}$$

$$= 666 \ \mu A$$

Figure 9.32

Parallel Circuit Current Law — You know that the total current flowing in the circuit is 1 milliamp. Since R_1 is passing 666 microamps and since resistor R_2 affords the only other path for the remainder of the current flow, at this point in the circuit, the current through R_2 may be found by subtracting the current through R_1 from the total current flowing in the circuit. Following the procedure shown in Figure 9.33, 1 milliamp plus 666 microamps equals 333 microamps of current for the current flowing through R_2.

$$I_{R2} = I_T - I_{R1}$$

$$= 1 \text{ mA} - 666 \ \mu A$$

$$= 333 \ \mu A$$

Figure 9.33

Ohm's Law (I_{R2}) — A second method may also be used to find the current flowing through R_2 which involves simply using Ohm's law. Twenty volts are being dropped across R_2 which is a 60-kilohm resistor. Ohm's law says that the current through R_2 equals 20 volts divided by 60 kilohms, as shown in Figure 9.34. Hence this calculation also yields a current of 333 microamps flowing through R_2, which agrees with the previous result.

$$I_{R2} = \frac{E_{R2}}{R_2}$$

$$= \frac{20 \text{ V}}{60 \text{ k}\Omega}$$

$$= 333 \ \mu A$$

Figure 9.34

Calculating Lower Branch Currents — Figure 9.35 shows the original circuit with all the calculated values for the upper parallel resistor cluster labeled. Ohm's law may be used once again to calculate the current flowing through the three individual resistors in the lower cluster, R_3, R_4, and R_5. From your previous analysis of the equivalent circuit, it was shown that 15 volts were being dropped across this lower parallel resistor combination.

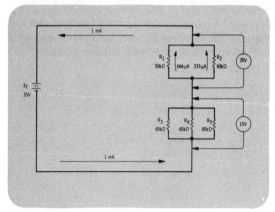

Figure 9.35

Ohm's Law (I_{R3}) — The current through R_3 can be calculated using Ohm's law in the form $I_{R3} = E_{R3}/R_3$, as shown in Figure 9.36. Plugging in the circuit values yields 15 volts across R_3, divided by the value of R_3, 45 kilohms, which equals 333 microamps of current for I_3.

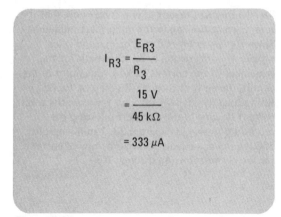

$$I_{R3} = \frac{E_{R3}}{R_3}$$

$$= \frac{15\ V}{45\ k\Omega}$$

$$= 333\ \mu A$$

Figure 9.36

Complete Circuit Solution — Determining the current flowing through resistors R_4 and R_5 is now relatively easy because these resistors all have the same value as R_3. Since 1 milliamp of current enters the parallel combination of the identical resistors R_3, R_4, and R_5, all the resistors must have one-third of 1 milliamp flowing through them. Thus each resistor in the lower parallel cluster has 333 microamps of current flowing through it, as shown in Figure 9.37.

   Keep in mind, however, that if the resistors in the lower parallel cluster had different resistive values, Ohm's law could have been used to solve for the current through each individual resistor.

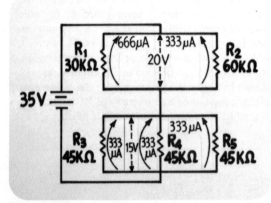

Figure 9.37

Review of Circuit Analysis — This is a good time to look back and review just how the analysis took place.

Originally, only the applied voltage and the values of the various resistors were known as shown in Figure 9.38(A).

By using parallel circuit laws, the parallel resistor combinations were reduced to their equivalent resistances illustrated in Figure 9.38(B).

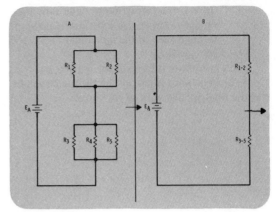

Figure 9.38

As you can see in Figure 9.39(A), the series circuit laws were then used to determine one single equivalent resistance for the circuit.

This equivalent resistance was then used to determine the total current flowing in the circuit as in Figure 9.39(B).

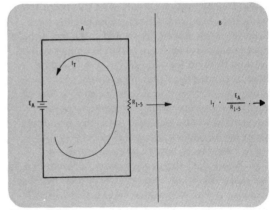

Figure 9.39

Once the total current was known, the individual resistances, and the series and parallel circuit laws were used to determine the voltage drops across the two parallel resistor combinations as you can see in Figure 9.40(A).

Once the voltage drops were known, this information was used to calculate the current flowing through each of the individual resistors in the circuit in Figure 9.40(B).

It is simply a process of:

1. Reducing the resistances
2. Determining the total current

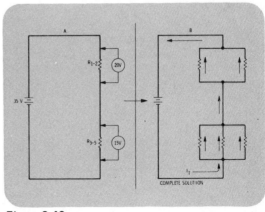

Figure 9.40

3. Using this information to determine the voltage drops across and current flowing through each of the individual components.

This procedure is basic to solving any series-parallel circuit where all the resistance values and the total applied voltage are known.

Lamp Demonstration Circuit — At this point the basic steps in solving series-parallel circuits have been set down and demonstrated and a few circuits have been analyzed. More circuit examples will be worked through for you but for now it will be helpful to consider a real-life example of a series-parallel circuit to watch and get the feel of them in actual operation. The demonstration circuit that will be discussed is an easy one to set up in the laboratory and will provide you with some insight as to how these circuits behave.

Figure 9.41 shows a schematic diagram of a series-parallel circuit that consists of four switches and four lamps. Notice that lamp 1 and switch 1 are in series with the total current, and that lamps 2, 3, and 4 are connected in parallel with each other. Each lamp is controlled by the switch in series with it.

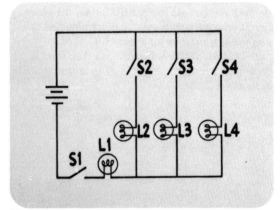

Figure 9.41

Switch S$_1$ Closed — Consider what would happen in this circuit if S$_1$ alone were closed. Would there be any current flow in the circuit? A quick examination of Figure 9.42 should tell you there would be no current flow. Trace the electron current flow out of the negative power supply terminal. With S$_1$ alone closed, current could flow through it, through lamp 1, and then as you can see, with S$_2$, S$_3$, and S$_4$ open, there is no complete path for current flow further.

If any one or all of these remaining switches are closed, the circuit will be completed and current will flow.

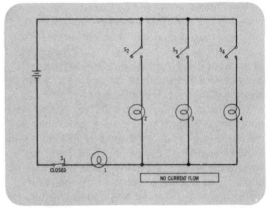

Figure 9.42

Picture of Circuit Operation — Consider what will happen if switches S$_1$ and S$_2$ are closed in Figure 9.43 which is a sketch of what this actual circuit might look like. With switches S$_1$ and S$_2$ closed, lamps L$_1$ and L$_2$ will be lit.

Electron current will flow from the power supply, through S$_1$, L$_1$, L$_2$, S$_2$, and then return to the power supply. With S$_1$ and S$_2$ closed, this circuit acts as a simple series circuit. Because these lamps are identical, they have the same amount of current flowing through them because they are connected in series, and they should produce the same amount of light as indicated in the figure. Equal currents through these identical lamps produce equal light output. (Also, because these two lamps are identical, they have the same resistance, and therefore the voltage drop across each is the same.)

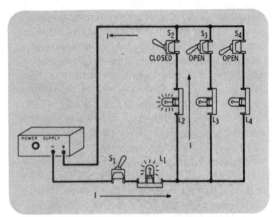

Figure 9.43

Circuit Operation: S_2 and S_3 Closed — If S_3 is now closed in this circuit, as shown in Figure 9.44, L_1 will glow much brighter than before, while L_2 and L_3 will not be as bright as L_1. Lamp 2 will not be as bright as it was before, when it alone was connected in series with L_1. Lamps 2 and 3 will be burning with equal (low) brightness.

   If this circuit were wired in front of you right now, that is the brightness picture you would be seeing. Keeping your series and parallel circuit laws in mind, along with some circuit sense, you can make some general statements about what is going on in this circuit before you actually analyze it mathematically. Focus your attention carefully on the brightness level of each lamp. Lamp 1 is burning more brightly than before; so since the lamp itself has not changed, there must be more current flowing through it than before. Since lamp 1 is in series in this circuit, this means that there is more total circuit current flowing now than there was before. This higher total circuit current flows on past L_1 where it *divides* to flow through L_2 and L_3. Since L_2 and L_3 are in parallel and have the same resistance (they are identical lamps, remember), they each must carry *one-half* the total circuit current. Now think, what does that tell you? First, this explains why L_2 and L_3 are of equal brightness, they each carry the same current. Second, L_2 is dimmer than it was before, so it is receiving *less* current than before in this new circuit situation. (This also means that one-half the new total circuit current is less than the old total circuit current.)

   Those are a few general deductions. Now consider how these can be backed up by circuit analysis.

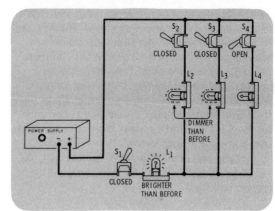

Figure 9.44

Lamp One Glows More Brightly — Why does L_1 glow more brightly when L_3 is switched on? As was mentioned, it must be because more current is flowing through it which means the total circuit current must increase. To calculate this increase, go back and consider the resistances in these two circuits as shown schematically in Figure 9.45.

When L_1 and L_2 were connected in series, the circuit contained a certain amount of resistance. For typical resistance in laboratory lamps you might use about 40 ohms. With two of these lamps in series, the total resistance of the first circuit would be 80 ohms.

When L_3 was switched into the circuit, the circuit's total resistance was reduced. Why? Notice that L_2 and L_3 are connected in *parallel*. Recall that one of the parallel resistance rules says that two resistors of equal value connected in parallel have a total resistance equal to half the value of one of the resistors.

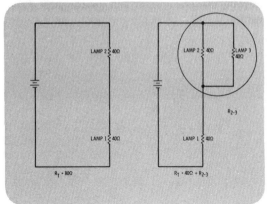

Figure 9.45

Calculating New Resistance — According to the parallel resistance rule, shown in Figure 9.46, the equivalent resistance of lamps L_2 and L_3 in parallel is equal to R_s over N, which equals 40 over 2 or 20 ohms. Now the total circuit resistance, R_T, equals 40 ohms plus $R_{2\text{-}3}$, or 40 ohms plus 20 ohms, which is 60 ohms.

Therefore, the circuit resistance dropped from 80 ohms to 60 ohms when L_3 was switched on. This is a decrease of 20 ohms or 20/80 (or 1/4 or 25%) *drop* in resistance. Since the resistance has dropped by 25%, the total current flowing will *rise* by 25%, because as you should recall, current and resistance are inversely related. So L_1, which is

$$R_{2\text{-}3} = \frac{R_S}{N}$$

$$R_{2\text{-}3} = \frac{40}{2} = 20\ \Omega$$

$$R_T = 40\ \Omega + R_{2\text{-}3}$$

$$R_T = 40\ \Omega + 20\ \Omega$$

$$R_T = 60\ \Omega$$

Figure 9.46

carrying the total current, will glow brighter than before.

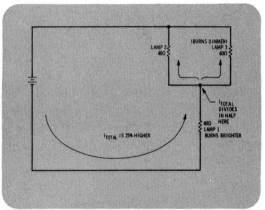

Figure 9.47

Lamps Two and Three Burn Dimmer — This new 25% higher current flows on to L_2 and L_3 where it divides, half flowing through each lamp (Figure 9.47). Although 25% more total current is flowing in this circuit, because it is divided by 2 as it flows through L_2 and L_3, these two lamps each receive less current than L_2 did previously. This is why L_2 and L_3 glow less brightly now than in the first circuit.

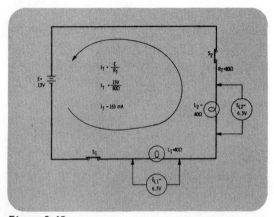

Figure 9.48

Circuit Calculation: Circuit 1 — To check this discussion, you could quickly calculate all the currents and voltages flowing in these circuits using the techniques covered in this lesson. In a laboratory demonstration of this type, the first circuit (shown in Figure 9.48) contains a 13-volt supply, and the two lamps L_1 and L_2 are in series. The total circuit resistance has already been calculated as 80 ohms, so the total circuit current is just 13 volts divided by 80 ohms which is 163 milliamps. The voltage dropped across each bulb equals this circuit current times the bulb's resistance, 163 milliamps times 40 ohms which is 6.5 volts. Each bulb drops one-half of the supply voltage.

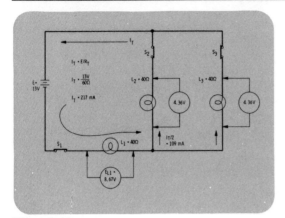

Figure 9.49

Circuit Calculation: Circuit 2 — In the second circuit everything is the same except another bulb, L_3, has been switched in across L_2, as shown in Figure 9.49. The total resistance in this case has been calculated as 60 ohms. The total circuit current is just 13 volts divided by 60 ohms or 217 milliamps. This is flowing through L_1, so the voltage across L_1 equals 217 milliamps times 40 ohms or 8.67 volts. This current divides so that half of it flows through L_2 and half through L_3, 217/2 is 109 milliamps. With 109 milliamps flowing through each of these 40-ohm bulbs, you can calculate that 109 milliamps times 40 ohms, or 4.36 volts is across each of L_2 and L_3. This gives you a complete picture of how these circuits differ in operation, and what happens when switch 3 is thrown bringing a third lamp into the circuit.

Fourth Lamp in Circuit — You should be able to predict what would happen if a fourth lamp were switched into the circuit, as shown in Figure 9.50. Lamp 1, which is in series, should glow more brightly, and L_2, L_3, and L_4 should glow less brightly than before.

This effect occurs because L_4 is put into the circuit in parallel, which *decreases* the total resistance of the circuit. This decrease again causes an increase in the total circuit current, which passes through lamp 1.

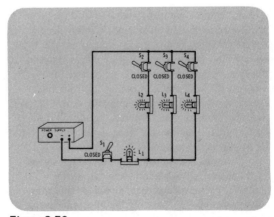

Figure 9.50

Fuse Action — The circuit has actually demonstrated an important circuit situation in electricity and electronics, that of the series-wired fuse. As has been discussed in earlier lessons, fuses are designed to be protective devices. They are usually made of a special resistive material that will melt if a prescribed current flows through it. If a circuit protected by a fuse were to draw too much current, the fuse would open the circuit.

In the demonstration circuit of Figure 9.51, L_1 is in the position that would normally be occupied by a fuse. All circuit current flows through it. In fact, L_1 may actually act as a fuse. If the current carrying capability of the lamp is exceeded, the lamp will burn out, opening the circuit just as a fuse would if the circuit exceeded its current rating.

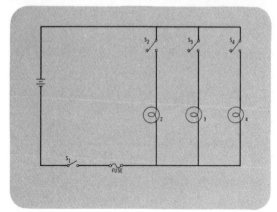

Figure 9.51

Circuit Switch Operation — Quite a lot has been reviewed concerning this demonstration circuit. Before leaving it completely, focus your attention on one final factor, the action of the switches in this circuit. By examining the schematic in Figure 9.52, you can see that S_1 controls the operation of all the circuit lamps because it is series with the total current. With S_1 closed, the other switches can then be used to activate L_2, L_3, and L_4 in the circuit branches.

Consider the behavior of the voltage across S_1. If this circuit were built for you in the laboratory, the voltage drop across S_1 could be measured simply by placing the voltmeter leads directly across the switch.

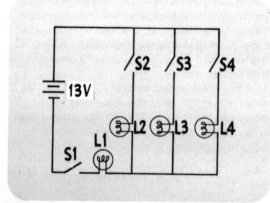

Figure 9.52

Open Switch: Maximum Voltage Drop — Let's check the operation of the circuit by closing switches S_1 and S_2. Lamps L_1 and L_2 will light. If S_1 is now opened, as in Figure 9.53, both lights will go out and no current will flow in the circuit. If you placed a voltmeter across this open switch, as shown in the figure, the meter would indicate approximately 13 volts, the total power supply voltage. This value is indicated because in the "open" position, the switch offers maximum opposition to current flow. Because of this, no current flows anywhere in the circuit, so there is no voltage dropped anywhere in this circuit. Across the point where the circuit is broken, the full supply voltage appears. It is not enough voltage, however, to force current to flow through the very high resistance of the open switch.

If the switch S_1 is closed, the voltmeter would read zero volts and the lamps will light.

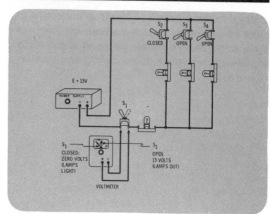

Figure 9.53

Ohm's Law and Switches — Figure 9.54 shows Ohm's law applied to a switch. The voltage drop across a device and the device's resistance vary directly. For an open switch, the resistance to current flow is infinite, so the voltage across it is at a maximum and it permits no current to flow. A closed switch offers zero resistance to current flow. Thus, if the resistance is zero ohms, the voltage drop across the resistance must be zero volts. The resistance of a closed switch is very close to zero ohms, so there's no voltage drop across it.

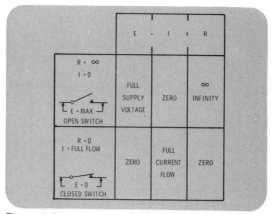

Figure 9.54

Voltage Across a Switch — This is an important point to remember: *voltage is always maximum or equal to the applied voltage across an open switch and minimum across a closed switch, as shown in Figure 9.55.*

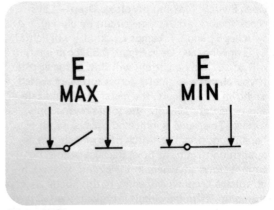

Figure 9.55

Before leaving this lesson, it will be helpful to extend your knowledge of and skill in handling series-parallel circuits by discussing a different type of example.

Up to this point, you have seen circuit reduction techniques used to determine voltages and currents in series-parallel circuits. In these examples you were given only the applied voltage and resistance values at the beginning of the problem. There will be times when you will know more about the circuit as you begin to analyze it than you did in the previous examples in this lesson. In some of these cases, the circuit analysis may proceed in a little more straightforward fashion.

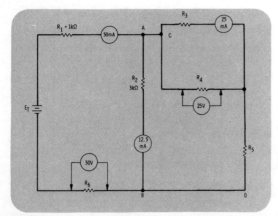

Figure 9.56

Series Parallel Circuit with Known Values — To see how this may be done, consider the circuit in Figure 9.56. Going through the solution to solve this circuit will help develop your circuit sense, because it is a tricky circuit. But, it is a circuit type often encountered and you should know how to handle it.

First of all, take a moment to analyze this circuit. You should recognize it as a series-parallel circuit, because in this case there are two resistors, R_1 and R_6, that are in series with the total circuit current.

You can also see that resistor R_2 is a parallel component. The function of resistors R_3, R_4, and R_5 in the circuit may not be clear to you at first.

This circuit can be redrawn in a slightly different fashion to clarify the operation of the right-hand side.

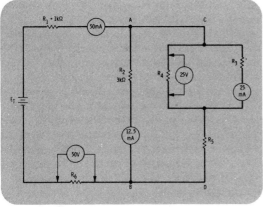

Figure 9.57

Redrawn Circuit — It is fairly easy to recognize in this redrawn circuit (Figure 9.57), that R_3 and R_4 are in parallel with each other, and that this parallel combination of R_3 and R_4 is in series with resistor R_5. In addition, the parallel combination of R_3 and R_4 *plus* the series resistor R_5 are all in *parallel* with resistor R_2. Hence, in this circuit you can see that there is a parallel-series element (R_3, R_4, and R_5) in parallel with another component, R_2. This entire resistor cluster is in series with the components R_1 and R_6.

Now focus your attention on the known quantities in this circuit as labeled on the schematic. You know that R_1 is 1 kilohm, and that the current flowing through this leg of the circuit is 50 milliamps. You know that resistor R_2 is 3 kilohms, and the current flowing through its branch is 12.5 milliamps. The current through the branch containing R_3 is 25 milliamps, and the voltage drop across R_4 is 25 volts. Finally, you are given that the voltage drop across R_6 is 50 volts.

- Known Circuit Values Chart
- Solution: Begin with R_1
- Calculation of E_{R1} and R_6

E	I	R
$E_{R1} =$	$I_{R1} = 50$ mA	$R_1 = 1$ kΩ
$E_{R2} =$	$I_{R2} = 12.5$ mA	$R_2 = 3$ kΩ
$E_{R3} =$	$I_{R3} = 25$ mA	$R_3 =$
$E_{R4} = 25$ V	$I_{R4} =$	$R_4 =$
$E_{R5} =$	$I_{R5} =$	$R_5 =$
$E_{R6} = 50$ V	$I_{R6} =$	$R_6 =$
$E_T =$	$I_T =$	$R_T =$

Figure 9.58

Known Circuit Values Chart — This example is really a little different from the others that have been analyzed in this lesson, except there are a few more knowns and unknowns, and they are sort of "scrambled." To help you keep track of the knowns and the unknowns as you determine them, you can use a chart such as the one shown in Figure 9.58. This chart lists all the possible voltages, currents, and resistances in the particular circuit you are going to solve. When you complete your calculations, all of the spaces will be filled in, and you can check your values since each voltage listed should equal the current times the resistance across each line of the chart.

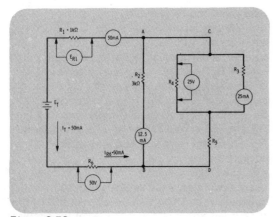

Figure 9.59

Solution: Begin with R_1 — There are several ways to begin the solution of this problem, but in general when solving series-parallel circuits, it is a good idea to determine the total current flowing in the circuit as soon as possible. Focus your attention on R_1 in Figure 9.59. Since R_1 is in series with the total current, and you are given that 50 milliamps is flowing through it, you immediately know that the total current flowing in this circuit, I_T, is also 50 milliamps. Also, since R_6 is in series with the total current, it too must have 50 milliamps flowing through it. With these values in mind, you can now use Ohm's law to calculate the voltage across R_1 (E_{R1}) and the resistance of R_6.

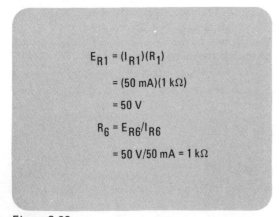

$$E_{R1} = (I_{R1})(R_1)$$
$$= (50 \text{ mA})(1 \text{ k}\Omega)$$
$$= 50 \text{ V}$$
$$R_6 = E_{R6}/I_{R6}$$
$$= 50 \text{ V}/50 \text{ mA} = 1 \text{ k}\Omega$$

Calculation of E_{R1} and R_6 — Using Ohm's law in the correct form, as shown in Figure 9.60 E_{R6} and R_6 can be calculated. E_{R1} is 50 volts and R_6 turns out to be identical to R_1, 1 kilohm.

Figure 9.60

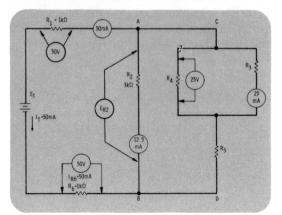

Figure 9.61

Solution: Branch Containing R_2 — Turn your attention to the branch containing resistor R_2 in Figure 9.61. You know two things about R_2: its resistance is 3 kilohms and the current flow through it is 12.5 milliamps. Using Ohm's law in the form $E_{R2} = I_{R2} \times R_2$, you can calculate the voltage across R_2, 12.5 milliamps times 3 kilohms is 37.5 volts for E_{R2}. Remember this important fact, the voltage across R_2 is also the voltage across the right-hand branch (between points C and D), since these two branches are connected in parallel. Therefore, E_{R2} (37.5 volts) is the voltage across the entire parallel-connected cluster of resistors. At this point you know:

1. The voltage across R_1 is 50 volts.
2. Connected in series with R_1 is the parallel cluster containing resistors R_2, R_3, R_4, and R_5, and the voltage across that cluster (E_{R2-5}) is 37.5 volts.
3. Connected in series with R_{2-5} is R_6 and the voltage across R_6 is 50 volts.

What do these three facts tell you? The total voltage, E_T, is the sum of these three voltages:

$$E_T = E_{R1} + E_{R2-5} + E_{R6}$$

or

$$E_T = 50 + 37.5 + 50$$
$$E_T = 137.5 \text{ V}$$

E	I	R
E_{R1} = 50 V	I_{R1} = 50 mA	R_1 = 1 kΩ
E_{R2} = 37.5 V	I_{R2} = 12.5 mA	R_2 = .3 kΩ
E_{R3} =	I_{R3} = 25 mA	R_3 =
E_{R4} = 25 V	I_{R4} =	R_4 =
E_{R5} =	I_{R5} =	R_5 =
E_{R6} = 50 V	I_{R6} = 50 mA	R_6 = 1 kΩ
E_T = 137.5 V	I_T = 50 mA	R_T =

Figure 9.62

Chart with Calculated Values — You are now quite well into the analysis of the operation of this circuit. Your chart of values at this point should look like the one in Figure 9.62. By inspecting the chart carefully, you can decide on your next step. First of all, since E_T and I_T, the total applied voltage and current, are known, you can calculate R_T, the total equivalent resistance of this circuit, using Ohm's law. Notice that *no* circuit reduction was involved in this case since you were given different types of known values at the start of this problem than in previous cases. Using Ohm's law in the form $R_T = E_T/I_T$, and substituting, yields R_T = 137.5 volts divided by 50 milliamps, which equals a value of 2.75 kilohms. Looking at the other blank spaces in your chart should tell you that it is time to move your attention to the right-hand branch of the circuit which contains R_3, R_4, and R_5.

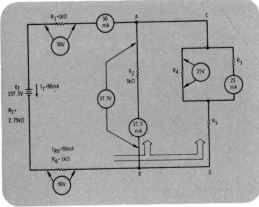

Figure 9.63

Circuit Solution: Right-Hand Branch — All you know about the branch in Figure 9.63 is that the voltage across it equals E_{R2}, 37.5 volts. What else can you calculate about it? The law for currents in a parallel circuit states that the total or main line current must equal the sum of the branch currents. Here, 50 milliamps is flowing through R_6 to point B, where it divides. You know that 12.5 milliamps out of this 50 milliamps flows through R_2, so the rest must flow in the right-hand branch. Fifty milliamps minus 12.5 milliamps yields 37.5 milliamps flowing in this branch. Automatically you know that this is I_{R5}, the current through R_5 since R_5 is in series in this branch and hence must carry all the branch current. Also, this branch current of 37.5 milliamps flowing past R_5 must again branch to go through R_3 and R_4. Twenty-five out of those 37.5 milliamps flow through R_3 so what is left must go through R_4. This 37.5 milliamps minus 25 milliamps leaves 12.5 milliamps for the current through R_4, I_{R4}. Immediately, your mind should "click." E_{R4} and I_{R4} are now known. Therefore, R_4 can be calculated with Ohm's law in the form $R_4 = E_{R4}/I_{R4}$. Substituting, $R_4 = 25$ volts/12.5 milliamps, or R_4 equals 2 kilohms.

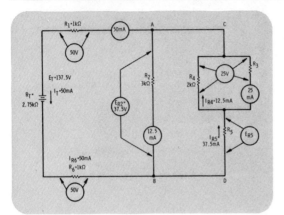

Figure 9.64

Circuit Solution R$_3$ and R$_5$ — Where do you proceed from here? Look at Figure 9.64. You need to "finish off" R$_3$ and R$_5$. What do you know about R$_3$? You are given that 25 milliamps of current flows through it. Also, notice that it is in *parallel* with R$_4$, so the voltage across R$_4$ is also the voltage across R$_3$, 25 volts. This enables you to calculate R$_3$:

$$R_3 = E_{R3}/I_{R3}$$
$$R_3 = 25 \text{ V}/25 \text{ mA}$$

which equals 1 kilohm.

Proceed to R$_5$. You have already calculated that I$_{R5}$ is 37.5 milliamps. You know that the total voltage across the right-hand branch, between points C and D, is 37.5 volts. Twenty-five volts of that is dropped across the parallel cluster containing R$_3$ and R$_4$, and the remainder must be dropped across R$_5$. This means that 37.5 volts minus 25 volts results in 12.5 volts for E$_{R5}$. With E$_{R5}$ and I$_{R5}$ known, you can now calculate

$$R_5 = E_{R5}/I_{R5}$$
$$= 12.5 \text{ V}/37.5 \text{ mA}$$

which equals 333 ohms.

Complete Analysis — Your analysis should be complete and the completed chart of values for this circuit is given in Figure 9.65 for your reference.

E	I	R
E$_{R1}$ = 50 V	I$_{R1}$ = 50 mA	R$_1$ = 1 kΩ
E$_{R2}$ = 37.5 V	I$_{R2}$ = 12.5 mA	R$_2$ = 3 kΩ
E$_{R3}$ = 25 V	I$_{R3}$ = 25 mA	R$_3$ = 1 kΩ
E$_{R4}$ = 25 V	I$_{R4}$ = 12.5 mA	R$_4$ = 2 kΩ
E$_{R5}$ = 12.5 V	I$_{R5}$ = 37.5 mA	R$_5$ = 333 Ω
E$_{R6}$ = 50 V	I$_{R6}$ = 50 mA	R$_6$ = 1 kΩ
E$_T$ = 137.5 V	I$_T$ = 50 mA	R$_T$ = 2.75 kΩ

Figure 9.65

Final Note-Power in Series-Parallel Circuits — As was the case for parallel-series circuits and *all* dc resistive circuits, the power dissipated by the circuit is an additive quantity. To find the total power a series-parallel circuit dissipates, you could:

1. Find the power dissipated by each resistor and add them all.
2. Apply a power formula to the total circuit quantities E_T, I_T, or R_T.

For the circuit that has just been analyzed then, as shown in Figure 9.66, the total dissipated power can be calculated using the formula $P_T = I_T \times E_T$. Substituting the circuit total values into the formula:

$$P_T = 50 \text{ mA} \times 137.5 \text{ V}$$
$$P_T = 6.88 \text{ W}$$

More information concerning the calculation of dissipated power in complex circuits will be covered in Lesson 10.

During this lesson quite a lot concerning the series-parallel circuit has been covered. These circuits were first discussed in general with attention to how they may be recognized and what the outstanding features of these operations are. Several simple circuits were analyzed where you were given the applied voltages and resistance values and where to calculate all voltages and currents. In cases such as this, three steps to follow in the circuit analysis were set down. A "live" example then demonstrated some of the peculiarities in the behavior of these circuits for you. Finally, an example involving a "detective" approach was covered for you. In cases such as the last one, you proceed through the circuit armed with your "circuit sense," building your analysis as you go along.

Practice, as usual, is the best way to develop the knack for handling these circuits. An extensive set of worked through examples and practice problems is included at the end of this lesson for your use. If you work with these problems, you will find yourself capable of solving these circuits very quickly. Also, you should begin to enjoy the "detective" aspect involved in predicting the behavior of voltages and currents in this complex circuit situation.

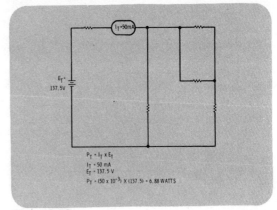

Figure 9.66

LESSON 9. SERIES-PARALLEL CIRCUITS

● **Worked Through Examples**

1. In the series-parallel circuit shown, calculate the total equivalent resistance and all unknown voltages and currents using Ohm's law and circuit reduction techniques.

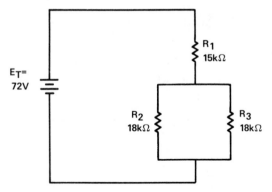

First, you can find R_T by circuit reduction techniques. Since R_2 and R_3 are of equal value and are connected in parallel, the equivalent resistance, $R_{2,3}$ can be found with the formula:

$$R_{eq} = \frac{R_s}{N}$$

R_s equals 18 kilohms and N equals 2, so:

$$R_{2,3} = R_{eq} = \frac{R_s}{N} = \frac{18\ k\Omega}{2}$$

$$R_{2,3} = 9\ k\Omega$$

After the first circuit reduction, the circuit now consists of R_1 in series with $R_{2,3}$ as shown.

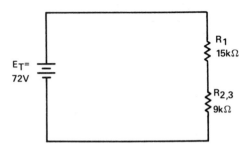

You can find the total resistance of the circuit by simply using the series circuit law which says that the total resistance of a series circuit equals the sum of the individual resistances. In formula form:

$$R_T = R_1 + R_2 + R_3 + \ldots$$

or in this case:

$$R_T = R_1 + R_{2,3}$$
$$R_T = 15 \text{ k}\Omega + 9 \text{ k}\Omega$$
$$R_T = 24 \text{ k}\Omega$$

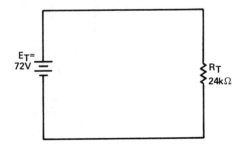

Once you know the total resistance, you can find the total current by using Ohm's law in the form $I_T = E_T/R_T$. Substituting the appropriate values in the formula gives:

$$I_T = \frac{E_T}{R_T} = \frac{72 \text{ V}}{24 \text{ k}\Omega}$$

$$I_T = 3 \text{ mA}$$

This total current can be used to find the voltage across R_1. Remember, since R_1 is in series with the rest of the circuit, the total current must flow through R_1. If you use Ohm's law in the form $E = I \times R$ and substitute the appropriate values, you get:

$$E_{R1} = I_T \times R_1$$

$$E_{R1} = 3 \text{ mA} \times 15 \text{ k}\Omega$$

$$E_{R1} = 45 \text{ V}$$

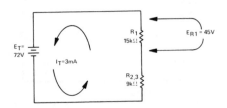

Remember that in a series circuit the total voltage equals the sum of the individual voltage drops. You know the total voltage and the voltage across R_1; the remainder of the voltage must be dropped across $R_{2,3}$. In formula form:

$$E_{R2,3} = E_T - E_{R1}$$

$$E_{R2,3} = 72 \text{ V} - 45 \text{ V}$$

$$E_{R2,3} = 27 \text{ V}$$

You can find the current through R_2 or R_3 by using Ohm's law in the form $I = E/R$. Remember, R_2 and R_3 are in parallel, so they have the same 27 volts dropped across them.

$$I_{R2} = \frac{E_{R2}}{R2} = \frac{27 \text{ V}}{18 \text{ k}\Omega}$$

$$I_{R2} = 1.5 \text{ mA}$$

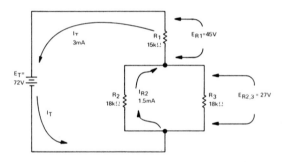

Since R_2 and R_3 have the same resistance value and the same voltage across them, they have the same current flow through them. You could have found the current through R_2 and R_3 by simply realizing that they must divide the total current of 3 milliamps equally between them.

$$I_{R2} = I_{R3} = \frac{I_T}{2} = \frac{3 \text{ mA}}{2}$$

$$I_{R2} = I_{R3} = 1.5 \text{ mA}$$

If R_2 and R_3 did not have the same resistance value, you could have found the current through R_3 by subtraction. You know the total current and you know the current through R_2, so the remainder of the current must flow through R_3.

$$I_{R3} = I_T - I_{R2}$$
$$I_{R3} = 3\text{ mA} - 1.5\text{ mA}$$
$$I_{R3} = 1.5\text{ mA}$$

and the circuit is completely solved.

2. In the series-parallel circuit shown, calculate the total equivalent resistance and all unknown voltages and currents using Ohm's law and circuit reduction techniques.

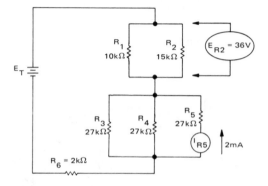

In order to keep track of all the knowns and unknowns, make a chart as shown below and fill in the known values. Then you can fill in the unknown values as you calculate them.

E	I	R
E_{R1} = 36 V	I_{R1} =	R_1 = 10 kΩ
E_{R2} = 36 V	I_{R2} =	R_2 = 15 kΩ
E_{R3} =	I_{R3} =	R_3 = 27 kΩ
E_{R4} =	I_{R4} =	R_4 = 27 kΩ
E_{R5} =	I_{R5} = 2 mA	R_5 = 27 kΩ
E_{R6} =	I_{R6} =	R_6 = 2 kΩ
E_T =	I_T =	R_T =

Notice that since R_1 and R_2 are in parallel, the voltage across them is the same.

You can use Ohm's law in the form $I = E/R$ to calculate I_{R1} and I_{R2}.

$$I_{R1} = \frac{E_{R1}}{R_1} \qquad\qquad I_{R2} = \frac{E_{R2}}{R_2}$$

$$I_{R1} = \frac{36\ V}{10\ k\Omega} \qquad\qquad I_{R2} = \frac{36\ V}{15\ k\Omega}$$

$$I_{R1} = 3.6\ mA \qquad\qquad I_{R2} = 2.4\ mA$$

You know that the total current in a parallel circuit equals the sum of the individual branch currents. In this circuit, the total current flows through the combination of R_1 and R_2; you can add I_{R1} and I_{R2} to get I_T.

$$I_T = I_{R1} + I_{R2}$$
$$I_T = 3.6\ mA + 2.4\ mA$$
$$I_T = 6.0\ mA$$

You can now fill in these calculated values on the chart as shown.

E	I	R
$E_{R1} = 36\ V$	$I_{R1} = 3.6\ mA$	$R_1 = 10\ k\Omega$
$E_{R2} = 36\ V$	$I_{R2} = 2.4\ mA$	$R_2 = 15\ k\Omega$
E_{R3}	I_{R3}	$R_3 = 27\ k\Omega$
E_{R4}	I_{R4}	$R_4 = 27\ k\Omega$
E_{R5}	$I_{R5} = 2\ mA$	$R_5 = 27\ k\Omega$
E_{R6}	I_{R6}	$R_6 = 2\ k\Omega$
E_T	$I_T = 6.0\ mA$	R_T

Looking at the chart or the circuit, you can see that you know two things about R_5, you know its resistance, and you know the current flow through it. You can use Ohm's law in the form $E = I \times R$ to find E_{R5}.

$$E_{R5} = I_{R5} \times R_5$$
$$E_{R5} = 2\ mA \times 27\ k\Omega$$
$$E_{R5} = 54\ V$$

Because R_3, R_4, and R_5 are in parallel, they have 54 volts dropped across them. If they all have the same voltage across them and they all have the same resistance value, then the current must be the same through all of them. Since I_{R5} equals 2 milliamps, then I_{R3} and I_{R4} also equal 2 milliamps each.

E	I	R
$E_{R1} = 36\ V$	$I_{R1} = 3.6\ mA$	$R_1 = 10\ k\Omega$
$E_{R2} = 36\ V$	$I_{R2} = 2.4\ mA$	$R_2 = 15\ k\Omega$
$E_{R3} = 54\ V$	$I_{R3} = 2\ mA$	$R_3 = 27\ k\Omega$
$E_{R4} = 54\ V$	$I_{R4} = 2\ mA$	$R_4 = 27\ k\Omega$
$E_{R5} = 54\ V$	$I_{R5} = 2\ mA$	$R_5 = 27\ k\Omega$
E_{R6}	I_{R6}	$R_6 = 2\ k\Omega$
E_T	$I_T = 6.0\ mA$	R_T

You could check your work at this point by adding I_{R3}, I_{R4} and I_{R5} to see that they do add up to the total current of 6 milliamps.

Because R_6 is in series with the rest of the circuit, the total current must flow through it. Thus I_{R6} equals 6 milliamps and you can now use this information to find E_{R6}.

$$E_{R6} = I_{R6} \times R_6$$
$$E_{R6} = 6\ mA \times 2\ k\Omega$$
$$E_{R6} = 12\ V$$

As shown, you know the voltage across and current flow through each portion of the circuit.

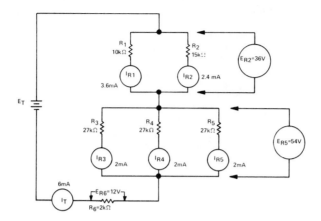

The voltage across R_1 and R_2 is the same; $E_{R1,2}$ equals 36 volts. The voltage is also the same across R_3, R_4, and R_5; $E_{R3,4,5}$ equals 54 volts. You also know the voltage across R_6; E_{R6} equals 12 volts. From series circuit laws, these voltages can be added to find the total voltage applied to the circuit.

$$E_T = E_{R1,2} + E_{R3,4,5} + E_{R6}$$

$$E_T = 36 \text{ V} + 54 \text{ V} + 12 \text{ V}$$

$$E_T = 102 \text{ V}$$

The only unknown quantity remaining to be calculated is the total resistance. This can be found in either of two ways. One way is to use Ohm's law in the form:

$$R_T = \frac{E_T}{I_T}$$

When you substitute the appropriate values in the formula, you obtain:

$$R_T = \frac{102 \text{ V}}{6 \text{ mA}}$$

$$R_T = 17 \text{ k}\Omega$$

Circuit reduction techniques can also be used to find R_T. First, consider R_1 in parallel with R_2. Using the product-over-the-sum formula:

$$R_{1,2} = \frac{R_1 \times R_2}{R_1 + R_2}$$

$$R_{1,2} = \frac{10 \text{ k}\Omega \times 15 \text{ k}\Omega}{10 \text{ k}\Omega + 15 \text{ k}\Omega}$$

$$R_{1,2} = \frac{(1 \times 10^{+4}) \times (1.5 \times 10^{+4})}{(1 \times 10^{+4}) + (1.5 \times 10^{+4})}$$

$$R_{1,2} = \frac{1.5 \times 10^{+8}}{2.5 \times 10^{+4}}$$

$$R_{1,2} = 0.6 \times 10^{+4} = 6 \text{ k}\Omega$$

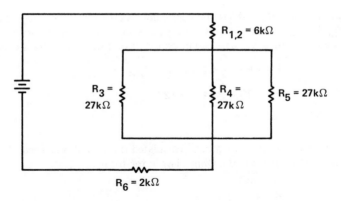

Because R_3, R_4 and R_5 all have the same resistance value, they can be reduced to an equivalent resistance by using the formula:

$$R_{eq} = \frac{R_s}{N}$$

$$R_{eq} = \frac{27 \text{ k}\Omega}{3}$$

$$R_{eq} = 9 \text{ k}\Omega$$

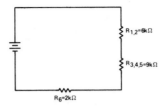

These three resistances are now in series and can be added to find R_T.

$$R_T = R_{1,2} + R_{3,4,5} + R_6$$
$$R_T = 6 \text{ k}\Omega + 9 \text{ k}\Omega + 2 \text{ k}\Omega$$
$$R_T = 17 \text{ k}\Omega$$

and this agrees with the previous calculation.

The chart can be filled in as shown, and the circuit is completely solved.

E	I	R
$E_{R1} = 36$ V	$I_{R1} = 3.6$ mA	$R_1 = 10$ kΩ
$E_{R2} = 36$ V	$I_{R2} = 2.4$ mA	$R_2 = 15$ kΩ
$E_{R3} = 54$ V	$I_{R3} = 2$ mA	$R_3 = 27$ kΩ
$E_{R4} = 54$ V	$I_{R4} = 2$ mA	$R_4 = 27$ kΩ
$E_{R5} = 54$ V	$I_{R5} = 2$ mA	$R_5 = 27$ kΩ
$E_{R6} = 12$ V	$I_{R6} = 6$ mA	$R_6 = 2$ kΩ
$E_T = 102$ V	$I_T = 6$ mA	$R_T = 17$ kΩ

LESSON 9. SERIES-PARALLEL CIRCUITS

● **Practice Problems**

___

The key objective of this lesson has been achieved if you can analyze any series parallel circuit in a variety of situations such as:

1. Given a series-parallel wired network of resistors, calculate their equivalent resistance, R_{eq}.
2. Given a series-parallel circuit with all of the resistor values and the applied voltage labeled, calculate any or all of the voltages across and currents through each resistor, as well as the total circuit current and equivalent resistance.
3. Given a series-parallel circuit schematic with several known values labeled, calculate any unknown values required.

The practice problems that follow are designed to give you as much practice as you may need in these areas. It is suggested that you work enough of these to enable you to approach and analyze any series-parallel circuit without referring back to the lesson. Fold over the page to check your answers

Depending upon the approach you use in solving these problems and how you round off intermediate results, your answers may vary slightly from those given here. However, any differences you encounter should only occur in the third significant digit of your answer. If the first two significant digits of your answers do not agree with those given here, recheck your calculations.

1.   Find R_{eq} for the following circuits.   **Fold Over**

a.

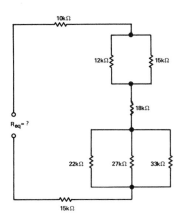

$R_{eq} =$ _____

Answers

1.a. $R_{eq} = 58.5$ kΩ

b.

Fold Over

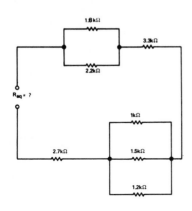

$R_{eq} =$ _____

c.

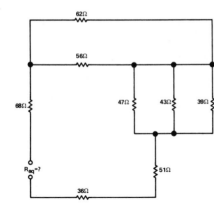

$R_{eq} =$ _____

Answers

1.b. $R_{eq} = 7.39 \ k\Omega$

1.c. $R_{eq} = 199 \ \Omega$

d.   **Fold Over**

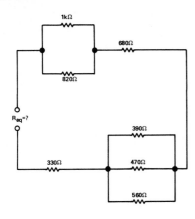

$R_{eq} =$ _____

e.

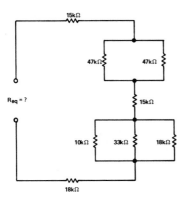

$R_{eq} =$ _____

Answers

1.d. R_{eq} = 1.61 kΩ

1.e. R_{eq} = 76.9 kΩ

f.   **Fold Over**

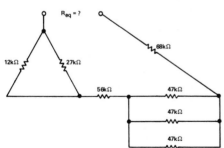

R_{eq} = _____

g.

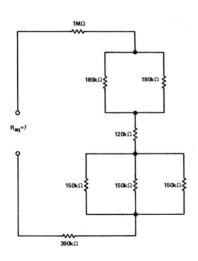

R_{eq} = _____

Answers

1.f. R_{eq} = 148 kΩ

1.g. R_{eq} = 1.65 MΩ

h. **Fold Over**

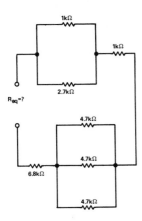

$R_{eq} = $ _____

i.

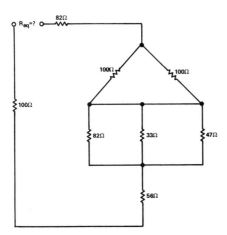

$R_{eq} = $ _____

Answers

1.h. $R_{eq} = 10.1\ k\Omega$

1.i. $R_{eq} = 304\ \Omega$

j.

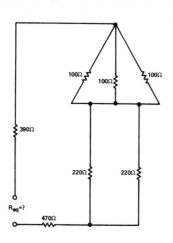

$R_{eq} =$ _____

Fold Over

k.

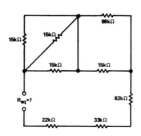

$R_{eq} =$ _____

Answers

1.j. R_{eq} = 1 kΩ

1.k. R_{eq} = 154 kΩ

l.

Fold Over

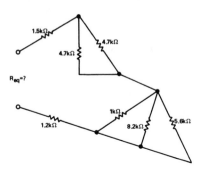

$R_{eq} =$ _____

m.

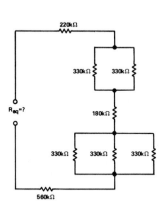

$R_{eq} =$ _____

Answers

1.l. R_{eq} = 5.82 kΩ

1.m. R_{eq} = 1.24 MΩ

n.

Fold Over

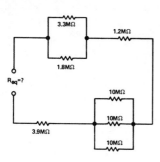

$$R_{eq} = \underline{\hspace{3cm}}$$

o.

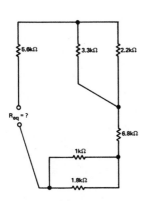

$$R_{eq} = \underline{\hspace{3cm}}$$

Answers

1.n. R_{eq} = 9.60 MΩ

1.o. R_{eq} = 14.4 kΩ

p.

Fold Over

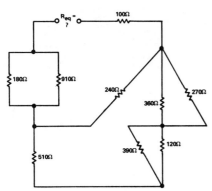

$R_{eq} =$ _____

Answers

1.p. $R_{eq} = 432 \ \Omega$

2. Using Ohm's law and circuit reduction techniques, calculate the requested voltages and currents in the following circuits.

Fold Over

a.

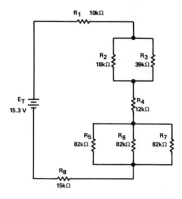

I_{R2} = _____

E_{R4} = _____

I_{R5} = _____

b.

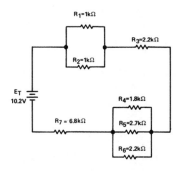

I_{R3} = _____

E_{R4} = _____

I_{R6} = _____

Answers

2.a. $I_{R2} = 137\ \mu A$

   $E_{R4} = 2.40\ V$

   $I_{R5} = 66.7\ \mu A$

2.b. $I_{R3} = 1\ mA$

   $E_{R4} = 724\ mV$

   $I_{R6} = 329\ \mu A$

c.

Fold Over

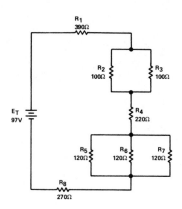

$E_{R1} =$ _____

$I_{R2} =$ _____

$E_{R6} =$ _____

d.

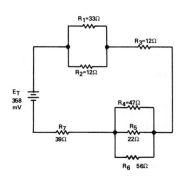

$E_{R1} =$ _____

$E_{R5} =$ _____

$I_{R7} =$ _____

Answers

2.c. $E_{R1} = 39$ V

$I_{R2} = 50$ mA

$E_{R6} = 4.0$ V

2.d. $E_{R1} = 44.0$ mV

$E_{R5} = 59.0$ mV

$I_{R7} = 5.0$ mA

e.

Fold Over

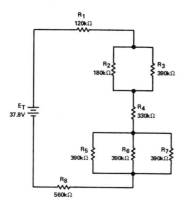

E_{R1} = _____

E_{R2} = _____

I_{R6} = _____

f.

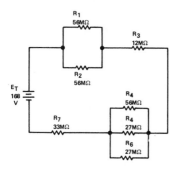

E_{R3} = _____

I_{R4} = _____

E_{R6} = _____

Answers

2.e. $E_{R1} = 3.6$ V

$E_{R2} = 3.69$ V

$I_{R6} = 10.0\ \mu A$

2.f. $E_{R3} = 24.0$ V

$I_{R4} = 389$ nA

$E_{R6} = 21.8$ V

g. **Fold Over**

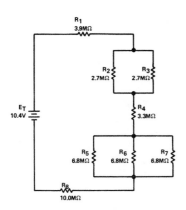

I_{R1} = _____

I_{R3} = _____

I_{R6} = _____

h.

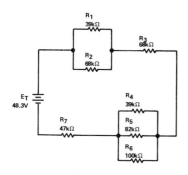

E_{R1} = _____

E_{R4} = _____

E_{R7} = _____

Answers

2.g. $I_{R1} = 500$ nA

$I_{R3} = 250$ nA

$I_{R6} = 167$ nA

2.h. $E_{R1} = 7.44$ V

$E_{R4} = 6.27$ V

$E_{R7} = 14.1$ V

Practice Problems

3. Calculate the requested voltages, currents, and resistances in the following circuits.

Fold Over

a.

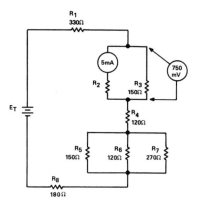

$R_2 =$ _____

$I_{R3} =$ _____

$E_{R7} =$ _____

$E_T =$ _____

b.

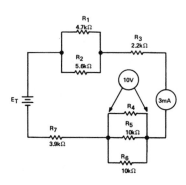

$I_{R1} =$ _____

$R_4 =$ _____

$E_{R7} =$ _____

$E_T =$ _____

Answers

3.a. $R_2 = 150 \ \Omega$

$I_{R3} = 5$ mA

$E_{R7} = 535$ mV

$E_T = 7.58$ V

3.b. $I_{R1} = 1.63$ mA

$R_4 = 10.0$ kΩ

$E_{R7} = 11.7$ V

$E_T = 36.0$ V

c. Fold Over

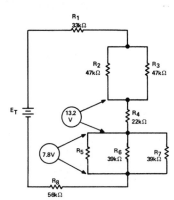

$I_{R2} =$ _____

$I_{R4} =$ _____

$R_5 =$ _____

$E_T =$ _____

d.

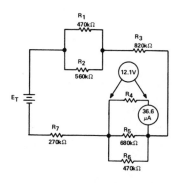

$I_{R1} =$ _____

$E_{R3} =$ _____

$R_4 =$ _____

$E_T =$ _____

Answers

3.c. $I_{R2} = 300\ \mu A$

$I_{R4} = 600\ \mu A$

$R_5 = 39\ k\Omega$

$E_T = 88.5\ V$

3.d. $I_{R1} = 43.6\ \mu A$

$E_{R3} = 65.8\ V$

$R_4 = 330\ k\Omega$

$E_T = 120\ V$

e.

Fold Over

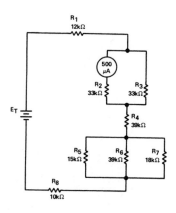

$E_{R1} =$ _____

$E_{R5} =$ _____

$I_{R6} =$ _____

$E_T =$ _____

f.

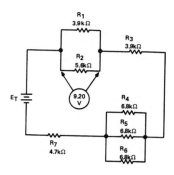

$I_{R1} =$ _____

$I_{R4} =$ _____

$E_{R5} =$ _____

$E_T =$ _____

Answers

3.e. $E_{R1} = 12$ V

$E_{R5} = 6.76$ V

$I_{R6} = 173\,\mu$A

$E_T = 84.3$ V

3.f. $I_{R1} = 2.36$ mA

$I_{R4} = 1.33$ mA

$E_{R5} = 9.04$ V

$E_T = 52.7$ V

Identify the following circuits as series (S), parallel (P), series-parallel (SP), or parallel series (PS).

1.

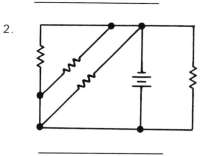

_____

2.

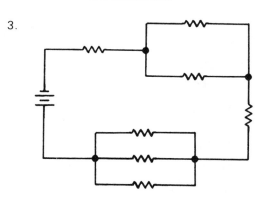

_____

3.

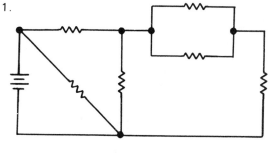

4.

5.

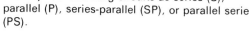

_____

Find R$_{eg}$ for the following circuits:

6.

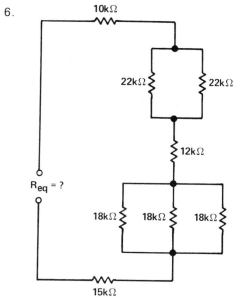

_____

7.

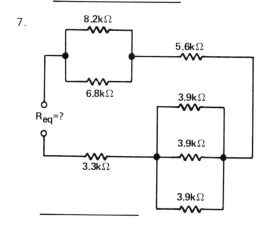

_____

Find the requested voltages and currents and the total resistance for the following circuit.

Find the requested voltages, currents, and resistances in the following circuit.

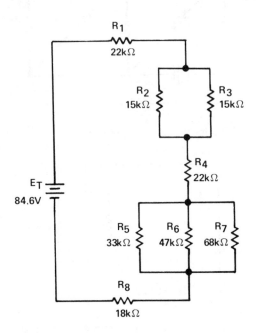

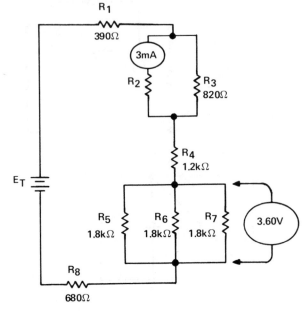

9. $R_T =$ _____
10. $I_T =$ _____
11. $I_{R2} =$ _____
12. $E_{R5} =$ _____
13. $I_{R7} =$ _____

14. $I_T =$ _____
15. $R_2 =$ _____
16. $E_{R4} =$ _____
17. $I_{R6} =$ _____
18. $E_T =$ _____

19. When solving Parallel-Series Networks:
   a. Each branch was examined first
   b. Series circuit laws were applied to each
   c. branch
   d. Parallel circuit laws reduced the elements to minimum
   All of the above

20. When solving Series-Parallel Networks:
   a. Parallel circuit clusters are reduced with parallel laws
   b. Series laws are used to reduce elements to minimum series elements
   c. The total current is found from the total equivalent series resistance
   d. All of the above

Lesson 10

Voltage, Current, and Power

This lesson introduces the loaded voltage divider as a particular application of series-parallel circuits. The concept of dividing voltage is illustrated with a simple voltage divider circuit and this circuit is used to examine the effect of selecting different ground points in the circuit. More practical voltage dividers with loads are discussed, and then the basic power formulas are reviewed and applied to a voltage divider circuit. The procedures for finding square roots with a calculator and with square root tables are also discussed.

LESSON 10. VOLTAGE DIVIDERS AND POWER

● Objectives

In this lesson your knowledge of series-parallel circuits will be extended to cover the voltage divider circuit configuration. In addition, the concept of power will be reviewed and calculation of power consumption for more complex circuits will be covered. Upon successful completion of this lesson, you should be able to:

1. *Design* a voltage divider circuit using a 100-volt power supply and five 10-kilohm resistors that will provide voltages of −20 V, −40 V, −60 V, +20 V, and +40 V to an external circuit that draws little or no current.

2. *Write* an explanation of the terms shown below, including diagrams where necessary:

    a.   Polarity
    b.   Ground reference
    c.   Load
    d.   Bleeder current

3. *Write* the "rules of thumb" by which bleeder currents may be selected for voltage dividers in simple applications.

4. *Design* a voltage divider that could be used to supply power to two loads requiring 45 volts at 20 milliamps, and 150 volts at 10 milliamps from a 200-volt power supply.

5. Using circle diagram (if necessary), *write* formulas that can be used to calculate power if

    a.   Voltage across and current through a load are known
    b.   Current through and resistance of a load are known
    c.   Voltage across and resistance of a load are known.

6. With what you have learned in previous lessons, *calculate* the voltage across, current through, or resistance of any resistor in a circuit schematic such as the one shown. At the end of this lesson, given the same schematic and using Ohm's law, the power formulas, and circuit reduction techniques, *calculate* the power each resistor dissipates.

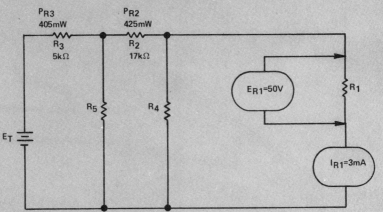

7. When performing square root calculations required in problems such as the above, *use a handheld calculator or square root tables to find the square roots*.

LESSON 10. VOLTAGE DIVIDERS AND POWER

● **Voltage Dividers**

This lesson, Lesson 10, introduces the topic of voltage dividers, and reviews and expands your knowledge of the power formulas used in dc circuits and how to use them. The review and discussion of the power formulas will be presented later on in the lesson, and they will be related to series-parallel circuits of the type that were covered in Lessons 8 and 9.

Voltage Dividers — The first part of this lesson gives you a look at what is called a voltage divider. You will learn what voltage dividers are, and why they are used (Figure 10.1). As you will see, a voltage divider is actually a series-parallel circuits, so in order to understand and work with voltage dividers, you need to know Ohm's law, the series voltage law, and the parallel current law.

VOLTAGE DIVIDERS

● WHAT THEY ARE
● WHY THEY ARE USED

Figure 10.1

Now you may be asking, "What are voltage dividers anyway, and why should I study them?" Often in your study of electricity or electronics, you may have an application or circuit situation that requires a voltage source which can provide several specified voltages and currents, but there is no power supply handy that provides just the voltages you need. (For example, suppose you need to operate devices that require 15, 25, and 150 volts, and maybe only a 200-volt supply is available.) Voltage dividers provide a fairly economical way to obtain one or several lower voltages from a single higher voltage supply. Since this requirement occurs quite often, it is convenient to know about voltage dividers and their use. It is also important to realize the limitations and tradeoffs involved in their use. They do waste some power and in general cannot be used to operate devices whose demand for current is very high or varies greatly.

Sample Voltage Divider — In order to better understand voltage dividers, it will be helpful to review some of the topics related to them that have already been covered. In Lesson 5, it was pointed out how it is possible to get both positive and negative voltages from one series string of resistors, depending on where the reference point is in the circuit. It may seem unusual that a circuit may require negative and positive voltages from a single source; however, circuits and devices requiring more than one polarity of voltage for their operation are quite common in many electronic applications, from your TV to the most advanced computers.

Figure 10.2 shows a simple series circuit with four resistors each having 25 ohms of resistance. This means that in this circuit there is a total of 100 ohms of resistance across a 100-volt supply.

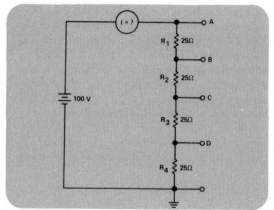

Figure 10.2

Calculations for Sample Circuit — Using Ohm's law as shown in Figure 10.3, you can calculate that the total current flowing in this circuit is 1 amp. You can use Ohm's law once again in the form $E = I$ times R to determine the voltage across R_4. Since you know the current is the same in all parts of a series circuit, the current through R_4 is 1 amp. Multiplying 1 amp times 25 ohms yields a calculated voltage drop of 25 volts across R_4.

$$I = E/R$$

$$I = \frac{100 \text{ VOLTS}}{100 \text{ OHMS}} \qquad I = 1 \text{ AMP}$$

$$E = I \times R$$

$$E = 1 \text{ AMP} \times 25 \text{ OHMS}$$

$$E = 25 \text{ VOLTS}$$

Figure 10.3

Sample Divider with Voltages — In the circuit of Figure 10.4, 25 volts is measured from ground to point D. If you were to measure the voltage from ground to point C, you would have an additional 25 volts across R_3 which when added to the voltage drop across R_4 would give you a total of 50 volts. If you were to measure the voltage from ground to point B, you would have 75 volts, and from ground to point A, you would measure the total supply voltage of 100 volts. You can see that at points A, B, C, and D, there are four different voltages increasing in 25-volt steps; the voltage has been *divided* by this circuit.

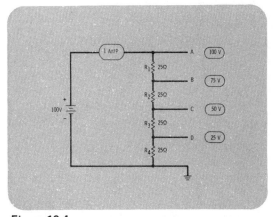

Figure 10.4

Figure 10.5

Voltage Divider Equation — The voltage output between any two points of a simple series circuit voltage divider such as this can be expressed in terms of a formula often called the *voltage divider* equation. Consider the simple circuit shown in Figure 10.5, where R_1, R_2, R_3, and R_4 can be any resistor values, and E_{supply} is the supply voltage connected in series with them. You know from Ohm's law that the total current flowing in this circuit, I, equals the supply voltage divided by the total series resistance of the circuit:

$$I = \frac{E_{supply}}{R_1 + R_2 + R_3 + R_4}$$

The voltage between any two points on this divider equals the product of this current times the resistance between these two points. For example, $E_{AB} = I \times R_1$. If you substitute for I, its equivalent expression as shown, you find that:

$$E_{AB} = \left(\frac{E_{supply}}{R_1 + R_2 + R_3 + R_4} \right) \times R_1$$

It turns out that for any two points on this divider the voltage output can be related directly to the supply voltage through a simple formula like this. Recognize that $R_1 + R_2 + R_3 + R_4$ is the total resistance of the circuit, R_{total}. Then the output voltage between any two points, E_{out}, can be expressed as

$$E_{out} = \left(\frac{E_{supply}}{R_{total}} \right) \times R_{out}$$

or, just moving these terms around:

$$E_{out} = \left(\frac{R_{out}}{R_{total}} \right) \times E_{supply}$$

This last expression is called the *voltage divider equation*. It states that the voltage output between any two points of a voltage divider equals the supply voltage times the ratio of the resistance between these two points, to the total resistance of the circuit.

This series circuit is one simple type of voltage divider. Suppose that you wanted to obtain a negative 25 volts from this same 100-volt supply. How could this circuit be altered to provide that? All you have to do is move the ground reference to point D, and move point D down to where the ground point was originally.

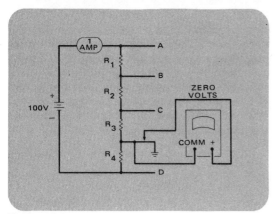

Figure 10.6

Review: Grounds and Voltage Reference — At this point it might be helpful to review the concept of a reference point for voltage measurement and ground. As has been mentioned previously, whenever a voltage is being measured, *two points* are involved. There are two probes on any voltmeter and the meter will measure the potential difference *between* these points. In circuits, voltage values are often stated with respect to one common reference point in the circuit. This reference point is given the general name "ground." Note that as shown in Figure 10.6, once this reference point is established for voltage, it is considered to be the zero voltage point in the circuit. This is because it is at zero volts with respect to *itself* as a reference. Think about this; *any* point is at zero volts with respect to *itself* as a reference. Also, *any* point in the circuit can be picked as a reference point for measuring voltage. If you place the voltmeter's reference probe at the reference point of the circuit, you can then measure all voltages and their polarities with respect to that point by touching the voltmeter's positive probe to other points in the circuit. Note that if you touch the positive probe to the reference point itself, you will read zero volts. The probes are effectively touched together.

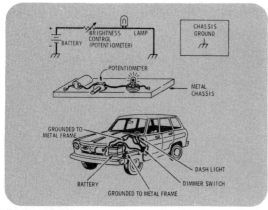

Figure 10.7

Return Path for Current — In addition to being used to specify the reference point from which voltage measurements are taken, the word *ground* is also often used to designate the *return path for current flow in a circuit*. In much electronic circuitry, the metal chassis upon which the circuit is assembled is used as the "return" path. In your car, the body and frame of the car are used as a return path for the current and are usually part of the path completing all circuits as shown in Figure 10.7. The great advantage of the common ground is realized in its contribution to the economy of wiring in circuits. Only one wire need be run to the part, and then the chassis completes the circuit. In addition, schematic diagrams are simplified, and voltage checks in equipment are made easier. Usually the technician is given a table of voltages with respect to the chassis of a piece of equipment. With the meter reference probe connected to the chassis, voltage readings may then be taken easily throughout the circuit with the other probe. The symbol used for a chassis ground is shown in Figure 10.7.

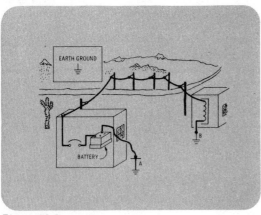

Figure 10.8

Earth Ground — As the use of electricity developed, one of the earliest reference points for voltage measurement and return paths for current flow was the earth itself. This is how the use of the word "ground" got started. In early telegraph systems, for example, one side of the system was connected directly into the earth. (A good long pipe or metal rod running into the earth is often used as a ground — good earth contact must be made.) The earth itself completed the circuit. (Actually, a good earth contact can be considered as a huge neutral reservoir containing both positive and negative charges. In the telegraph circuit shown in Figure 10.8, electrons are pushed into the ground at point A, and pulled out at point B). The use of earth grounding eliminated the need to string *two* telegraph wires over long distances and saved considerable cost.

- Grounding
- Three-Prong Plug and Outlet
- Don't Ground Yourself

Figure 10.9

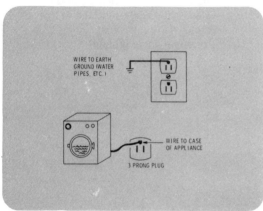

Figure 10.10

Grounding — Earth ground is an important concept to review because in most homes and laboratories the structural metal surfaces are at earth ground potential. Water pipes, gratings, electric conduit boxes and plates should all be connected to earth ground. Since it is easy to touch these things as you move about your home or the lab, it is important that none of the other surfaces you can easily touch have a greatly different potential from ground. If, for example, through faulty wiring, the outside of the refrigerator in your kitchen was raised to 110 volts, look at the possibilities in Figure 10.9. If you touched a faucet with one hand and the refrigerator with the other, you would have a 110-volt potential difference right across your body. This is very often a lethal situation.

Three-Prong Plug and Outlet — For this reason most appliances now come with a special three-prong plug. The third (long) prong is connected to a wire which should be connected to nothing but the outside surfaces and case of the appliance. In your wall outlet, the third socket in the receptacle should be connected directly to a good earth ground (Figure 10.10). In this way, all the appliances and fixtures in your kitchen and the rest of your home have their outside conducting surfaces at ground potential. If through faulty wiring some appliance developed a high potential on its outside case, a large current would flow through the ground wire, usually enough to blow a fuse and alert you of trouble.

Don't Ground Yourself — Electrical chassis are often connected to earth ground for safety reasons. In schematics you may see the earth ground symbol used to indicate an earth grounded point. But often electrical chassis grounds are NOT earth grounded. *This means a metal chassis may be at a much higher (or lower) potential than earth ground*, in which case touching the chassis and touching a good ground could result in an unhealthy shock. Always be careful when handling electrical equipment, and, in general, avoid grounding your body while working in the lab. If the floor and your shoes are wet, for example, you could become well grounded through the water.

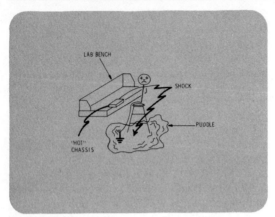

Figure 10.11

Touching a "hot" wire could then cause quite a bit of current flow through your body (Figure 10.11).

In review:

1. Voltages are measured between two points, one of these is usually referred to as a reference point or "ground."

2. A chassis ground (⌐⟋⟋) refers to an electrical chassis or metal frame that is used as part of the current pathway for circuits wired on it. The chassis is usually also used as the reference point from which voltages in these circuits are measured. A chassis ground may or may not be an earth ground.

3. An earth ground (⟂) refers to a point that is at the same potential as the earth itself, and most usually is a point that is electrically well connected to a set of metal water pipes that run for long distances under ground.

Another Similar Voltage Divider — To get back to voltage dividers, consider a circuit similar to the previous voltage divider circuit. This circuit (Figure 10.12) is essentially the same as the first voltage divider shown in Figure 10.2. All that has been done is that the ground reference point has been moved. If you were to measure the voltage from the ground reference to point C, you would measure a positive 25 volts.

If you were to measure from ground to point B, you would get a plus 50 volts, and from ground to point A, you would get a plus 75 volts. However, if you were to measure from ground to point D, you would get a negative voltage, a

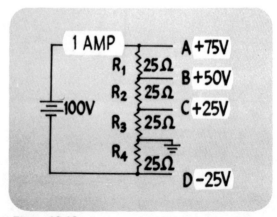

Figure 10.12

negative 25 volts. Remember that the term "polarity" is used when speaking of positive or negative voltage. For instance, in this circuit, the polarity of voltage from points A through C is positive with respect to ground; the polarity of voltage at point D is negative with respect to ground. Notice that this circuit still divides the applied voltage into four equal "chunks," and that the same current will flow as before. The maximum positive voltage available is 75 volts, but note a −25 volts is available from the reference point to point D.

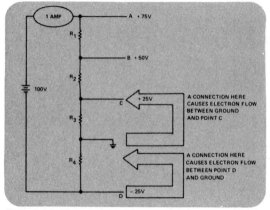

Figure 10.13

Polarity and Ground — Notice that when a voltage is said to be positive with respect to ground, this means that if a circuit were connected between that point and ground, electrons would flow out of ground to that point. If a point is negative with respect to ground, connecting the point and ground would result in a flow of electrons from that point to ground.

This situation is pictured for you in Figure 10.13. If ground and point C were connected, electron current would flow from ground to point C. If ground and point D were connected, electron current would flow from point D to ground.

A Load — Notice that so far in this discussion of voltage dividers, no external circuits that would draw any current have been connected to the voltage divider. If a current carrying circuit were attached to one of the voltage points, the whole operation of the circuit would change. Such an external circuit is called a *"load"*.

Concentrate your attention for a moment on this word, load. *A load is defined as any circuit or device that draws current and/or has resistance, requires voltage, or dissipates power.* As shown in Figure 10.14, when there is a load on a voltage source or power supply, current is drawn. The load may be a simple resistor.

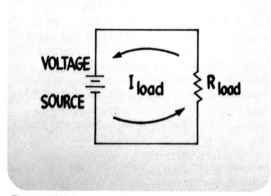

Figure 10.14

Large and Small Loads — Different devices will create different loading effects when attached to a power supply or circuit. Figure 10.15 shows an important point, a lower resistance draws a greater current and is, therefore, a greater load. A higher resistance draws less current, and is a lighter load. So a larger load causes a large current drain on the power supply, while a small load causes a small current drain. In some voltage divider applications, a voltage reference may be needed that will not have to drive any load. However, in many practical applications, a load drawing significant current will be attached to the divider, and the current drawn must be considered in its design.

A **LOWER RESISTANCE** DRAWS A GREATER CURRENT AND IS THEREFORE A GREATER LOAD.

A **HIGHER RESISTANCE** DRAWS LESS CURRENT AND IS A LIGHTER LOAD.

Figure 10.15

Voltage Divider Problem — Figure 10.16 shows a more typical type of voltage divider problem. Consider that you have to operate a load that draws 20 milliamps of current and requires 100 volts to operate. In this situation, also imagine that all that is available in your lab is a 150-volt power source. The problem is to design a voltage divider circuit to supply the correct voltage and current to the load, using what you have learned about series-parallel circuits in previous lessons.

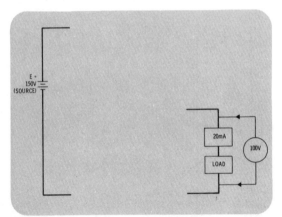

Figure 10.16

Practical Voltage Divider Circuit — Figure 10.17 shows the schematic of a basic voltage divider circuit to do the job, with the specified load connected to points A and B. Notice that this voltage divider is basically just a series-parallel circuit. There are two parallel paths for current to flow through. One branch current flows through R_2 and the other branch current flows through the load. The sum of these two currents equals the total current, which flows through R_1. The dotted lines indicate the paths of current flow.

To make this voltage divider supply the correct voltage and current for the load, you must select the correct resistances for R_1 and R_2. How

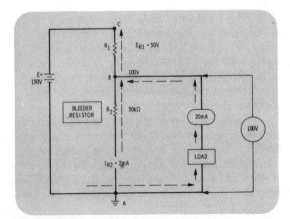

Figure 10.17

do you do this? This question leads to several interesting points concerning voltage dividers. The voltage delivered to the load depends on the relative sizes of R_1 and R_2, their ratio. Notice again that resistor R_1 will carry the total current in the circuit but that resistor R_2 carries a smaller current, called the *bleeder current*. Resistor R_2, through which the bleeder current flows, is called the *bleeder resistor*. The bleeder resistor is what develops the voltage that is delivered to the load, and also will function to stabilize the voltage to the load.

Choosing Bleeder Current — The design of a voltage divider begins by choosing a value of bleeder current. A large bleeder current will tend to keep the load voltage constant, even if the load current should vary for some reason. Variations in load current may occur for a variety of reasons, which depend on the nature of the load. The load attached to this circuit may be a motor or other device whose current demand may be continually varying. A high bleeder current causes a lot of power to be wasted in the voltage divider.

A compromise is made and as a general rule of thumb, shown in Figure 10.18, for simple voltage divider applications, bleeder current is selected to be 10% or one-tenth of the total load current. This conserves power, while allowing some measure of stability for the voltage divider.

"RULE OF THUMB" FOR VOLTAGE DIVIDERS

THE BLEEDER CURRENT EQUALS 10% OR ONE-TENTH OF THE TOTAL LOAD CURRENT

Figure 10.18

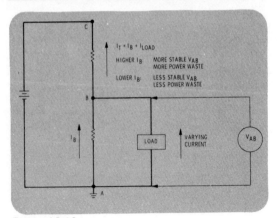

Figure 10.19

Voltage Divider Bleeder Current — This "rule of thumb" for voltage dividers is one that achieves a good compromise for most low-power, constant-load demand situations. To accommodate other loads and other power situations, you may see a variety of "rules of thumb" listed in various textbooks. Any rule, however, is working around the same compromise. The higher the bleeder current, the more stable the load voltage will be if the load current should vary. This is illustrated in Figure 10.19. However, bleeder current causes wasted power, all of which must be dissipated in the bleeder resistor. So in reality, the 10% rule offers a good compromise for loads drawing a moderate amount of current (generally around 100 milliamps or less) that will not vary drastically. For heavier loads, this rule is sometimes modified to read as follows: For loads from zero to 99 milliamps, select the bleeder current as 10% of the load current. For loads drawing greater than 100 milliamps, choose a bleeder current of 10 milliamps. This modified rule is primarily designed to keep the power dissipated by the bleeder resistor to a level that can be handled by fairly common components.

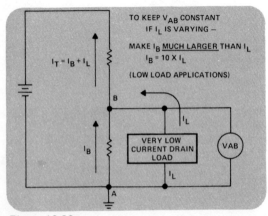

Figure 10.20

Low Power Highly Varying Load — The choice of bleeder current might take another direction in a case where you have a small load drawing very little current but whose current drain is varying wildly. If a load such as this required a constant voltage to operate, you might choose to make the bleeder current much higher than the load current, say ten times the load current. In this case (Figure 10.20), the voltage between points A and B is primarily determined by the bleeder current. Actually, if the bleeder current is very high and the load current is low and varying, the varying load acts as a slight disturbance to the total current. The resulting change in V_{AB} will be quite small. The amount of bleeder current, as mentioned earlier, is limited by how much power loss can be tolerated and whether or not the bleeder resistor can safely dissipate this power.

What about loads that need a lot of current and power and vary as well? For loads of this type, a voltage divider is probably not the best choice of circuitry to provide power. A regulated power supply will be needed to handle a large varying load. The word "regulated" means that the power supply contains special circuitry designed to keep its output voltage constant, even if the current drain on it is changing. Some regulation circuits are quite sophisticated and complex and their design often incorporates circuit components called capacitors, as will be discussed in Lesson 13.

Note again that the 10% rule for bleeder current will work for simple loads and voltage divider applications and that is what has been assumed here. The example voltage dividers you will see in this lesson will be confined to handling load currents of about 100 milliamps or less.

Calculation of Bleeder Current, I_{R2} — As shown in Figure 10.21, the value of the current through R_2 is chosen to be 10% or one-tenth of the total load current, or 0.1 times 20 milliamps, which equals 2 milliamps.

$$I_{R2} = \frac{1}{10} \times I_{LOAD}$$

$$I_{R2} = 0.1 \times 20 \text{ mA}$$

$$I_{R2} = 2 \text{ mA}$$

Figure 10.21

Calculation of R_2 — You know that you want 2 milliamps of current to flow through R_2, and you know that the voltage across R_2 must equal 100 volts, since it is connected in parallel with the load which requires 100 volts. To find the resistance of R_2, you can use Ohm's law as shown in Figure 10.22. When you substitute the values of voltage and current in the formula, you should have 100 volts divided by 2 milliamps, which equals 50 kilohms.

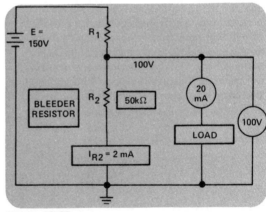

$$R_2 = \frac{E_{R2}}{I_{R2}}$$

$$R_2 = \frac{100\,V}{2\,mA} = \frac{100\,V}{2 \times 10^{-3}\,A}$$

$$R_2 = 50 \times 10^{+3}\,\Omega$$

$$R_2 = 50\,k\Omega$$

Figure 10.22

Partially Completed Voltage Divider — Figure 10.23 shows the voltage divider circuit with the bleeder resistance and bleeder current labeled. You are ready to calculate the value of R_1. From the series circuit laws, you know that the voltage across R_2 plus the voltage across R_1 must equal the total source voltage of 150 volts. Since you know the voltage across R_2 is 100 volts, the remainder of 50 volts must be dropped across R_1. Note also that the *total circuit current* flows through R_1. So before you calculate R_1, it will be necessary to calculate the total circuit current.

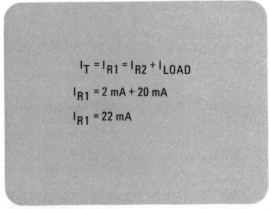

Figure 10.23

Total Circuit Current — As shown in Figure 10.24, this is equal to the current through R_2, plus the current through the load, which simply equals 2 milliamps plus 20 milliamps. Thus the current through R_1 equals 22 milliamps.

$$I_T = I_{R1} = I_{R2} + I_{LOAD}$$

$$I_{R1} = 2\,mA + 20\,mA$$

$$I_{R1} = 22\,mA$$

Figure 10.24

Calculation of R_1 — Knowing the current flowing through R_1 and the voltage that must appear across it, you can use Ohm's law in the form $R = E/I$ to find the resistance of R_1. As shown in Figure 10.25, R_1 equals E_{R1} divided by I_{R1}. When you substitute the values into the formula, you should have 50 volts divided by 22 milliamps which equals 2.27 kilohms.

$$R_1 = \frac{E_{R1}}{I_{R1}}$$

$$R_1 = \frac{50 \text{ V}}{22 \text{ mA}}$$

$$R_1 = 2.27 \text{ k}\Omega$$

Figure 10.25

Final Circuit — Figure 10.26 shows the completed voltage divider design with all values labeled.

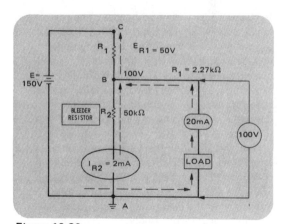

Figure 10.26

VOLTAGE DIVIDER DESIGN BASICS

1) EXAMINE POWER SUPPLY AND LOAD REQUIREMENTS
2) SELECT BLEEDER CURRENT
3) CALCULATE BLEEDER RESISTANCE
4) CALCULATE RESISTANCE OF OTHER DIVIDER RESISTOR(S)

Figure 10.27

Basic Voltage Divider Design — The basics of voltage divider design have been reviewed in this example and are shown in abbreviated form in Figure 10.27. In the rest of this lesson several examples will be used to illustrate and reinforce this procedure. Basically, the procedure is:

1. Examine your available power supply output voltage and the requirements of your load carefully. (Your power supply must provide a higher voltage and current capability than your load in order for a voltage divider to be designed between them.)

2. Select a value of bleeder current for the divider. For most voltage dividers, one tenth or 10% of the load current is selected.

3. Calculate the value needed for the bleeder resistor. (In building a divider, be sure to use a resistor with a high enough *power* rating to handle the bleeder current at the load voltage.)

4. Calculate the value of the other resistor or resistors needed for the circuit by calculating the total current flow through it, and then using Ohm's law in the form R = E/I.

Voltage Divider with Two Loads — The next logical step in considering the design of voltage divider circuits is to consider the design of a voltage divider that will supply the voltage and current needed to power two or more loads. This type of circuit basically involves a fairly straightforward extension of what has already been covered and the problem being considered is shown in Figure 10.28.

In this situation there are two loads; one load (load 1) requires 45 volts and draws 20 milliamps of current, while the other load (load 2) requires 150 volts and draws 10 milliamps of current. Suppose that the only power supply readily available is a 200-volt supply.

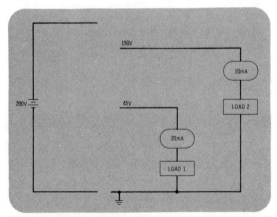

Figure 10.28

Resistor Voltage Divider — In order to deliver the required voltages and currents you will need a voltage divider with three resistors of the type shown in Figure 10.29. The problem boils down to one of calculating the values of R_1, R_2, and the bleeder resistor, R_3.

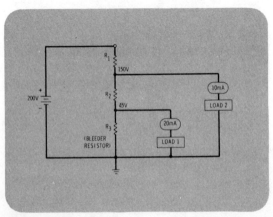

Figure 10.29

- **Calculation of Bleeder Current**
- **Calculation of R_3**

Calculation of Bleeder Current — To solve this problem you will follow the same basic steps that were followed previously (Figure 10.30). First of all, assume that the power supply has ample current capability to handle the job of powering the 10-milliamp and 20-milliamp loads, plus providing the bleeder current. Your next step is to select a suitable bleeder current to flow through R_3.

If you follow the rule of thumb for the bleeder current, you will want the current through R_3 to be about 10% of the total load current. In this case, since the load currents are 10 milliamps and 20 milliamps, the total load current is 10 milliamps plus 20 milliamps, which equals 30 milliamps. The current through R_3 then equals 10% of 30 milliamps, which is 3 milliamps.

$$I_{BLEEDER} = I_{R3} = \frac{1}{10} I_{LOAD}$$

$$I_{LOAD} = 10\text{ mA} + 20\text{ mA} = 30\text{ mA}$$

$$I_{R3} = (0.1) \times (30\text{ mA}) = 3\text{ mA}$$

Figure 10.30

Calculation of R_3 — Load 1 must have a voltage of 45 volts across it to operate correctly. Since R_3, the bleeder resistor, is in parallel with load 1, you know that the voltage across R_3 must also equal 45 volts. So at this point you know the current flowing through R_3, the bleeder current, and the voltage across it, 45 volts. Therefore, you can use Ohm's law in the form R_3 equals E_{R3} divided by I_{R3} to find R_3 as shown in Figure 10.31. Substituting the values of voltage and current into the formula, you have 45 volts divided by 3 milliamps, which in scientific notation equals $4.5 \times 10^{+1}$ divided by 3×10^{-3}. When you divide, you get $1.5 \times 10^{+4}$, which is 15 kilohms.

$$R_3 = \frac{E_{R3}}{I_{R3}} = \frac{45\text{ V}}{3\text{ mA}}$$

$$R_3 = \frac{4.5 \times 10^{+1}}{3 \times 10^{-3}}$$

$$R_3 = 1.5 \times 10^{+4} = 15\text{ k}\Omega$$

Figure 10.31

Determination of I_{R2} and R_2 — With R_3 determined, move up the divider to consider the next resistor, R_2. First you should analyze the currents flowing through R_2 as shown in Figure 10.32. From the parallel circuit laws, the current through R_2 equals the sum of the currents flowing through R_3 and load 1. This equals 3 milliamps for the bleeder current plus 20 milliamps from load 1 for a total current through R_2 of 23 milliamps.

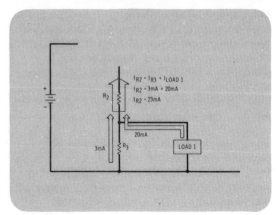

Figure 10.32

Calculation of E_{R2} — You need to determine the voltage across R_2. As shown in Figure 10.33, the voltage between ground and point B equals the voltage required by load 2, which is 150 volts. The voltage between ground and point A is 45 volts. The voltage across R_2 equals the difference between these two voltages, so subtract 45 volts from 150 volts and you get 105 volts. Now that you know the voltage across R_2 is 105 volts, and the current through R_2 is 23 milliamps, you can use Ohm's law to calculate the resistance of R_2.

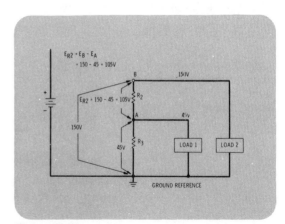

Figure 10.33

Calculation of R_2 — As shown in Figure 10.34, R_2 equals E_{R2} divided by I_{R2}. Substituting the values of voltage and current in the formula yields 105 volts divided by 23 milliamps. In scientific notation that equals $1.05 \times 10^{+2}$ divided by 2.3×10^{-2} which equals about 4.6 kilohms for R_2.

$$R_2 = \frac{E_{R2}}{I_{R2}} = \frac{105\ V}{23\ mA}$$

$$R_2 = \frac{1.05 \times 10^{+2}}{2.3 \times 10^{-2}}$$

$$R_2 = 4.6\ k\Omega$$

Figure 10.34

Determination of I_{R1} — Move up the divider one more time and consider how to calculate the resistance needed for R_1. This procedure is the same as for R_2 and would be the same for a fourth or fifth resistor in this circuit. First determine the current flowing through the resistor, then the voltage across it, and finally use Ohm's law in the form $R = E/I$ to calculate the resistance.

First consider the flow of current as shown in Figure 10.35. The current through R_1 equals the 23 milliamps of current flowing through R_2 plus the 10 milliamps of current flowing through load 2. The sum of these two currents is 33 milliamps.

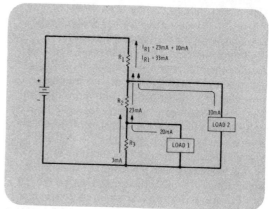

Figure 10.35

Calculation of E_{R1} — You need to calculate the voltage across R_1. As shown in Figure 10.36, the voltage from ground to point C is the full supply voltage of 200 volts. The voltage between ground and point B is 150 volts. The voltage across R_1 is the difference between the two, or 200 volts minus 150 volts which equals 50 volts across R_1.

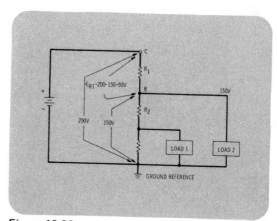

Figure 10.36

Calculation of R_1 — To find R_1 you can use Ohm's law as shown in Figure 10.37. R_1 equals E_{R1} divided by I_{R1}. When you substitute the circuit values in the formula, you have 50 volts divided by 33 milliamps. In scientific notation this equals $5 \times 10^{+1}$ divided by 3.3×10^{-2}. When you divide, you should get $1.515 \times 10^{+3}$ which is about 1.5 kilohms for R_1.

$$R_1 = \frac{E_{R1}}{I_{R1}} = \frac{50\ V}{33\ mA}$$

$$R_1 = \frac{5 \times 10^{+1}}{3.3 \times 10^{-2}} = 1.515 \times 10^{+3}$$

$$R_1 = 1.5\ k\Omega$$

Figure 10.37

Completed Voltage Divider — As shown in Figure 10.38, the voltage divider design is now complete with the resistance values of R_1, R_2, and R_3 found. As long as the current drawn by the load remains fairly constant, this will provide a reliable power source for them. It is interesting to note that in a voltage divider with more than one load, such as this one, one stable load will keep the voltage to an unstable load constant. If load 1 was a very constant load, it would help stabilize the voltage at point B if load 2 should start to vary.

A more advanced discussion of the specific applications of voltage dividers is beyond the scope of this course. However, for those of you who are interested, some reference books on this subject are listed at the end of this lesson.

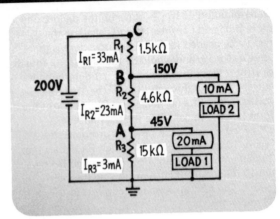

Figure 10.38

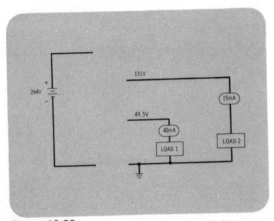

Figure 10.39

Additional Voltage Divider Problem — Consider two additional examples of voltage divider design. Figure 10.39 shows the requirements for the first problem. Load 1 draws 40 milliamps of current and operates at 49.5 volts while load 2 draws 15 milliamps of current and needs 131 volts applied to operate properly. Suppose that the only readily available power supply has an output voltage of 264 volts.

Basic Voltage Divider Circuit — The voltage divider circuit necessary to supply the correct voltages and currents to these two loads is shown in Figure 10.40. As you saw in the previous case, the design of this voltage divider boils down to the calculation of three resistors, R_1, R_2 and the bleeder resistor, R_3.

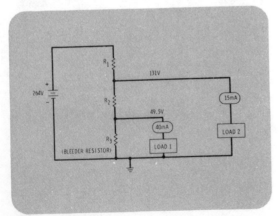

Figure 10.40

- Calculation of I_B
- Calculation of R_3
- Determination of I_{R2} and R_2

$$I_B = 1/10TH\ OF\ I_{TOTAL\ LOAD}$$

$$I_B = 0.1 \times (40\ mA + 15\ mA)$$

$$I_B = 0.1 \times (55\ mA)$$

$$I_B = 5.5\ mA$$

Figure 10.41

Calculation of I_B — To begin the design of this voltage divider, you need to select a value for the bleeder current, I_B. From the rule of thumb for bleeder current, I_B equals 10% or one-tenth of the total load current. As shown in Figure 10.41, I_B equals one-tenth of 40 milliamps plus 15 milliamps. If you add the two load currents and multiply by 0.1, you get 5.5 milliamps for the bleeder current.

$$R_3 = \frac{E_3}{I_b}$$

$$R_3 = \frac{49.5\ V}{5.5\ mA}$$

$$R_3 = \frac{4.95 \times 10^{+1}}{5.5 \times 10^{-3}}$$

$$R_3 = 0.9 \times 10^{+4} = 9.0\ k\Omega$$

Figure 10.42

Calculation of R_3 — You can use that information to find R_3 by using Ohm's law in the form $R = E/I$. When you substitute the appropriate values in the formula as shown in Figure 10.42, you get 49.5 volts for E_{R3} divided by 5.5 milliamps for I_B. When you divide, you get 9.0 kilohms for R_3.

Here's an interesting point. If you were to look at a chart of preferred resistance values you would not find 9.0 listed. In the actual construction of the circuit you could use two 18-kilohm resistors in parallel in order to obtain the 9 kilohms needed for R_3.

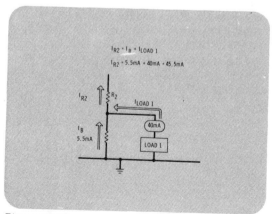

Figure 10.43

Determination of I_{R2} and R_2 — To calculate the value of R_2, you need to know the current flow through R_2 (I_{R2}) and the voltage across R_2 (E_{R2}). As shown in Figure 10.43, I_{R2} is equal to the sum of I_B and the current through load 1. When these are added, you get 45.5 milliamps for I_{R2}.

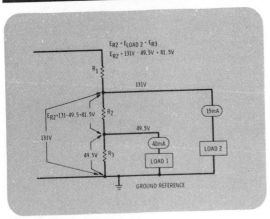

Figure 10.44

Calculation of E_{R2} — Figure 10.44 illustrates the calculation of E_{R2}. The voltage across R_2 equals the difference between the two load voltages since R_2 is essentially connected between the two loads. The voltage required by load 2 is 131 volts, and the voltage required by load 1 is 49.5 volts. The difference between these two is 81.5 volts for E_{R2}.

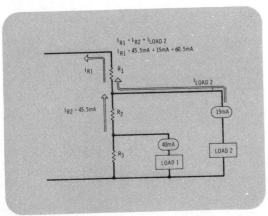

Figure 10.45

Calculation of R_2 — You can use Ohm's law in the form $R = E/I$ to find the value of R_2. As shown in Figure 10.45, when you substitute 81.5 volts for E_{R2} and 45.5 milliamps in the formula and divide, you get 1.79 kilohms for R_2. In actual circuit construction you would use a 1.8-kilohm resistor.

The formulas shown in Figure 10.45:

$$R_2 = \frac{E_{R2}}{I_{R2}}$$

$$R_2 = \frac{81.5\ V}{45.5\ mA}$$

$$R_2 = \frac{8.15 \times 10^{+1}}{4.55 \times 10^{-2}}$$

$$R_2 = 1.79 \times 10^{+3} = 1.79\ k\Omega$$

Determination of I_{R1} and R_1 — To finish the design of this voltage divider circuit, you need to find the value of R_1. First you can find the current flow through R_1 as illustrated in Figure 10.46 by adding I_{R2} and the current through load 2. Their sum is 60.5 milliamps for I_{R1}.

Figure 10.46

- Calculation of E_{R1}
- Calculation of R_1
- Completed Voltage Divider

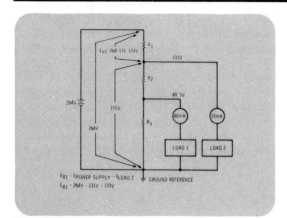

Figure 10.47

Calculation of E_{R1} — You also need to know the voltage across R_1 in order to calculate the resistance value of R_1. As illustrated in Figure 10.47, E_{R1} is the difference between the supply voltage of 264 volts and the voltage required by load 2 of 131 volts. This difference is 133 volts for E_{R1}.

$$R_1 = \frac{E_{R1}}{I_{R1}}$$

$$R_1 = \frac{133 \text{ V}}{60.5 \text{ mA}}$$

$$R_1 = \frac{1.33 \times 10^{+2}}{6.05 \times 10^{-2}}$$

$$R_1 = 0.2198 \times 10^{+4} = 2.2 \text{ k}\Omega$$

Figure 10.48

Calculation of R_1 — Now you can use Ohm's law in the form $R = E/I$ to find R_1. As shown in Figure 10.48, when you substitute the appropriate values in the formula, you get 133 volts over 60.5 milliamps, and when you divide, the result is 2.2 kilohms for R_1.

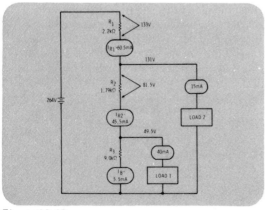

Figure 10.49

Completed Voltage Divider — The design of this voltage divider is complete, the values of R_1, R_2 and R_3 have been calculated and are shown in Figure 10.49 along with the values of voltage and current used to calculate the resistances.

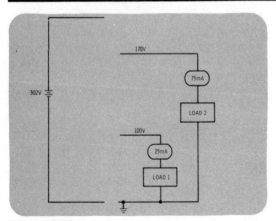

Figure 10.50

Another Voltage Divider Design — In order to become thoroughly familiar with the design of voltage dividers, consider one more example. The requirements for the circuit are shown in Figure 10.50. Load 1 draws 25 milliamps of current and requires 100 volts to operate properly, and load 2 draws 75 milliamps of current and requires 170 volts. Suppose that the only available power supply has an output voltage of 302 volts and assume that it is capable of delivering the required currents.

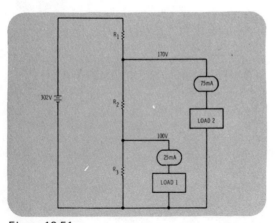

Figure 10.51

Basic Voltage Divider Circuit — As before, the voltage divider circuit necessary to supply these voltages and current to the two loads consists of three resistors, R_1, R_2, and the bleeder resistor, R_3 shown in Figure 10.51. Again, the design of this voltage divider simply involves the calculation of the three resistances.

$$I_B = 10\% \text{ OR } 1/10\text{TH OF } I_{TOTAL\ LOAD}$$

$$I_B = 0.1 \times (25\ mA + 75\ mA)$$

$$I_B = 0.1 \times 100\ mA$$

$$I_B = 10\ mA$$

Determination of I_B and R_3 — In order to calculate R_3, you need to know the current flowing through it, which is the bleeder current, I_B, and the voltage across it, which is 100 volts. (R_3 is in parallel with load 1, which requires 100 volts.) From your rule of thumb for bleeder current, I_B equals 10% or one-tenth of the total load current, or, as shown in Figure 10.52, I_B equals one-tenth of 25 milliamps plus 75 milliamps, therefore I_B equals one-tenth of 100 milliamps, which is 10 milliamps.

Figure 10.52

- Calculation of R_3
- Determination of I_{R2} and E_{R2}
- Calculation of R_2

$$R_3 = \frac{E_{R3}}{I_B}$$

$$R_3 = \frac{100 \text{ V}}{10 \text{ mA}}$$

$$R_3 = \frac{1 \times 10^{+2}}{1 \times 10^{-2}}$$

$$R_3 = 1 \times 10^{+4} = 10 \text{ k}\Omega$$

Figure 10.53

Calculation of R_3 — If you substitute the values of E_{R3} and I_B in the Ohm's law formula as shown in Figure 10.53 and divide, you get 10 kilohms for R_3.

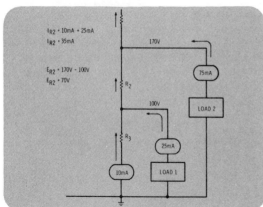

Figure 10.54

Determination of I_{R2} and E_{R2} — Moving up the voltage divider to R_2, you will need to determine I_{R2} and E_{R2} in order to calculate R_2. I_{R2} is simply the sum of I_B and $I_{load\ 1}$, which is 10 milliamps plus 25 milliamps. When you add these two currents, you obtain 35 milliamps for I_{R2}.

To find E_{R2} you need to subtract the voltage across load 1 from the voltage across load 2. Since R_2 is between these loads, E_{R2} is the difference between their two voltage requirements. As shown in Figure 10.54, this is 170 volts minus 100 volts or 70 volts for E_{R2}.

$$R_2 = \frac{E_{R2}}{I_{R2}}$$

$$R_2 = \frac{70 \text{ V}}{35 \text{ mA}}$$

$$R_2 = \frac{7.0 \times 10^{+1}}{3.5 \times 10^{-2}}$$

$$R_2 = 2 \times 10^{+3} = 2 \text{ k}\Omega$$

Figure 10.55

Calculation of R_2 — Now that you know E_{R2}, you can use Ohm's law in the form $R = E/I$ to calculate R_2. As shown in Figure 10.55, when you substitute the values in the formula and divide, the result is 2 kilohms for R_2.

- Determination of I_{R1} and E_{R1}
- Calculation of R_1
- **Power**

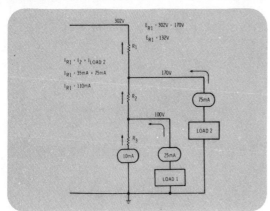

Figure 10.56

Determination of I_{R1} and E_{R1} — Following the same procedure for R_1, you should find the voltage across R_1 is 132 volts and the current through R_1 is 110 milliamps, as illustrated by Figure 10.56.

$$R_1 = \frac{E_{R1}}{I_{R1}}$$

$$R_1 = \frac{132 \text{ V}}{110 \text{ mA}}$$

$$R_1 = \frac{1.32 \times 10^{+2}}{1.1 \times 10^{-1}}$$

$$R_1 = 1.2 \times 10^{+3} = 1.2 \text{ k}\Omega$$

Figure 10.57

Calculation of R_1 — In order to complete the design of this voltage divider all you need to do is to use Ohm's law in the form $R = E/I$ as shown in Figure 10.57. When you substitute the values of voltage and current in the formula, the result is 132 volts divided by 110 milliamps, which equals 1.2 kilohms and the design is complete.

Power — Thus far you have seen how voltage dividers can be used to provide different polarities of voltage, and how to calculate voltages, currents, and resistances in a voltage divider circuit. Loads and how the voltage or current of a load affects a voltage divider were also discussed. There is another factor used as a measure of a load, however, and that is its power or wattage rating. The concept of power was introduced in Lesson 4 along with a discussion of Ohm's law. At this point, it will be helpful to review this important subject and introduce some additional topics of interest that will expand your useful knowledge in this area.

Basic Power Formula in Circle Form — As you may recall from Lesson 4, the basic formula for power states that power equals current times voltage, and remember the unit for power is the watt. This formula may be put into a circle form as shown in Figure 10.58, which can be used in the same way that you used the Ohm's law circle. This gives you a convenient device for handling problems involving power calculations. Simply cover the quantity you want to find with your thumb, and the position of the remaining letters tells you the procedure to follow. Remember, a vertical line means multiply the quantities on either side of the line, and a horizontal line means divide the quantity on the top by the quantity on the bottom. For example, if you cover P, the position of the remaining letters gives you the basic power formula, power equals current times voltage. If you cover up the I, the formula is $I = P/E$, and covering E yields $E = P/I$.

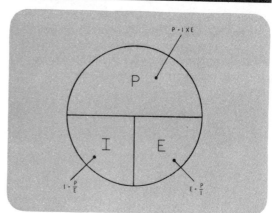

Figure 10.58

Derived Power Formula — In Lesson 4 another formula was derived for power using the first power formula and Ohm's law. That formula was $P = I_2R$. Figure 10.59 shows that formula in a circle form similar to the one above. This time by covering the P, you can see that power equals the current squared times the resistance. In other words, to find power, multiply current times itself and then multiply by the resistance.

You can also use this circle to find current when you know the power dissipated by a resistor, and the value of resistance. If you cover I^2, you can see that the square of the current equals power

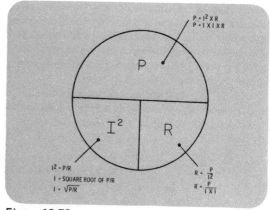

Figure 10.59

divided by resistance. How then do you find the current, if you know the square of the current? As you may recall, you must take the square root of the answer. In this case you can find the current by taking the square root of power divided by resistance.

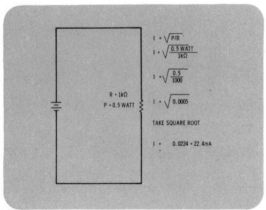

Figure 10.60

Example: $I = \sqrt{P/R}$ — From the circle formula shown in Figure 10.59, covering the I^2 gives P/R so you can easily write $I^2 = P/R$. This means that P/R is the square of I, or equal to I times I, not equal to I itself. You can find I by taking the square root of P/R. In the simple circuit of Figure 10.60, the 1-kilohm resistor is dissipating one-half of a watt. How would you calculate the current flowing through it? Substituting in the formula as shown, I equals the square root of P/R, which is the square root of 0.0005. This square root can be found either using a calculator or square root tables (both methods will be reviewed later in the lesson). When the square root is computed, you find that I = 22 milliamps.

Circle Diagram for Calculating R — The third way you can use this circle is in finding resistance when you know the power dissipated by a resistor and the current flowing through the resistor. By covering the R, you can see that resistance equals power divided by the square of current (Figure 10.61). In this case, you must square the current or multiply it by itself before you divide it into the power, to calculate the resistance.

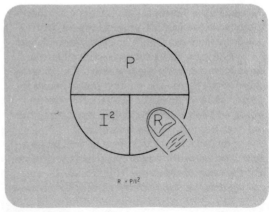

Figure 10.61

- Example $R = P/I^2$
- **Third Power Formula**
- **Third Circle Diagram**

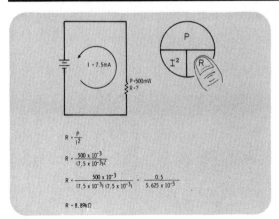

Figure 10.62

Example $R = P/I^2$ — For example, if you know that a resistor in a circuit is dissipating 500 milliwatts and is handling a current of 7.5 milliamps, how would you calculate its resistance? Use the formula $R = P/I^2$ as shown in Figure 10.62 and substitute. Remember that the 7.5 milliamps must be multiplied by itself (squared) before it is divided into the power. When you perform the calculations, you find that $R = 8.89$ kilohms.

Third Power Formula — A third helpful power circle can be developed by using Ohm's law and the first power formula that was introduced. Figure 10.63 shows the first power formula, P equals I times E, and also shows Ohm's law in the form I equals E over R. What you can do is substitute the equivalent of current from Ohm's law into the power formula, that is, you can replace current in the power formula with E over R. This enables you to write power as the product of voltage times voltage all divided by resistance, and this is equivalent to voltage squared divided by resistance.

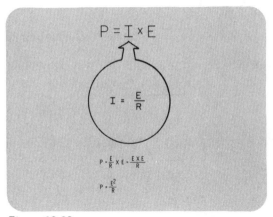

Figure 10.63

Third Circle Diagram — This relationship can now be put in circle form. You should have voltage squared on the top with power and resistance on the bottom as shown in Figure 10.64. This circle diagram can be used to give you three formulas in the same way as all the others. If you cover P, you should find that power equals voltage squared divided by resistance. In order to find the power dissipated by a resistor, when you know the voltage across the resistor and the value of resistance, you must square the voltage, and then divide by the resistance.

A second way this circle can be used is to find resistance when you know the power dissipated by

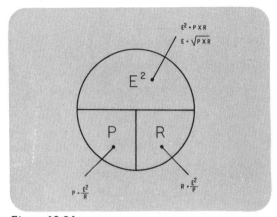

Figure 10.64

- Example $P = E^2/R$
- Example $R = E^2/P$

a resistor and the voltage drop across the resistor. By covering the R, you can see that resistance equals the square of voltage divided by power.

You can also use this circle formula to find voltage when you know the power dissipated by a resistor and the value of resistance. If you cover E squared, all you have to do is multiply power times resistance. However, that does not give you voltage, it gives you the square of voltage. You must take the square root of that to find the voltage. Thus, voltage equals the square root of power times resistance.

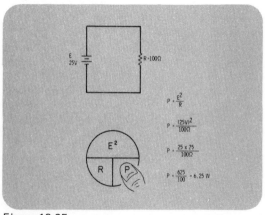

Figure 10.65

Example $P = E^2/R$ — In order for you to become more familiar with this third circle formula for power, consider some examples of its use. The 100-ohm resistor in Figure 10.65 has 25 volts applied to it. How much power is it dissipating? Cover the P in the circle to find that $P = E^2/R$. Then substitute 25 volts in the formula for E and 100 ohms for R. When you square 25 and divide by 100, you get 6.25 watts for the power dissipated by the resistor.

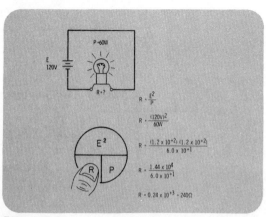

Figure 10.66

Example $R = E^2/P$ — A light bulb is rated at 60 watts and the required voltage is 120 volts. What is the resistance value? In Figure 10.66, cover the R in the circle to get $R = E^2/P$, then substitute 120 volts in the formula for E and 60 watts for P. When you square 120 and divide by 60, you find that the resistance of the bulb is 240 ohms.

- Example $E = \sqrt{P \times R}$
- **Power in Series-Parallel Circuits**

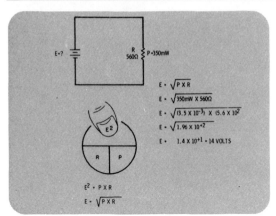

Figure 10.67

Example $E = \sqrt{P \times R}$ — The power dissipated by a 460-ohm resistor is 350 milliwatts. What is the voltage drop across the resistor? Cover the E in the circle as shown in Figure 10.67 to get $E^2 = P \times R$. Remember that this means $E = \sqrt{P \times R}$. Then substitute 560 ohms for R and 350 milliwatts for P, multiply, and take the square root of that to get $1.4 \times 10^{+1}$ which equals 14 volts dropped across the resistor.

Power in Series-Parallel Circuits — Now that three separate circle formulas for power have been introduced and discussed, it will be helpful to show how they all can be applied to the analysis of a series-parallel circuit of the type shown in Figure 10.68.

In this circuit diagram, all the known quantities have been labeled and the unknown quantities have been labeled with question marks. This circuit can be analyzed in a variety of ways by focusing your attention on each component and each unknown, one at a time. You can begin your analysis by using the first power formula to find the power dissipated by R_1. First of all, what do you know about this resistor? You know that the current through R_1 is 3 milliamps, and you know that the voltage across R_1 is 50 volts. You have to calculate the power.

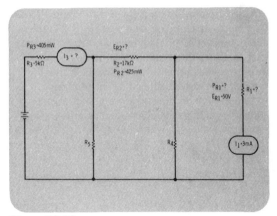

Figure 10.68

Calculation of P_{R1} — The first circle formula that was introduced gave you the relationship between P, I, and E, as shown in Figure 10.69. If you cover the P, you can see that power dissipated in a resistor equals current through it times voltage across it. Substituting the values of voltage and current in the formula gives you 3 milliamps times 50 volts. In powers of ten form, 3 milliamps equals 3×10^{-3} and 50 volts equals $5 \times 10^{+1}$. When you multiply, you should have 15×10^{-2} which equals 150 milliwatts for the power being dissipated by R_1.

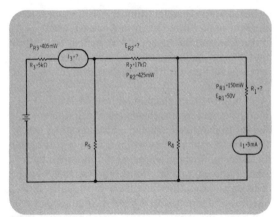

$$P = I \times E$$

$$P_{R1} = I_{R1} \times E_{R1}$$

$$P_{R1} = 3\,mA \times 50\,V$$

$$P_{R1} = (3 \times 10^{-3}) \times (5 \times 10^{+1})$$

$$P_{R1} = 15 \times 10^{-2} = 150\,mW$$

Figure 10.69

Circuit Diagram: Study R_3 — From Figure 10.70 you can see that the resistance of R_3 is 5 kilohms and the power dissipated by R_3 is 405 milliwatts. You know resistance and power dissipated and need to calculate the current flowing in the resistor. The second circle diagram that was introduced relates power, current, and resistance and can be used to help in this calculation.

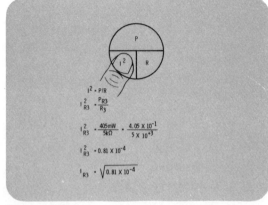

Figure 10.70

Calculation of I_{R3} — If you look at this circle diagram, reproduced in Figure 10.71, and cover I_2, you can see that the square of current equals power divided by resistance. Substituting the values into the formula as shown, you should have 405 milliwatts divided by 5 kilohms. In scientific notation, this equals 4.05×10^{-1} divided by $5 \times 10^{+3}$. When you divide, you can see that the square of the current through R_3 equals 0.81×10^{-4}.

You must take the square root of that to find the current through R_3. In previous lessons it was shown how a calculator such as the TI SR-50 can

$$I^2 = P/R$$

$$I_{R3}^2 = \frac{P_{R3}}{R_3}$$

$$I_{R3}^2 = \frac{405\,mW}{5k\Omega} = \frac{4.05 \times 10^{-1}}{5 \times 10^{+3}}$$

$$I_{R3}^2 = 0.81 \times 10^{-4}$$

$$I_{R3} = \sqrt{0.81 \times 10^{-4}}$$

Figure 10.71

be used to help simplify calculations of this sort. You can also use it to find the square root in this example.

Use of Calculator for Square Root — The chart in Figure 10.72 will help you keep track of the steps needed and what the display should read as you go along. First enter 0.81. Then press the $\boxed{\text{EE}}$ or enter exponent key. (The $\boxed{\text{EE}}$ or enter exponent key is what tells the calculator that the number has an exponent; the exponent is displayed to the right of the number.) Next press the $\boxed{+/-}$ key to make the sign of the exponent negative. Then enter four and the calculator should display 0.81×10^{-4}. In order to find the square root of this number, you simply press the square root key. The calculator should display 9×10^{-3} which is the current through R$_3$ expressed in scientific notation. In abbreviated form, this is 9 milliamps of current through R$_3$.

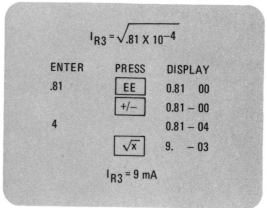

Figure 10.72

Circuit Diagram: Consider R$_1$ — Figure 10.73 shows the circuit diagram with the value of I$_{R3}$ and P$_{R1}$ labeled. You might recall that there are other methods that can be used to determine square roots. The use of tables and a manual calculation method were introduced in Lesson 4. The tabular method will be reviewed later as you proceed to analyze this circuit. Right now, focus your attention on R$_1$. At this point you still need to calculate its resistance value.

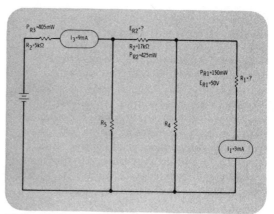

Figure 10.73

Calculation of R_1 — Since you know the voltage across and the current through R_1, you could use Ohm's law to find the resistance. However, from a previous calculation, you know that the power dissipated by R_1 is 150 milliwatts. The last circle diagram for power relates power, voltage, and resistance. In this case you can use this circle to help calculate R_1. As shown in Figure 10.74, if you cover R, you can see that resistance equals voltage squared divided by power. Substituting the values of voltage and power into the formula as shown, you should have 50 volts squared divided by 150 milliwatts. In powers of ten form, this equals $5 \times 10^{+1}$ times $5 \times 10^{+1}$ divided by 1.5×10^{-1}. This works out to be $16.7 \times 10^{+3}$ which equals 16.7 kilohms; thus the resistance of R_1 is 16.7 kilohms.

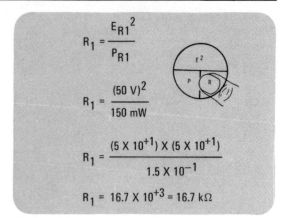

$$R_1 = \frac{E_{R1}^2}{P_{R1}}$$

$$R_1 = \frac{(50 \text{ V})^2}{150 \text{ mW}}$$

$$R_1 = \frac{(5 \times 10^{+1}) \times (5 \times 10^{+1})}{1.5 \times 10^{-1}}$$

$$R_1 = 16.7 \times 10^{+3} = 16.7 \text{ k}\Omega$$

Figure 10.74

Calculation of E_{R2} — If you look back at the circuit and scan it for additional unknown quantities, you see that E_{R2} needs to be calculated. You can find the voltage across R_2 using the last circle formula. You know the resistance of R_2 is 17 kilohms, and you know that the power dissipated by R_2 is 415 milliwatts. If you look at the third circle formula for power as shown in Figure 10.75 and cover E^2, you can see that the square of voltage equals power times resistance. Substituting the values into the formula as shown, you should have 425 milliwatts times 17 kilohms. In scientific notation, this is

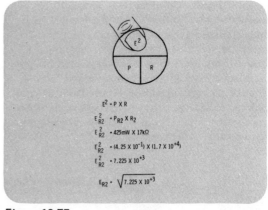

$E^2 = P \times R$

$E_{R2}^2 = P_{R2} \times R_2$

$E_{R2}^2 = 425 \text{mW} \times 17 \text{k}\Omega$

$E_{R2}^2 = (4.25 \times 10^{-1}) \times (1.7 \times 10^{+4})$

$E_{R2}^2 = 7.225 \times 10^{+3}$

$E_{R2} = \sqrt{7.225 \times 10^{+3}}$

Figure 10.75

4.25×10^{-1} times $1.7 \times 10^{+4}$. When you multiply, you can see that the square of the voltage across R_2 equals $7.225 \times 10^{+3}$.

Square Root Tables — You must take the square root of that to get the voltage. Before you do this, however, consider how you can find square roots using square root tables. As shown in Figure 10.76, many of these tables list numbers from one to a thousand, and then across the page to the right are columns listing the squares and square roots. Some tables also list cubes and cube roots, but these will not be discussed here.

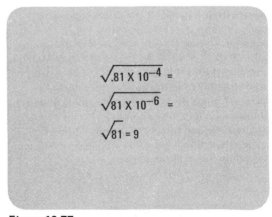

SQUARE ROOT TABLE

n	n^2	$\sqrt{n}$	n^3	$\sqrt[3]{n}$
1	1	1	1	1
1000	1,000,000	31.62278		

Figure 10.76

Square Root Example 1 — As a review of using tables to find square roots, go back to your first example and try to find the square root of 0.81×10^{-4}. When using tables, a good way to proceed is to try to alter the number you are working with until it is in a form that is easier to find in the tables. You will find that you can find the square root of this number much more easily by converting it to 81×10^{-6} as shown in Figure 10.77. (All this operation does is move the decimal point two places to the right and make the exponent more negative to make up for it.) Looking in the square root table, you see that the square root of 81 is 9.

$$\sqrt{.81 \times 10^{-4}} =$$

$$\sqrt{81 \times 10^{-6}} =$$

$$\sqrt{81} = 9$$

Figure 10.77

- Square Root of Exponent
- Example 1 Complete
- Square Root Example 2

Square Root of Exponent – Here's the key point on taking the square root of the exponent. As stated in Figure 10.78, *to find the square root of the power of ten, simply divide the power or exponent by 2.* You want to avoid powers of ten that are not whole numbers for the results, so you must make sure that the power of ten you are dividing is an even number, or one that is a multiple of 2. If not, the power of ten in your result may come out uneven like $10^{3.91}$ power. Powers such as this are difficult to handle and should be avoided.

TO FIND THE SQUARE ROOT OF A POWER OF TEN, DIVIDE THE POWER OR EXPONENT BY 2.

Figure 10.78

Example 1 Complete – For example, in Figure 10.79, you know that the square root of 81 is 9 from the tables. You must now take the square root of the power of ten which is minus 6. To do this, you just divide the exponent by 2, giving you a minus 3. So, you can see that the square root of 81×10^{-6} equals 9×10^{-3}.

$$\sqrt{81 \times 10^{-6}}$$

$$= \sqrt{81} \times \sqrt{10^{-6}}$$

$$= 9 \times 10^{-3}$$

Figure 10.79

Square Root Example 2 – As a second example, you can finish the circuit problem you were working. To find R_2 you need to take the square root of $7.225 \times 10^{+3}$. As shown in Figure 10.80, you could convert this number to $72.25 \times 10^{+2}$ and then find 71 in the square root tables. However, since the tables list only whole numbers, this would not be very accurate. There is a more accurate method which can be used to find the square roots of numbers that are larger or smaller than those listed in the square root table.

$$E_{R2} = \sqrt{7.225 \times 10^{+3}}$$

$$E_{R2} = \sqrt{72.25 \times 10^{2}}$$

Figure 10.80

Alternate Method — As shown in Figure 10.81, convert the number to decimal form and you should have 7225. Then there is a little mathematical trick you can use to find its square root. Find two numbers that when multiplied together give 7225, both of which are listed in your square root table. You can then take the square root of each of these numbers and multiply them together to get your answer. For example, select some number, say 25, and divide it into 7225. You find that 7225 is equal to 289 times 25. Looking in the tables you can find the square root of each of these numbers. When you look up the square roots, you find that the square root of 289 is 17 and the square root of 25 is 5. Then you multiply these two square roots together, 17 times 5 equals 85. Thus, the square root of 7225 is 85, and this is also the result for E_{R2}.

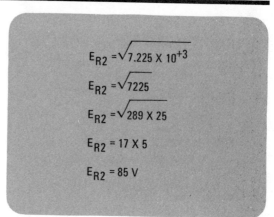

$$E_{R2} = \sqrt{7.225 \times 10^{+3}}$$

$$E_{R2} = \sqrt{7225}$$

$$E_{R2} = \sqrt{289 \times 25}$$

$$E_{R2} = 17 \times 5$$

$$E_{R2} = 85 \text{ V}$$

Figure 10.81

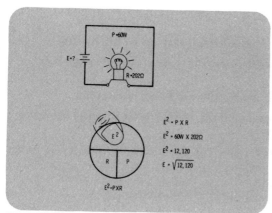

Figure 10.82

Additional Square Root Example — Suppose you wanted to know the proper voltage to apply to a light bulb rated at 60 watts, and you have measured its resistance and found it to be 202 ohms. Here again, you can use the third circle for power as shown in Figure 10.82. First, cover E^2 and the formula is $E^2 = R \times P$. When you substitute the appropriate values in the formula and multiply, you get 12,120 for E^2. You must find the square root of that to get the voltage. As shown in Figure 10.83, you must first find two numbers that when multiplied together equal 12,120. To do this, divide 12,120 by, say 60, and you get 202. Now find the square root of each of

- Additional Square Root Example
- **Additional Reference Material on Voltage Dividers**

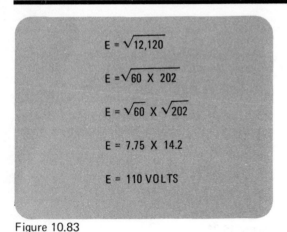

$$E = \sqrt{12,120}$$

$$E = \sqrt{60 \times 202}$$

$$E = \sqrt{60} \times \sqrt{202}$$

$$E = 7.75 \times 14.2$$

$$E = 110 \ VOLTS$$

Figure 10.83

these numbers in the square root tables. The square root of 60 is 7.75 and the square root of 202 is 14.2. When you multiply these two square roots together, you find that the voltage required by the 60 watt bulb is 110 volts.

In this lesson, several new concepts have been discussed, along with a review and expansion of some that were covered previously. The voltage divider circuit (both with and without loading) was covered. At this point you should be able to design a voltage divider circuit to power several small loads from a single adequate power supply. The concept of power has been reviewed and expanded, and at this point, you should be able to use the three power circle diagrams as an aid in analyzing complex circuits. You are now ready to discuss multiple source dc circuits and some of the methods used to analyze them.

Additional Reference Material on Voltage Dividers

Grob, B., *Basic Electronics*, 3rd edition (New York: McGraw-Hill, 1971), pp. 61-64, 106-108.

Herrick, Clyde N., *Unified Concepts of Electronics* (Englewood Cliffs, New Jersey: Prentice-Hall, 1970), pp. 173-178.

Korneff, T., *Introduction to Electronics* (New York: Academic Press, 1966), pp. 160-164.

Tocci, R. J., *Introduction to Electric Circuit Analysis* (Columbus, Ohio: Charles E. Merrill Publishing Co., 1974), p. 168 (unloaded), p. 238 (loaded).

Weick, Carl B., *Principles of Electronic Technology* (Toronto, Canada: McGraw-Hill, 1969), pp. 79-82.

LESSON 10. VOLTAGE DIVIDERS AND POWER

- **Worked Through Examples**

In these examples and in the practice problems that follow, you will be designing voltage divider circuits for various loads which operate at specific values of voltage and current. You will find that quite often the resistances you calculate for your voltage divider circuits will not be "preferred values" of resistance. Therefore, if you were to actually build any of these circuits, you would more than likely have to combine various resistors in parallel or series to obtain the value of resistance required for the voltage divider circuit.

1. Design a voltage divider to supply power to a load which draws 30 milliamps of current and requires 50 volts to operate properly. Assume that the only readily available power supply capable of delivering the required current has an output voltage of 80 volts.

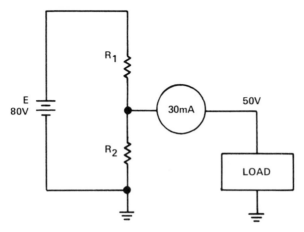

Since there is only one load, this voltage divider requires only two resistors to divide the 80 volts down to the 50 volts required by the load. The design of this voltage divider then simply requires you to calculate the values of R_1 and R_2.

The first step is to find the value of the bleeder current which flows through R_2, the bleeder resistor. The rule of thumb for finding bleeder current says that the bleeder current should be 10% or one-tenth of the total load current.

$$I_b = 1/10\text{th of } I_{total\ load}$$

$$I_b = I_{R2} = 0.1 \times 30 \text{ mA}$$

$$I_b = I_{R2} = 3 \text{ mA}$$

Now, since R_2 is in parallel with the load it will have the same voltage of 50 volts dropped across it.

You now know the current through and the voltage across R_2. and can therefore use Ohm's law in the form $R = E/I$ to find the value of R_2.

$$R_2 = \frac{E}{I}$$

$$R_2 = \frac{50 \text{ V}}{3 \text{ mA}}$$

$$R_2 = \frac{5 \times 10^{+1}}{3 \times 10^{-3}}$$

$$R_2 = 1.67 \times 10^{+4} = 16.7 \text{ k}\Omega$$

Now to find the value of R_1 you first need to know the current through it. As shown in the schematic below, the current through R_1 is the sum of I_b and the current through the load.

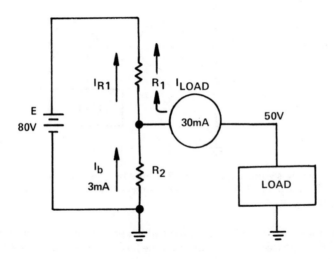

$$I_{R1} = I_b + I_{load}$$

$$I_{R1} = 3 \text{ mA} + 30 \text{ mA}$$

$$I_{R1} = 33 \text{ mA}$$

Next you need to determine the voltage across R_1.

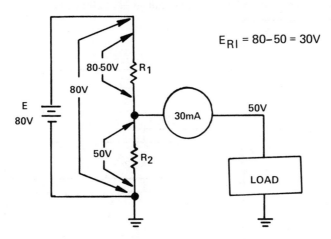

$$E_{RI} = 80-50 = 30V$$

As shown, E_{R1} is the difference between the supply voltage of 80 volts and the load voltage of 50 volts and consequently, E_{R1} is 30 volts.

Now you can use Ohm's law in the form R = E/I to find R_1.

$$R_1 = \frac{E_{R1}}{I_{R1}} = \frac{30 \text{ V}}{33 \text{ mA}}$$

$$R_1 = \frac{3.0 \times 10^{+1}}{3.3 \times 10^{-2}}$$

$$R_1 = 0.909 \times 10^{+3}$$

$$R_1 = 909 \ \Omega$$

This calculation completes your design of the voltage divider for this load.

2. Design a voltage divider circuit for a 250-volt power supply. The loads to be connected to the voltage divider require 40 milliamps at 150 volts and 50 milliamps at 200 volts.

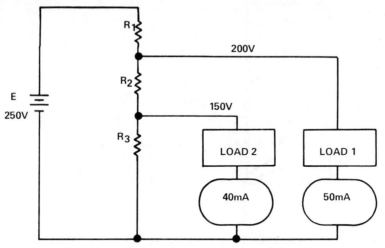

First find the bleeder current through R_3.

$$I_b = 10\% \text{ or } 1/10\text{th } I_{total\ load}$$

$$I_b = 0.1 \times (40 \text{ mA} + 50 \text{ mA})$$

$$I_b = 0.1 \times 90 \text{ mA} = 9 \text{ mA}$$

Since R_3 is in parallel with load 2, E_{R3} equals 150 volts, and you can now use Ohm's law to calculate R_3.

$$R_3 = \frac{E_3}{I_b} = \frac{150 \text{ V}}{9 \text{ mA}}$$

$$R_3 = \frac{1.5 \times 10^{+2}}{9 \times 10^{-3}} = 0.167 \times 10^{+5} = 16.7 \text{ k}\Omega$$

Moving up the voltage divider to R_2, you need to determine the voltage across R_2 and the current through it.

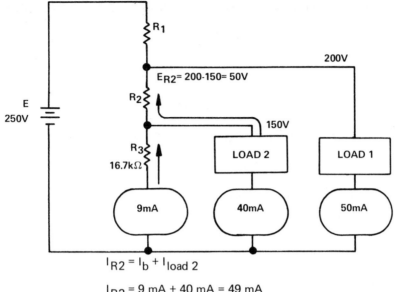

$$I_{R2} = I_b + I_{load\ 2}$$

$$I_{R2} = 9\ mA + 40\ mA = 49\ mA$$

As shown in the above sketch, E_{R2} is the difference between the two load voltages or 50 volts, and I_{R2} is the sum of the bleeder current and the current through load 2, which equals 49 milliamps.

You can use Ohm's law as shown to calculate R_2.

$$R_2 = \frac{E_{R2}}{I_{R2}} = \frac{50\ V}{49\ mA}$$

$$R_2 = \frac{5.0\ X\ 10^{+1}}{4.9\ X\ 10^{-2}} = 1.02\ X\ 10^{+3} = 1.02\ k\Omega$$

In order to calculate R_1 you need to find its voltage and current. E_{R1} is the difference between the supply voltage and the voltage required by load 1, or $250 - 200$, which equals 50 volts. I_{R1} is the total current in the circuit, which equals the sum of I_{R2} and the current through load 1. Here, 49 milliamps plus 50 milliamps equals 99 milliamps for I_{R1}.

You can use Ohm's law to find R_1.

$$R_1 = \frac{E_{R1}}{I_{R1}} = \frac{50\ V}{99\ mA}$$

$$R_1 = \frac{5.0\ X\ 10^{+1}}{9.9\ X\ 10^{-2}} = 0.505\ X\ 10^{+3}$$

$$R_1 = 505\ \Omega$$

and your design is complete.

3. Design a voltage divider circuit for a 125-volt power supply that will supply voltage and current to the following loads.

    Load 1: 0.0 milliamps at −25 volts

    Load 2: 5.0 milliamps at +25 volts

    Load 3: 15 milliamps at 50 volts

    Load 4: 30 milliamps at 100 volts

The voltage divider circuit consists of R_1, R_2, R_3, and R_4 with loads connected as shown in the sketch.

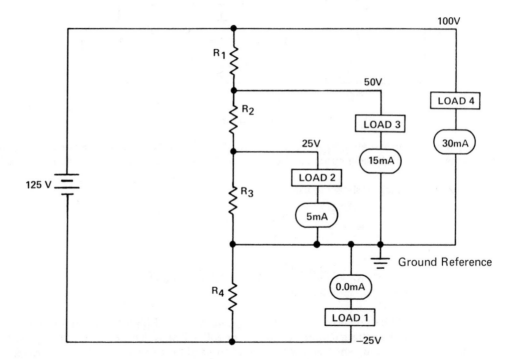

Since load 1 requires a negative voltage, it is connected from ground to the negative side of the source. All that is necessary to complete the design of this voltage divider circuit is to calculate the values of resistors 1 through 4 using the same procedures and techniques for the proceding examples.

In this case, R_3 is the bleeder resistor and R_4 is simply a voltage dropping resistor necessary to produce the −25 volts required by load 1.

To find I_b, take one-tenth of the total load current.

$$I_b = 10\% \text{ or } 1/10\text{th } I_{\text{total load}}$$

$$I_b = 0.1 \times (5 \text{ mA} + 15 \text{ mA} + 30 \text{ mA})$$

$$I_b = 0.1 \times (50 \text{ mA}) = 5 \text{ mA}$$

E_{R3} equals plus 25 volts since R_3 is in parallel with load 2. Now you can calculate R_3.

$$R_3 = \frac{E_{R3}}{I_b} = \frac{25 \text{ V}}{5 \text{ mA}}$$

$$R_3 = \frac{2.5 \times 10^{+1}}{5 \times 10^{-3}}$$

$$R_3 = 0.5 \times 10^{+4} = 5 \text{ k}\Omega$$

Moving up the divider to R_2, you need to know I_{R2} and E_{R2}. The current through R_2 is simply the sum of I_b and the current through load 2.

$$I_{R2} = I_b + I_{\text{load 2}}$$

$$I_{R2} = 5 \text{ mA} + 5 \text{ mA} = 10 \text{ mA}.$$

The voltage across R_2 is the difference between the voltages required by load 2 and load 3 since R_2 is connected between these two loads.

$$E_{R2} = E_{\text{load 3}} - E_{\text{load 2}}$$

$$E_{R2} = 50 \text{ V} - 25 \text{ V} = 25 \text{ V}$$

Using Ohm's law in the form $R = E/I$, you can calculate the value of R_2.

$$R_2 = \frac{E_{R2}}{I_{R2}} = \frac{25 \text{ V}}{10 \text{ mA}}$$

$$R_2 = \frac{2.5 \times 10^{+1}}{1.0 \times 10^{-2}}$$

$$R_2 = 2.5 \times 10^{+3} = 2.5 \text{ k}\Omega$$

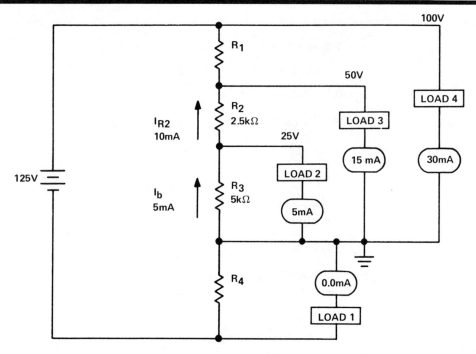

Proceeding up the voltage divider to R_1, you need to determine E_{R1} and I_{R1}. The current through R_1 is simply the sum of I_{R2} and the current through load 3.

$$I_{R1} = I_{R2} + I_{load\ 3}$$

$$I_{R1} = 10\ mA + 15\ mA$$

$$I_{R1} = 25\ mA$$

E_{R1} is the difference between the voltages required by load 3 and load 4 because R_1 is connected between these two loads.

$$E_{R1} = E_{load\ 4} - E_{load\ 3}$$

$$E_{R1} = 100\ V - 50\ V$$

$$E_{R1} = 50\ V$$

Using Ohm's law in the same form as before, you can calculate the value of R_1.

$$R_1 = \frac{E_{R1}}{I_{R1}} = \frac{50\ V}{25\ mA}$$

$$R_1 = \frac{5.0 \times 10^{+1}}{2.5 \times 10^{-2}}$$

$$R_1 = 2 \times 10^{+3} = 2\ k\Omega$$

All that remains to finish the design is to calculate the resistance of R_4. Because it is in parallel with load 1, you know that E_{R4} is 25 volts. You also know that no current flows through load 1 ($I_{load\ 1} = 0.0$ mA), therefore the total circuit current must flow through R_4.

$$I_{R4} = I_b + I_{load\ 1} + I_{load\ 2} + I_{load\ 3}$$

$$I_{R4} = 5\ mA + 5\ mA + 15\ mA + 30\ mA$$

$$I_{R4} = 55\ mA$$

Using Ohm's law in the form $R = E/I$, you can calculate the value of R_4.

$$R_4 = \frac{E_{R4}}{I_{R4}} = \frac{25\ V}{55\ mA}$$

$$R_4 = \frac{2.5 \times 10^{+1}}{5.5 \times 10^{-2}} = 0.455 \times 10^{+3}$$

$$R_4 = 455\ \Omega$$

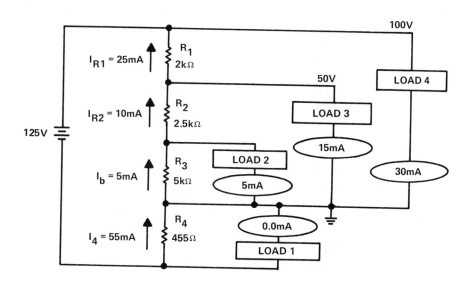

4. An electric iron dissipates 300 watts of power, and its measured resistance is 50 ohms. Using the appropriate power circle memory aid, determine the correct formula for finding the operating voltage of the iron and then solve for the voltage.

Cover E^2 in the $E^2 - R - P$ circle to see that $E^2 = P \times R$, and then take the square root to get:

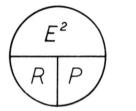

$$E^2 = P \times R$$
$$E = \sqrt{P \times R}$$

Then substitute the appropriate values in the formula and multiply.

$$E = \sqrt{300 \times 50}$$
$$E = \sqrt{15000}$$

In order to find the voltage required by the iron, you must take the square root of 15,000. If you have a calculator with a square root key, simply enter 15,000 and press the square root key to get 122 volts. If you don't have a calculator, you may use the square root tables. Since 15,000 is too large to be listed listed in the tables, you must find two numbers that when multiplied equal 15,000 and both of which are listed in the tables.

$$
\begin{array}{r}
150 \\
100 \overline{\smash{)}15000} \\
100 \\
\hline
500 \\
500 \\
\hline
\end{array}
$$

Dividing 15,000 by 100 gives you 150. Then

$$E = \sqrt{100 \times 150}$$
$$E = \sqrt{100} \times \sqrt{150}$$

Look up the square root of 100 and 150, and multiply to get your final answer.

$$E = \sqrt{100} \times \sqrt{150}$$
$$E = 10 \times 12.2$$
$$E = 122 \text{ volts.}$$

The voltage required by the iron to produce 300 watts of power is 122 volts, which agrees with the previous answer.

LESSON 10. VOLTAGE DIVIDERS AND POWER

● **Practice Problems**

The key objectives of this lesson have been achieved if you can now:
1. Design a simple voltage divider for use in any typical low-power application.
2. Apply the power formulas in analyzing circuit configurations of the type covered up to this point in the course.

The following problems are divided into two sets to enable you to get some practice in these areas. Fold over the page to check your progress and the accuracy of your calculations.

Depending upon the approach you use in solving these problems and how you round off intermediate results, your answers may vary slightly from those given here. However, any differences you encounter should only occur in the third significant digit of your answer. If the first two significant digits of your answers do not agree with those given here, recheck your calculations.

Calculate the following values: **Fold Over**

a.

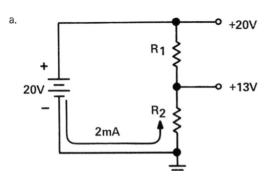

$R_1 =$ _____

$R_2 =$ _____

b.

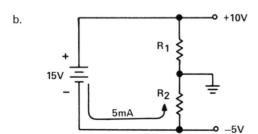

$R_1 =$ _____

$R_2 =$ _____

Hint: (Bleeder current should be one-tenth of the total load current)

c.

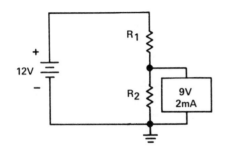

$I_{R1} =$ _____

$I_{R2} =$ _____

$R_1 =$ _____

$R_2 =$ _____

Answers

1.

a. $R_1 = 3.5$ kΩ

       $R_2 = 6.5$ kΩ

b. $R_1 = 2$ kΩ

       $R_2 = 1$ kΩ

c. $I_{R1} = 2.2$ mA

       $I_{R2} = 0.2$ mA

       $R_1 = 1.36$ kΩ

       $R_2 = 45$ kΩ

d.

Fold Over

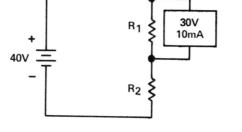

I_{R1} = _____

I_{R2} = _____

R_1 = _____

R_2 = _____

e.

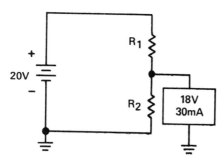

I_{R1} = _____

I_{R2} = _____

R_1 = _____

R_2 = _____

P_{R1} = _____

P_{R2} = _____

f.

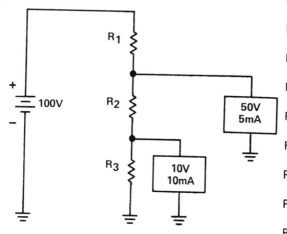

I_{R1} = _____

I_{R2} = _____

I_{R3} = _____

R_1 = _____

R_2 = _____

R_3 = _____

P_{R1} = _____

P_{R2} = _____

P_{R3} = _____

Answers

d. I_{R1} = 1 mA

I_{R2} = 11 mA

R_1 = 30 kΩ

R_2 = 909 Ω

e. I_{R1} = 33 mA

I_{R2} = 3 mA

R_1 = 60.6 Ω

R_2 = 6 kΩ

P_{R1} = 66 mW

P_{R2} = 54 mW

f. I_{R1} = 16.5 mA

I_{R2} = 11.5 mA

I_{R3} = 1.5 mA

R_1 = 3.03 kΩ

R_2 = 3.48 kΩ

R_3 = 6.67 kΩ

P_{R1} = 8 25 mW

P_{R2} = 460 mW

P_{R3} = 15 mW

Fold Over

g.

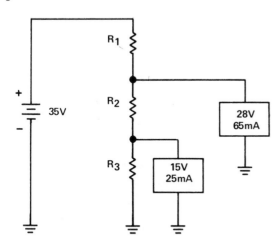

I_{R1} = _____

I_{R2} = _____

I_{R3} = _____

R_1 = _____

R_2 = _____

R_3 = _____

P_{R1} = _____

P_{R2} = _____

P_{R3} = _____

h.

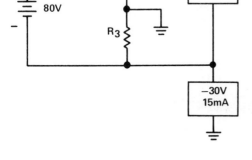

I_{R1} = _____

I_{R2} = _____

I_{R3} = _____

R_1 = _____

R_2 = _____

R_3 = _____

P_{R1} = _____

P_{R2} = _____

P_{R3} = _____

Answers

g. $I_{R1} = 99$ mA

$I_{R2} = 34$ mA

$I_{R3} = 9$ mA

$R_1 = 70.7\ \Omega$

$R_2 = 382\ \Omega$

$R_3 = 1.67$ kΩ

$P_{R1} = 693$ mW

$P_{R2} = 442$ mW

$P_{R3} = 135$ mW

h. $I_{R1} = 44$ mA

$I_{R2} = 19$ mA

$I_{R3} = 4$ mA

$R_1 = 909\ \Omega$

$R_2 = 526\ \Omega$

$R_3 = 7.5$ kΩ

$P_{R1} = 1.76$ W

$P_{R2} = 190$ mW

$P_{R3} = 120$ mW

Fold Over

i.

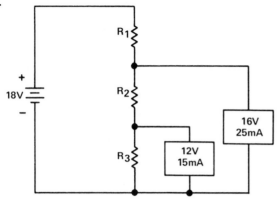

$I_{R1} =$ _____

$I_{R2} =$ _____

$I_{R3} =$ _____

$R_1 =$ _____

$R_2 =$ _____

$R_3 =$ _____

$P_{R1} =$ _____

$P_{R2} =$ _____

$P_{R3} =$ _____

j.

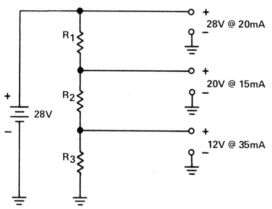

$I_{R1} =$ _____

$I_{R2} =$ _____

$I_{R3} =$ _____

$R_1 =$ _____

$R_2 =$ _____

$R_3 =$ _____

$P_{R1} =$ _____

$P_{R2} =$ _____

$P_{R3} =$ _____

Answers

i. $I_{R1} = 44$ mA

     $I_{R2} = 19$ mA

     $I_{R3} = 4$ mA

     $R_1 = 45$ Ω

     $R_2 = 211$ Ω

     $R_3 = 3$ kΩ

     $P_{R1} = 88$ mW

     $P_{R2} = 76$ mW

     $P_{R3} = 48$ mW

j. $I_{R1} = 57$ mA

     $I_{R2} = 42$ mA

     $I_{R3} = 7$ mA

     $R_1 = 140$ Ω

     $R_2 = 190$ Ω

     $R_3 = 1.71$ kΩ

     $P_{R1} = 456$ mW

     $P_{R2} = 336$ mW

     $P_{R3} = 84$ mW

Fold Over

k.

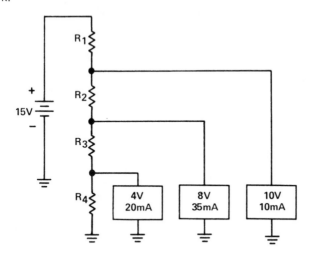

I_{R1} = _____

I_{R2} = _____

I_{R3} = _____

I_{R4} = _____

R_1 = _____

R_2 = _____

R_3 = _____

R_4 = _____

P_{R1} = _____

P_{R2} = _____

P_{R3} = _____

P_{R4} = _____

l.

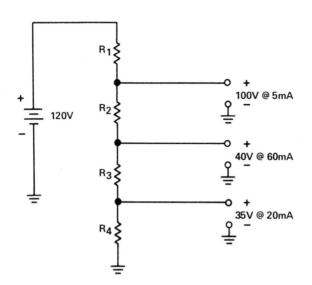

I_{R1} = _____

I_{R2} = _____

I_{R3} = _____

I_{R4} = _____

R_1 = _____

R_2 = _____

R_3 = _____

R_4 = _____

P_{R1} = _____

P_{R2} = _____

P_{R3} = _____

P_{R4} = _____

Answers

k. $I_{R1} = 71.5$ mA

   $I_{R2} = 61.5$ mA

   $I_{R3} = 26.5$ mA

   $I_{R4} = 6.5$ mA

   $R_1 = 70 \ \Omega$

   $R_2 = 32.5 \ \Omega$

   $R_3 = 151 \ \Omega$

   $R_4 = 615 \ \Omega$

   $P_{R1} = 358$ mW

   $P_{R2} = 123$ mW

   $P_{R3} = 106$ mW

   $P_{R4} = 26$ mW

l. $I_{R1} = 93.5$ mA

   $I_{R2} = 88.5$ mA

   $I_{R3} = 28.5$ mA

   $I_{R4} = 8.5$ mA

   $R_1 = 214 \ \Omega$

   $R_2 = 678 \ \Omega$

   $R_3 = 175 \ \Omega$

   $R_4 = 4.12$ kΩ

   $P_{R1} = 1.87$ W

   $P_{R2} = 5.31$ W

   $P_{R3} = 143$ mW

   $P_{R4} = 298$ mW

Fold Over

m.

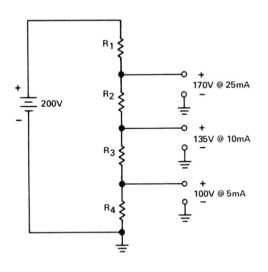

$I_{R1} =$ _____

$I_{R2} =$ _____

$I_{R3} =$ _____

$I_{R4} =$ _____

$R_1 =$ _____

$R_2 =$ _____

$R_3 =$ _____

$R_4 =$ _____

$P_{R1} =$ _____

$P_{R2} =$ _____

$P_{R3} =$ _____

$P_{R4} =$ _____

n.

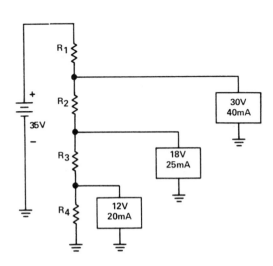

$I_{R1} =$ _____

$I_{R2} =$ _____

$I_{R3} =$ _____

$I_{R4} =$ _____

$R_1 =$ _____

$R_2 =$ _____

$R_3 =$ _____

$R_4 =$ _____

$P_{R1} =$ _____

$P_{R2} =$ _____

$P_{R3} =$ _____

$P_{R4} =$ _____

Answers

m. $I_{R1} = 44$ mA

$I_{R2} = 19$ mA

$I_{R3} = 9$ mA

$I_{R4} = 4$ mA

$R_1 = 682$ Ω

$R_2 = 1.84$ kΩ

$R_3 = 3.89$ kΩ

$R_4 = 25$ kΩ

$P_{R1} = 1.32$ W

$P_{R2} = 665$ mW

$P_{R3} = 315$ mW

$P_{R4} = 400$ mW

n. $I_{R1} = 93.5$ mA

$I_{R2} = 53.5$ mA

$I_{R3} = 28.5$ mA

$I_{R4} = 8.5$ mA

$R_1 = 53.5$ Ω

$R_2 = 224$ Ω

$R_3 = 211$ Ω

$R_4 = 1.41$ kΩ

$P_{R1} = 468$ mW

$P_{R2} = 642$ mW

$P_{R3} = 171$ mW

$P_{R4} = 102$ mW

Fold Over

o.

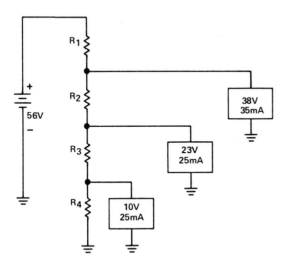

I_{R1} = _____

I_{R2} = _____

I_{R3} = _____

I_{R4} = _____

R_1 = _____

R_2 = _____

R_3 = _____

R_4 = _____

P_{R1} = _____

P_{R2} = _____

P_{R3} = _____

P_{R4} = _____

2. Calculate the following unknown values using the three power formulas:

a. P = 5 W

 I = 25 mA

 E = _____

b. P = 1320 W

 I = _____

 E = 12 V

c. P = _____

 I = 6 A

 E = 5 V

Answers

o. $I_{R1} = 93.5$ mA

$I_{R2} = 58.5$ mA

$I_{R3} = 33.5$ mA

$I_{R4} = 8.5$ mA

$R_1 = 193 \ \Omega$

$R_2 = 256 \ \Omega$

$R_3 = 388 \ \Omega$

$R_4 = 1.18$ kΩ

$P_{R1} = 1.68$ W

$P_{R2} = 878$ mW

$P_{R3} = 436$ mW

$P_{R4} = 85$ mW

2.

a. E = 200 V

b. I = 110 A

c. P = 30 W

Fold Over

2. d. P = 6 W

    I = 0.39 A

    E = _____

    e. P = 100 W

    I = _____

    E = 38 V

    f. P = 30 W

    I = 3 A

    R = _____

    g. P = 6 W

    I = _____

    R = 500 Ω

    h. P = _____

    I = 0.036 A

    R = 15 kΩ

    i. P = 2 W

    I = 0.009 A

    R = _____

    j. P = 12 W

    I = _____

    R = 320 Ω

Answers

d.   E = 15.4 V

e.   I = 2.63 A

f.   R = 3.33 Ω

g.   I = 110 mA

h.   P = 19.4 W

i.   R = 24.7 kΩ

j.   I = 194 mA

Fold Over

2. k. E = _____

R = 33 kΩ

P = 20 W

l. E = 120 V

R = 400 kΩ

P = _____

m. E = 12 V

R = _____

P = 0.5 W

n. E = _____

R = 10 kΩ

P = 0.25 W

o. E = 1000 V

R = 5 MΩ

P = _____

●   **Practice Problems**

Answers

k.   E = 812 V

l.   P = 36 mW

m.   R = 288 Ω

n.   E = 50 V

o.   P = 200 mW

1. Any circuit or device that draws current and/or has resistance, requires voltage, or dissipates power is defined as:

   a. A voltage generator
   b. A load
   c. An alternator
   d. A short circuit

2. A point that is at the same potential as the earth itself is called:

   a. Chassis ground
   b. The minus terminal
   c. An earth ground
   d. The positive terminal

3. A rule of thumb for voltage dividers: _____% of the load current is selected as bleeder current.

   a. 10%
   b. 25%
   c. 50%
   d. 90%

4. The next step in designing a voltage divider after examining the power supply and load requirements is:

   a. To draw a schematic
   b. To select the bleeder current
   c. To turn on the switch
   d. To calculate the current

5. A 100 watt light bulb requires 120 volts; what is its resistance?

   a. 14.4 K
   b. 120 ohms
   c. 1.2 ohms
   d. 144 ohms

Two quantities are given for each of the five resistors drawn below. In each case, calculate the quantity indicated with a question mark.

6.

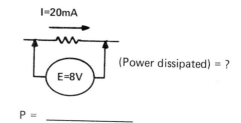

I=20mA

E=8V

(Power dissipated) = ?

P = _____

7.

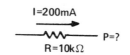

I=200mA

R=10kΩ   P=?

P = _____

8.

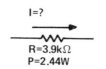

I=?

R=3.9kΩ
P=2.44W

I = _____

9.

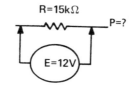

R=15kΩ

P=?

E=12V

P = _____

10.

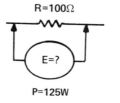

R=100Ω

E=?

P=125W

E = _____

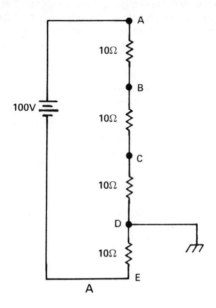

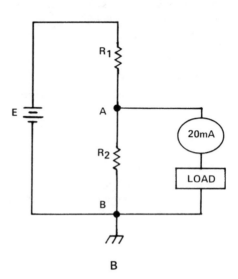

The questions that follow refer to schematics A and B.

11. The series network of resistors from A thru E in schematic A is known as a:

    a. Series-parallel circuit
    b. Voltage divider
    c. Parallel - series circuit
    d. Open circuit

12. Point E in schematic A is_____with respect to chassis ground (point D).

    a. Negative
    b. Positive
    c. Neutral
    d. The same voltage

13. Point A in Schematic A is_____with respect to chassis ground (Point D).

    a. Negative
    b. Positive
    c. Neutral
    d. The same voltage

14. Point C of Schematic A has a voltage with respect to point E that is_____of the applied voltage from A to E.

    a. 10%
    b. 20%
    c. 60%
    d. 50%

15. R_2 of Schematic B is called:

    a. The load
    b. A needless bypass
    c. A bleeder resistor
    d. A power saver

16. R_1 of Schematic B has a current through it that should be at least_____of the load current:

    a. 5%
    b. 50%
    c. 110%
    c. 90%

17. Increasing the current through R_2 of Schematic B will make V_{AB} more stable:

    a. But wastes more power
    b. But doesn't help the circuit.
    c. And requires changing R_1
    d. All of above
    e. a and c above

18. If the load of Schematic B is changed to a smaller resistor the current drain from the power supply:

    a. Decreases
    b. Increases
    c. Remains the same
    d. Goes negative

Lesson 11

Introduction to Kirchhoff's Laws

This lesson introduces *Kirchhoff's Current Law* and *Kirchhoff's Voltage Law* and explains all new terminology relating to these two laws. Discussion includes an examination of how to *write* and *solve loop equations* for the unknown current or currents. This includes all algebraic manipulations necessary to solve the loop equations for *multiple source* circuits.

LESSON 11. INTRODUCTION TO KIRCHHOFF'S LAWS

- Objectives

This lesson introduces two new and powerful laws that you will find useful in analyzing dc circuits where you cannot use Ohm's law alone. This lesson and the next explore Kirchhoff's laws and several advanced methods of circuit analysis. At the end of this lesson, you should be able to:

1.   *Write* Kirchhoff's current law, using diagrams to explain it.

2.   In any circuit of the type illustrated in the schematic diagram below

    a.   *Identify* and label the loops and nodes (or junctions)

    b.   *Identify* three typical circuit paths that are not loops.

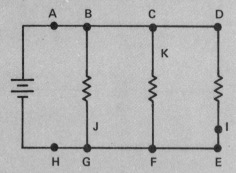

3.   At any node of the type illustrated below, *use Kirchhoff's current law to calculate* the unknown current labeled with a question mark.

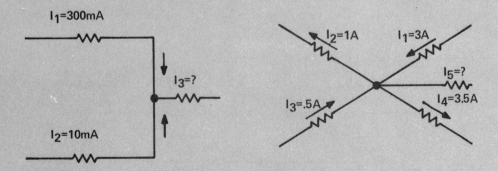

4.   *Write* Kirchhoff's voltage law, using diagrams to explain it.

5.   *Write* the six rules governing series circuit and parallel circuit operation. *Explain* which of these rules are related to Kirchhoff's voltage law, which to Kirchhoff's current law, and why.

6.   *Write* an explanation of the significance of a negative current value solution in a Kirchhoff's law problem.

7. In any loop, such as the type represented in the schematic below, *write* a Kirchhoff's voltage law or "loop" equation, being careful to observe the correct rules with respect to the signs of the voltages. *Write* one equation using electron current and another equation using conventional current.

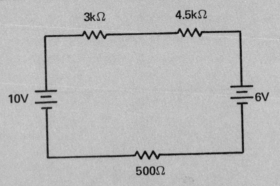

8. *Solve* any loop equation of the type shown for a single unknown quantity such as the current, *correctly using the procedures of*:

   a. Transposing

   b. Combining "like" terms

   c. Multiplying or dividing both sides of the equation by the same quantity

   d. Adding or subtracting the same quantity on both sides of an equation

   e. Changing the signs of all parts of an equation.

   $$10\text{ V} - 3\text{ kI} - 4.5\text{ kI} - 6\text{ V} - 500\text{ kI} = 0$$

9. In a multiple source circuit of the type shown, using Kirchhoff's laws, *solve* for all unknown currents (including direction) and voltages (including polarity). You should be able to write loop equations for this circuit using either conventional or electron current.

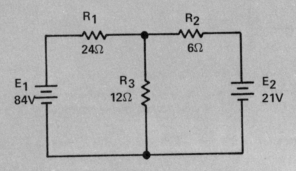

LESSON 11. INTRODUCTION TO KIRCHHOFF'S LAWS

● **Gustav Kirchhoff**

An important objective of this course on dc circuits is to provide you with the tools you will need to make electricity work for you, or to predict how it will behave in any given circuit situation. Up to this point, you have been using Ohm's law, the power formulas, and some basic circuit rules as tools for analyzing dc circuits (Figure 11.1).

Figure 11.1

In this lesson and the next, you will be introduced to some new tools you can use to help simplify and analyze more complex circuits (Figure 11.2). The key difference between the methods that will be covered in this lesson and methods you have seen earlier is that the methods introduced here will allow you to analyze dc circuits which cannot be solved using Ohm's law alone. This includes certain types of single power supply circuits, as well as those with more than one voltage source. You will find that in certain circuits, the solution is not possible using Ohm's law alone.

Figure 11.2

Gustav Kirchhoff — Some of the basic and general methods for analyzing circuits that will be introduced and analyzed in this lesson, were developed in 1874 by the German physicist, Gustav Kirchhoff (pronounced Kirk'hawf). A photo of Kirchhoff is shown as Figure 11.3.

Figure 11.3

Lesson Objectives — The key objective of this lesson and the next is to enable you to write down and use what are called Kirchhoff's laws to completely analyze a complex, multiple source dc circuit, such as the one in Figure 11.4. At the end of Lessons 11 and 12, you should be able to calculate any voltage drop and any current flowing in a multiple source complex circuit, such as this.

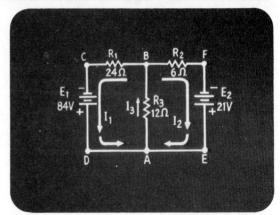

Figure 11.4

Basic Circuit Rules: Series Circuit — You have already seen Kirchhoff's laws at work in some specific cases, such as when the rules describing the operation of series and parallel circuits were introduced. Take a minute to briefly review these rules. As listed in Figure 11.5 for series circuits, you have seen that:

1. The current is the same in all parts of a series circuit.
2. The sum of the individual voltage drops in a series circuit equals the total applied voltage.
3. The total resistance of a series circuit is equal to the sum of the individual resistances.

SERIES CIRCUIT RULES

1. CURRENT IS THE SAME IN ALL PARTS OF A SERIES CIRCUIT.

2. THE SUM OF THE INDIVIDUAL VOLTAGE DROPS IN A SERIES CIRCUIT EQUALS THE TOTAL APPLIED VOLTAGE.

3. THE TOTAL RESISTANCE OF A SERIES CIRCUIT EQUALS THE SUM OF THE INDIVIDUAL RESISTANCES.

Figure 11.5

Basic Circuit Rules: Parallel Circuit — For parallel circuits (Figure 11.6) you have seen that:

1. The total or main line current is the sum of the individual branch currents.
2. The voltage is the same across all branches.
3. The total resistance is always less than, or approximately equal to, the smallest branch resistance.

As you proceed through this lesson, you will see how Kirchhoff's laws really are just more general statements of these circuit rules you have been using.

PARALLEL CIRCUIT RULES

1. TOTAL CURRENT IS THE SUM OF THE INDIVIDUAL BRANCH CURRENTS.

2. VOLTAGE IS THE SAME ACROSS ALL BRANCHES.

3. TOTAL RESISTANCE IS ALWAYS LESS THAN OR APPROXIMATELY EQUAL TO THE SMALLEST BRANCH RESISTANCE.

Figure 11.6

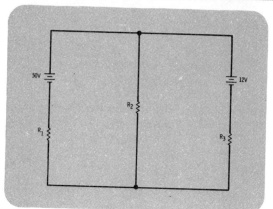

Figure 11.7

Why Do You Need Kirchhoff's Laws? — You may be wondering just why these new laws are needed. The reason is that the laws you already know won't always work with complex circuits such as the one shown in Figure 11.7.

Why? Suppose that you need to find the voltage dropped across R_2. There are two voltage sources in the circuit, and as you see, each one acting on R_2.

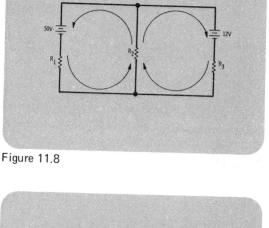

Figure 11.8

Which Way Does Current Flow Through R_2 — There are several possibilities for this question. Will the currents flow as shown in Figure 11.8, where electron current from both sources combines and flows through R_2?

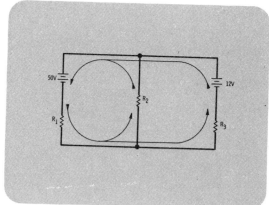

Figure 11.9

Or, will the larger battery "overpower" the smaller battery and push current backwards through the smaller source as shown in Figure 11.9?

If this is the case, what effect does the 12-volt battery have on the circuit? Should it be treated as a resistance, an open circuit, a short circuit, or something else?

These are all good questions. Unfortunately, they cannot be answered by applying Ohm's law and the series and parallel circuit laws you already know. For this reason, some other circuit analysis methods, including methods such as Kirchhoff's laws must be studied. There are *two* of these laws, and they will be discussed one at a time.

Kirchhoff's Current Law — Kirchhoff's first law, sometimes called *Kirchhoff's current law*, simply states that the sum of the currents *arriving* at any point in a circuit must equal the sum of the currents *leaving* that point (Figure 11.10). You may see this law stated several different ways, and you will be shown several alternate statements of this law. But, basically, any statement of Kirchhoff's current law means the same thing: whatever current arrives at any point in a circuit must equal the total current that leaves.

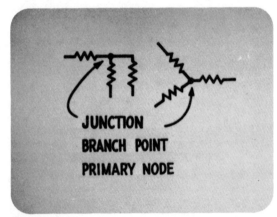

Figure 11.10

Junctions or Nodes — This law applies to any point in a circuit but is most frequently used to analyze points in circuits where three or more components are joined together. As shown in Figure 11.11, points with three or more connections are called "junction points" or *junctions*. They may also be called *branch points* or *nodes* as well.

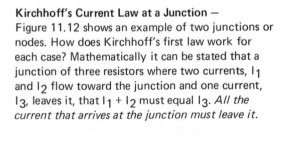

Figure 11.11

Kirchhoff's Current Law at a Junction — Figure 11.12 shows an example of two junctions or nodes. How does Kirchhoff's first law work for each case? Mathematically it can be stated that a junction of three resistors where two currents, I_1 and I_2 flow toward the junction and one current, I_3, leaves it, that $I_1 + I_2$ must equal I_3. *All the current that arrives at the junction must leave it.*

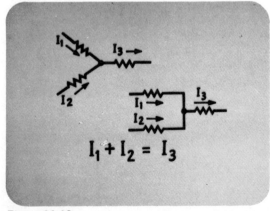

Figure 11.12

Kirchhoff's Current Law at a Junction —
Figure 11.13 shows another example set of two junctions. In each case there is one current, I_1, entering the junction and two currents, I_2 and I_3, leaving the junction. By Kirchhoff's first law, I_1 must equal $I_2 + I_3$. For example, suppose you know that I_1 equals 7 amps, and I_2 equals 5 amps. How would you determine I_3? Recall again that the current leaving the junction must be equal to the current entering. In order for this to be true, I_3 must be 2 amps.

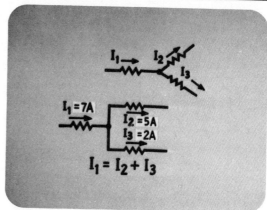

Figure 11.13

If Kirchhoff's Current Law Were Not Followed —
Consider for a second what would happen if Kirchhoff's current law were not followed. Suppose, as shown in Figure 11.14, that more current arrives at a point in a circuit than leaves it. Remember that electron current is a flow of electrons, and if there is more electron current arriving at a point than leaving, what would the result be? Electrons would have to be building up somehow at a point since more negative charges are coming in than are leaving. This type of circuit behavior — with a wire blowing up like a balloon just doesn't happen anywhere in nature.

Figure 11.14

If More Electrons Leave Than Arrive — In the opposite vein, as shown in Figure 11.15, suppose more electrons were leaving a circuit point than arrived. If this is the case, electrons have to be magically created somehow, or secretly "snuck" into the circuit and this doesn't happen either. Kirchhoff's first law states that all the electrons that enter any point in a circuit, leave it, no more, no less. The total current arriving at any point, must leave that point.

Figure 11.15

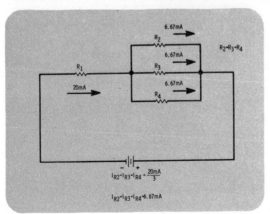

Figure 11.16

Example: Kirchhoff's First Law — Say that you have the circuit pictured in Figure 11.16. The current flowing through R_1 is 20 milliamps. Resistors 2, 3, and 4 are of equal size. Kirchhoff's current law says that the current leaving a junction must equal the current flowing into that junction. The 20 milliamps flowing into the junction will be divided three ways by the three equal-sized resistors. You could then calculate that the current flowing through each resistor will be 6.67 milliamps, for a total of 20 milliamps.

Kirchhoff's Current Law in Parallel Circuits — You have already seen Kirchhoff's first law at work in parallel and series circuits. In parallel circuits, recall that "the total or main line current equals the sum of the branch currents." Basically, you have seen this law expressed as the formula $I_T = I_1 + I_2 + I_3$, for parallel circuits. If you focus your attention on the points where the circuit branches in Figure 11.17, you will see that the law for currents in a parallel circuit is actually just a specific statement of Kirchhoff's current law that applies to parallel circuits.

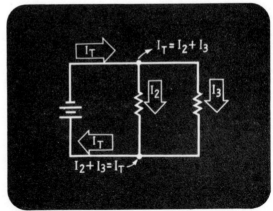

Figure 11.17

Kirchhoff's Current Law in Series Circuits — In series circuits you have seen that the "current is the same in all parts of the circuit." This just means that if you closely examine any point in the circuit, as shown in Figure 11.18, the current entering that point always equals the current leaving it. This is, again, just a specific statement of Kirchhoff's first law or Kirchhoff's current law that applies to series circuits.

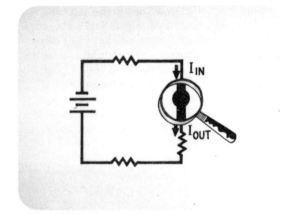

Figure 11.18

11-10

Kirchhoff's Voltage Law — Moving on, Kirchhoff's second law, sometimes called Kirchhoff's voltage law will be discussed. Then the discussion will proceed to how Kirchhoff's current and voltage laws may be used together to analyze complex dc circuits. Figure 11.19 shows one way that Kirchhoff's voltage law can be stated: the total *voltage applied* to any *closed circuit path* is always equal to the sum of the *voltage drops* across the individual parts of the path. To really understand the meaning of this law and how to use it, you will have to concentrate your attention on all the parts of the law one at a time. First, consider the concept of a *"closed circuit path"* for a minute.

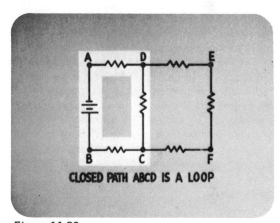

KIRCHHOFF'S VOLTAGE LAW

THE TOTAL VOLTAGE APPLIED TO ANY "CLOSED CIRCUIT PATH" IS ALWAYS EQUAL TO THE SUM OF THE VOLTAGE DROPS IN THAT PATH.

Figure 11.19

Closed Circuit Path — A "closed circuit path" simply means any continuous path you trace in a circuit that starts and ends at the same point. In electricity, these closed circuit pathways are usually called *loops*, as they will be referred to from now on in this course. In Figure 11.20, the pathway ABCD and back to A is a loop.

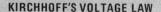

CLOSED PATH ABCD IS A LOOP

Figure 11.20

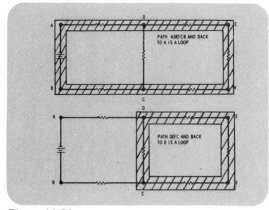

PATH ADEFCB AND BACK TO A IS A LOOP

PATH DEFC AND BACK TO D IS A LOOP

Figure 11.21

Other Loops in this Circuit — Actually in most any circuit, it is possible to trace out several loops. A loop is any closed circuit pathway, regardless of what kinds of components it may contain. Since one of the objectives of this lesson requires that you be able to identify loops in a circuit, it is important that you spend a little time familiarizing yourself with loops. For example, in Figure 11.21, two other simple loops in the circuit are shown. To trace a loop in your mind, start at any point and mentally move around (or *traverse*) a path that returns to the same point. In doing this, you will have traversed a closed path, or loop.

Paths That Are Not Loops — Circuit pathways that are *not closed* are *not loops*. This means that if a path in a circuit doesn't start and end at the same point, it is not a loop. In Figure 11.22, paths going through points BAD and CFE are not loops. Now remember that Kirchhoff's voltage law holds for *loops* only, and again states that the total voltage applied to a loop equals the sum or all the voltage drops in the loop.

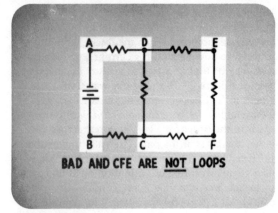

Figure 11.22

Kirchhoff's Voltage Law in Series Circuits — You have already seen Kirchhoff's voltage law in operation in series circuits. Series circuits are circuits with only one closed circuit path, or only one loop, as shown in Figure 11.23. You have seen that in any series circuit such as this, the sum of the voltage drops across the circuit resistors must equal the applied voltage. This is just another way of stating Kirchhoff's second law. This law tells us that in any loop the algebraic sum of the voltage applied by generators, batteries, etc., must always equal the sum of the voltage drops across all resistances.

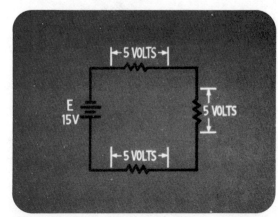

Figure 11.23

Loops in Parallel Circuits — The really *new* thing about Kirchhoff's second law is that it applies to loops anywhere in any type of circuit. For example, Figure 11.24 shows a simple parallel circuit. In this circuit you can "move around" or "traverse" several loops, starting at any point. You can start at point A and find several possible loops in this circuit. There is one loop around the outer path (points ABCDHGFE and back to A); there is another loop through the central branch (points ABCGFE and back to A); and there is a third loop through points ABFE and back to A. Note in Figure 11.24(D) that loop CDHG is also a loop; this particular loop contains only resistors, no

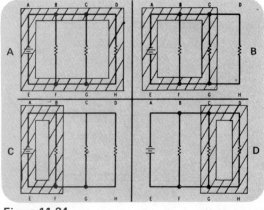

Figure 11.24

voltage sources. Kirchhoff's laws hold in loops such as this, also.

Kirchhoff's Voltage Law: Alternate Statement — In each of these loops, and in any closed circuit path, Kirchhoff's voltage law holds true. As this lesson proceeds, this law will be stated for you in several different ways to help clarify its meaning. One alternate way it may be stated is: *for any loop the sum of all the voltages that aid current flow in the loop, must equal the sum of all those voltages that oppose it* (Figure 11.25).

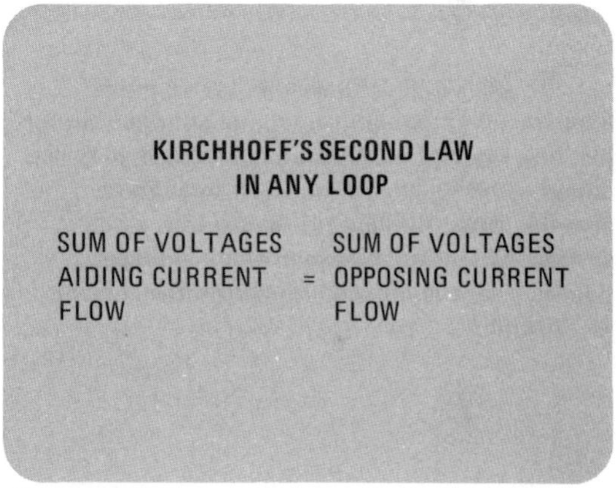

KIRCHHOFF'S SECOND LAW
IN ANY LOOP

SUM OF VOLTAGES   SUM OF VOLTAGES
AIDING CURRENT   =   OPPOSING CURRENT
FLOW   FLOW

Figure 11.25

Voltages Aiding and Opposing Current — As you study Kirchhoff's law, you will be asked to keep track of several facts concerning loops. Normally, as you analyze a loop you will start at one point in the loop and mentally "walk around" or *traverse* the loop. As you traverse the loop you will be writing down each voltage you come across along with its correct sign. That is, you will write down all the positive and negative voltages you come to as you go around the loop. Anytime you come to a battery or voltage source, you will write down its voltage, and any time you come to a resistor, you will write down a voltage expressed as a current times a resistance (IR drop). Now a key point. The *sign* (positive or negative) that you put in front of

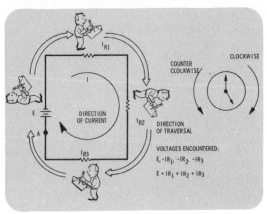

Figure 11.26

these voltages will be important. How do you determine these signs? Basically, it is a matter of considering what effect each voltage you come to has on the current flow in the circuit. Consider Figure 11.26; you see a single loop with electron current flowing clockwise as shown. If you traverse this circuit in the same direction as the current, then you mark all voltage sources *positive* that *push current in that same direction*. When you come to a *resistor* and traverse it in the same direction as current flow through it, you write down a negative voltage, $-IR$. This is because the resistors act to *oppose* current flow; a voltage *drop* develops across them. So starting at point A in this simple loop and going around the circuit, you would encounter voltages $+E$, $-IR_1$, $-IR_2$ and $-IR_3$. Kirchhoff's second law states that the applied voltage E, equals IR_1 plus IR_2 plus IR_3 or $E = IR_1 + IR_2 + IR_3$.

To use Kirchhoff's second law in more complex circuit situations, you need to be careful just how you add up the applied voltages and the voltage drops in circuit loops. In some more complex cases, it might not be obvious which voltages aid current flow and which voltages oppose it, or even in what direction the currents are flowing.

Kirchhoff's Voltage Law: Alternate Form — In cases such as this, it is often more convenient to rewrite Kirchhoff's second law in another form. If you label all the voltages aiding current flow in a loop with a plus sign and all voltages hindering or opposing current flow in the same loop with a minus sign, then Kirchhoff's second law says that their *algebraic sum* must equal zero (Figure 11.27).

KIRCHHOFF'S VOLTAGE LAW

| SUM OF VOLTAGES + AIDING CURRENT FLOW | VOLTAGES OPPOSING − CURRENT FLOW | = 0 |

Figure 11.27

Kirchhoff's Voltage Law Example — The voltage law tells you that the sum of these voltages aiding current flow minus those that oppose it equals zero. This is an alternate (but equivalent) statement of Kirchhoff's voltage law. You will find that writing this law in this second way may make it easier to simplify solutions of more complex circuits.

To use Kirchhoff's second law, what you really need to do is go around or traverse the loop and focus on the voltage sources and resistors you encounter. Voltages which aid current flow in the direction you are moving get a plus sign. Voltages which are opposing current flow in the direction you are moving get a minus sign. Put all these plus and minus voltages on one side of an equal sign. Kirchhoff's voltage law tells you there must be a zero on the other side. This then gives you a Kirchhoff's voltage law equation.

Consider the expression containing all the voltages and their signs on the left-hand side of the equal sign in Figure 11.28. This is called the *algebraic sum* of these voltages. To take the algebraic sum of these voltages, you add all of the positive voltages together, and subtract each negative voltage from that result. Kirchhoff's second law states that for any loop you have as many positive as negative volts; so that when you finish adding and subtracting, the result will be exactly zero.

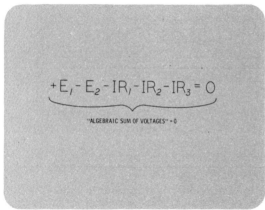

$$+E_1 - E_2 - IR_1 - IR_2 - IR_3 = 0$$

"ALGEBRAIC SUM OF VOLTAGES" = 0

Figure 11.28

Kirchhoff's Voltage Law: Third Form — If you keep careful track of the signs of all the voltages in the circuit, you can rewrite Kirchhoff's voltage law once again, this time in a simpler form as shown in Figure 11.29. The algebraic sum of all the voltages encountered in any loop equals zero.

Again, algebraic sum means the sum of the voltages taking into account whether each is positive or negative. Positive voltages are simply added, and negative voltages are subtracted to get the total.

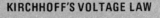

KIRCHHOFF'S VOLTAGE LAW

THE ALGEBRAIC SUM OF ALL THE VOLTAGES ENCOUNTERED **IN ANY LOOP** EQUALS ZERO.

Figure 11.29

Loop Equation — When the algebraic sum of all the voltages in a loop is set to equal zero, the result is called a loop equation. A loop equation (Figure 11.30) is simply a mathematical expression of Kirchhoff's voltage law. As you will see, loop equations can be used to help solve for unknown currents and voltages in circuits, where Ohm's law and basic circuit rules can't always be used.

LOOP EQUATION

THE ALGEBRAIC SUM OF THE VOLTAGES IN A LOOP EQUALS ZERO.

$$+ \begin{matrix} \text{SUM OF} \\ \text{APPLIED} \\ \text{VOLTAGES} \end{matrix} - \begin{matrix} \text{SUM OF} \\ \text{VOLTAGE} \\ \text{DROPS} \end{matrix} = 0$$

Figure 11.30

Example: Kirchhoff's Voltage Law — As an example, examine the single-loop circuit shown in Figure 11.31, and go through the steps that you would need in writing a loop equation for it. In this circuit, a loop equation could be used to allow you to solve for the total current. (Note that you could solve this circuit using series circuit rules, but this type of example will be a good place for you to begin learning the use of Kirchhoff's laws.)

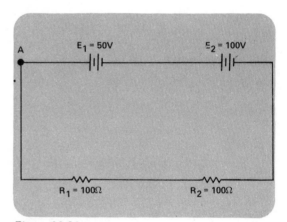

Figure 11.31

Assign Current Directions — The first step in writing a loop equation is to mark all the current directions. In some circumstances, you may not know the direction of the current flow. Here's a trick. In those *cases where the current flow is unknown, simply guess a direction.*

A key point to remember is that if you happen to guess the current direction incorrectly, the current you calculate using Kirchhoff's laws will turn out to be negative. That's a key to using Kirchhoff's laws. *If any current you calculate using Kirchhoff's laws turns out to be negative, you then know it's flowing opposite to the direction you've guessed* (Figure 11.32).

DON'T KNOW CURRENT DIRECTION?

GUESS A DIRECTION!

IF ANY CURRENT YOU CALCULATE IS **NEGATIVE**, IT IS FLOWING IN THE OPPOSITE DIRECTION.

Figure 11.32

- Selected Clockwise Electron Current Flow
- Step 2: Traverse the Circuit
- Determining Signs: Voltage Sources

Selected Clockwise Electron Current Flow — In this loop you could just blindly assume that electron current is flowing clockwise. An arrow has been inserted in Figure 11.33 to help you keep this in mind. Remember that if you calculate a negative current, this means that this assumed current direction is incorrect. If that's the case, you will know that electron current is actually flowing in the opposite direction in this circuit.

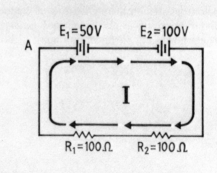

Figure 11.33

Step 2: Traverse the Circuit — The second step in writing a loop equation is to mentally get inside your circuit and walk around or "traverse" it. Start at one point and write down all the voltages you encounter along the way, with their correct signs.

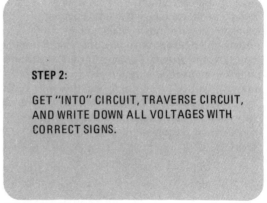

STEP 2:

GET "INTO" CIRCUIT, TRAVERSE CIRCUIT, AND WRITE DOWN ALL VOLTAGES WITH CORRECT SIGNS.

Figure 11.34

Determining Signs: Voltage Sources — How do you determine the correct signs? The trick is to pay careful attention to the direction in which you are going through or traversing your circuit. Source voltages (like batteries) are considered *positive* if you go through (or traverse) them in the *same* direction that they normally push current (Figure 11.35).

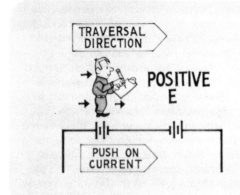

Figure 11.35

Negative Voltage Sources — Correspondingly, source voltages are considered *negative* in the loop equation if the source is pushing current *against* you as you traverse it, as shown in Figure 11.36.

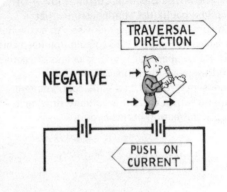

Figure 11.36

Signs of Source Voltages — To determine the right sign for a voltage source in a loop equation, compare the direction in which you are going through the source, to the direction that the source is pushing on current. If you go through the source in the *same direction* as it pushes on current, write down the voltage in your loop equation with a *plus sign*. If the source is pushing current *against you* as you traverse it, write down the voltage with a *negative sign*. These two rules are illustrated in Figure 11.37.

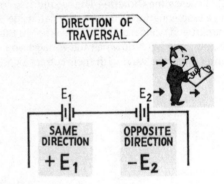

Figure 11.37

Example — Mentally traverse the example circuit and write down the two source voltages according to these rules. Starting at point A, traverse the circuit clockwise. This way you will be moving through the top of this circuit from left to right. The first thing you come to is a voltage source which you will go through from plus to minus. Now notice that this is the same direction this source pushes on electron current. This gives you a plus 50 volts. Proceed through a second source from minus to plus. Notice that this source is pushing electron current against you as you go through it, so you get a minus 100 volts. Next as you move clockwise around this circuit, you come to resistor R$_2$. So now consider how to handle the

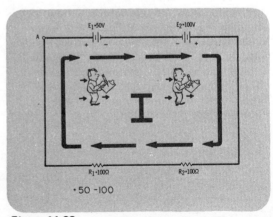

Figure 11.38

voltage across a resistor when writing loop
equations.

Size of Voltage Across Resistors — First of all,
from Ohm's law you know that the *size* of the
voltage across any resistor equals the product of
the current through it times its resistance. The
voltage across a resistor equals I times R, and, as
you know by now, is called an IR drop, as shown
in Figure 11.39.

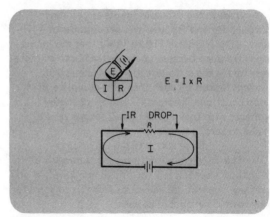

Figure 11.39

Sign of Voltage Across Resistors — How do you
determine the correct sign (positive or negative) for
an IR voltage in your loop equation? To do this,
concentrate on the direction you are moving
through or traversing the resistor and how current
is flowing through it.

As shown in Figure 11.40, if you go through a
resistor *in the direction of assumed current flow*,
you write down its voltage with a *negative* sign.

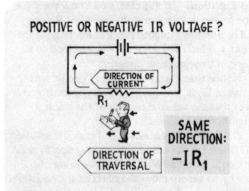

Figure 11.40

Sign of Voltage Across Resistors — If you go through a resistor *opposite to the direction of current* through it, the IR voltage across it is considered *positive*, as shown in Figure 11.41.

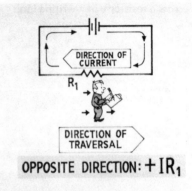

Figure 11.41

Back in the example circuit as you move through R_2 and R_1, you will get voltages of $-IR_2$ and $-IR_1$. Notice in Figure 11.42 that the minus signs are there because you traversed both of these resistors in the same direction as the electron current flowed through them. To complete the loop equation, algebraically add all of these voltages with their correct signs, set the result equal to zero.

$$+50 - 100 - IR_2 - IR_1 = 0$$

This is, in equation form, the statement of Kirchhoff's second law for this simple loop. As you will see, you can use equations like this to solve for unknown quantities in circuits.

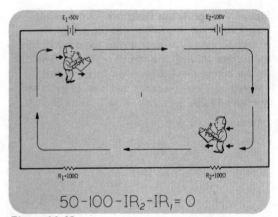

Figure 11.42

Loop Equation — In this case, you know the value of the source voltages, and the values of R_1 and R_2. The unknown quantity in this equation is the current, I, which as you recall was assumed to be flowing in the clockwise direction.

Notice that if you use a little "circuit sense" here, since E_2 is twice as large as E_1, you should expect that the actual direction of the electron current flow is counterclockwise.

What steps are necessary to finish the solution for I? Since I is the only quantity you don't know in this equation, one way to find the value of I would be to simply move the parts of the equation around until you have I all alone on one side of the equal sign. Then the current will simply equal all the known quantities on the other side of the equal sign as shown in Figure 11.43.

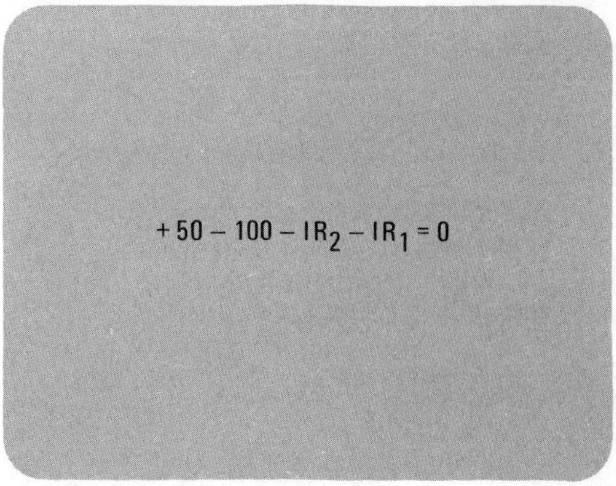

$$+50 - 100 - IR_2 - IR_1 = 0$$

Figure 11.43

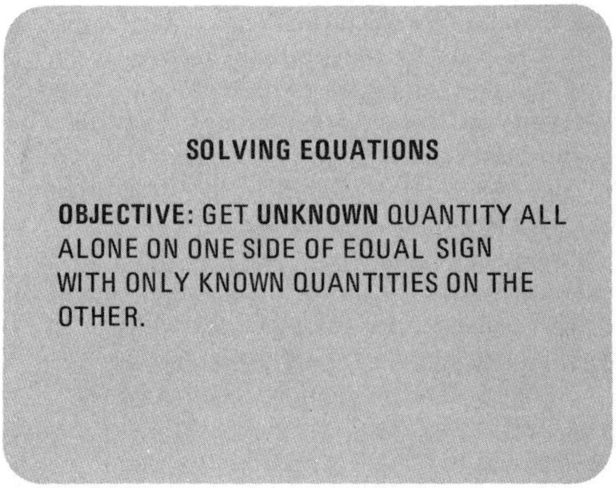

SOLVING EQUATIONS

OBJECTIVE: GET UNKNOWN QUANTITY ALL ALONE ON ONE SIDE OF EQUAL SIGN WITH ONLY KNOWN QUANTITIES ON THE OTHER.

Figure 11.44

Solution of Equations — Quite often in electronic problems you may encounter situations where you must "solve" an equation (a mathematical statement with an equal sign). In this equation may be several known quantities and an unknown quantity that you need to find. To "solve the equation" means to rearrange it using correct procedures, until the unknown quantity is all by itself on one side of the equal sign, and only known quantities are on the other (Figure 11.44). When that has been accomplished, you know that your unknown quantity is equal to a combination of known quantities, and you have your answer.

Whenever you are manipulating equations in solving for unknowns, you must follow the correct procedures which will allow you to rearrange an equation without actually changing its equality. These procedures are fairly easy to learn, and will be reviewed for you.

- Basic Principle
- One Equation for Each Unknown

BASIC PRINCIPLE IN SOLVING EQUATIONS

ANY MATHEMATICAL MANIPULATION PERFORMED ON ONE SIDE OF THE EQUAL SIGN **MUST** BE PERFORMED ON THE OTHER SIDE OF THE EQUAL SIGN.

Figure 11.45

EQUATIONS AND UNKNOWNS

YOU NEED **ONE** EQUATION FOR EVERY UNKNOWN FOR WHICH YOU MUST SOLVE.

Figure 11.46

Basic Principle – The basic principle to follow when solving any equation is this: Any mathematical operation that's performed on one side of an equal sign must be performed on the other. If this principle is not followed, as you work with an equation, the equation will no longer be an equation and you will start getting wrong answers. The operations you are allowed to perform on an equation to solve it are listed below. You can apply these to an equation as many times as you need to; and in any order until the unknown is alone on one side of the equal sign. That is always your objective as you work on an equation to solve it: Get the unknown all by itself on one side of the equal sign with known quantities on the other side of the equal sign (Figure 11.45).

One Equation for Each Unknown – Also, note one more point, before the procedures are reviewed. One equation can be solved for only one unknown. You will need one equation for each unknown quantity that shows up in any problem (Figure 11.46). This is why Kirchhoff's laws come in handy in circuits where there are several unknowns. For example, if there are three unknown currents, you can use Kirchhoff's laws to write three equations and then solve these equations together for the three unknowns.

Solving several equations together will be discussed in a moment. For now, consider the rules you can use to solve one equation for one unknown quantity. In the example equations that are discussed for you, the letter "X" will be used to mean "the unknown quantity." In your work, the "X" might be an "I" for an unknown current or an "E" for an unknown voltage.

RULE: YOU CAN **ADD** OR **SUBTRACT** THE SAME QUANTITY FROM BOTH SIDES OF THE EQUAL SIGN.

EXAMPLE: YOU HAVE X − 300 AND NEED TO FIND X

ADD 30 TO BOTH SIDES

$$X - (30 + 30) = (300 + 30)$$

$$(-30 + 30 = 0; \ 300 + 30 = 330 \rightarrow)$$

$$X = 330.$$

UNKNOWN ⟋   ⟍ KNOWN

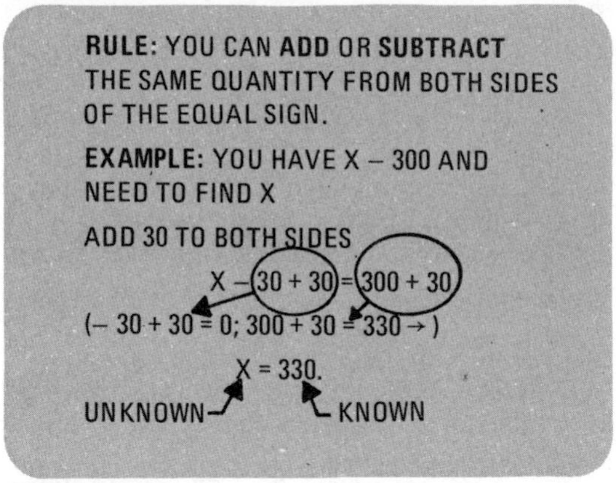

Figure 11.47

First Rule — What are the procedures you can use with equations in solving them? First of all, you can add or subtract the same quantity from both sides of the equal sign in any equation, as shown in Figure 11.47. For example, if you have an equation such as $X - 30 = 300$ and need to calculate X, you can add 30 to both sides. Here's where some plain old common sense comes into play. Why would you add 30 to both sides? Because you know that you want X all alone on one side of the equal sign. If you add 30 to both sides of this equation, you get

$$X - 30 + 30 = 300 + 30.$$

The −30 and +30 on the left-hand side of the equal sign will then cancel and you will have

$$X = 300 + 30$$

or

$$X = 330$$

and the equation is solved.

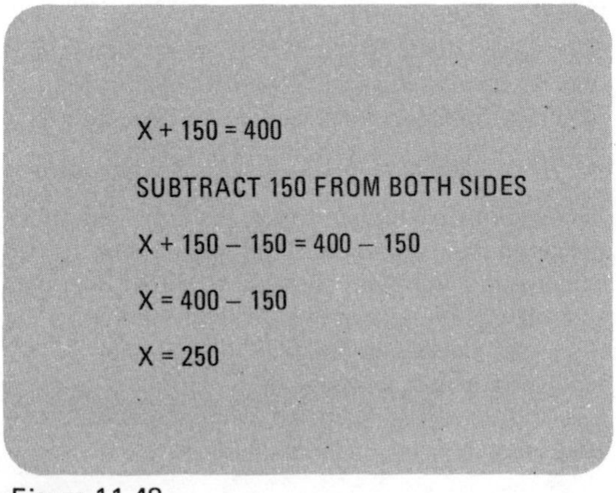

$$X + 150 = 400$$

SUBTRACT 150 FROM BOTH SIDES

$$X + 150 - 150 = 400 - 150$$

$$X = 400 - 150$$

$$X = 250$$

Example — Suppose you had an equation like $X + 150 = 400$, as shown in Figure 11.48. How would you solve this? Subtracting 150 from both sides will put X all alone on the left-hand side of the equal sign, and 250 on the right-hand side.

Figure 11.48

RULE: YOU CAN **MOVE** A QUANTITY FROM ONE SIDE OF THE EQUAL SIGN TO THE OTHER; **IF YOU CHANGE ITS SIGN** (THIS IS CALLED **TRANSPOSING**).

EXAMPLE: YOU HAVE +50 +X −30 = 300 AND NEED TO FIND X

TRANSPOSE +50 AND −30 − (CHANGE SIGNS AND MOVE TO RIGHT OF =)

X = 300 + 30 − 50

(300 + 30 − 50 = 280)

X = 280

UNKNOWN ⤒   ⤒ KNOWN

Figure 11.49

Rule 2 — Another rule you can use (that arises directly from the previous rule) states that you can move any quantity from one side of the equal sign to the other, *if you change its sign*. This procedure is called *transposing*. Transposing is really just the same as adding or subtracting the same quantity from both sides of the equation. However, transposing saves you some steps. For example, as shown in Figure 11.49, if you have

$$+50 + X - 30 = 300$$

and you want to find X, you can transpose the −30 and the +50 to the right-hand side. Remember when you do, to *change the signs of the terms you transpose*. This will give you

$$X = 300 + 30 - 50$$

or

$$X = 280$$

RULE: YOU CAN **MULTIPLY OR DIVIDE** BOTH SIDES OF THE EQUAL SIGN BY THE SAME QUANTITY.

EXAMPLE: YOU HAVE: 300X = 30 AND NEED X DIVIDE BOTH SIDES BY 300

$$\frac{300X}{300} = \frac{30}{300}$$

SINCE 300/300 =1

X = 30/300

UNKNOWN ⟋   ⟍ KNOWN

Figure 11.50

Rule 3 — Following the basic principle of "whatever you do to one side of an equation you must also do to the other" leads to an additional rule you can use in solving equations. As stated in Figure 11.50, you can multiply or divide the quantities on both sides of the equal sign by the same quantity. This is the rule you use when your unknown in an equation is multiplied by or divided by some number and you want to get the unknown all by itself.

For example, if you have the equation 300X = 30, how would get X all by itself? If you divide this equation (that means all of the terms in it) by 300, you would have (Figure 11.50):

$$\frac{300X}{300} = \frac{30}{300}$$

Now 300/330 equals 1, and 1 times X equals X, so you would then have

$$X = \frac{30}{300}$$

or

$$X = 0.1$$

The trick in cases such as this is to divide both sides of the equation by whatever is multiplying X. This will give you X all by itself.

$$\frac{X}{6000} = .003$$

MULTIPLY BOTH SIDES BY 6000

$$6000 \; \frac{X}{6000} = .003 \,(6000)$$

$$X = 18$$

Figure 11.51

Example — Suppose the unknown is divided by some quantity as in Figure 11.51. How would you get X all by itself in this situation? In cases like this you can multiply both sides of the equation by the quantity by which X is divided. In this case you would get X = 0.003 times 6000 or X = 18.

RULE: YOU CAN CHANGE THE SIGN OF EVERY TERM IN AN EQUATION, AS LONG AS YOU CHANGE THEM ALL AT THE SAME TIME.

EXAMPLE: YOU HAVE −X = 30 + 400 − 10 AND NEED X. CHANGE **ALL** SIGNS

$$X = -30 - 400 + 10$$
$$X = -420$$

Figure 11.52

Rule 4 — Growing out of the last rule is an additional one. In any equation you can change the signs of all of the terms if you need to. (This is actually equivalent to multiplying the entire equation by −1). The thing to remember when doing this is that *all* the terms' signs must be changed. As an example, if you have −X = 30 + 400 − 10, as shown in Figure 11.52, you could change all the signs to give you X (actually +X) all by itself.

RULE: IN ANY EQUATION YOU CAN COMBINE LIKE TERMS.

EXAMPLE: YOU HAVE 30X + 60X = 90 AND NEED X

COMBINE 30X
$$\underline{+60X}$$

TO GET   90X

$90X = 90$

DIVIDE BY 90

$X = 1$

Figure 11.53

Rule 5 — One additional and useful rule concerns equations where the unknown appears more than once, as shown in Figure 11.53. If you have a situation such as

$$30X + 60X = 90$$

you can *combine the terms that each contain the unknown*, the "like" terms. Then $30X + 60X$ equals $90X$, and your equation reads $90X = 90$.

Now you could use an earlier rule and divide both sides of this equation by 90 to get $X = 1$.

With these rules in mind, you can get back to the solution of basic loop equations.

Solution for the Current, I — How do you proceed to solve the equation underlined at the top of Figure 11.54 for the current? One thing you can do is to move the $50 - 100$ to the right side of the equation. Remember this process is called *transposing* and when you do this you must change the sign of *each* quantity.

Algebraically add -50 and $+100$ to get $+50$. Then, since you know that the values of R_1 and R_2 are each 100 ohms, you can substitute that into the equation.

You can combine like terms: $-100\,I$ combined with $-100\,I$ gives you a $-200\,I$.

Next you can multiply both sides of the equation by -1, remembering that when two minus numbers are multiplied, they yield a plus number. This reduces the equation further to: $200\,I = -50$.

You want I, your unknown, all by itself. Since it is multiplied by 200, you can divide both sides of the equation by 200. When you do, you will find that the current, I, equals -0.25 amp. Notice that this answer is *negative*, which as expected tells you that the *current is actually flowing in the opposite direction*.

$$50 - 100 - IR_1 - IR_2 = 0$$
$$-IR_1 - IR_2 = -50 + 100$$
$$-IR_1 - IR_2 = 50$$
$$-100\,I - 100\,I = 50$$
$$-200\,I = 50$$
$$(-1)(-200\,I) = (50)(-1)$$
$$200\,I = -50$$
$$200\,I/200 = -50/200$$
$$I = -.25\ A$$

Figure 11.54

Kirchhoff's Laws in More Complex Circuts — At this point you have seen how to use Kirchhoff's laws and some basic tools to solve a simple type of circuit problem, one which you really know how to solve with simpler techniques. As mentioned, however, this problem was presented as an introduction to Kirchhoff's laws. Their real power comes into play when you are faced with the analysis of a more complex circuit, which may contain several voltage sources in a configuration like that shown in Figure 11.55.

In circuits such as this where more than one branch contains a voltage source, you cannot use Ohm's law and basic circuit rules all by themselves, since they are designed to take into accout only one source at a time. Here is where you must use more powerful tools such as Kirchhoff's laws. All the procedures involved in using these tools will be outlined. To keep things simple, electron current examples will be used at first. Keep in mind, however, that all of these procedures will be applicable, whether you are considering conventional or electron current. Examples using each will be worked through for you.

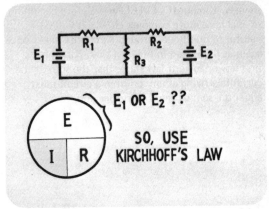

Figure 11.55

Electron and Conventional Current — Remember that all of the effects of electron and conventional current are the same as shown in Figure 11.56. They just flow in opposite directions.

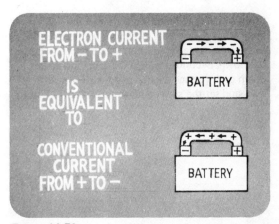

Figure 11.56

Review: Kirchhoff's First Law — Before beginning to analyze a more complex circuit, it will be helpful to briefly go through Kirchhoff's laws once again as illustrated in Figure 11.57. Kirchhoff's first law or current law states that the sum of the currents arriving at any point in a circuit must equal the sum of the currents leaving.

KIRCHHOFF'S FIRST LAW OR KIRCHHOFF'S CURRENT LAW

THE SUM OF THE CURRENTS ARRIVING AT ANY POINT IN A CIRCUIT MUST EQUAL THE SUM OF THE CURRENTS LEAVING THAT POINT.

Figure 11.57

Review: Kirchhoff's Second Law — Kirchhoff's second law or voltage law states that the algebraic sum of the voltages around any closed loop must be zero (Figure 11.58). Again, when writing loop equations, the following steps must be kept in mind.

KIRCHHOFF'S VOLTAGE LAW

THE TOTAL VOLTAGE APPLIED TO ANY "CLOSED CIRCUIT PATH" IS ALWAYS EQUAL TO THE SUM OF THE VOLTAGE DROPS IN THAT PATH.

Figure 11.58

Steps in Using Kirchhoff's Voltage Law — First, label all of the current directions in the circuit, *assuming* a current direction if one happens to be unknown. Second, traverse the loop and write down each voltage encountered with the correct sign, and set the sum equal to zero (Figure 11.59).

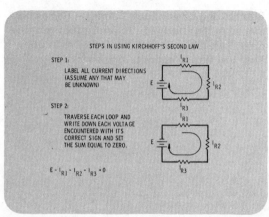

Figure 11.59

Rules for Voltage Signs — If you traverse a voltage source in the *same direction* that it pushes on the type of current you are using, write down its voltage with a *plus* sign (Figure 11.60).

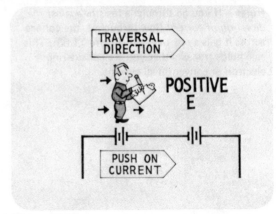

Figure 11.60

Rules — If you traverse a voltage source *opposite* to the direction it pushes the type of current you are using, write down its voltage with a *negative sign* (Figure 11.61).

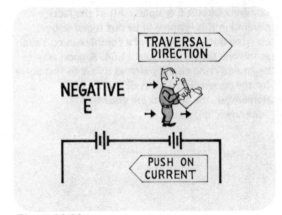

Figure 11.61

Rules — If you traverse a resistor *in the direction of current* flow through it, the voltage across it gets a *negative* sign in the loop equation (Figure 11.62).

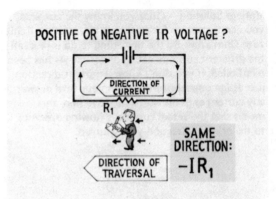

Figure 11.62

Rules — If you go through a resistor *against the direction of current* flow through it, the voltage across it gets a *positive* sign (Figure 11.63). This rule holds true whether you are considering electron or conventional current.

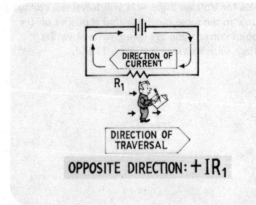

Figure 11.63

Complex Circuit Example — All of the facts covered in this lesson will be put together by carrying out the analysis of a complex circuit such as the one shown in Figure 11.64. A good way to begin analyzing circuits such as this is to first solve for the *current* in all parts of the circuit. Remember, the currents are your first "unknown" quantities.

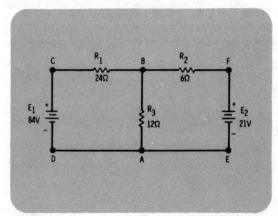

Figure 11.64

Voltage Solution — Once you know the currents, you can then solve for all the voltages in the circuit with Ohm's law. So the first thing to do is label all the different currents with a direction. As has been mentioned, if you don't know a current direction, just assign one arbitrarily. If, in your final answer, any current ends up with a negative sign, this means that the actual current is flowing opposite to the original direction you assumed.

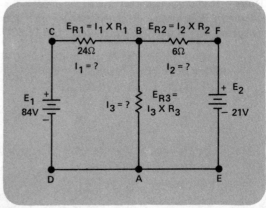

Figure 11.65

Circuit Solution: Step 1 — Label all the currents with their directions. In the same problem the current flowing in the left-hand branch will be labeled I_1, the current in the right-hand branch I_2, and the current through the central resistor I_3, as shown in Figure 11.66.

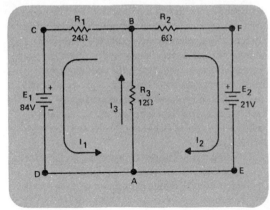

Figure 11.66

Kirchhoff's Current Law Equation — Your "circuit sense" should begin "tingling" a little as soon as you set up the problem. Right away you can see in Figure 11.67 that at junction point A that I_1 and I_2 enter the junction and I_3 leaves. You immediately see that according to Kirchhoff's current law: $I_3 = I_1 + I_2$.

You have three unknown quantities to find: I_1, I_2, and I_3. An important mathematical rule states that you need one equation for each unknown quantity you are trying to find. Since you already have one equation, you now need to find two more. Then these three equations can be solved altogether (simultaneously) for the unknown quantities I_1, I_2, and I_3. The methods you will need will be outlined for you; however, you need two more equations.

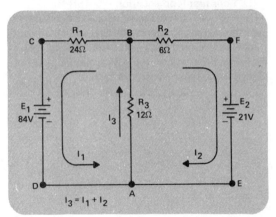

Figure 11.67

Kirchhoff's Voltage Law Equation 1 — The two other equations you need are where Kirchhoff's voltage law applies. Now you simply pick two loops and write a loop equation for each. In the example circuit (Figure 11.68), you can write equations for the left-hand and right-hand loops.

If you go around the left-hand loop counterclockwise starting at point A, and write down all the IR voltages and the source voltage, you will get:

Minus I_3R_3 (since you go through R_3 in the direction of the current I_3)

Minus I_1R_1 (since you go through R_1 in the direction of I_1)

Plus E_1 (since you go through E_1 in the direction that it is pushing current)

Your first loop equation is: $-I_3R_3 - I_1R_1 + E_1 = 0$. On the schematic all the resistances and the source voltages are labeled so you can put them into this equation where they belong. You know that $R_3 = 12\ \Omega$, $R_1 = 24\ \Omega$, and $E_1 = 84$ V. Substituting these values gives you:

$$-12\,I_3 - 24\,I_1 + 84 = 0.$$

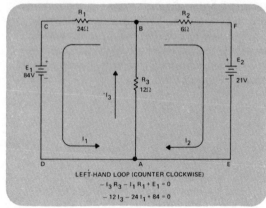

LEFT-HAND LOOP (COUNTER CLOCKWISE)
$-I_3R_3 - I_1R_1 + E_1 = 0$
$-12\,I_3 - 24\,I_1 + 84 = 0$

Figure 11.68

Right-Hand Loop Equation — Follow the same procedure in the right-hand loop. Going around the circuit counterclockwise starting at point A, you get (Figure 11.69):

$-E_2$ (since E_2 is pushing current against the direction in which you go through it).

$+I_2R_2$ (since you go through R_2 against the current direction).

$+I_3R_3$ (since you go through R_3 against the current direction).

So for this loop, you have $-E_2 + I_2R_2 + I_3R_3 = 0$. Again, putting in all of the known values labeled on the circuit diagram you have:

$$-21 + 6\,I_2 + 12\,I_3 = 0.$$

This is your second loop equation.

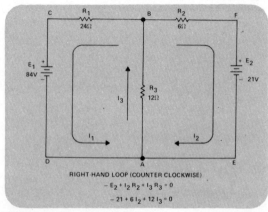

RIGHT-HAND LOOP (COUNTER CLOCKWISE)
$-E_2 + I_2R_2 + I_3R_3 = 0$
$-21 + 6\,I_2 + 12\,I_3 = 0$

Figure 11.69

Three Equations to Solve — Figure 11.70 shows the current equation and two loop equations together. Now you want to solve for I_1, I_2, and I_3. Your aim in using three equations like this with three unknown quantities is to solve for one unknown quantity at a time.

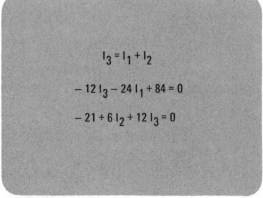

$$I_3 = I_1 + I_2$$

$$-12 I_3 - 24 I_1 + 84 = 0$$

$$-21 + 6 I_2 + 12 I_3 = 0$$

Figure 11.70

METHODS USED IN SOLVING SIMULTANEOUS EQUATIONS WHICH ALLOW YOU TO ELIMINATE UNKNOWNS FROM ONE OR MORE OF THESE EQUATIONS:

1) ADDITION METHOD
2) SUBSTITUTION METHOD

Figure 11.71

ADDITION METHOD

ANY TWO EQUATIONS CAN BE ADDED TOGETHER

Figure 11.72

Solving Simultaneous Equations — At this point you encounter the need for another technique for analyzing complex circuits. You now have three equations to solve for three unknown quantities, in this case I_1, and I_2, and I_3. Your goal is the same as before. You want to get each unknown all by itself alone on one side of the equals sign, and only known quantities on the other side. Since each of the three equations contains more than one unknown at this point, the procedures that were described earlier won't enable you to solve these equations. You need some methods that allow you to eliminate some of the unknowns from these equations (Figure 11.71). If you could do this, you could manipulate your three equations so that each equation contained only one unknown quantity.

Addition Method — What methods can you use to eliminate some of the unknowns from these equations? The first is simple addition; any two equations can be added together (Figure 11.72).

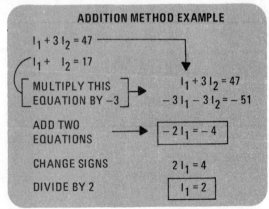

Figure 11.73

Addition Method Example — To see how this works, consider the example shown in Figure 11.73. Suppose you have two equations, each of which contains the two unknowns I_1 and I_2, and you need to solve for I_1:

$$I_1 + 3I_2 = 47$$
$$I_1 + I_2 = 17$$

Notice that each of these equations contains *both* unknowns I_1 and I_2. You can't solve for I_1 until you have an equation whose only unknown quantity is I_1.

Focus your attention on these two equations and consider this point. Any of the procedures outlined in the rules you were given earlier can be performed on either one of the equations. You now know that these two equations can be added together. The secret to getting rid of I_2 in these two equations is to make the I_2 terms equal in value but opposite in sign. For example, if you take the second equation in Figure 11.73 and multiply both sides by -3, it will become:

$$(I_1 + I_2) \times (-3) = (17) \times (-3)$$
$$-3I_1 - 3I_2 = -51.$$

Add this equation and your first equation together. See what happens? The $+3I_2$ in the upper equation and $-3I_2$ in the lower equation cancel when you add them, and the resulting equation is now

$$-2I_1 = -4$$

You can now solve this easily for I_1. If you change the signs of both terms in this equation and then divide by 2 you get $I_1 = 2$ and the equation is solved for I_1.

Steps in Using Addition Method — One method you can use in solving simultaneous equations is to eliminate unknowns using addition. To do this, follow these steps, as reviewed in Figure 11.74:

1. Decide which unknown you want to eliminate, say X.
2. Make the terms containing this unknown equal in value but opposite in sign.
3. Add the equations together: X is eliminated.
4. Solve for the other unknown.

STEPS IN USING ADDITION METHOD

DECIDE WHICH UNKNOWN TO ELIMINATE

MAKE THE TERMS CONTAINING THIS UNKNOWN EQUAL IN VALUE BUT OPPOSITE IN SIGN

ADD THE EQUATIONS TOGETHER; ONE UNKNOWN IS ELIMINATED

SOLVE FOR REMAINING UNKNOWN QUANTITY

Figure 11.74

SUBSTITUTION METHOD

SUBSTITUTE ONE EQUATION INTO THE OTHER EQUATION TO ELIMINATE AN UNKNOWN.

Figure 11.75

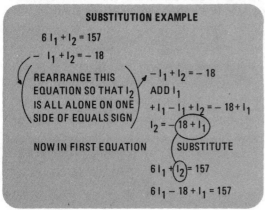

SUBSTITUTION EXAMPLE

$6I_1 + I_2 = 157$

$-I_1 + I_2 = -18$

REARRANGE THIS EQUATION SO THAT I_2 IS ALL ALONE ON ONE SIDE OF EQUALS SIGN

$-I_1 + I_2 = -18$

ADD I_1

$+I_1 - I_1 + I_2 = -18 + I_1$

$I_2 = -(18 + I_1)$

NOW IN FIRST EQUATION

SUBSTITUTE

$6I_1 + (I_2) = 157$

$6I_1 - 18 + I_1 = 157$

Figure 11.76

Substitution Method — Another method exists that will be helpful in solving several equations together. This is called the *substitution* method. In this situation you take one equation and actually substitute it into the other to eliminate an unknown quantity (Figure 11.75).

Substitution Example — Consider the two equations shown in Figure 11.76:

$$6I_1 + I_2 = 157$$
$$-I_1 + I_2 = -18$$

and you need to calculate I_1. Here is another way to proceed that will allow you to eliminate one unknown in one of these equations. First, rearrange one equation, the second one, so that I_2 is all alone on one side of the equals sign. In this case, if you add I_1 to both sides, you will get

$$+I_1 - I_1 + I_2 = -18 + I_1$$

or

$$I_2 = -18 + I_1.$$

You know that I_2 equals, or is exactly the same as, the expression, $-18 + I_1$. You can go back to your first equation and substitute or replace I_2 by the expression $-18 + I_1$. Therefore your first equation

$$6I_1 + I_2 = 157$$

becomes

$$6I_1 - 18 + I_1 = 157.$$

This equation now has only one unknown, I_1.

SUBSTITUTION EXAMPLE (CONTINUED)

$$6 I_1 - 18 + I_1 = 157$$

COMBINE TERMS

$$7 I_1 - 18 = 157$$

ADD 18 TO BOTH SIDES

$$7 I_1 - 18 + 18 = 157 + 18$$

$$7 I_1 = 175$$

DIVIDE BY 7

$$\frac{7 I_1}{7} = \frac{175}{7} ; \quad I_1 = 25$$

Figure 11.77

Substitution Example — To finish solving for I_1 combine terms

$$7 I_1 - 18 = 157$$

Then add 18 to both sides

$$+18 + 7 I_1 - 18 = 157 + 18$$

which gives you $7 I_1 = 175$. Next, divide both sides by 7 to get

$$\frac{7 I_1}{7} = \frac{175}{7}$$

or

$$I_1 = 25$$

These two methods, addition and substitution may be used together and repetitively to solve for any number of unknown quantities. Remember though, when using Kirchhoff's laws that you must have one equation for each unknown. Now you can return to your previous circuit problem.

Circuit Example — Your first goal is to get I_1 on one side of an equals sign, and only things you know on the other side. Examine your equations again carefully. Notice that the second and third equations contain I_1, I_2, and I_3. You know from your first equation that $I_3 = I_1 + I_2$, so now you can substitute the expression $I_1 + I_2$ wherever you see I_3 in your last two equations. Then your last two equations will only contain the two unknowns I_1 and I_2. This will reduce the number of unknowns in the last two equations and bring you a little closer to your goal.

$$I_3 = I_1 + I_2$$

$$-12 I_3 - 24 I_1 + 84 = 0$$

$$-21 + 6 I_2 + 12 I_3 = 0$$

Figure 11.78

- Substitution in First Loop Equation
- Substitution in Second Loop Equation
- Two Loop Equations

Substitution in First Loop Equation — The first loop equation is shown in Figure 11.79. When you substitute $I_1 + I_2$ for I_3, as shown, you get

$$-12(I_1 + I_2) - 24 I_1 + 84 = 0.$$

Now notice that when $I_1 + I_2$ are each multiplied by -12, you have $-12 I_1 - 12 I_2 - 24 I_1 + 84 = 0$. You can algebraically add all the terms in this equation that contain the same unknown. (Again, this is called *combining like terms*.) You have a $-12 I_1$ and $-24 I_1$ on the same side of the equation, so add them to get $-36 I_1 - 12 I_2 + 84 = 0$.

FIRST LOOP EQUATION

$$-12\, I_3 - 24 I_1 + 84 = 0$$
$$I_3 = I_1 + I_2$$
$$-12(I_1 + I_2) - 24 I_1 + 84 = 0$$
$$-12 I_1 - 12 I_2 - 24 I_1 + 84 = 0$$
$$-36 I_1 - 12 I_2 + 84 = 0$$

Figure 11.79

Substitution in Second Loop Equation — Examine your second loop equation, which is underlined in Figure 11.80. You can follow a similar procedure to eliminate one unknown in this equation. First, substitute $I_1 + I_2$ for I_3 as shown, and this gives you

$$-21 + 6 I_2 + 12(I_1 + I_2) = 0.$$

Multiply the 12 times each of the terms in parentheses to get

$$-21 + 6 I_2 + 12 I_1 + 12 I_2 = 0.$$

Now combine like terms, and you get

$$-21 + 18 I_2 + 12 I_1 = 0.$$

SECOND LOOP EQUATION

$$-21 + 6 I_2 + 12 I_3 = 0$$
$$I_3 = I_1 + I_2$$
$$-21 + 6 I_2 + 12 (I_1 + I_2) = 0$$
$$-21 + 6 I_2 + 12 I_1 + 12 I_2 = 0$$
$$-21 + 18 I_2 + 12 I_1 = 0$$

Figure 11.80

Two Loop Equations — Now that both loop equations have been simplified, you want to manipulate them to eliminate one of the unknown loop currents so that you can solve for the other. You can solve for either unknown first. The analysis will be aimed at solving for I_2, and then I_1.

In Figure 11.81 both simplified loop equations are written one right under the other, and the terms have been moved around a little so that all of the terms of the same kind (that is, containing the same unknown) are aligned one under another. A rule of mathematics says that any two equations can be added together to help eliminate unknowns. The idea is to do this in such

2 LOOP EQUATIONS

$$-21 + 18 I_2 + 12 I_1 = 0$$
$$+84 - 12 I_2 - 36 I_1 = 0$$

Figure 11.81

a way that the sum of the two equations contains only one loop current. So how do you do this?

Manipulation of First Equation — Focus your attention on the first equation. If this equation is multiplied by 3 you get $-63 + 54\,I_2 + 36\,I_1 = 0$, as shown in Figure 11.82. Now, with this change, rewrite the equation above the second equation.

$$-21 + 18I_2 + 12I_1 = 0$$

$$\underline{\qquad\qquad\qquad \times 3 \qquad \times 3 \qquad}$$

$$-63 + 54I_2 + 36I_1 = 0$$

Figure 11.82

Solution by Addition — If this equation is added to the second equation, term by term, you will have $21 + 42\,I_2 = 0$. Notice that the two I_1 terms have canceled out. You now have an equation with only one unknown current. Transpose and divide as shown in Figure 11.83 to get $I_2 = -21/42$ which equals -0.5 amp. Notice that this answer came out negative; that means that the wrong direction was originally assumed for I_2. So you now know that I_2 (electron current) is actually flowing in the opposite direction in the right-hand loop.

$$-63 + 54I_2 + 36I_1 = 0$$

$$+84 - 12I_2 - 36I_1 = 0$$

$$\underline{\qquad\qquad\qquad\qquad\qquad\qquad}$$

$$21 + 42I_2 \qquad\quad = 0$$

$$42I_2 = -21$$

$$I_2 = -21/42 = -.5\text{ A}$$

Figure 11.83

- Solution for I_1
- Solution for I_3
- Solution for Voltages

Solution for I_1 — Now that you know I_2, you can substitute its value in either of the two loop equations to solve for I_1. For example, if you take the second loop equation and replace I_2 with -0.5 amp, as shown in Figure 11.84, you now get $84 + 6 - 36 I_1 = 0$, or $90 - 36 I_1 = 0$. Transposing and dividing by -36, you have $I_1 = -90/-36$ or 2.5 amps. Notice that this current came out positive which means the initially assumed direction of flow in this loop was correct.

$$84 - 12 I_2 - 36 I_1 = 0$$
$$84 - 12(-.5) - 36 I_1 = 0$$
$$84 + 6 - 36 I_1 = 0$$
$$90 - 36 I_1 = 0$$
$$-36 I_1 = -90$$
$$I_1 = -90/-36 = 2.5 \text{ A}$$

Figure 11.84

Solution for I_3 — Now that you know I_1 and I_2, you can go back to the current equation and solve for I_3 as shown in Figure 11.85. You simply substitute your known values of I_1 and I_2 in the equation, and you get $I_3 = 2.5 - 0.5$ or 2 amps, which is positive, so it's in the correct direction. Now that all of the currents have been calculated, you can use Ohm's law to find the voltage across each resistor.

$$I_3 = I_1 + I_2$$
$$I_3 = 2.5 - .5$$
$$I_3 = 2 \text{ AMPS}$$

Figure 11.85

Solution for Voltages — Simply multiply the appropriate currents by the appropriate resistances as shown in Figure 11.86.

$$E_1 = 60 \text{ volts}$$
$$E_2 = 3 \text{ volts}$$
$$E_3 = 24 \text{ volts}$$

and the problem is completely solved.

$$E_1 = I_1 \times R_1 = 2.5 \times 24 = 60 \text{ VOLTS}$$
$$E_2 = I_2 \times R_2 = .5 \times 6 = 3 \text{ VOLTS}$$
$$E_3 = I_3 \times R_3 = 2.0 \times 12 = 24 \text{ VOLTS}$$

Figure 11.86

Complete Solution — Figure 11.87 shows the circuit schematic with all of the answers labeled. Notice also that the voltage polarities have been labeled. Recall that if you wanted to solve this problem using conventional currents, you would follow exactly the same steps and you would get the same answers, except all the current directions would be opposite.

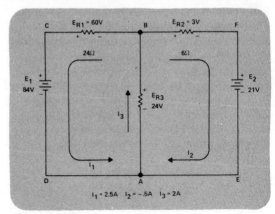

Figure 11.87

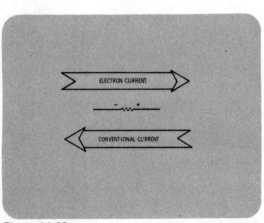

Figure 11.88

How to Determine Voltage Polarities — In simple circuits your rule for determining the polarity of a resistor's voltage drop simply states that the side of a resistor closest to the positive terminal of the battery or voltage source is positive and the side closest to the negative terminal is negative. This is understood when you only have one voltage source or battery, but what do you do when you are working with more complex circuits which contain more than one source?

You use this simple rule (Figure 11.88):

1. First determine the correct direction for the current (either electron or conventional current)
2. Electron current flows through resistors from minus (−) to plus (+)
3. Conventional current flows through resistors from plus (+) to minus (−).

Then you can label your voltage drops accordingly.

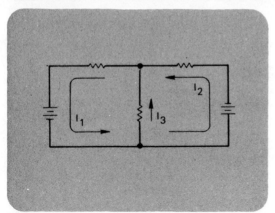

Figure 11.89

Example — Suppose you have determined the correct directions for all the electron currents in a circuit and they are as shown in Figure 11.89. What then are the correct polarities for the voltage drops? Following the above rule you should find that the polarities are as shown in Figure 11.90.

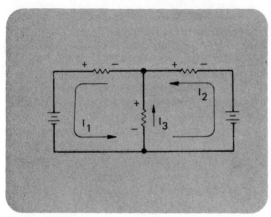

Figure 11.90

This has been a busy lesson. The next lesson will continue to give you some practice using some of what has been covered here; and more importantly, you will be introduced to some shortcuts to make using these laws easier. In the meanwhile, it is suggested that you carefully go through the worked examples set out on the next few pages. These will help you in understanding the techniques involved in using Kirchhoff's laws.

LESSON 11. INTRODUCTION TO KIRCHHOFF'S LAWS

● **Worked Through Examples**

1. Write a node equation for the diagram shown below, substitute the appropriate currents and solve the equation for I_5. Also indicate the direction of I_5.

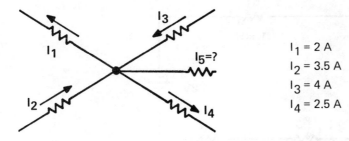

$I_1 = 2$ A

$I_2 = 3.5$ A

$I_3 = 4$ A

$I_4 = 2.5$ A

From Kirchhoff's current law you know that whatever current arrives at a junction must equal the current that leaves the junction. Write down the currents entering the junction on one side of an equals sign, and then write down the currents that leave the junction on the other side of the equals sign.

$$\text{Leaving} = \text{Entering}$$
$$I_1 + I_4 = I_2 + I_3$$

On which side of the equals sign does I_5 belong? If you substitute the values for I_1 through I_4 in the equation, you will see.

$$\text{Leaving} = \text{Entering}$$
$$2\text{ A} + 2.5\text{ A} = 3.5\text{ A} + 4\text{ A}$$
$$4.5\text{ A} = 7.5\text{ A}$$

Obviously, 4.5 amps does not equal 7.5 amps, so I_5 must belong with the 4.5 amp leaving the junction.

$$\text{Leaving} = \text{Entering}$$
$$4.5\text{ A} + I_5 = 7.5\text{ A}$$

In order for the currents leaving to equal the currents entering, I_5 must be the right value so that there will be 7.5 amps leaving *and* entering the junction. I_5 should be 3 amps *leaving* the junction. You can prove this by subtracting 4.5 amps from each side of the equation.

$$
\begin{aligned}
4.5\text{ A} + I_5 &= 7.5\text{ A}\\
-4.5\text{ A} \qquad\ \ &\quad -4.5\text{ A}\\
\hline
I_5 &= \ \ 3\text{ A}
\end{aligned}
$$

Thus, I_5 does equal 3 amps and it must leave the junction.

2. Write a loop equation for the circuit shown below using electron current, and write another loop equation using conventional current.

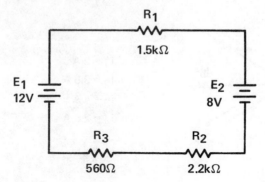

Step One: Assign a current direction. Any direction is fine but more than likely the actual direction of electron current is counterclockwise since E_1 is larger than E_2. Assume that the electron current is flowing in the counterclockwise direction and label it accordingly.

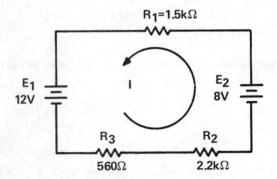

Step Two: Traverse the circuit and write down all the source voltages and IR voltages according to the rules presented in this lesson. If you start at the positive terminal of E_1 and move through the circuit counterclockwise, you should get:

$+E_1$ (since you go through E_1 in the *same* direction it pushes electron current)

$-IR_3$ (since you traverse R_3 *in* the direction of electron current)

$-IR_2$ (since you traverse R_2 *in* the direction of electron current)

$-E_2$ (since you go through E_2 *against* the direction it is pushing electron current)

$-IR_1$ (since you traverse R_1 *in* the direction of electron current).

When you set this equal to zero, the loop equation for this circuit, considering electron current, is:

$$E_1 - IR_3 - IR_2 - E_2 - IR_1 = 0.$$

To write a loop equation for conventional current, traverse the loop again and write down the voltages according to your rules. Assume the same direction for current as before. If you start at the same point (the positive terminal of E_1) and move through the circuit counterclockwise, you should get:

$-E_1$ (since you go through E_1 *against* the direction it pushes conventional current)

$-IR_3$ (since you traverse R_3 *in* the assumed direction for conventional current)

$-IR_2$ (since you traverse R_3 *in* the assumed direction for conventional current)

$+E_2$ (since you go through E_2 in the *same* direction it pushes conventional current)

$-IR_1$ (since you traverse R_1 *in* the assumed direction for conventional current).

The loop equation for this circuit, considering conventional current, is:

$$-E_1 - IR_3 - IR_2 + E_2 - IR_1 = 0.$$

3. Solve each of the equations from the previous example for the current.

Electron Current Equation

$$E_1 - IR_3 - IR_2 - E_2 - IR_1 = 0$$

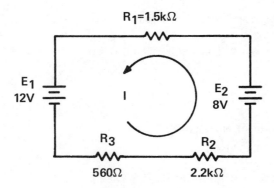

First, substitute the appropriate values from the circuit into the equation.

$$12 - 0.56 \text{ kI} - 2.2 \text{ kI} - 8 - 1.5 \text{ kI} = 0$$

When the two source voltages are added algebraically, they yield 4.

$$4 - 0.56 \text{ kI} - 2.2 \text{ kI} - 1.5 \text{ kI} = 0$$

You can combine the I terms to get:

$$4 - 4.26 \text{ kI} = 0$$

Transpose the 4, remembering to change its sign.

$$-4.26 \text{ kI} = -4$$

Divide both sides of the equation by -4.26 k.

$$\frac{-4.26 \text{ kI}}{-4.26 \text{ k}} = \frac{-4}{-4.26 \text{ k}}$$

$$I = 0.939 \text{ mA or } 939 \text{ } \mu\text{A}$$

Since this answer is positive, the assumed direction for the electron current (counterclockwise) is correct.

Conventional Current Equation

$$-E_1 - IR_3 - IR_2 + E_2 - IR_1 = 0$$

First, substitute the appropriate values from the circuit into the equation.

$$-12 - 0.56 \text{ kI} - 2.2 \text{ kI} + 8 - 1.5 \text{ kI} = 0$$

When the two source voltages are added algebraically, they yield -4. This, as you will see, will make a difference in your answer.

$$-4 - 0.56 \text{ kI} - 2.2 \text{ kI} - 1.5 \text{ kI} = 0$$

Combine the I terms to get:

$$-4 - 4.26 \text{ kI} = 0$$

Transpose the 4, remembering to change its sign.

$$-4.26 \text{ kI} = 4$$

Divide both sides of this equation by -4.26 k.

$$\frac{-4.26 \text{ kI}}{-4.26 \text{ k}} = \frac{4}{-4.26 \text{ k}}$$

$$I = -0.939 \text{ mA or } -939 \text{ } \mu\text{A}$$

Since this answer is *negative* the assumed direction for the conventional current was wrong, and so you know that the conventional current is actually flowing *clockwise*.

You know, if you thought about this answer for a minute, it makes a great deal of sense. The solution to the electron current equation told you that the electron current was flowing *counter-clockwise*. Recall that electron and conventional current have the *same* effect in a circuit; they just flow in *opposite* directions. Thus, you know that conventional current for this circuit must flow in the *clockwise* direction.

4. Write the loop and node equations for the following circuit using electron current. Then solve the equations for the branch currents, including their directions, and use these currents to find the voltage drop across each resistor. Also indicate the polarity of each voltage drop.

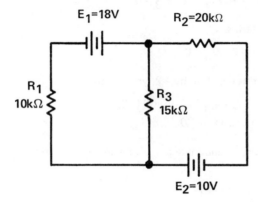

First Step: Assign a direction for each current and label it accordingly.

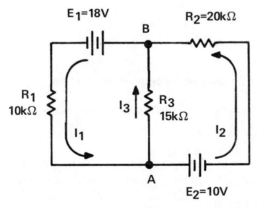

Immediately, you can see from Kirchhoff's current law that at junction point A:

$$I_1 = I_3 + I_2$$

Second Step: Traverse each loop and write down all the voltages you encounter with their correct signs.

Loop 1	$18 - 10 \, kI_1 - 15 \, kI_3 = 0$	(counterclockwise from point B)
Loop 2	$10 - 20 \, kI_2 + 15 \, kI_3 = 0$	(counterclockwise from point A)

Third Step: Simplify the equations. If you substitute $I_3 + I_2$ for I_1 in the first equation, you will then have only two unknowns, and you will have two equations with which to find the two unknowns.

$$18 - 10 \, k \, (I_3 + I_2) - 15 \, kI_3 = 0$$
$$10 - 20 \, k \, I_2 + 15 \, k \, I_3 = 0$$

In the first equation, multiply I_3 and I_2 by $-10k$.

$$18 - 10 \, kI_3 - 10 \, kI_2 - 15 \, kI_3 = 0$$

You can now combine the I_3 terms.

$$18 - 10 \, kI_2 - 25 \, kI_3 = 0.$$

If you multiply both sides of this equation by -2, you can then add it to your equation for loop 2.

$$(-2) \, (18 - 10 \, kI_2 - 25 \, kI_3) = (0) \, (-2)$$
$$-36 + 20 \, kI_2 + 50 \, kI_3 = 0$$

Fourth Step: Add the equations to eliminate one of the unknown currents, thus enabling you to calculate the other current.

$$\begin{array}{rl} -36 + 20 \, kI_2 + 50 \, kI_3 &= 0 \\ 10 - 20 \, kI_2 + 15 \, kI_3 &= 0 \\ \hline -26 \qquad\qquad + 65 \, kI_3 &= 0 \end{array}$$

$$65 \, kI_3 = 26$$

$$I_3 = \frac{26}{65 \, k} = 0.4 \text{ mA} = 400 \, \mu A$$

Since this answer is positive, you know that the assumed direction for I_3 is correct.

Fifth Step: Substitute the value of I_3 in one of the previous loop equations to find I_1 or I_2.

Loop 2	$10 - 20 \, kI_2 + 15 \, kI_3 = 0$
	$10 - 20 \, kI_2 + 15 \, k \, (0.4 \text{ mA}) = 0$

When 15 k is multiplied by 4 mA, the result is 6, which can then be added to the 10.

$$10 - 20 \, kI_2 + 6 = 0$$
$$16 - 20 \, kI_2 = 0$$

Transpose and divide.

$$-20\ kI_2 = -16$$

$$\frac{-20\ kI_2}{-20\ k} = \frac{-16}{-20\ k}$$

$$I_2 = 0.8\ mA = 800\ \mu A$$

This answer is also positive, so the assumed direction for I_2 is correct.

Sixth Step: Substitute I_2 and I_3 in the node current equation to find I_1.

$$I_1 = I_3 + I_2$$

$$I_1 = 0.4\ mA + 0.8\ mA$$

$$I_1 = 1.2\ mA$$

The answer is again positive, so its assumed direction is also correct. If any of the answers for the branch currents were negative, you would know that the assumed direction for the current was wrong and that the actual direction was *opposite* to the assumed direction for that current.

Seventh Step: Use Ohm's law to calculate the voltage drops across the resistors.

$$E_{R1} = I_1 \times R_1$$
$$E_{R1} = 1.2\ mA \times 10\ k\Omega$$
$$E_{R1} = 12\ V$$

$$E_{R2} = I_2 \times R_2$$
$$E_{R2} = 0.8\ mA \times 20\ k\Omega$$
$$E_{R2} = 16\ V$$

$$E_{R3} = I_3 \times R_3$$
$$E_{R3} = 0.4\ mA \times 15\ k\Omega$$
$$E_{R3} = 6\ V$$

Recall the rule for determining the polarity of the voltage across a resistor, which states that electron current flows through a resistor from minus to plus or from the negative side to the positive side. Thus, the voltage drops and their polarities are as shown below.

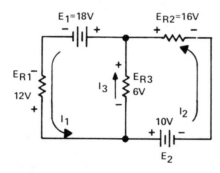

● **Worked Through Examples**

5. Write the loop and node equations for the circuit shown in example 4 using *conventional* current. Then solve the equations for the branch currents, including their directions. Also indicate the polarities of the voltage drops produced by these conventional currents.

First Step: Assign a direction for each current and lable it accordingly.

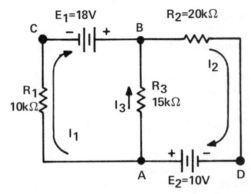

Then, from Kirchhoff's current law, the node current equation for node A is:

$$I_2 = I_3 + I_1$$

Second Step: Traverse each loop and write down all the voltages you encounter with their correct signs.

Loop 1 $18 + 15\,kI_3 - 10\,kI_1 = 0$ (clockwise from point C)

Loop 2 $10 - 15\,kI_3 - 20\,kI_2 = 0$ (clockwise from point D)

Third Step: Simplify the equations. If you substitute $I_3 + I_1$ for I_2 in the second equation, you will have two equations with two unknowns. You can then easily solve the equations for the unknown currents.

$$10 - 15\,kI_3 - 20\,kI_2 = 0$$
$$10 - 15\,kI_3 - 20\,k\,(I_3 + I_1) = 0$$

Multiply I_3 and I_1 by $-20\,k$.

$$10 - 15\,kI_3 - 20\,kI_3 - 20\,kI_1 = 0$$

You can combine the I_3 terms.

$$10 - 35\,kI_3 - 20\,kI_1 = 0$$

If you divide both sides of this equation by -2, you can add it to the equation for loop 1.

$$(10 - 35\,kI_3 - 20\,kI_1) \div (-2) = (0) \div (-2)$$
$$-5 + 17.5\,kI_3 + 10\,kI_1 = 0$$

Fourth Step: Add the equations to eliminate one of the unknown currents, thus enabling you to find the other current.

Loop 1 $\quad 18 + 15\,kI_3 - 10\,kI_1 = 0$

Loop 2 $\quad -5 + 17.5\,kI_3 + 10\,kI_1 = 0$

$$\overline{13 + 32.5\,kI_3 = 0}$$

$$32.5\,kI_3 = -13$$

$$I_3 = \frac{-13}{32.5\,k}$$

$$I_3 = -0.4\ mA = -400\ \mu A$$

Since this answer is negative, you know that the assumed direction for I_3 is wrong and that the conventional current I_3 actually flows down through R_3.

Fifth Step: Substitute the value of I_3 in one of the previous loop equations to find I_1 or I_2.

Loop 1 $\quad 18 + 15\,kI_3 - 10\,kI_1 = 0$

$ 18 + 15\,k\,(-0.4\ mA) - 10\,kI_1 = 0$

When 15 k is multiplied by −0.4 mA, the result is −6, which can then be added algebraically to the 18.

$$18 - 6 - 10\,kI_1 = 0$$

$$12 - 10\,kI_1 = 0$$

Transpose and divide.

$$-10\,kI_1 = -12$$

$$\frac{-10\,kI_1}{-10\,k} = \frac{-12}{-10\,k}$$

$$I_1 = 1.2\ mA$$

This answer is positive, so you know that the assumed direction for I_1 is correct.

Sixth Step: Substitute I_1 and I_3 in the node current equation to find I_2;

$$I_2 = I_3 + I_1$$

$$I_2 = -0.4\ mA + 1.2\ mA$$

$$I_2 = 0.8\ mA \text{ or } 800\ \mu A$$

The answer is positive so the assumed direction for I_2 is correct.

Seventh Step: Use Ohm's law to find the voltage drops across the resistors. Since the answers for the currents have the same numerical value as in the previous example, the voltage drops will be the same as they were before, or:

$$E_{R1} = 12 \text{ V}$$
$$E_{R2} = 16 \text{ V}$$
$$E_{R3} = 6 \text{ V}$$

In determining the correct polarities of these voltage drops, remember two things:

1. Conventional current flows through resistors from plus to minus.

2. I_3 is actually flowing down through R_3.

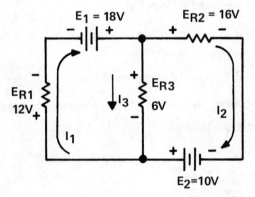

LESSON 11. INTRODUCTION TO KIRCHHOFF'S LAWS

● **Practice Problems**

Depending upon the approach you use in solving these problems and how
you round off intermediate results, your answers may vary slightly from those
given here. However, any differences you encounter should only occur in the
third significant digit of your answer. If the first two significant digits of your
answers do not agree with those given here, recheck your calculations. Fold
over the page to check your answers.

Fold Over

1. Write the node equations for the following diagrams.

a.

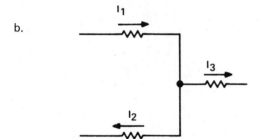

b.

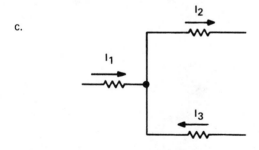

c.

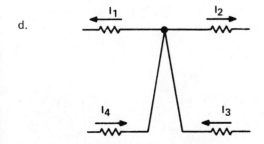

d.

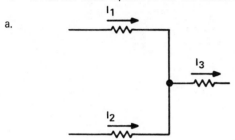

• **Practice Problems**

Answers

1.a. $I_1 + I_2 = I_3$

1.b. $I_1 = I_2 + I_3$

1.c. $I_1 + I_3 = I_2$

1.d. $I_4 + I_3 = I_1 + I_2$

e.

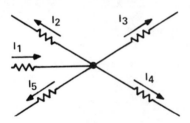

2. Write the loop equations for the following diagrams.

a.

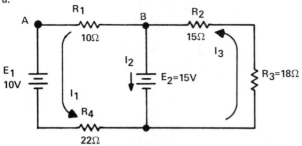

b.

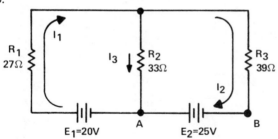

c.

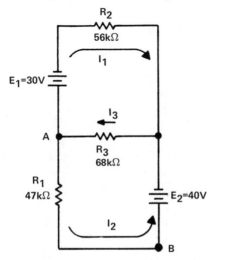

Answers

1.e. $I_1 = I_2 + I_3 + I_4 + I_5$

2.a. Loop 1 — Start at point A and trace the loop ccw.
$$10 - 22I_1 - 15 - 10I_1 = 0$$
Loop 2 — Start at point B and trace the loop ccw.
$$15 - 18I_3 - 15I_3 = 0$$

2.b. Loop 1 — Start at point A and trace the loop cw.
$$20 - 27I_1 - 33I_3 = 0$$
Loop 2 — Start at point B and trace the loop cw.
$$25 + 33I_3 - 39I_2 = 0$$

2.c. Loop 1 — Start at point A and trace the loop cw.
$$30 - 56 kI_1 - 68 kI_3$$
Loop 2 — Start at point B and trace the loop ccw.
$$40 - 68 kI_3 - 47 kI_2$$

d.

Fold Over

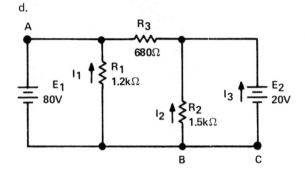

e.

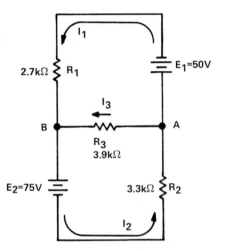

3.   Solve the following circuits for all currents and voltage drops. Indicate the polarity of the voltage drops and the direction of the currents.

a.

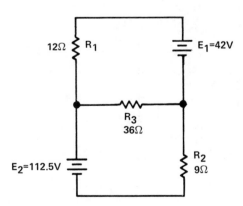

Answers

2.d. Loop 1 — Start at point A and trace the loop ccw.
$$80 - 1.2\,k\,I_1 = 0$$
Loop 2 — Start at point B and trace the loop ccw.
$$-1.5\,k\,I_2 - .68\,k\,(I_2 + I_3) + 1.2\,k\,I_1 = 0$$
Loop 3 — Start at point C and trace the loop ccw.
$$-20 + 1.5\,k\,I_2 = 0$$

2.e. Loop 1 — Start at point A and trace the loop ccw.
$$50 - 2.7\,k\,I_1 + 3.9\,k\,I_3 = 0$$
Loop 2 — Start at point B and trace the loop ccw.
$$75 - 3.3\,k\,I_2 - 3.9\,k\,I_3 = 0$$

3.a.

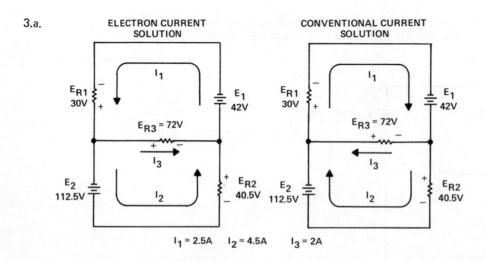

ELECTRON CURRENT SOLUTION   CONVENTIONAL CURRENT SOLUTION

$I_1 = 2.5A$ $I_2 = 4.5A$ $I_3 = 2A$

b.

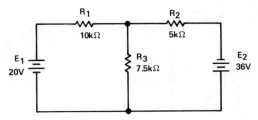

c.

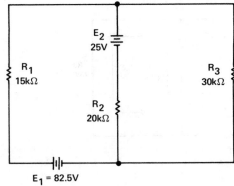

d.

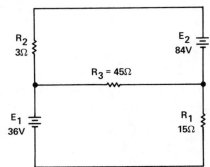

Answers

3.b.

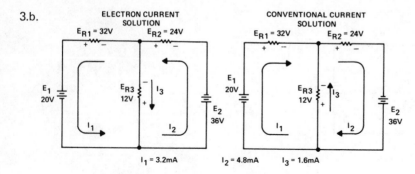

$I_1 = 3.2mA$ $I_2 = 4.8mA$ $I_3 = 1.6mA$

3.c.

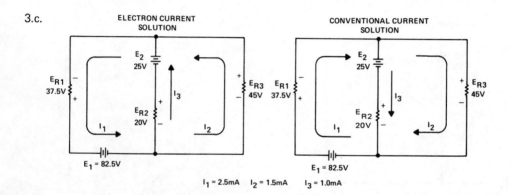

$I_1 = 2.5mA$ $I_2 = 1.5mA$ $I_3 = 1.0mA$

3.d.

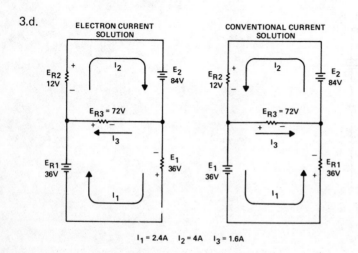

$I_1 = 2.4A$ $I_2 = 4A$ $I_3 = 1.6A$

Fold Over

e.

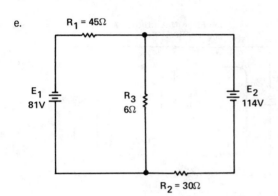

f.

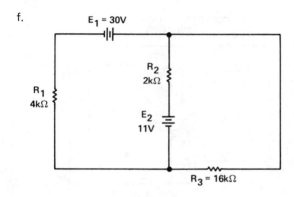

g.

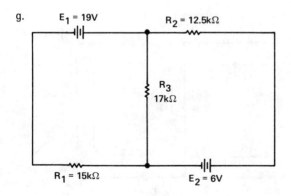

Answers

3.e.

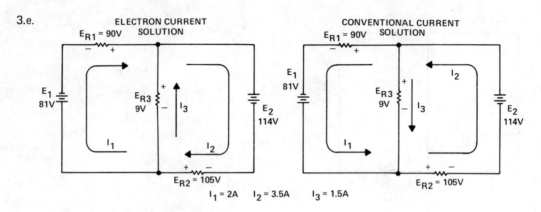

$I_1 = 2A$ $I_2 = 3.5A$ $I_3 = 1.5A$

3.f.

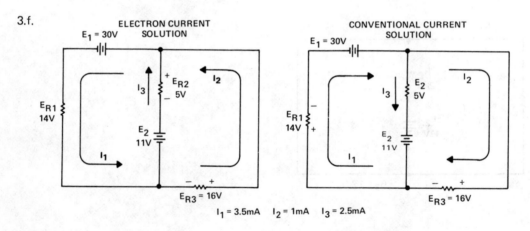

$I_1 = 3.5mA$ $I_2 = 1mA$ $I_3 = 2.5mA$

3.g.

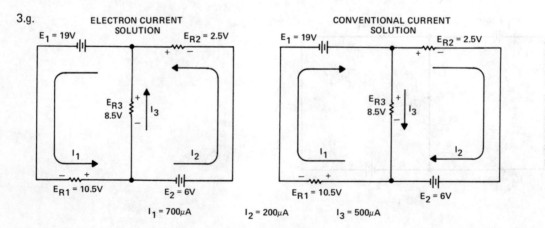

$I_1 = 700\mu A$ $I_2 = 200\mu A$ $I_3 = 500\mu A$

Fold Over

h.

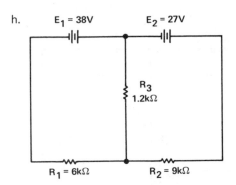

i.

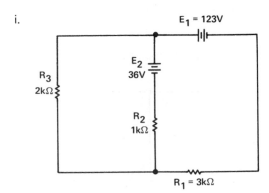

Answers

3.h.

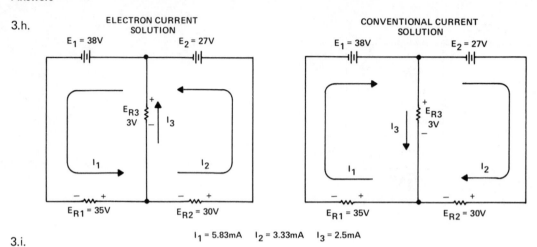

$I_1 = 5.83mA$ $I_2 = 3.33mA$ $I_3 = 2.5mA$

3.i.

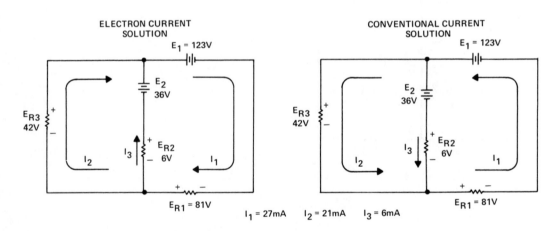

$I_1 = 27mA$ $I_2 = 21mA$ $I_3 = 6mA$

Fold Over

j.

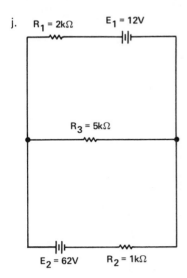

$R_1 = 2k\Omega$ $E_1 = 12V$

$R_3 = 5k\Omega$

$E_2 = 62V$ $R_2 = 1k\Omega$

Answers

3.j.

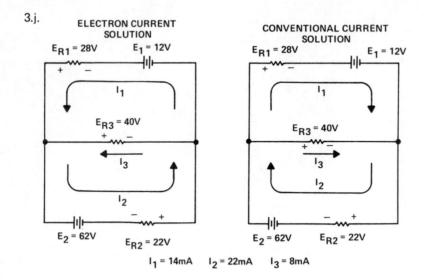

ELECTRON CURRENT
SOLUTION

$E_{R1} = 28V$ $E_1 = 12V$

I_1

$E_{R3} = 40V$

I_3

I_2

$E_2 = 62V$ $E_{R2} = 22V$

CONVENTIONAL CURRENT
SOLUTION

$E_{R1} = 28V$ $E_1 = 12V$

I_1

$E_{R3} = 40V$

I_3

I_2

$E_2 = 62V$ $E_{R2} = 22V$

$I_1 = 14mA$ $I_2 = 22mA$ $I_3 = 8mA$

1. "The sum of currents arriving at any point in a circuit must be equal to the sum of the currents leaving the point" is called:

   a. The law of Electrostatics
   b. Kirchhoff's current law
   c. The product over sum law
   d. The loop equation

2. Junctions or branch points in a circuit are commonly called:

   a. High points
   b. Terminals
   c. Extra paths
   d. Nodes

3. "The total circuit current equals the sum of the branch currents" is:

   a. The first law of electrostatics
   b. Coulomb's law
   c. Kirchhoff's current law for parallel circuits
   d. A series circuit law

4. "The current is the same at any point in the circuit" is:

   a. The first law of electrostatics
   b. A parallel circuit law
   c. Coulomb's law
   d. Kirchhoff's current law for series circuits

5. Kirchhoff's second law is:

   a. For series circuits
   b. For parallel circuits
   c. Kirchhoff's voltage law
   d. For a closed circuit path
   e. All of the above
   f. None of the above

6. "The total voltage applied to any closed circuit path is always equal to the sum of the voltage drops in that path" is:

   a. To be used with series-parallel circuits
   b. To be used with parallel-series circuits
   c. Kirchhoff's voltage law
   d. a, b, c above
   e. a only

7. A continuous path that is traced in a circuit that starts and ends at the same point is called:

   a. A junction
   b. A loop
   c. A closed path
   d. A branch
   e. b and c above
   f. All of above

8. Kirchhoff's voltage law _____ hold for circuit loops that have no voltage sources.

   a. Does
   b. Doesn't
   c. May
   d. May not

9. "For any loop, the sum of all the voltages that aid current flow in the loop, must equal the sum of all those voltages that oppose it" is:

   a. Kirchhoff's current law
   b. An alternate statement of Kirchhoff's voltage law
   c. Applies only to series circuits
   d. Applies only to parallel circuits

10. "The algebraic sum of the voltages aiding current and the voltages opposing current in any circuit loop is equal to zero" is:

    a. To be used only for series circuits
    b. To be used only for parallel circuits
    c. An alternate of Kirchhoff's current law
    d. An alternate statement of Kirchhoff's voltage law.

11. "The algebraic sum of all voltages in a loop equals zero" is:

    a. To be used only for series circuits
    b. To be used only for parallel circuits
    c. An alternate of Kirchhoff's current law
    d. An alternate of Kirchhoff's voltage law

12. When the algebraic sum of all voltages in a loop is set equal to zero the result is called:

    a. A series circuit
    b. A loop equation
    c. The total current
    d. A branch

13. When analyzing circuits, using Kirchhoff's laws for circuit branches where the current direction is unknown, _____ a direction.

    a. Guess
    b. The current dictates
    c. The voltage dictates
    d. All of the above

14. When calculating currents in a circuit using Kirchhoff's laws, if the current is negative it:

    a. Must not be flowing
    b. Is too great in value
    c. Is flowing opposite to the direction chosen
    d. Is very small in value

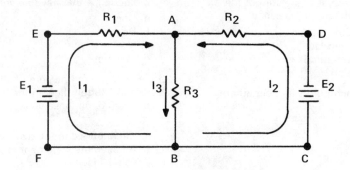

The questions that follow use the schematic above.

15. Point A and Point B are:
    a. Junctions
    b. Nodes
    c. Form a branch
    d. All of above
    e. b only

16. Circuit paths A B F E A and A B C D A are called:
    a. Circuit loops
    b. Nodes
    c. Junctions
    d. Voltage sources

17. The direction of I_1 and I_2 shown is for _____ flow.
    a. Electron
    b. Conventional current
    c. Branch current
    d. Load current

18. If E_1 and E_2 and R_1, R_2 and R_3 are known, solving for I_1, I_2, and I_3 requires_____ equations.
    a. One
    b. Two
    c. Three
    d. Four

19. Kirchhoff's current law can be used to write an equation for:
    a. The loop A B F E A
    b. Point A
    c. Node B
    d. a and b above
    e. b and c above
    f. c only

20. Kirchhoff's voltage law can be used to write equations around:
    a. Loop A B F E A
    b. Loop A B C D A
    c. Loop B F E A D C B
    d. All of the above
    e. a and b only

Lesson 12

Advanced Methods of DC Circuit Analysis

In this lesson, Kirchhoff's laws and the procedures for using them will be reviewed, and a new circuit analysis technique called the superposition theorem will be introduced. The use of the superposition theorem in analyzing multiple source dc circuits will be covered. Several additional advanced methods of dc circuit analysis will also be introduced. A brief synopsis of the use of these methods will be covered at the end of this lesson.

LESSON 12. ADVANCED METHODS OF DC CIRCUIT ANALYSIS

● Objectives

1. Using Kirchhoff's voltage and current laws correctly, *write* the loop and junction equations for any dc circuit such as the bridge circuit shown below, using either electron or conventional current to analyze the circuit.

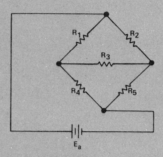

2. *Write* the superposition theorem, including a statement of its limitations.

3. Using the superposition theorem, *solve* for the unknown voltages and currents in multiple source dc circuits of the type illustrated in the schematic diagram below.

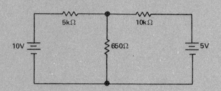

4. With the aid of this lesson summary and other reference material, *state* how any of the following circuit analysis techniques may be used in analyzing dc circuits:
   a. Method of mesh currents
   b. Node voltage analysis
   c. Thevenin's theorem
   d. Norton's theorem
   e. Millman's theorem

-
- **Kirchhoff's Voltage Law**
- **Series Circuit as a Closed Loop**

In the previous lesson, Kirchhoff's laws were introduced and you were shown how to apply them in analyzing multiple source dc circuits. This lesson goes on to examine the use of Kirchhoff's laws to analyze a complex series-parallel circuit, and also discusses what is called the *superposition theorem*. You will see that the superposition theorem is a very important tool, because you can use it to greatly simplify the analysis of multiple source circuits. This lesson then goes on to briefly familiarize you with some more methods used in circuit analysis. These advanced methods will not be discussed at great length but they will be concisely outlined for you at the end of the lesson for your later reference.

Kirchhoff's Voltage Law — In the last lesson Kirchhoff's laws were first introduced to you. Before covering any new material in this lesson, it will be helpful to review these laws and the rules you were given on how to use them. Remember there are two laws: Kirchhoff's current law and Kirchhoff's voltage law. It will be helpful to review Kirchhoff's voltage law which is stated as shown in Figure 12.1; *the algebraic sum of the voltages around any closed loop in a circuit must equal zero.*

    Since quite a few new terms and concepts are involved in using this law, a brief review of some of its basic elements will be helpful.

KIRCHHOFF'S VOLTAGE LAW

THE ALGEBRAIC SUM OF THE VOLTAGES AROUND ANY CLOSED LOOP MUST EQUAL ZERO.

Figure 12.1

Series Circuit as a Closed Loop — A closed path, or *loop* as it was called in the preceding lesson, is determined by traversing a path through a circuit in a particular direction until the starting point of the path is reached. As you can see in the simple series circuit of Figure 12.2, if you start at the starting point labeled in this circuit and proceed in a counterclockwise direction, you must pass through R_1, R_2, R_3 and the voltage source before you return to the starting point. This path through the circuit is a closed path or loop.

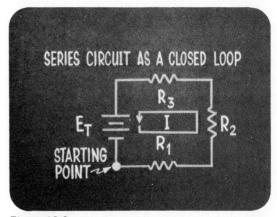

Figure 12.2

- Use of Kirchhoff's Laws
- Positive Source Voltage
- Negative Source Voltage

Use of Kirchhoff's Laws — You must remember the steps involved in using Kirchhoff's second law. In order to write down a loop equation as shown in Figure 12.3, the first step is to label all the currents flowing in the circuit and assign them a direction. Then traverse the loop you are analyzing, adding the voltages algebraically, and set the algebraic sum of the voltages equal to zero. In the loop equation, a voltage source gets a positive sign if you traverse it in the same direction as it normally pushes current.

TO USE KIRCHHOFF'S LAWS

1. LABEL ALL CURRENTS AND ASSIGN THEM A DIRECTION.
2. TRAVERSE THE LOOP, ADD ALL VOLTAGES ALGEBRAICALLY, AND SET THE SUM EQUAL TO ZERO.
3. A VOLTAGE SOURCE RECEIVES A POSITIVE SIGN, IF YOU TRAVERSE IT IN THE SAME DIRECTION, IT NORMALLY PUSHES CURRENT.
4. A VOLTAGE SOURCE RECEIVES A NEGATIVE SIGN, IF YOU TRAVERSE IT OPPOSITE TO THE DIRECTION IT NORMALLY PUSHES CURRENT.

Figure 12.3

Positive Source Voltage — Stop and consider for a moment what that means. If you are considering *electron current* (Figure 12.4), a *source voltage* is considered *positive* if you traverse it from *plus to minus*. If you are considering *conventional current*, a *source voltage* is considered *positive* if you traverse it from *minus to plus*. In either convention, a *voltage source* gets a *positive* sign in the loop equation if the direction you traverse it is the *same* as the direction it pushes current.

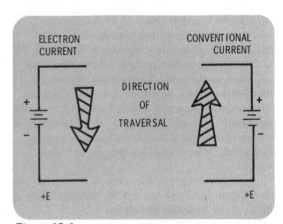

Figure 12.4

Negative Source Voltage — Conversely, a voltage source gets a *negative sign* (Figure 12.5) if you traverse it *opposite* to the direction it normally pushes current. In other words, if you are considering *electron current*, a source voltage receives a *negative sign* when you traverse the source from negative to positive. For *conventional current*, the source voltage is *negative* when you traverse the source from *positive to negative*. In either current convention, a *voltage source* gets a *negative sign* in the loop equation if you traverse it *opposite* to the direction it normally carries current.

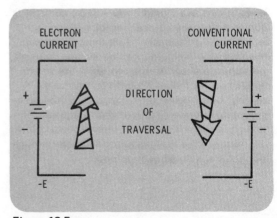

Figure 12.5

- **Signs of Voltages Across Resistors**
- **Voltages Around a Closed Loop**
- **Series Circuit Rule**

Signs of Voltages Across Resistors — Two rules governing the signs of voltage terms across resistors were also covered in Lesson 11 (Figure 12.6). If you traverse a *resistor in the direction of current flow through it*, the *voltage across it* is considered *negative*. If you go through a *resistor against the current direction*, the *voltage across it* is considered *positive*.

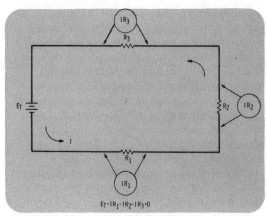

TO USE KIRCHHOFF'S LAWS (CONTINUED)

5. IF YOU TRAVERSE A RESISTOR IN THE DIRECTION OF ASSUMED CURRENT FLOW, THE VOLTAGÉ ACROSS IT IS CONSIDERED **NEGATIVE.**
6. IF YOU TRAVERSE A RESISTOR **AGAINST** THE DIRECTION OF ASSUMED CURRENT FLOW, THE VOLTAGE ACROSS IT IS CONSIDERED POSITIVE.

Figure 12.6

Voltages Around a Closed Loop — In the single loop shown in Figure 12.7, consider that electron current is flowing counterclockwise as labeled. To write the loop equation for this circuit, begin with the voltage source and traverse the loop, writing down all the voltages you encounter along the way. It is normally convenient in loops such as this to traverse in the same direction as the current flowing in the circuit. If you were traversing counterclockwise, beginning with the voltage source, you would get $E_T - IR_1 - IR_2 - IR_3 = 0$. Remember that the voltages across each resistor are equal to I times R, from Ohm's law.

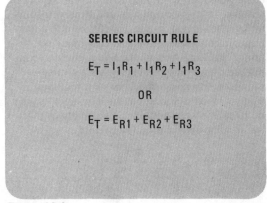

Figure 12.7

Series Circuit Rule — As mentioned in Lesson 11, the loop equation above is simply another way of stating an earlier rule about voltages in a series circuit. If the resistor's IR drops are transposed to the other side of the equation (remember to change the signs), the result is a mathematical expression for the series circuit rule. As shown in Figure 12.8, the total applied voltage must be equal to the sum of the individual drops. Here E_T equals the sum of the voltage across the resistors or E_{R1} plus E_{R2} plus E_{R3}.

SERIES CIRCUIT RULE

$$E_T = I_1 R_1 + I_1 R_2 + I_1 R_3$$

OR

$$E_T = E_{R1} + E_{R2} + E_{R3}$$

Figure 12.8

Kirchhoff's Current Law — Go back a moment to consider Kirchhoff's first law or Kirchhoff's *current* law, which states that the sum of the currents *into* any point of a circuit must equal the sum of the currents *out* of that point (Figure 12.9). This law is generally applied where there is a branch in a circuit, where the current will either divide or recombine. Recall that such circuit points are called *nodes*.

KIRCHHOFF'S CURRENT LAW

THE SUM OF THE CURRENTS INTO ANY POINT OF A CIRCUIT, MUST EQUAL THE SUM OF THE CURRENTS OUT OF THAT POINT.

Figure 12.9

Example Circuit — For example, in the circuit of Figure 12.10, you can see that the current will divide at point A and combine at point B. So, if there is a total of, say, 5 milliamps coming into point A, there must be a total of 5 milliamps leaving that point and dividing into I_1 and I_2. Thus, you can see that this is just another way of saying that the total or main line current in a parallel circuit is equal to the sum of the individual branch currents, or $I_T = I_1 + I_2$. This, you recall, is the rule for currents in parallel circuits.

Remember, Kirchhoff's laws are really just another more general way of stating the series and parallel circuit rules.

In the last lesson it was shown how these laws can be applied to the solution of *multiple source* dc circuits, in configurations where Ohm's law alone would not be enough to complete the circuit analysis. This lesson will help you expand the use of Kirchhoff's two laws and apply them to another class of complex series-parallel circuits called bridge circuits. Although the bridge circuits to be discussed have only one voltage source, they cannot be analyzed using Ohm's law methods alone. After this discussion, the lesson will move on to cover several additional new methods that are useful in solving complex circuits.

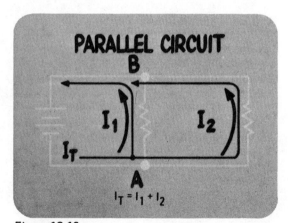

Figure 12.10

Lesson Objectives — So, at the end of this lesson you should:
1. Know and be able to use Kirchhoff's laws to write the loop and node equations for a bridge circuit
2. Know the superposition theorem, its limitations, and how to use it to analyze multiple source dc circuits
3. Be familiar with more advanced circuit analysis methods, and know where to look for reference material necessary to implement them (Figure 12.11).

- KIRCHHOFF'S LAWS
- SUPERPOSITION THEOREM
- ADVANCED CIRCUIT ANALYSIS METHODS

Figure 12.11

Bridge Circuit — In the previous lesson you saw how to use Kirchhoff's laws to analyze multiple source circuits. This lesson first considers a complex series-parallel circuit which (although it has only one voltage source) cannot be analyzed by ordinary means; that is, with Ohm's law and simple circuit rules.

A circuit of the type shown in Figure 12.12 is called a *bridge* circuit. As you will see in the laboratory portion of this course, a specialized version of this circuit called a *Wheatstone bridge* can be used in making very accurate resistance measurements. At this point in this lesson, Kirchhoff's laws will be used to *begin* the analysis of this circuit. That is, all the loop equations and node equations necessary to find all currents and voltages in the circuit will be written down, going through all procedures step by step. At this point, if you feel ready to write down these equations and solve them on your own, please do so. This would be a good point to see how far you can go on your own in handling Kirchhoff's laws. If you have any difficulty, the complete analysis of this circuit will be worked out step by step for you right here.

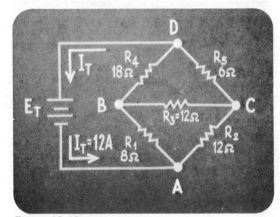

Figure 12.12

Label Current Direction — The first step in analysing this circuit is to label all the electron currents and assign them a direction, as shown in Figure 12.13. (Note, at this point if you do not know the direction of current flow, just guess a direction. Remember that, if any current you solve for turns out *negative*, the current is flowing *opposite* to the direction you guessed.) Notice in the circuit that the total main line electron current is 12 amps, and that it flows from the negative terminal of the source to point A. At point A, the total current divides into two parts which you can label I_1 and I_2, which flow up through R_1 and R_2 respectively. Assume that I_3, the current flowing through R_3, is flowing from left to right. Now I_4 and I_5 flow up through R_4 and R_5 respectively, and then join at point D to produce the total current, which flows back to the voltage source.

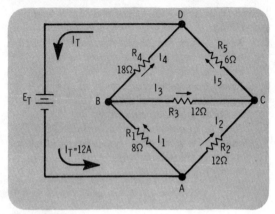

Figure 12.13

Node Current Equations: Node A — Now that all the currents flowing in the circuit have been labeled and a direction assigned to them, you can use Kirchhoff's current law to write the *node* current equations at points A, B, C, and D.

At point A in Figure 12.14, the sum of I_1 and I_2 must equal the total current. Since you know the total current is 12 amps, you can say that I_1 plus I_2 equals 12 amps.

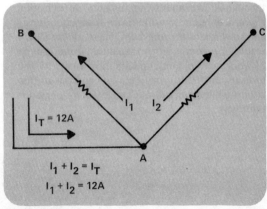

Figure 12.14

- Node B
- Node C
- Node D

Node B — At point B in the circuit, as shown in Figure 12.15, I_1 must equal I_3 plus I_4. This gives you another node equation to use in analyzing this circuit.

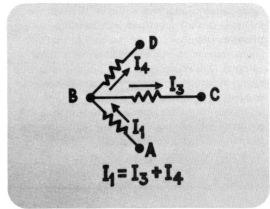

$$I_1 = I_3 + I_4$$

Figure 12.15

Node C — Likewise, at point C in Figure 12.16, I_5 must equal I_3 plus I_2.

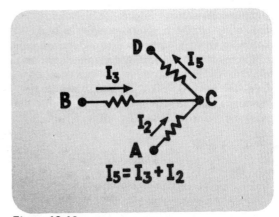

$$I_5 = I_3 + I_2$$

Figure 12.16

Node D — At point D in Figure 12.17, I_4 plus I_5 equals I_T or 12 amps.

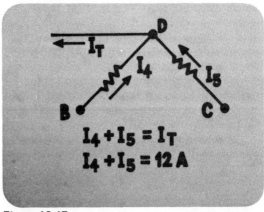

$$I_4 + I_5 = I_T$$
$$I_4 + I_5 = 12\,A$$

Figure 12.17

12-11

Node Current Equations — Figure 12.18 shows all the node current equations. As you will be seeing, these equations can be manipulated so that any one current can be expressed in terms of other currents. This is usually done to simplify the loop equations which are developed next. To carry the circuit analysis further, three loop equations will be written for it. The loop equations when analyzed with the node equations, will allow you to completely analyze this circuit.

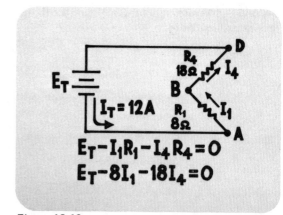

$$I_1 + I_2 = 12$$

$$I_1 = I_3 + I_4$$

$$I_5 = I_3 + I_2$$

$$I_4 + I_5 = 12$$

Figure 12.18

First Loop Equation — Figure 12.19 shows one loop of the circuit. If you traverse the loop beginning with the voltage source and move around in the counterclockwise direction in the assumed direction of electron current, the loop equation is $E_T - I_1 R_1 - I_4 R_4 = 0$. When you substitute the resistance values written on the schematic into the equation, you should have $E_T - 8 I_1 - 18 I_4 = 0$.

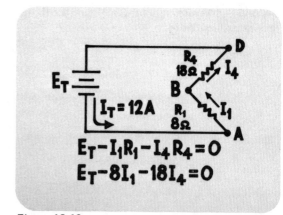

$$E_T - I_1 R_1 - I_4 R_4 = 0$$
$$E_T - 8I_1 - 18I_4 = 0$$

Figure 12.19

Second Loop Equation — Figure 12.20 shows a second loop in the circuit. If you traverse this loop in the clockwise direction starting at point A, the loop equation is $-I_1 R_1 - I_3 R_3 + I_2 R_2 = 0$. Note that since you traversed R_2 *against* the direction of current flowing through it, the last term in the equation has a *plus* sign. When you substitute the resistance values in the equation, you should have $-8 I_1 - 12 I_3 + 12 I_2 = 0$.

Now consider one more loop and write the equation for it.

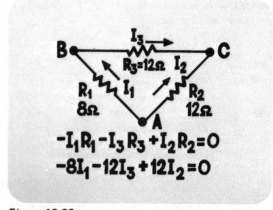

$$-I_1 R_1 - I_3 R_3 + I_2 R_2 = 0$$
$$-8I_1 - 12I_3 + 12I_2 = 0$$

Figure 12.20

Third Loop Equation — If you traverse the loop in Figure 12.21 in the counterclockwise direction, starting at point B, the equation is $-I_3R_3 - I_5R_5 + I_4R_4 = 0$. Again, substitute the resistance values in the equation, and you should have $-12\,I_3 - 6\,I_5 + 18\,I_4 = 0$. The last term in this equation also has a plus sign since you traverse R_4 against the direction of current flowing through it. (Note, any three loops could be used in arriving at the loop equations you need for this problem.)

As you will see, the remainder of this problem involves manipulations and substitutions with the node current equations and the loop equations to find the individual branch currents. Once all the branch currents are known, you simply multiply them by the appropriate resistances to find the individual voltages in the circuit.

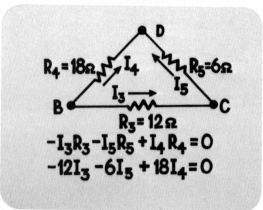

$$-I_3R_3 - I_5R_5 + I_4R_4 = 0$$
$$-12I_3 - 6I_5 + 18I_4 = 0$$

Figure 12.21

For those of you who want to follow the analysis of this circuit, its step-by-step solution is included here.

To start, write all of the node current equations in simplified form:

$$I_1 + I_2 = 12 \rightarrow I_1 = 12 - I_2$$
$$I_1 = I_3 + I_4 \rightarrow I_4 = I_1 - I_3 = 12 - I_2 - I_3$$
$$I_5 = I_3 + I_2 \rightarrow I_3 = I_5 - I_2$$
$$I_4 + I_5 = 12 \rightarrow I_4 = 12 - I_5 = 12 - I_2 - I_3$$

Note that in writing a simplified expression for I_4 in the second equation, the simplified value of I_1 from the first equation was used. This is also done in the fourth equation. By having several alternate expressions for I_4, the problem can be solved more easily.

Write the three loop equations.
1. $E_T - 8\,I_1 - 18\,I_4 = 0$
2. $-8\,I_1 - 12\,I_3 + 12\,I_2 = 0$
3. $-12\,I_3 - 6\,I_5 + 18\,I_4 = 0$

Pick two equations that share a common term so that that term may be canceled out. In this case, equations 2 and 3 are used, because they both contain the term "I_3". The two equations may now be simplified. Equation 2 is divisible by 4, and equation 3 is divisible by 6.

$$\frac{-8\,I_1 - 12\,I_3 + 12\,I_2 = 0}{4} = 2\,I_1 - 3\,I_3 + 3\,I_2 = 0\,(*)$$

$$\frac{-12\,I_3 - 6\,I_5 + 18\,I_4 = 0}{6} = -2\,I_3 - I_5 + 3\,I_4 = 0$$

In order to simplify the problem further, the equations may be reduced to two unknowns. This may be done by substituting a node current equivalent value into the loop equations. For example, I_1 may also be expressed as $12 - I_2$. If this value is substituted into the simplified loop equation (* above) you have:

$$-2\,I_1 - 3\,I_3 + 3\,I_2 = 0$$
$$-2(12 - I_2) - 3\,I_3 + 3\,I_2 = 0$$
$$-24 + 2\,I_2 - 3\,I_3 + 3\,I_2 = 0$$

or

$$5\,I_2 - 3\,I_3 \; 24$$

Next, reduce the other equation to two unknowns (I_2 and I_3)

$$-2\,I_3 - I_5 + 3\,I_4 = 0$$

From the node current equations you know that I_5 is equal to $I_2 + I_3$. I_4 is equal to $12 - I_2 - I_3$. When these values are substituted into the loop equation, the equation will contain only the unknowns I_2 and I_3, and can then be solved.

$$-2\,I_3 - I_5 + 3\,I_4 = 0$$
$$-2\,I_3 - (I_2 + I_3) + 3(12 - I_2 - I_3) = 0$$
$$-2\,I_3 - I_2 - I_3 + 36 - 3\,I_2 - 3\,I_3 = 0$$
$$+36 - 4\,I_2 - 6\,I_3 = 0$$
$$-4\,I_2 - 6\,I_3 = -36 \text{ or}$$
$$4\,I_2 + 6\,I_3 = 36$$

Now the loop equations have been reduced to two unknowns, I_2 and I_3. One of the currents may now be found.

$$5\,I_2 - 3\,I_3 = 24$$
$$4\,I_2 + 6\,I_3 = 36$$

In order to cancel the I_3 term, the first equation above should be multiplied by 2, and then the equations can be added.

$$2\,(5\,I_2 - 3\,I_3) = (24)\,2$$

$$10\,I_2 - 6\,I_3 = 48$$

$$\underline{4\,I_2 + 6\,I_3 = 48}$$

$$14\,I_2 \qquad = 84$$

$$\frac{14\,I_2}{14} = \frac{84}{14}$$

$$I_2 \qquad = 6\text{ A}$$

Now that one of the currents is known, this value may be substituted into the node current and loop equations to solve for the remainder of the circuit currents.

Node current equation:

$$I_1 = 12 - I_2$$
$$I_1 = 12 - 6$$
$$I_1 = 6 \text{ A}$$

Loop equation:

$$-3\,I_3 + 5\,I_2 = 24$$
$$-3\,I_3 + 5(6) = 24$$
$$-3\,I_3 + 30 = 24$$
$$-3\,I_3 = 24 - 30$$
$$-3\,I_3 = -6$$
$$I_3 = 2 \text{ A}$$

Node current equation:

$$I_4 = I_1 - I_3$$
$$I_4 = 6 \text{ A} - 2 \text{ A}$$
$$I_4 = 4 \text{ A}$$
$$I_5 = I_2 + I_3$$
$$I_5 = 6 \text{ A} + 2 \text{ A}$$
$$I_5 = 8 \text{ A}$$

To find the voltages across all resistors in the circuit, Ohm's law may be used.

$$E_{R1} = I_1 \times R_1 = 6 \text{ A} \times 8\,\Omega = 48 \text{ V}$$
$$E_{R2} = I_2 \times R_2 = 6 \text{ A} \times 12\,\Omega = 72 \text{ V}$$
$$E_{R3} = I_3 \times R_3 = 2 \text{ A} \times 12\,\Omega = 24 \text{ V}$$
$$E_{R4} = I_4 \times R_4 = 4 \text{ A} \times 18\,\Omega = 72 \text{ V}$$
$$E_{R5} = I_5 \times R_5 = 8 \text{ A} \times 6\,\Omega = 48 \text{ V}$$

To determine E_T, go back to the loop equation in Figure 12.19 and substitute the appropriate current values in the equation.

$$E_T - 8\,I_1 - 18\,I_4 = 0$$
$$E_T - 8(6) - 18(4) = 0$$
$$E_T - 48 - 72 = 0$$
$$E_T - 120 = 0$$
$$E_T = 120 \text{ V}$$

Superposition Theorem — At this point you have seen how to use Kirchhoff's laws to analyze several different types of circuits. Several examples have now been worked through for you to show you how Kirchhoff's laws may be used in analyzing multiple source dc circuits and bridge circuits, which cannot be worked if Ohm's law methods alone are used. This lesson now goes on to examine another powerful method which simplifies the analysis of circuits with more than one voltage source. This method essentially involves carefully examining the effect that each source has on the circuit *by itself*. For example, consider the simple circuit of Figure 12.22. The basic circuit consists of a 10-ohm resistor connected to a 10-volt source. The current produced in the resistor is 10 volts divided by 10 ohms or 1 amp.

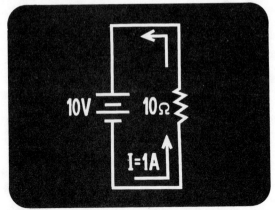

Figure 12.22

Second Source — Figure 12.23 shows this same 10-ohm resistor connected to another 10-volt source. The current produced in this resistor by this second source acting alone is again 1 amp.

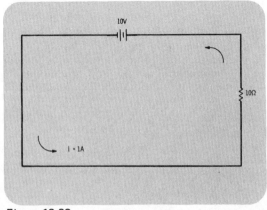

Figure 12.23

Circuit with Both Sources Acting — When both voltage sources are connected to the resistor as shown in Figure 12.24, the current flowing through the resistor is now 2 amps. The action of this circuit is a simplified illustration of what is called the *superposition theorem*.

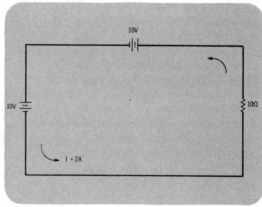

Figure 12.24

Statement of Superposition Theorem — The superposition theorem is stated as shown in Figure 12.25. *In a network with two or more sources, the current or voltage for any component is equal to the algebraic sum of the effects produced by each source acting separately.*

The superposition theorem is easy to use. As was illustrated in the simplified example, when you analyze a circuit considering only one source at a time, you can solve the remaining circuit by simply using Ohm's law and the rules you have learned for the behavior of series and parallel circuits.

SUPERPOSITION THEOREM

IN A NETWORK WITH TWO OR MORE SOURCES, THE CURRENT OR VOLTAGE FOR ANY COMPONENT IS EQUAL TO THE ALGEBRAIC SUM OF THE EFFECTS PRODUCED BY EACH SOURCE ACTING SEPARATELY.

Figure 12.25

Limitations of the Superposition Theorem — The superposition theorem, however, does have two limitations (Figure 12.26). First, all components must be *linear*, which means that as the input to the component increases or decreases, the output must increase or decrease in direct proportion. Second, all components must be *bilateral*, which means that current must flow equally well in either direction. Because of these limitations, the superposition theorem is most effectively used in analyzing the behavior of resistive circuits.

LIMITATIONS OF THE SUPERPOSITION THEOREM

1. LINEAR COMPONENTS

2. BILATERAL COMPONENTS

Figure 12.26

- Sample Solution Using the Superposition Theorem
- Trick in Using Superposition: Short Out E_2
- Redrawn Circuit

Sample Solution Using the Superposition

Theorem — To see how the superposition theorem works, it will be used to solve for the voltage from point A to ground, across R_3, in a circuit with two voltage sources such as the one shown in Figure 12.27. Notice you could use Kirchhoff's laws to solve this problem, but not Ohm's law and the simple circuit rules alone. (Ohm's law methods alone would not allow you to determine what was happening in this circuit because two sources are interacting here to produce current flow.)

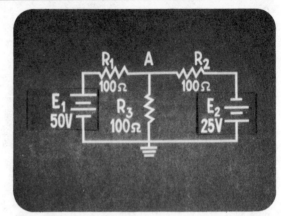

Figure 12.27

Trick in Using Superposition: Short Out E_2 — The key trick in using the superposition theorem to analyze a circuit like this is to just mentally replace one of the voltage sources on your schematic, say E_2, with a short. With E_2 replaced by a short, as shown in Figure 12.28, you can now use simple circuit rules and Ohm's law to first find the voltage from point A to ground with E_1 acting alone. This voltage will be called E_{A1}.

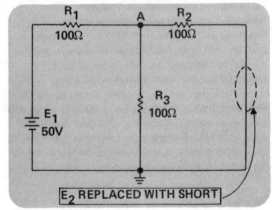

Figure 12.28

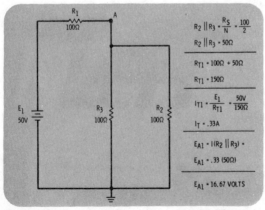

Figure 12.29

$$R_2 \parallel R_3 = \frac{R_S}{N} = \frac{100}{2}$$

$$R_2 \parallel R_3 = 50\Omega$$

$$R_{T1} = 100\Omega + 50\Omega$$

$$R_{T1} = 150\Omega$$

$$I_{T1} = \frac{E_1}{R_{T1}} = \frac{50V}{150\Omega}$$

$$I_T = .33A$$

$$E_{A1} = I(R_2 \parallel R_3) =$$

$$E_{A1} = .33 (50\Omega)$$

$$E_{A1} = 16.67 \text{ VOLTS}$$

Redrawn Circuit — This circuit can be redrawn slightly to indicate clearly how you would solve for E_{A1} (Figure 12.29). With E_2 shorted out, this circuit consists of the single power source, E_1, powering a series-parallel circuit. In the circuit, R_1 is in series with the parallel combination of R_2 and R_3. Remember that a shorthand notation that can be used to express "R_2 in parallel with R_3" is $R_2 \parallel R_3$. The first step in analyzing our original two-source circuit is to find the voltage from point A to ground in this circuit, with E_1 acting alone. This is a straightforward series-parallel circuit problem.

Begin by finding the total resistance of the circuit. To do that you need to calculate the total resistance of R_2 in parallel with $R_3(R_2\|R_3)$. Since R_2 and R_3 are both 100-ohm resistors, use the formula $R_{eq} = R_S/N$. Substituting, you have 100 ohms divided by 2 or 50 ohms for $R_2\|R_3$. Now the total resistance of this circuit is just $R_1 + (R_2\|R_3)$ or 100 plus 50 or 150 ohms. The total circuit current I_{T1} just equals E_1/R_{T1} or 50 volts/150 ohms which equals 333 milliamps. Finally, E_{A1} now equals this total current times the resistance between point A and ground. $E_{A1} = I_{T1} \times (R_2\|R_3)$ which equals 0.333 × 50 ohms or 16.67 volts.

Solution for E_{A1} — If you were to calculate this voltage with the series-parallel circuit rules and ohm's law, you would find that it is 16.67 volts as shown in Figure 12.30. The polarity of E_{A1} as shown is negative at ground and positive at point A.

At this point you know the voltage across R_3, from point A to ground, produced by one of the two sources acting alone. If you find the voltage across R_3 produced by the other source acting alone, you can add the two voltages algebraically to find the total voltage across R_3.

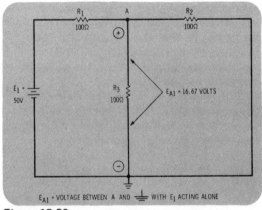

Figure 12.30

Next Step: "Short" E_1 — Your next step is to mentally reconnect E_2, and short E_1 as shown in Figure 12.31. Now you can find the voltage from point A to ground with E_2 acting alone; call it E_{A2}.

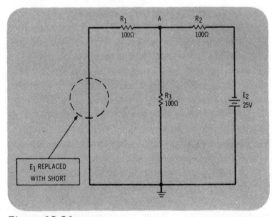

Figure 12.31

- Solving for E_{A2}
- **Solution for E_{A2}**
- **Add E_{A1} and E_{A2} to Get Final Result**

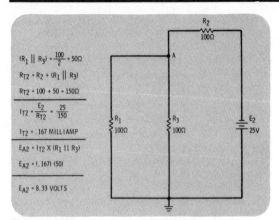

$(R_1 \parallel R_3) = \frac{100}{2} = 50\Omega$

$R_{T2} = R_2 + (R_1 \parallel R_3)$

$R_{T2} = 100 + 50 = 150\Omega$

$I_{T2} = \frac{E_2}{R_{T2}} = \frac{25}{150}$

$I_{T2} = .167$ MILLIAMP

$E_{A2} = I_{T2} \times (R_1 \parallel R_3)$

$E_{A2} = (.167)(50)$

$E_{A2} = 8.33$ VOLTS

Figure 12.32

Solving for E_{A2} — This circuit of Figure 12.31 can be redrawn as shown in Figure 12.32 to show that this too is a series-parallel circuit. The equivalent resistance of $(R_1 \parallel R_3)$ is now 50 ohms, similar to the previous case, and the total circuit resistance is again 150 ohms. This time the total current I_{T2} is equal to E_2/R_{T2} or 5 volts divided by 150 ohms or 167 milliamps. E_{A2} is then equal to I_{T2} times $(R_1 \parallel R_3)$ which equals 0.167 amp times 50 ohms or 8.33 volts.

Solution for E_{A2} — If you were to calculate this voltage with the simple circuit rules and Ohm's law, you would find that it is 8.33 volts. The polarity of E_{A2}, as shown in Figure 12.33, is opposite to the polarity of E_{A1}, that is, positive at ground and negative at point A.

To find the *total voltage* from point A to ground, *algebraically add up the effects of each of the sources acting separately*, remembering that the voltages are of opposite polarity.

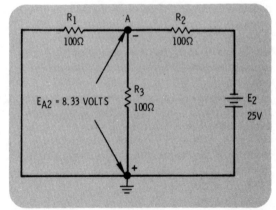

Figure 12.33

Add E_{A1} and E_{A2} to Get Final Result — In Figure 12.34, if you consider E_{A1} positive and E_{A2} negative, you should have 16.67 minus 8.33, and you get 8.34 volts for the total voltage between point A and ground. The polarity will be that of the larger voltage: positive at point A and negative at ground.

A word of caution may be in order here. Although you can talk about shorting out voltage sources in schematic diagrams when analyzing circuits with the superposition theorem, you should *never just short out an actual voltage source in a real circuit*. This can damage the source, cause electrical burns, as well as a variety of other undesirable effects. When the term "shorting out a

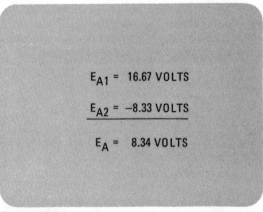

$E_{A1} = 16.67$ VOLTS

$E_{A2} = -8.33$ VOLTS

$E_A = 8.34$ VOLTS

Figure 12.34

source'' is used in circuit analysis, it refers to a *mental* process only.

Another Sample Solution Using the Superposition Theorem — To provide you with some additional practice in using the superposition theorem, it will now be used with a slightly different circuit. This time the superposition theorem will be used to find the current flowing in one of the resistors in the complex circuit situation shown in Figure 12.35.

This is a complex two-source circuit in which you want to find the current flowing through the resistor, R_1. The two sources, E_1 which equals 10 volts, and E_2 which equals 20 volts, are pushing current through R_1 in opposite directions. Using the superposition theorem to solve this problem, you first want to find I_1, the current flowing in R_1 produced by E_1 acting alone, so you short out E_2. Then you want to find I_2, the current flowing through R_1 with E_2 acting alone — so you short out E_1. Once I_1 and I_2 are calculated, algebraically add the results to find the total current flow through R_1 with both sources acting together.

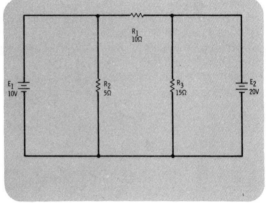

Figure 12.35

- Short E$_2$: E$_1$ Acting Alone
- Redrawn Circuit
- Short E$_1$: E$_2$ Acting Alone

Short E$_2$: E$_1$ Acting Alone — Begin by shorting out E$_2$ so that E$_1$ is acting alone in the circuit (Figure 12.36). Notice that when E$_2$ is shorted out this creates a short circuit directly across R$_3$. There is now essentially a path with zero resistance right across R$_3$. This means that no current will flow through R$_3$ and the circuit essentially consists of R$_1$ in parallel with R$_2$ with 10 volts applied.

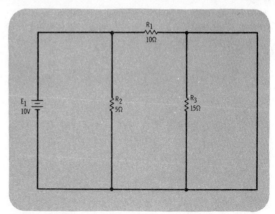

Figure 12.36

Redrawn Circuit — To clarify this, the equivalent circuit has been redrawn in Figure 12.37 with R$_3$ replaced by a short. The position of I$_1$ in the circuit is now more easily seen. To find I$_1$, use Ohm's law in the form I$_1$ = E$_1$/R$_1$, as shown in Figure 12.37. E$_1$ divided by R$_1$ equals 1 amp. I$_1$ flows in the direction shown in the circuit diagram.

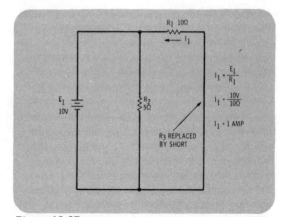

Figure 12.37

Short E$_1$: E$_2$ Acting Alone — Next you want to find I$_2$, the current produced in R$_1$ by E$_2$ acting alone, so you short out E$_1$ as shown in Figure 12.38. This puts a zero resistance path right across R$_2$. Now the circuit essentially consists of R$_1$ in parallel with R$_3$, since no current will flow through R$_2$.

To find I$_2$, you can use Ohm's law just as before. As shown in the figure, E$_2$ divided by R$_1$ equals 2 amps. The direction of I$_2$ is opposite to that of I$_1$, as indicated on the circuit diagram.

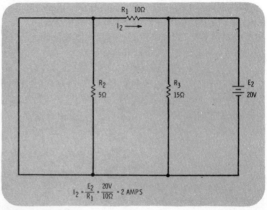

Figure 12.38

Final Solution — Back in the original circuit (Figure 12.39), the two currents I_1 and I_2 have been labeled along with their directions. Since they are flowing in opposite directions, the total current through R_1 is the difference between the two currents which is equal to 1 amp. The direction of I_{R1} is in the direction of I_2, since I_2 is the larger of the two currents.

     This problem is fairly easy to work using the superposition theorem, especially when you compare it to the previous methods. If you were to use loop and node equations to analyze this circuit, you would have to write three equations, simplify them, and then manipulate the three equations in order to solve for the current you want. Using the superposition theorem, a complex two-source problem can be reduced to two single-source circuits that can be solved using methods you already know.

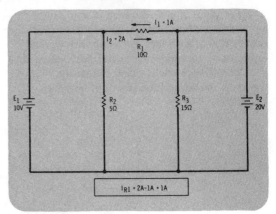

Figure 12.39

Other Methods of Circuit Analysis — The objective of the remainder of this lesson is to very briefly introduce you to some other methods of dc circuit analysis which may be useful in certain specific circuit applications.

     Figure 12.40 is a list of methods and theorems which will be discussed. Relax as a general survey of these methods is presented for you. The intent is not to explain each method of analysis in depth, but to give a brief description of each method so that you will be familiar with the language and terminology of each and know generally where the different methods can be applied most effectively.

MESH CURRENT ANALYSIS

NODE VOLTAGE ANALYSIS

THEVENIN'S THEOREM

NORTON'S THEOREM

MILLMAN'S THEOREM

Figure 12.40

For those of you who wish to study these methods more fully, there is a more thorough explanation of each theorem along with worked examples at the end of this lesson. Also included is a list of reference books which discuss the various theorems and advanced methods of dc circuit analysis.

The presentation of this material is deliberately intended to serve only as an introduction for several reasons. First of all, detailed coverage of these advanced methods will be taken up and discussed in detail in more advanced courses in electricity and electronics. Secondly, most circuits you will encounter can be solved by applying the methods that have already been discussed, although some of the solutions may become quite lengthy. The methods presented here are more powerful tools that can aid in circuit analysis or greatly simplify the prediction of circuit behavior.

Mesh Current Analysis — First consider what is called the *mesh current method* of analyzing dc circuits. This method is handy for analyzing circuits with many, many branches such as the one shown in Figure 12.41. A *mesh* is defined as the *simplest form of a loop*. In a circuit such as that shown, a typical mesh looks sort of like a single window pane. Other larger paths in this circuit are loops, but not meshes. The current flowing in a mesh is called, aptly enough, a mesh current. When working with mesh currents, you usually assume that all of the currents in all of the circuit meshes flow in the same direction, usually clockwise. Also, it is assumed that the mesh current flows all the way around the mesh and doesn't break up at

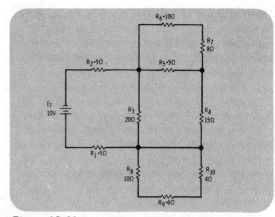

Figure 12.41

circuit nodes. If you get a negative sign in your final solution for the circuit currents, that indicates the assumed direction of the current is opposite to the actual direction of current.

Circuit with Drawn-In Mesh Currents — An important thing to remember about mesh currents is that they are *assumed currents*. Figure 12.42 shows the circuit with all the assumed branch currents drawn in. Notice that all of the branch currents are assumed to be flowing in the clockwise direction. The *actual current* flowing in any branch of the circuit may be *opposite* in direction to the *assumed* mesh current. Also, in some branches of a mesh, the current may consist of *two* mesh currents flowing in opposite directions.

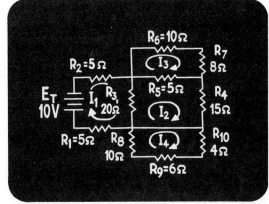

Figure 12.42

Close-Up on Central Resistor — Figure 12.43, for example, is a "close-up" of this circuit centered on the branch containing resistor R_3. Note that the current flow through R_3 is the algebraic sum of I_1 and I_2. Since these two currents flow through R_3 in different directions, $I_{R3} = I_1 - I_2$.

To carry out a complete mesh current analysis of a complex circuit like this, write mesh equations for each mesh the same way you write loop equations, except that now you use mesh currents.

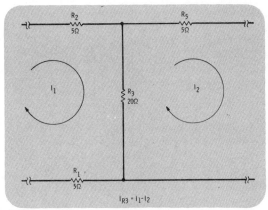

Figure 12.43

- **Left-Hand Mesh**
- **Central Mesh**
- Complete Solution

Left-Hand Mesh — For example, in the left-hand mesh of Figure 12.44, you can start with the voltage source and traverse the circuit clockwise to get: $10\,V - 5\,I_1 - 20\,I_1 - 5\,I_1 + 20\,I_2 = 0$. Notice the 20-ohm resistor also has I_2 flowing through it (opposite to I_1) so the $20\,I_2$ term in the mesh equation has a plus sign.

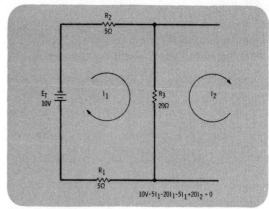

Figure 12.44

Central Mesh — In the mesh in which I_2 is flowing, the mesh equation of Figure 12.45 is $-5\,I_2 - 15\,I_2 - 20\,I_2 + 20\,I_1 + 5\,I_3 = 0$. Since I_1 also flows in R_3, the $20\,I_1$ term is positive, and because I_3 flows in R_5, the $5\,I_3$ term is also positive.

To finish the analysis of this circuit completely, you would write one equation for each mesh and solve them for all of the unknown currents. With the currents known, you could then solve for all the unknown voltages in the circuit.

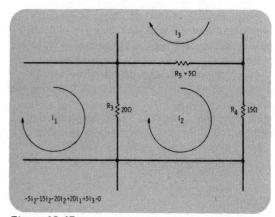

Figure 12.45

Complete Solution — The mesh equation for the top mesh, shown in Figure 12.45, starting to the left of R_5 and proceeding clockwise is: $-10\,I_3 - 8\,I_3 - 5\,I_3 + 5\,I_2 = 0$.

The bottom mesh equation may be written as: $-4\,I_4 - 6\,I_4 - 10\,I_4 = 0$. As you can see, this equation contains only terms for I_4. When added up, this equation is: $-20\,I_4 = 0$. No current flows in this loop! This fact can be seen by examining the circuit diagram. Notice that a short circuit is connected across the three resistors R_8, R_9, and R_{10}, thus no current can flow through them.

The three remaining mesh equations may now be written and simplified.

1. $10 - 5 I_1 - 20 I_1 - 5 I_1 + 20 I_2 = 0$
    $10 - 30 I_1 + 20 I_2 = 0$
    $- 30 I_1 + 20 I_2 = -10$

2. $-20 I_2 - 5 I_2 - 15 I_2 + 20 I_1 + 5 I_3 = 0$
    $- 40 I_2 + 20 I_1 + 5 I_3 = 0$

3. $-10 I_3 - 8 I_3 - 5 I_3 + 5 I_2 = 0$
    $- 23 I_3 + 5 I_2 = 0$

In order to solve for any single current value, a pair of equations containing the same current values must be set up. This can be done by expressing I_3 in terms of either I_1 or I_2. The third mesh equation allows this to be done rather easily.

The equation states that $5 I_2 - 23 I_3 = 0$. Changing the equation around a bit, it can be seen that $23 I_3 = 5 I_2$. Divide both sides of the equation by 23 and you see that:

$$\frac{23 I_3}{23} = \frac{5 I_2}{23}$$

$$I_3 = 0.217 I_2$$

Now that I_3 can be expressed in terms of I_2, a simultaneous equation may be set up using mesh equations 1 and 2. First, the I_3 term in mesh equation 2 must be written in terms of I_2.

2. $-40 I_2 + 20 I_1 + 5 I_3 = 0$
    $-40 I_2 + 20 I_1 + 5 (0.217 I_2) = 0$
    $-40 I_2 + 20 I_1 + 1.09 I_2 = 0$
    $+20 I_1 - 38.9 I_2 = 0$

This equation may be solved by canceling out a term through addition to equation 1.

1. $-30 I_1 + 20 I_2 = -10$
    $20 I_1 - 38.9 I_2 = 0$

Equation 2 must be multiplied by 1.5 in order to cancel the I_1 term.

$1.5 (20 I_1 - 38.9 I_2) = 0 (1.5)$
$30 I_1 - 58.4 I_2 = 0$

Now, equations 1 and 2 may be added.

1. $\begin{aligned} -30\,I_1 + 20\,I_2 &= -10 \\ 30\,I_1 - 58.4\,I_2 &= 0 \\ \hline -38.4\,I_2 &= -10 \end{aligned}$

$$\frac{-38.4\,I_2}{-38.4} = \frac{-10}{-38.4}$$

$$I_2 = 0.26\ A = 260\ mA$$

The answer is positive, which means the originally assumed direction of current flow was correct.

To solve for I_1, equation 1 may be used, and the value of I_2 simply inserted.

$$-30\,I_1 + 20\,I_2 = -10$$
$$-30\,I_1 + 20\,(0.26) = -10$$
$$-30\,I_1 + 5.2 = -10$$
$$-30\,I_1 = -10 - 5.2$$
$$-30\,I_1 = -15.2$$
$$\frac{-30\,I_1}{-30} = \frac{-15.2}{-30}$$
$$I_1 = 0.507\ A = 507\ mA$$

The positive answer for I_1 indicates that the originally assumed direction of current flow was again correct.

To find I_3, the third equation may be used, and the value of I_2 substituted in.

$$5\,I_2 = 23\,I_3$$
$$23\,I_3 = 5\,I_2$$
$$23\,I_3 = 5\,(0.26)$$
$$23\,I_3 = 1.30$$
$$\frac{23\,I_3}{23} = \frac{1.3}{23}$$
$$I_3 = 0.057\ A = 57\ mA$$

Now that all the currents are known, the individual voltages may be found by using Ohm's law.

$$I_1 = 0.507\ A$$
$$I_2 = 0.26\ A$$
$$I_3 = 0.57\ A$$

$E_{R1} = I_1 \times R_1$
$E_{R1} = 0.507\,A \times 5\,\Omega$
$E_{R1} = 2.54\,V$
$E_{R2} = I_1 \times R_2$
$E_{R2} = 0.507\,A \times 5\,\Omega$
$E_{R2} = 2.54\,V$
$E_{R3} = (I_1 - I_2)\,R_3$
$E_{R3} = (0.507 - 0.26)\,20$
$E_{R3} = 0.247 \times 20$
$E_{R3} = 4.94\,V$
$E_{R4} = I_2 \times R_4$
$E_{R4} = 0.26\,A \times 15\,\Omega$
$E_{R4} = 3.9\,V$
$E_{R5} = (I_2 - I_3)\,R_5$
$E_{R5} = (0.26\,A - 0.057\,A)\,5\,\Omega$
$E_{R5} = 0.203\,A \times 5\,\Omega$
$E_{R5} = 1.02\,V$
$E_{R6} = I_3 \times R_6$
$E_{R6} = 0.057\,A \times 10\,\Omega$
$E_{R6} = 0.57\,V$
$E_{R7} = I_3 \times R_7$
$E_{R7} = 0.057\,A \times 8\,\Omega$
$E_{R7} = 0.456\,V$

Since no current flows through R_8, R_9 or R_{10}, the voltage dropped across each of these resistors is zero volts.

The real advantage of mesh current analysis is that, even though a circuit may have many loops or voltage sources, all mesh currents are assumed to flow in the same direction continuously around each mesh. This can make writing the equations much simpler; and if the *assumed direction* for one of the currents is *wrong*, this is indicated by a *negative sign* in the answer.

Node Voltage Analysis — Another method for analyzing dc circuits is called *node voltage* analysis. When using this method (Figure 12.46), you no longer express the voltage drops around the circuit in terms of IR drops; now you express them in terms of the *voltage at a node*. A node is just a common connection of two or more circuit components. A *principal node* is defined as a point where *three or more* components are connected.

NODE VOLTAGE ANALYSIS

- VOLTAGE DROPS ARE EXPRESSED IN TERMS OF THE VOLTAGE AT A **NODE**

- A NODE IS A COMMON CONNECTION OF TWO OR MORE COMPONENTS

Figure 12.46

Sample Solution with Node Voltage Analysis — In the circuit of Figure 12.47, points A and G are principal nodes. In this circuit, point G has been selected as the ground or reference point. The voltage, E_N, between points A and G (with G used as the reference) can be used to find all other voltages in the circuit, and is called the *node voltage*.

Figure 12.47

Use of Node Equation — In the node voltage analysis of this circuit, first write a Kirchhoff's current law equation for point A in the circuit. As shown in Figure 12.48, which is a close-up of this circuit at point A, $I_3 = I_1 + I_2$.

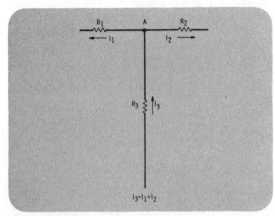

Figure 12.48

- Express Each Current as a Voltage
- Express All Voltages in Terms of Source and Node Voltages
- Substitute in Previous Equation

Express Each Current as a Voltage — Now each of these three currents can be expressed in terms of voltage and resistance as shown in Figure 12.49. The current flowing through R_3 can be put into an Ohm's law formula and expressed in terms of E_N and R_3: $I_3 = E_N/R_3$. Similarly, you can write that $I_1 = E_{R1}/R_1$ and $I_2 = E_{R2}/R_2$.

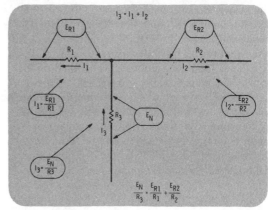

Figure 12.49

Express All Voltages in Terms of Source and Node Voltages — The next step is to take this equation and express all the voltages in it in terms of the source voltages (E_1 and E_2) and the node voltage E_N. You can do this by going around the loops containing the voltage sources, the node A and the ground reference as shown in Figure 12.50 and writing the loop equations. In the left-hand loop (starting at R_1 and traversing clockwise), you get $E_{R1} + E_N - E_1 = 0$. Transposing E_N and E_1 to the right-hand side of the equals sign (remembering to change their signs), you get $E_{R1} = E_1 - E_N$. For the right-hand loop (starting at R_2 and traversing counterclockwise), you get $E_{R2} + E_N - E_2 = 0$. Transposing this equation becomes $E_{R2} = E_2 - E_N$.

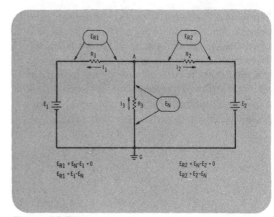

Figure 12.50

Substitute in Previous Equation — If you substitute this information into the current equation, it should appear as shown in Figure 12.51. Notice that in this final equation you know the values of all the resistances and source voltages, so you could solve the equation for E_N. Once E_N is known, all other voltages in the circuit can be found, using the equations you have just derived.

$$I_3 = I_1 + I_2$$

$$\frac{E_N}{R_3} = \frac{E_{R1}}{R_1} + \frac{E_{R2}}{R_2}$$

$$E_{R1} = E_1 - E_N \quad E_{R2} = E_2 - E_N$$

$$\frac{E_N}{R_3} = \frac{E_1 - E_N}{R_1} + \frac{E_2 - E_N}{R_2}$$

Figure 12.51

- **Summary: Node Voltage Analysis**
- **Thevenin's Theorem**
- **Sample Circuit**

Summary: Node Voltage Analysis — Figure 12.52 briefly summarizes the node voltage method of analyzing dc circuits. First, determine which node you want to analyze and which you want to use as a ground. Then write a node current equation for the node you have selected to analyze. Next, express each current in the node equations in terms of voltage and resistance. When these voltages are then expressed in terms of the node voltage and source voltages, the equation can be solved for the node voltage. Once the node voltage is determined, all other voltages can be found.

NODE VOLTAGE ANALYSIS

1. CHOOSE A NODE
2. WRITE A NODE CURRENT EQUATION
3. EXPRESS EACH CURRENT IN TERMS OF VOLTAGE AND RESISTANCE
4. EXPRESS THE VOLTAGES IN TERMS OF THE NODE VOLTAGE AND THE SOURCE VOLTAGES
5. SOLVE THE EQUATION FOR THE NODE VOLTAGE

Figure 12.52

Thevenin's Theorem — Another method of circuit analysis that is useful in simplifying the calculation of voltages in certain special situations is called Thevenin's Theorem (Figure 12.53). Thevenin's theorem states that *a complex circuit or network can be reduced to an equivalent series circuit with a single voltage source and a single series resistance*, as long as all components are linear.

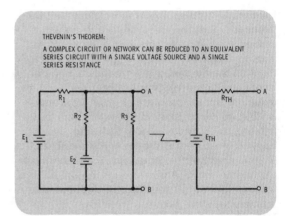

Figure 12.53

Sample Circuit — One type of problem (Figure 12.54) where Thevenin's theorem is useful is if you have a complex circuit and you wish to connect several different loads between two selected points, say points A and B. If you want to calculate the voltage between points A and B, you would normally have to analyze the complete series-parallel circuit every time you changed loads.

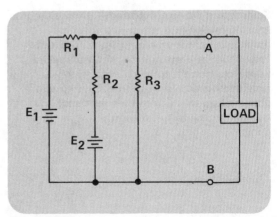

Figure 12.54

Thevenin Equivalent — When using Thevenin's theorem, you replace the complex circuit by an *equivalent voltage*, which is called the *Thevenin voltage* or E_{TH}, in series with a *resistance* which is called the *Thevenin resistance*, or R_{TH}, as shown in Figure 12.55. Then to find the voltage and current for each load, you connect to points A and B, and need analyze only this simple series circuit shown in Figure 12.55.

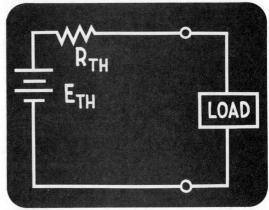

Figure 12.55

Calculation of Thevenin Voltage — To find E_{TH}, the Thevenin voltage as shown in Figure 12.56, calculate the voltage between the two selected points, when the *load is removed*.

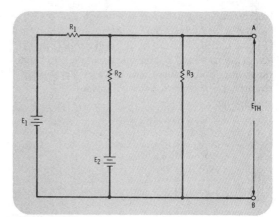

Figure 12.56

Calculation of Thevenin Resistance — The Thevenin resistance, R_{TH}, is the resistance you calculate between points A and B with *no load connected, and all the voltage sources replaced by short circuits*. So as shown in Figure 12.57 you would usually use series-parallel circuit reduction techniques to help you find R_{TH}.

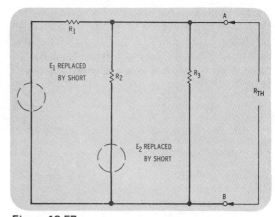

Figure 12.57

12-33

- **Norton's Theorem**
- **Norton Equivalent Current**
- **Norton Equivalent Resistance = Thevenin Equivalent Resistance**

Norton's Theorem — Another technique that you can use to analyze similar circuit situations is called Norton's theorem. Norton's theorem (Figure 12.58) states that *any complex circuit can be replaced by an equivalent parallel circuit consisting of a single current source, I_N, and a single shunt or parallel resistance, R_N.* Then, in a circuit where various loads are to be connected across points A and B, as shown, the circuit analysis is reduced to that of a simple parallel circuit shown.

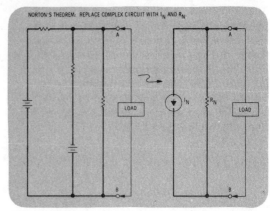

Figure 12.58

Norton Equivalent Current — The value of the Norton equivalent *current source*, I_N (Figure 12.59), is found by putting a *short circuit across points A and B*, and finding the current that flows through this short. The value of this short circuit current is the Norton equivalent current, I_N.

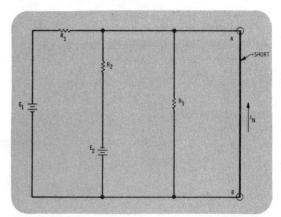

Figure 12.59

Norton Equivalent Resistance = Thevenin Equivalent Resistance — The Norton equivalent resistance, R_N (Figure 12.60), is found in the same way as the Thevenin resistance, R_{TH}. Thus, R_N equals R_{TH} for any one circuit between any two points.

THE NORTON EQUIVALENT RESISTANCE, R_N, IS FOUND IN THE SAME WAY AS THE THEVENIN RESISTANCE, R_{TH}

$$R_N = R_{TH}$$

Figure 12.60

- **Norton Equivalent Circuit**
- **Constant Current Source**
- **Millman's Theorem**

Norton Equivalent Circuit — Once the Norton equivalent circuit (Figure 12.61) is known, you can easily calculate the voltage across and current through any resistance you place between points A and B. It should be obvious that Norton's theorem is useful in design of voltage divider circuits, where you might want to connect different loads at different times to the same two points.

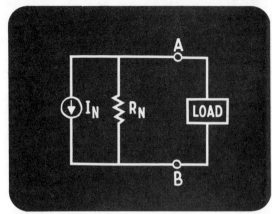

Figure 12.61

Constant Current Source — Notice that the circle with the arrow inside it used in the Norton equivalent circuit is the symbol for a *current source* (Figure 12.62). This is a special idealized source that would provide a *constant current output to a load*, but whose output voltage would vary depending on the load put across it. (This is different from the normal power supplies discussed throughout this course whose *voltage output is constant* and whose current output would vary depending on the load connected to it.)

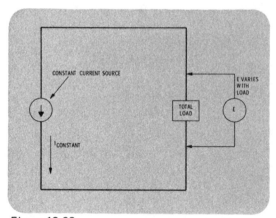

Figure 12.62

Millman's Theorem — The final circuit analysis technique discussed in this lesson is called *Millman's theorem*. This theorem is similar to Thevenin's and Norton's theorems in that it is used to help simplify a circuit by analyzing circuit behavior between two selected points.

More specifically, Millman's theorem gives you a formula to find the total voltage across several parallel branches (Figure 12.63), each of which may contain a different voltage source and several resistors. The details of *deriving* Millman's equation are not shown here, but a complete derivation is included at the end of this lesson. The intent is to show you the results of Millman's

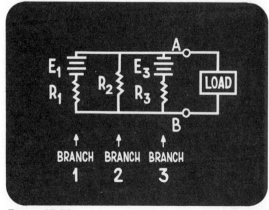

Figure 12.63

theorem and how it can help in simplifying circuit analysis problems.

Millman's Formula — Figure 12.64 shows the final formula that expresses Millman's theorem. I_1, I_2, and I_3 are the branch currents and each of these can be calculated using the branch voltages and branch resistances. Once you use this formula to calculate the voltage between points A and B, you can use that voltage to find the current through any resistive load you may place between A and B. Millman's theorem can therefore be used to greatly simplify repetitive calculations involving a parallel circuit where different loads are to be applied in circuits where there are *no series resistances between the parallel branches*.

    In this lesson you have reviewed Kirchhoff's laws and seen these laws along with the superposition theorem used to analyze some typical circuits. Also, a very brief survey of some of the other more advanced methods of circuit analysis has been presented. Remember, at the end of this lesson there is additional information, including a review chart covering the advanced methods along with a list of reference texts, should you care to study these methods of analysis in more depth. The complete details of these advanced methods introduced are properly covered in more advanced courses in circuit analysis.

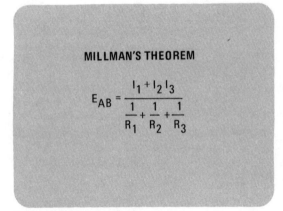

MILLMAN'S THEOREM

$$E_{AB} = \frac{I_1 + I_2\, I_3}{\dfrac{1}{R_1} + \dfrac{1}{R_2} + \dfrac{1}{R_3}}$$

Figure 12.64

● **Worked Through Examples**

1. Solve for all voltages and currents in the following circuit using Kirchhoff's laws:

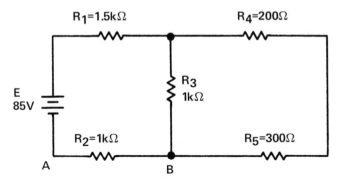

Notice first that this type of circuit could be solved using Ohm's law and circuit reduction techniques, but can also be analyzed using the more general Kirchhoff's law techniques, presented in Lessons 11 and 12. As a first step in solving this circuit, label all the currents flowing in the circuit, assigning each a direction as shown below. (This circuit will be analyzed here using *electron current*.)

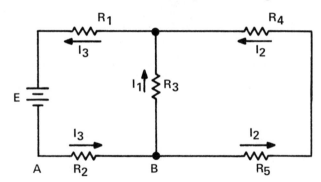

You can immediately apply Kirchhoff's current law at point B in this circuit: at that point I_3 enters, and I_1 and I_2 leave. So by Kirchhoff's current law $I_3 = I_1 + I_2$. Now proceed to use Kirchhoff's voltage law to write two loop equations for the circuit.

Starting at point A and traversing the left-hand loop counterclockwise, the voltages are: $-R_2 I_3 - R_3 I_1 - R_1 I_3 + E = 0$. Traversing the right-hand loop counterclockwise and starting at point B, the voltages are: $-R_5 I_2 - R_4 I_2 + R_3 I_1 = 0$. Now, these loop equations may be simplified by substituting $I_1 + I_2$ for I_3, and combining terms.

$$-R_2 I_3 - R_3 I_1 - R_1 I_3 + E = 0 \qquad \text{(substitute } I_1 + I_2 \text{ for } I_3)$$
$$-R_2(I_1 + I_2) - R_3 I_1 - R_1 (I_1 + I_2) + E = 0$$
$$-R_2 I_1 - R_2 I_2 - R_3 I_1 - R_1 I_1 - R_1 I_2 + E = 0$$

Now substitute in the circuit values:

$$-1k I_1 - 1k I_2 - 1k I_1 - 1.5k I_1 - 1.5k I_2 + 85 = 0$$

And combine terms:

$$-3.5kI_1 - 2.5kI_2 + 85 = 0$$
$$-3.5kI_1 - 2.5kI_2 = -85$$

or

$$3.5kI_1 + 2.5kI_2 = 85$$

Now, write the second loop equation:

$$-R_5I_2 - R_4I_2 + R_3I_1 = 0$$

Substitute circuit values:

$$-300I_2 - 200I_2 + 1kI_1 = 0$$

Combine terms:

$$-500I_2 + 1kI_1 = 0.$$

Now the two equations may be solved by canceling out one of the terms through addition.

$$3.5kI_1 + 2.5kI_2 = 85$$
$$1kI_1 - 500I_2 = 0$$

To cancel out the I_1 term, multiply the lower equation by -3.5.

$$-3.5(1kI_1 - 500I_2) = 0(-3.5)$$
$$-3.5kI_1 + 1.75kI_2 = 0.$$

Now add the two equations.

$$3.5kI_1 + 2.5kI_2 = 85$$
$$\underline{-3.5kI_1 + 1.75kI_2 = 0}$$
$$4.25kI_2 = 85$$
$$\frac{4.25kI_2}{4.25k} = \frac{85}{4.25k}$$
$$I_2 = 20 \text{ mA}$$

Substitute this value back into one of the equations and solve for I_1.

$$1kI_1 - 500I_2 = 0$$
$$1kI_1 - 500 (20 \text{ mA}) = 0$$
$$1kI_1 - 10,000 \text{ mA} = 0$$
$$1kI_1 = 10,000 \text{ mA}$$
$$\frac{1kI_1}{1k} = \frac{10,000 \text{ mA}}{1k}$$

$$I_1 = 10 \text{ mA}$$
$$I_3 = I_1 + I_2 = 10 \text{ mA} + 20 \text{ mA} = 30 \text{ mA}$$

All of the individual circuit voltage drops may now be found by using Ohm's law.

$E_{R1} = I_3 \times R_1$

$E_{R1} = 30 \text{ mA} \times 1.5 \text{ k}\Omega$

$E_{R1} = 45 \text{ V}$

$E_{R2} = I_3 \times R_2$

$E_{R2} = 30 \text{ mA} \times 1 \text{ k}\Omega$

$E_{R2} = 30 \text{ V}$

$E_{R3} = I_1 \times R_3$

$E_{R3} = 10 \text{ mA} \times 1 \text{ k}\Omega$

$E_{R3} = 10 \text{ V}$

$E_{R4} = I_2 \times R_4$

$E_{R4} = 20 \text{ mA} \times 200 \text{ }\Omega$

$E_{R4} = 4 \text{ V}$

$E_{R5} = I_2 \times R_5$

$E_{R5} = 20 \text{ mA} \times 300 \text{ }\Omega$

$E_{R5} = 6 \text{ V}$

2.   Using Kirchhoff's laws, solve for all voltages and currents in the circuit shown below.

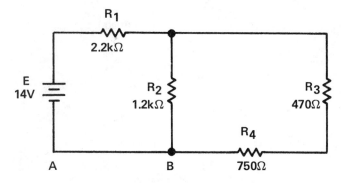

Label all currents and assign them a direction as shown in the figure below. Assuming electron current, I_3, flows from the source to point B where it splits up into I_1 and I_2 as shown.

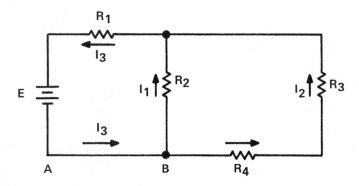

Kirchhoff's current law can therefore be applied at point B: it simply states that $I_3 = I_1 + I_2$. Now traverse two loops, and apply Kirchhoff's voltage law to each to obtain two loop equations. This will yield a total of three equations, which will allow you to solve for the three unknown currents in the problem.

First, the left-hand loop will be traversed from point A in a counterclockwise direction. The resulting loop equation is: $-R_2I_1 - R_1I_3 + E = 0$. The right-hand loop may be traversed from point B in a counterclockwise direction also. This yields $-R_4I_2 - R_3I_2 + R_2I_1 = 0$. These two equations may now be simplified by substituting in $(I_1 + I_2)$ for I_3, and combining like terms.

$$-R_2I_1 - R_1I_3 + E = 0$$
$$-R_2I_1 - R_1(I_1 + I_2) + E = 0$$
$$-R_2I_1 - R_1I_1 - R_1I_2 + E = 0$$

Now substitute the circuit values into the equations.

$$-1.2kI_1 - 2.2kI_1 - 2.2kI_2 + 14 = 0$$
$$-3.4kI_1 - 2.2kI_2 + 14 = 0$$
$$-3.4kI_1 - 2.2kI_2 = -14 \qquad \text{(1st simplified loop equation)}$$

$$-R_4I_2 - R_3I_2 + R_2I_1 = 0$$
$$-750\,I_2 - 470\,I_2 + 1.2kI_1 = 0$$
$$-1220\,I_2 + 1200\,I_1 = 0$$
$$+1200\,I_1 = 1220\,I_2$$
$$\frac{1200\,I_1}{1200} = \frac{1220\,I_2}{1200}$$

$$I_1 = 1.02\,I_2 \qquad \text{(2nd simplified loop equation)}$$

Now that I_1 has been expressed in terms of I_2, the I_1 terms in the first simplified loop equation may all be replaced with I_2 terms. The equation may then be solved for I_2 as shown below.

$$-3.4kI_1 - 2.2kI_2 = -14$$
$$-3.4k\,(1.02\,I_2) - 2.2kI_2 = -14$$
$$-3.47kI_2 - 2.2kI_2 = -14$$
$$-5.67kI_2 = -14$$
$$\frac{-5.67kI_2}{-5.67k} = \frac{-14}{-5.67k}$$

$$I_2 = 2.47 \text{ mA}$$

To find I_1, the value of I_2 may be substituted into the second simplified loop equation.

$$I_1 = 1.02 \, I_2$$
$$I_1 = 1.02 \, (2.47 \text{ mA})$$
$$I_1 = 2.52 \text{ mA}$$

Finally, since $I_1 + I_2 = I_3$, you can calculate that:

$$I_3 = (2.47 \text{ mA} + 2.52 \text{ mA})$$
$$I_3 = 4.99 \text{ mA}$$

The three current values may now be used to find the voltage dropped across each resistor in the circuit.

$$E_{R1} = R_1 \times I_3$$
$$E_{R1} = 2.2 \text{ k}\Omega \times 4.99 \text{ mA}$$
$$E_{R1} = 11.0 \text{ V}$$

$$E_{R2} = R_2 \times I_1$$
$$E_{R2} = 1.2 \text{ k}\Omega \times 2.52 \text{ mA}$$
$$E_{R2} = 3.02 \text{ V}$$

$$E_{R3} = R_3 \times I_2$$
$$E_{R3} = 470 \, \Omega \times 2.47 \text{ mA}$$
$$E_{R3} = 1.16 \text{ V}$$

$$E_{R4} = R_4 \times I_2$$
$$E_{R4} = 750 \, \Omega \times 2.47 \text{ mA}$$
$$E_{R4} = 1.85 \text{ V}$$

3. Solve for all of the circuit *currents* in the multiple-source dc circuit shown below.

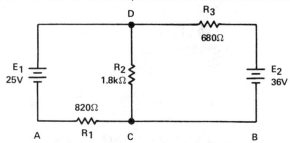

First, label all of the currents and assign them a direction, as shown below. In this analysis, electron current directions are being assumed.

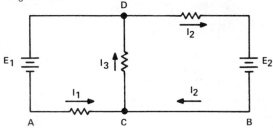

Kirchhoff's current law can be immediately applied at point C in the circuit: $I_1 + I_2 = I_3$.

Proceed to use Kirchhoff's voltage law on two of the circuit loops to get two loop equations. The two loop equations, together with the current law equation, will give you the three equations necessary to solve for the unknown currents.

First, traverse the left-hand loop counterclockwise, starting from point A, going through points C and D, and returning to point A. Algebraically adding each voltage you encounter (remembering the rules for using the voltage law) you get:

$$-R_1 I_1 - R_2 I_3 + 25 = 0.$$

Traversing the right-hand loop clockwise from point B (through points C and D and back to B) yields the equation:

$$-R_2 I_3 - R_3 I_2 + 36 = 0.$$

These two equations may now be reduced to simpler forms by substituting the expression $(I_1 + I_2)$ for I_3 each time it appears, and then combining like terms.

$$-R_1 I_1 - R_2 I_3 + 25 = 0$$
$$-R_1 I_1 - R_2 (I_1 + I_2) + 25 = 0$$
$$-R_1 I_1 - R_2 I_1 - R_2 I_2 + 25 = 0$$
$$-820 I_1 - 1.8k I_1 - 1.8k I_2 + 25 = 0$$
$$-2.62k I_1 - 1.8k I_2 + 25 = 0$$
$$-2.62k I_1 - 1.8k I_2 = -25$$

$$-R_2 I_3 - R_3 I_2 + 36 = 0$$
$$-R_2 (I_1 + I_2) - R_3 I_2 + 36 = 0$$
$$-R_2 I_1 - R_2 I_2 - R_3 I_2 + 36 = 0$$
$$-1.8k I_1 - 1.8k I_2 - 680 I_2 + 36 = 0$$
$$-1.8k I_1 - 2.48k I_2 + 36 = 0$$
$$-1.8k I_1 - 2.48k I_2 = -36$$

These two equations may now be solved simultaneously by manipulating one of the equations so that one of the "I" terms cancels out the pair.

$$(*) \; -2.62k I_1 - 1.8k I_2 = -25$$
$$-1.8k I_1 - 2.48k I_2 = -36$$

Now to arrange for one of the "I" terms to cancel, the lower equation may be multiplied by -1.46:

$$(-1.46)(-1.8k I_1 - 2.48k I_2) = -36(-1.46)$$
$$2.62k I_1 + 3.62k I_2 = +52.6$$

Now this equation may be added to the first loop equation (*) to yield:

$$-2.62kI_1 - 1.8kI_2 = -25$$
$$2.62kI_1 + 3.62kI_2 = +52.6$$
$$\overline{\hspace{2.5cm}1.82kI_2 = 27.6}$$

$$\frac{1.82kI_2}{1.82k} = \frac{27.6}{1.82k}$$

$$I_2 = 15.2 \text{ mA}$$

Substitute this value back into the first equation:

$$-2.62kI_1 - 1.8k\,(15.2m) = -25$$
$$-2.62kI_1 - 27.36 = -25$$
$$-2.62kI_1 = -25 + 27.36$$
$$-2.62kI_1 = 2.36$$
$$\frac{-2.62kI_1}{-2.62k} = \frac{2.36}{-2.62k}$$
$$I_1 = -0.9 \text{ mA}$$

The negative sign in front of the 0.9 indicates that this current flows in the opposite direction from that direction originally assumed. The circuit, redrawn with the correct current appears as shown below.

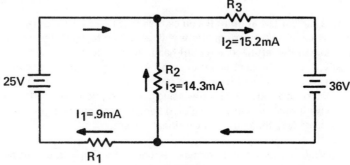

As can be seen from the figure, the 36-volt battery is actually pushing current through the 25-volt battery in a backward direction. The value of I_3 may now easily be calculated:

$$I_1 + I_2 = I_3$$
$$-0.9 + 15.2 = I_3$$
$$14.3 \text{ mA} = I_3$$

The result is positive, indicating that the correct direction was assumed for the current flow at the start of the problem.

4. Solve for all the voltages and currents in the following circuit using Kirchhoff's laws. Use *conventional current* in your analysis.

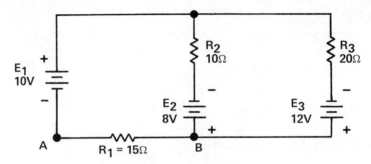

First, label all currents and assign them a direction, as shown in the circuit diagram below. The directions shown are assumed *conventional current directions*.

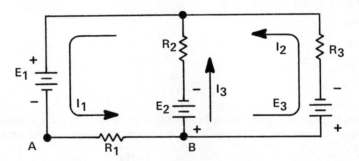

Applying Kirchhoff's current law at point B you can see that $I_1 = I_2 + I_3$, or $I_3 = I_1 - I_2$. Since these *three* currents are unknown, two more equations will be needed to finish the circuit solution. Apply Kirchhoff's voltage law to the left-hand and right-hand loops, that is, traverse the loops, add all voltages algebraically, and set the sums equal to zero. Assuming *conventional current* and starting at point A, the voltages are: In the right-hand loop, traversing clockwise, $10 + 10 I_3 + 8 + 15 I_1 = 0$. In the right-hand loop, starting at point B and traversing clockwise, you have: $-8 - 10 I_3 + 20 I_2 + 12 = 0$. Remember, since conventional current is being used here, voltage *source* terms are positive if you traverse them in the same way they push conventional current.

Now, these equations may be rewritten in terms of I_1 and I_2 by simply substituting $(I_1 - I_2)$ for I_3 each time it appears. In the first equation, this is equal to:

$$10 + 10 I_3 + 8 + 15 I_1 = 0$$
$$10 + 10 (I_1 - I_2) + 8 + 15 I_1 = 0$$
$$10 + 10 I_1 - 10 I_2 + 8 + 15 I_1 = 0$$
$$10 + 25 I_1 - 10 I_2 + 8 = 0$$
$$25 I_1 - 10 I_2 + 18 = 0$$
$$25 I_1 - 10 I_2 = -18$$

Following the same procedure in the second equation, you have:

$$-8 - 10 I_3 + 20 I_2 + 12 = 0$$
$$-8 - 10 (I_1 - I_2) + 20 I_2 + 12 = 0$$
$$-8 - 10 I_1 + 10 I_2 + 20 I_2 + 12 = 0$$
$$-8 - 10 I_1 + 30 I_2 + 12 = 0$$
$$-10 I_1 + 30 I_2 + 4 = 0$$
$$-10 I_1 + 30 I_2 = -4$$

These loop equations may be solved by addition after multiplying the top equation by 3 (in order to make the I_2 terms cancel).

$$3 (25 I_1 - 10 I_2) = (-18) (3)$$
$$-10 I_1 + 30 I_2 = -4$$

$$75 I_1 - 30 I_2 = -54$$
$$-10 I_1 + 30 I_2 = -4$$

$$65 I_1 \qquad = -58$$

$$\frac{65 I_1}{65} = \frac{-58}{65}$$

$$I_1 = -0.892 \text{ A}$$

Since this answer is negative, you know that the *conventional current*, I_1, is flowing *opposite* to the assumed direction in the left-hand loop.

I_2 may now be found by substituting the I_1 value into one of the original loop equations.

$$-10 I_1 + 30 I_2 = -4$$
$$-10 (-0.892) + 30 I_2 = -4$$
$$+8.92 + 30 I_2 = -4$$
$$30 I_2 = -12.92$$

$$\frac{30 I_2}{30} = \frac{-12.92}{30}$$

$$I_2 = -0.431 \text{ A}$$

The negative answer for I_2 again indicates that the originally assumed direction was incorrect and that the conventional current I_2 actually flows clockwise in the right-hand loop. To find the value of I_3, go back to the equation that expresses I_3 in terms of I_1 and I_2. I_3 is equal to $I_1 - I_2$.

$$I_3 = I_1 - I_2$$
$$I_3 = (-0.892 \text{ A}) - (-0.431 \text{ A})$$
$$I_3 = -0.461 \text{ A}$$

This answer is also negative, indicating that I_3 actually flows down through R_2, so conventional current in this circuit flows as shown below.

To go on and find the individual voltage drops in this problem, simply substitute the values of current and resistance into Ohm's law for each resistor in the circuit.

$$E_{R1} = I_1 \times R_1$$
$$E_{R1} = 0.892 \text{ A} \times 15$$
$$E_{R1} = 13.38 \text{ V}$$

$$E_{R2} = I_3 \times R_2$$
$$E_{R2} = 0.461 \times 10$$
$$E_{R2} = 4.61 \text{ V}$$

$$E_{R3} = I_2 \times R_3$$
$$E_{R3} = 0.431 \times 20$$
$$E_{R3} = 8.62 \text{ V}$$

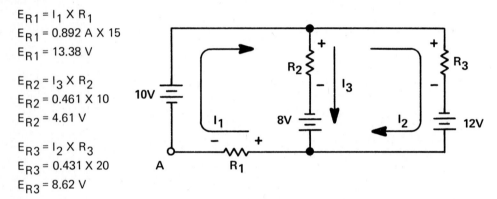

The voltage polarities are also labeled on the diagram. Remember that conventional current always flows through resistors from + to −; while electron current flows through resistors from − to +.

5. Solve for all voltages and currents in the following circuit using the *superposition theorem*.

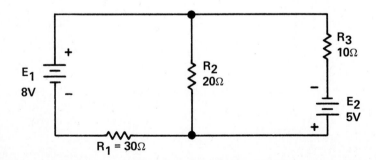

The superposition theorem says that "In a network with two or more sources, the current or voltage for any component is the algebraic sum of the effects produced by each source acting separately. The first step in solving this problem, is to remove one source and replace it with a short, or conducting path.

The circuit with E_1 "shorted out" looks like this:

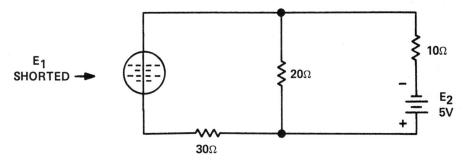

Circuit reduction laws may be applied, and the circuit reduced down to one equivalent resistance.

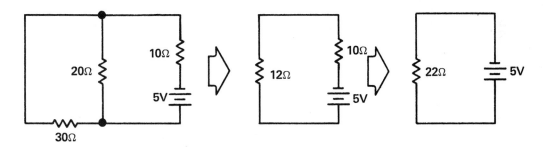

The total current may now be found by applying Ohm's law:

$$I_T = E_T/R_T$$
$$I_T = 5/22$$
$$I_T = 0.227 \text{ A} = 227 \text{ mA}$$

This "total current" flows through resistor R_3. E_{R3} may be found by applying Ohm's law.

$$E_{R3} = I_{R3} \times R_3$$
$$E_{R3} = 0.227 \times 10$$
$$E_{R3} = 2.27 \text{ V}$$

The balance of the circuit voltage is dropped across the parallel combination of resistors R_1 and R_2 (R_{1-2}). The voltage value may be found by applying Kirchhoff's voltage law for series circuits.

$$E_2 = E_{R1-2} + E_{R3}$$
$$5 \text{ V} = E_{R1-2} + 2.27 \text{ V}$$
$$-E_{R1-2} = -5 \text{ V} + 2.27 \text{ V}$$
$$E_{R1-2} = 2.73 \text{ V}$$

Knowing the voltage across the parallel resistors R_1 and R_2, their currents may be found by using Ohm's law.

$I_{R1} = E_1/R_1$ $I_{R2} = E_2/R_2$

$I_{R1} = 2.73/30$ $I_{R2} = 2.73/20$

$I_{R1} = 91$ mA $I_{R2} = 137$ mA

The circuit, completely analyzed with E_1 shorted, looks like this:

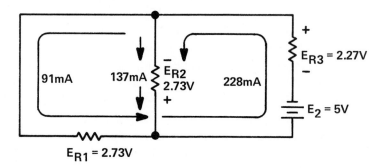

Short E_2, and solve the circuit for all voltages and currents using E_1 as the source voltage.

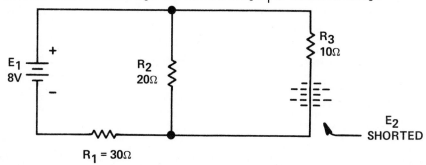

First reduce the circuit to find its equivalent resistance.

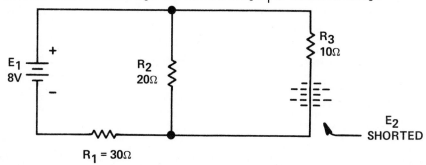

Now, use the equivalent resistance and the total voltage to find the total current.

$$I_T = E_T/R_{eq}$$
$$I_T = 8/36.67$$
$$I_T = 0.218 \text{ A} = 218 \text{ mA}$$

This current value flows through R_1. Ohm's law may be used to find the voltage dropped across R_1.

$$E_{R1} = I_T \times R_1$$
$$E_{R1} = 0.218 \text{ A} \times 30 \, \Omega$$
$$E_{R1} = 6.54 \text{ V}$$

According to the series circuit voltage law, the rest of the applied voltage must be dropped across the other circuit resistance, which in this case is the parallel combination of resistors R_2 and R_3 ($R_{2\text{-}3}$).

$$E_{R2\text{-}3} = E_1 - E_{R1}$$
$$E_{R2\text{-}3} = 8 \text{ V} - 6.54 \text{ V}$$
$$E_{R2\text{-}3} = 1.46 \text{ V}$$

Since the 20-ohm and 10-ohm resistors are connected in parallel, the voltage across each of them is equal to 1.46 volts. The current flowing through these resistors may now be found by using Ohm's law and the known values of resistance and voltage:

$$I_{R2} = \frac{E_{R2}}{R_2}$$

$$I_{R2} = \frac{1.46}{20}$$

$$I_{R2} = 73 \text{ mA}$$

$$I_{R3} = \frac{E_{R3}}{R3}$$

$$I_{R3} = \frac{1.46}{10}$$

$$I_{R3} = 146 \text{ mA}$$

Here is the circuit with all circuit values shown, with the source, E_1, acting alone.

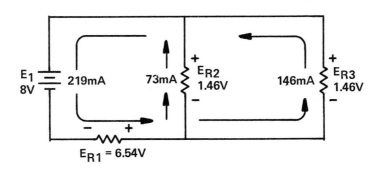

Now, superimpose this circuit over the first circuit you worked and take the algebraic sum of the values of current and voltage in the circuit.

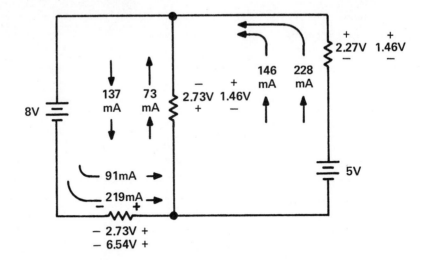

Algebraically add all the voltage and current values for each resistor to get the final circuit values as labeled below:

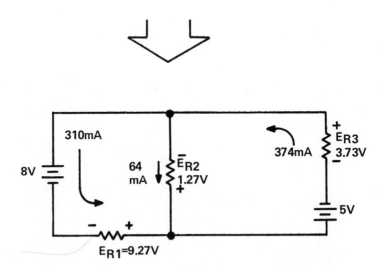

6. Solve the following circuit for all values of current and voltage using the superposition theorem:

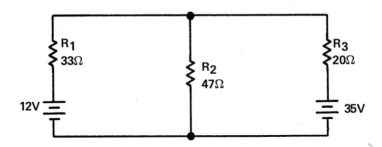

First, short one source and reduce the circuit to its equivalent resistance.

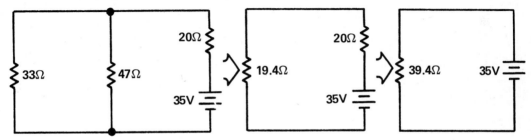

The total circuit current may be found now by using Ohm's law.

$$I_T = \frac{E_T}{R_{eq}}$$

$$I_T = \frac{35 \text{ V}}{39.4 \text{ }\Omega}$$

$$I_T = 0.888 \text{ A} = 888 \text{ mA}$$

Looking back to the simplified circuit diagrams, you can now see that the total current I_T flows through the 20-ohm resistor and throgh the parallel combination of resistors R_1 and R_2 ($R_{1\text{-}2}$), which is 19.4 ohms. The voltage across these resistors may be calculated by using Ohm's law.

$$E_{R3} = I_T \times R_3$$
$$E_{R3} = 0.888 \times 20$$
$$E_{R3} = 17.8 \text{ V}$$

$$E_{R1\text{-}2} = I_T \times R_{1\text{-}2}$$
$$E_{R1\text{-}2} = 0.888 \times 19.4$$
$$E_{R1\text{-}2} = 17.2 \text{ V}$$

Now the current through resistors R_1 and R_2 may be found by using Ohm's law.

$$I_{R1} = \frac{E_{R1}}{R_1} \qquad\qquad I_{R2} = \frac{E_{R2}}{R_2}$$

$$I_{R1} = \frac{17.2\ V}{33\ \Omega} \qquad\qquad I_{R2} = \frac{17.2\ V}{47\ \Omega}$$

$$I_{R1} = 521\ mA \qquad\qquad I_{R2} = 366\ mA$$

Here is the circuit with E_1 shorted and all currents and voltages listed.

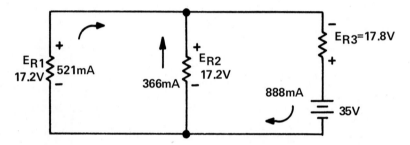

Now, short E_2 and calculate all voltages and currents in the circuit.

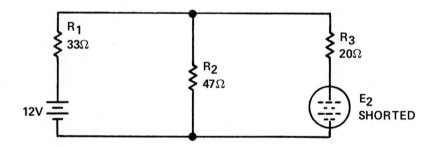

The circuit may now be reduced to its equivalent resistance.

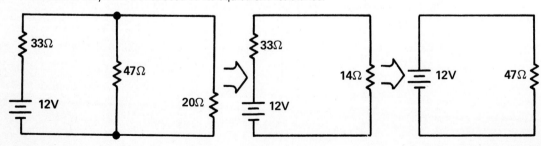

The total current may be calculated using Ohm's law.

$$I_T = \frac{E_T}{R_{eq}}$$

$$I_T = \frac{12\ V}{47\ \Omega}$$

$$I_T = 0.255\ A$$

This total current flows through the 33-ohm resistor producing a voltage equal to $I_T \times R_1$.

$$E_{R1} = I_T \times R_1$$
$$E_{R1} = 0.255\ A \times 33\ \Omega$$
$$E_{R1} = 8.42\ V$$

The balance of the 12 volts is dropped across the resistor combination $R_{2\text{-}3}$, as stated by the series circuit voltage law $E_T = E_1 + E_2$. Since, in this case, $E_T = 12$ volts, $12 - 8.42$ is equal to 3.58 V. This is the voltage dropped across the parallel combination of resistors $R_{2\text{-}3}$.

Now that E_{R2} and E_{R3} are known, the current through the resistors may be calculated using Ohm's law.

$$I_{R2} = \frac{E_{R2}}{R_2}$$

$$I_{R3} = \frac{E_{R3}}{R_3}$$

$$I_{R2} = \frac{3.58\ V}{47}$$

$$I_{R3} = \frac{3.58\ V}{20\ \Omega}$$

$$I_{R2} = 76\ mA$$

$$I_{R3} = 179\ mA$$

Here is the circuit with E_2 shorted and all voltages and currents listed.

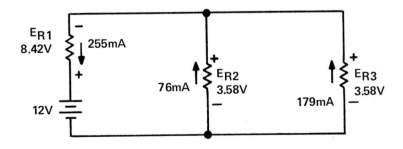

Now, superimpose the values of this circuit over the values of the circuit that had E_1 shorted.

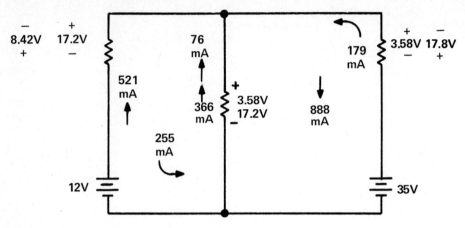

Now take the algebraic sum of all these voltages and currents.

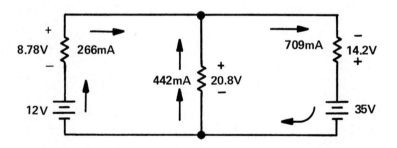

• **Practice Problems**

Depending upon the approach you use in solving these problems and how   **Fold Over**
you round off intermediate results, your answers may vary slightly from those
given here. However, any differences you encounter should only occur in the
third significant digit of your answer. If the first two significant digits of your
answers do not agree with those given here, recheck your calculations.

1.

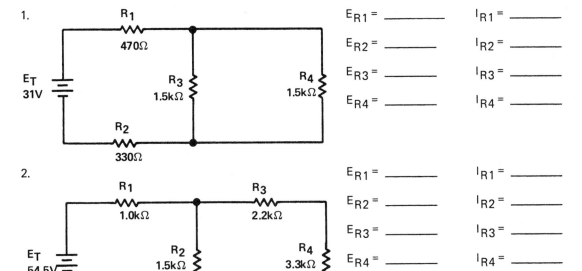

$E_{R1} =$ _____ $I_{R1} =$ _____

$E_{R2} =$ _____ $I_{R2} =$ _____

$E_{R3} =$ _____ $I_{R3} =$ _____

$E_{R4} =$ _____ $I_{R4} =$ _____

2.

$E_{R1} =$ _____ $I_{R1} =$ _____

$E_{R2} =$ _____ $I_{R2} =$ _____

$E_{R3} =$ _____ $I_{R3} =$ _____

$E_{R4} =$ _____ $I_{R4} =$ _____

Find the following values for these circuits

3.   Thevenin's theorem

$R_{TH} =$ _____

$E_{TH} =$ _____

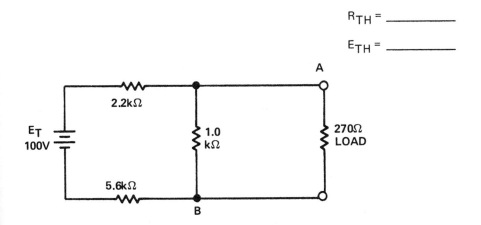

Answers

1. $E_{R1} = 9.4$ V $I_{R1} = 20$ mA

   $E_{R2} = 6.6$ V $I_{R2} = 20$ mA

   $E_{R3} = 15$ V $I_{R3} = 10$ mA

   $E_{R4} = 15$ V $I_{R4} = 10$ mA

2. $E_{R1} = 25$ V $I_{R1} = 25$ mA

   $E_{R2} = 29.5$ V $I_{R2} = 19.7$ mA

   $E_{R3} = 11.7$ V $I_{R3} = 5.33$ mA

   $E_{R4} = 17.6$ V $I_{R4} = 5.33$ mA

3. $R_{TH} = 886$ Ω

   $E_{TH} = 11.4$ V

4.

Fold Over

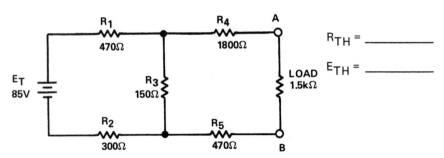

$R_{TH} =$ _____

$E_{TH} =$ _____

5.

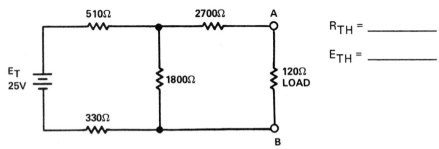

$R_{TH} =$ _____

$E_{TH} =$ _____

6. Norton's theorem

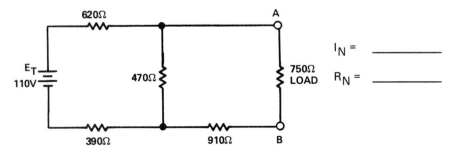

$I_N =$ _____

$R_N =$ _____

7.

$I_N =$ _____

$R_N =$ _____

Answers

4. $R_{TH} = 2.40 \text{ k}\Omega$

   $E_{TH} = 13.9 \text{ V}$

5. $R_{TH} = 3.27 \text{ k}\Omega$

   $E_{TH} = 17 \text{ V}$

6. $I_N = 28.4 \text{ mA}$

   $R_N = 1.23 \text{ k}\Omega$

7. $I_N = 28 \text{ mA}$

   $R_N = 600 \ \Omega$

8. Millman's theorem

Fold Over

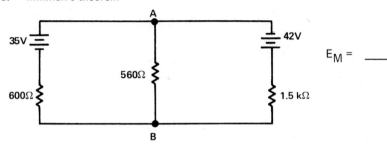

$E_M =$ _____

9.

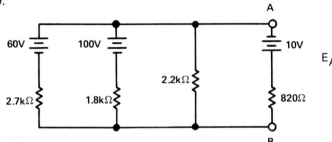

$E_{AB} =$ _____

10. Mesh currents

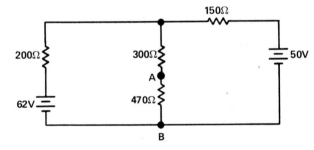

$E_{AB} =$ _____

$I_{AB} =$ _____

11.

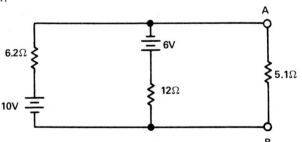

$E_{AB} =$ _____

$I_{AB} =$ _____

Answers

8. $E_M = 7.36$ V

9. $E_{AB} = 25.2$ V

10. $E_{AB} = 30.1$ V

   $I_{AB} = 64.0$ mA

11. $E_{AB} = 2.53$ V

   $I_{AB} = 496$ mA

12. Node voltage

Fold Over

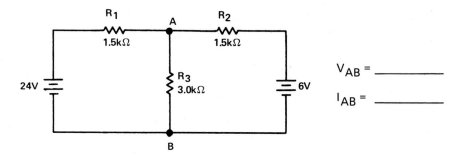

V_{AB} = _____

I_{AB} = _____

13.

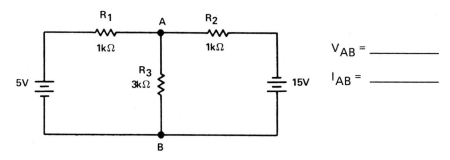

V_{AB} = _____

I_{AB} = _____

14. Superposition theorem

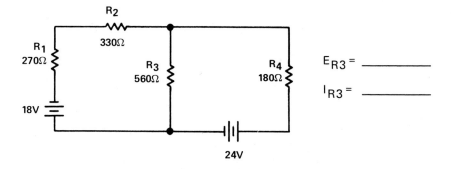

E_{R3} = _____

I_{R3} = _____

Answers

12. V_{AB} = 12 V

   I_{AB} = 4 mA

13. V_{AB} = 4.29 V

   I_{AB} = 1.43 mA

14. E_{R3} = 11.5 V

   I_{R3} = 20.5 mA

15. **Fold Over**

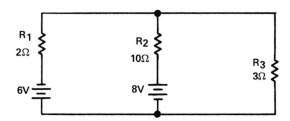

E_{R2} = ⎯⎯⎯⎯⎯⎯

I_{R2} = ⎯⎯⎯⎯⎯⎯

Answers

15. $E_{R2} = 10.4$ V

    $I_{R2} = 1.04$ A

1. A voltage source receives a _____ sign in a loop equation if one passes over it in the same direction as it normally pushes current.

   a. positive
   b. negative
   c. current
   d. voltage

2. If a voltage source is passed over when writing a loop equation in the opposite direction to that which it normally pushed current, the voltage receives:

   a. a negative sign
   b. a positive sign
   c. a current sign
   d. a voltage sign

3. When writing loop equations, if a resistor is passed over in the direction of the assumed current flow, the voltage across it is considered:

   a. negative
   b. positive
   c. zero
   d. opposite

4. The voltage across a resistor in a loop equation is considered positive if the resistor is passed over in _____ direction as the assumed current.

   a. The same
   b. A like
   c. An upward
   d. The opposite

5. $I_1 + I_2 + I_3 = 0$ is an equation using Kirchhoff's _____ law.

   a. Ohms
   b. Voltage
   c. Current
   d. Resistance

   Solve the following for the missing current. Give a value and direction. + is away from the node; − is toward the node.

6. $I_1=300mA$
   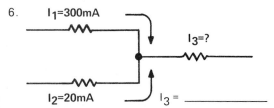
   $I_3=?$
   $I_2=20mA$ $I_3 =$ _____

7.
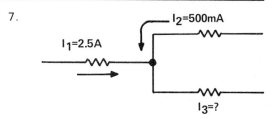
$I_1=2.5A$
$I_2=500mA$
$I_3=?$

$I_3 =$ _____

8. $I_1=30mA$

$I_3=.05A$

$I_2=?$

$I_2 =$ _____

9.

$I_1=20mA$

$I_3=15.5mA$ $I_2=5mA$

$I_4=?$

$I_4 =$ _____

10.

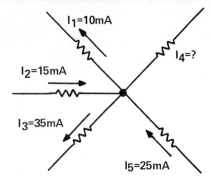

$I_4 =$ _____

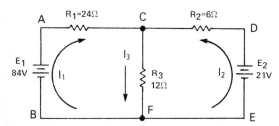

Use the above schematic for questions 11, 12, 13, 14 and 15.

11. Write the conventional current loop equation for loop A B C F A:

a. $84 - 24I_1 - 12I_3 - 12I_2 = 0$
b. $84 - 24I_1 + 12I_3 = 0$
c. $84 - 24I_1 - 12I_3 = 0$
d. $-84 - 24I_1 - 12I_3 = 0$

12. Write the conventional current loop equation for loop D E F C D:

a. $21 - 6I_2 - 12I_3 = 0$
b. $21 - 6I_2 - 12I_3 - 12I_1 = 0$
c. $-21 + 6I_2 - 12I_3 = 0$
d. $21 - 6I_2 + 12I_3 = 0$

13. Write the Kirchoff's current law for node C:

a. $I_1 + I_2 + I_3 = 0$
b. $I_1 = I_2 + I_3$
c. $I_1 = -I_2 + I_3$
d. $I_1 + I_2 = I_3$
e. d only
f. c and d above

14. Solve for I_1, I_2, and I_3:

a. $+3.0, -2.0, +1.0$
b. $+2.5, +0.5, +3.0$
c. $+2.29, -0.36, +1.93$
d. $+2.5, -0.5, +2.0$

15. What are the voltages across R_1, R_2, and R_3?

a. 81, 24, 3
b. 55, 8, 29
c. 60, 3, 24
d. None of above

16. When using the Superposition Theorem, all components in the circuits must be:

a. Voltage dependent
b. Linear and bilateral
c. Have low temperature coefficients
d. Be stable

17. When analyzing a circuit using mesh currents, it is usually assumed that all mesh currents:

a. Flow backwards
b. Flow in the same direction
c. Are indefinitely small
d. Can't flow opposite to the direction assumed

18. In the node voltage analysis method for circuit analysis, the voltage drops around a circuit are expressed in terms:

a. Of a node voltage
b. Familiar to all
c. That requires power dissipation
d. Of superposition

19. When a complex circuit is replaced with an equivalent voltage and an equivalent source resistance for circuit analysis, _____ theorem is being used.

a. Ampere
b. Coulombs
c. Thevenin's
d. Ohm's

20. Norton's Theorem is similar to Thevenin's theorem except _____ source is used and a parallel equivalent resistance.

a. A voltage
b. A current
c. A resistor
d. A series

LESSON 12. SUPPLEMENT
AN OUTLINE OF ADVANCED METHODS OF DC CIRCUIT ANALYSIS
AND HOW THEY ARE USED

In the previous lessons, Kirchhoff's voltage and current laws have been thoroughly discussed. This included not only a complete discussion of how they are applied in solving some fairly complex circuits, but also an introduction into some more specific applications of these rules as applied to complex circuit analysis. The methods discussed included node and loop equations and the superposition theorem; and also briefly introduced mesh current analysis, node voltage analysis, Thevenin's theorem, Norton's theorem, and finally Millman's theorem.

With this introduction as a background, the purpose of this additional material is to give a more complete treatment of each of these methods of circuit analysis. Specifically, each method will be discussed as to *where it is useful*, as well as the *specific steps to follow* in using the method in solving problems.

Then, each will have a *complete worked through example* illustrating details of the methods used.

The methods of circuit analysis will be discussed in this order:
1. Mesh currents
2. Node voltages
3. Thevenin's theorem
4. Norton's theorem
5. Millman's theorem

These five methods can be placed into two similar groups.

The first group consists of the mesh and node analysis methods. These two are actually used as extensions of Kirchhoff's current and voltage laws and are used primarily to analyze a circuit *completely* by finding all of the various current and voltage values throughout the circuit.

The second group, Thevenin's, Norton's, and Millman's theorems, is used to analyze circuits with a specific solution as an objective. For example, Thevenin's theorem is useful for analyzing the effect of substituting various loads across two specific test terminals in a circuit. Norton's theorem allows analysis similar to Thevenin's; however, the specific test point in the circuit is analyzed with respect to current rather than the Thevenin voltage consideration. Finally, Millman's theorem provides a shortcut method of finding the voltage across any number of parallel circuit branches, where each branch may contain different voltage sources, as well as resistances.

If further information is desired concerning these analysis methods, refer to the reference list following this supplement.

Where Useful

Mesh current solutions are particularly useful in solving for currents and voltages in complex circuits with several branch points, such as the circuit shown in Figure 12S.1.

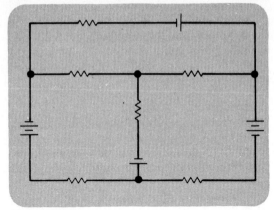

Figure 12S.1

Definitions and Methods of Use

The concept of mesh currents analyses of a circuit is similar to the analysis of loops and equations using Kirchhoff's laws. The basic differences center around the initial rules, or assumptions, used *before* writing the equations to solve for the currents in the circuits.

Rule — Assign mesh currents to all meshes in the circuit so that each element in the circuit is included at least once. A *mesh* is defined as the simplest closed path, or loop, in a circuit. Within a circuit, a mesh resembles a single window pane, and may or may not include a voltage source. Notice in Figure 12S.2 that the circuit shown has only two meshes. (Other closed circuit pathways will be *loops*, but not meshes.)

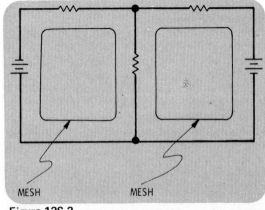

MESH   MESH

Figure 12S.2

Now, choose a current convention for use in the circuit (either electron current or conventional current) and assign all of the mesh currents the same direction (either all clockwise or all counterclockwise), as shown in Figure 12S.3. Note that with a resistor such as R_2 there are *two* mesh currents flowing through it. The actual value of the current flowing through R_2 will be the *difference* of the two mesh currents, with the *direction* being that of the *larger* current.

After these rules are employed, mesh equations are written in the same way as loop equations, using Kirchhoff's voltage law. Note: When writing the equations, shared components, like R_2 in Figure 12S.3, produce two terms in each mesh equation, one positive and one negative. (See sample solution.)

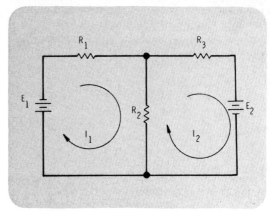

Figure 12S.3

Problem

Solve the circuit shown in Figure 12S.4 for the currents through R_1, R_2, and R_3, and voltages across R_1, R_2, and R_3.

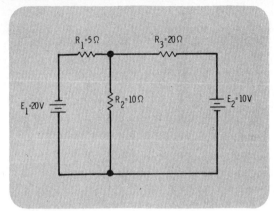

Figure 12S.4

Solution

First choose a current convention (here electron current is chosen), and assign mesh currents to each mesh, all in the same direction, as shown in Figure 12S.5. Then, write the mesh equation for each mesh current following Kirchhoff's voltage law:

Mesh 1. Beginning at Point A and traversing clockwise in the direction of I_1, you get: $-5 I_1 - 10 I_1 + 10 I_2 + 20 = 0$ or $-15 I_1 + 10 I_2 + 20 = 0$. Note the $(-10 I_1 + 10 I_2)$ for R_2, due to the opposite mesh currents through R_2.

Mesh 2. Begin at point B and follow clockwise in the direction of I_2. This results in: $-10 I_2 + 10 I_1 - 20 I_2 + 10 = 0$, or $+10 I_1 - 30 I_2 + 10 = 0$. Again note the R_2 term. The polarities reverse for the second mesh for $+10 I_1 - 10 I_2$.

Then solve the two resulting equations shown in Figure 12S.6. There will be one equation for each mesh so that there is one equation for each unknown mesh current.

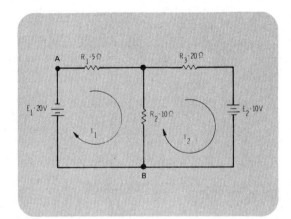

Figure 12S.5

EQUATION (1) $- 15 I_1 + 10 I_2 + 20 = 0$

EQUATION (2) $+ 10 I_1 - 30 I_2 + 10 = 0$

Figure 12S.6

Figure 12S.7 begins the solution. The first thing to do is to try to eliminate one of the unknowns, and the procedure followed here is aimed at eliminating I_2. To do this, multiply the first equation by +3 to get common coefficients for I_2 in both mesh equations. Then (Figure 12S.8), add the equations to eliminate I_2. This leaves the equation at the top of Figure 12S.9 with only one unknown. Solving this equation, I_1 is found to be 2 amps.

FIRST – MULTIPLY EQUATION (1) BY 3

$$3X\,(-15\,I_1 + 10\,I_2 + 20 = 0)$$

THIS GIVES

$$-45\,I_1 + 30\,I_2 + 60 = 0$$

Figure 12S.7

EQUATIONS (1) AND (2) NOW HAVE THE SAME COEFFICIENT FOR I_2, THUS I_2 CAN BE ELIMINATED BY ADDITION:

$$(1) \quad -45\,I_1 + 30\,I_2 + 60 = 0$$
$$+\,(2) \quad +10\,I_1 - 30\,I_2 + 10 = 0$$

$$\overline{\qquad -35\,I_1 \qquad\qquad +70 = 0 \qquad}$$

Figure 12S.8

$$-35\,I_1 + 70 = 0$$

$$-35\,I_1 = -70$$

$$I_1 = \frac{-70}{-35}$$

$$I_1 = 2A$$

Figure 12S.9

Once I_1 is known, substitute I_1 into either of the two equations and solve for I_2 as shown in Figure 12S.10. I_2 is found to be 1 amp.

EQUATION (2)

$$+ 10\,I_1 - 30\,I_2 + 10 = 0$$
(SUBSTITUTE $I_1 = 2$)
$$+ 10\,(2) - 30\,I_2 + 10 = 0$$
$$20 - 30\,I_2 + 10 = 0$$
$$- 30\,I_2 + 30 = 0$$
$$- 30\,I_2 = 0$$

$$I_2 = \frac{-30}{-30} = 1A$$

Figure 12S.10

Then solve for the current through R_2 by subtracting the smaller mesh current from the larger mesh current, giving the 1 amp shown in Figure 12S.11. This current flows *down*, through the central resistor R_2, in the direction of the larger mesh current I_1.

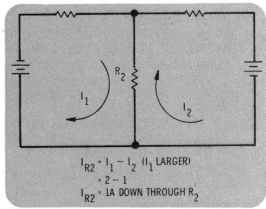

$$I_{R2} = I_1 - I_2 \ (I_1 \ \text{LARGER})$$
$$= 2 - 1$$
$$I_{R2} = 1A \ \text{DOWN THROUGH } R_2$$

Figure 12S.11

The rest of the solution is shown in Figure 12S.12 and involves finding each component's voltage using Ohm's law.

$$E_{R1} = I_1 \times R_1$$
$$E_{R1} = 2A \times 5\,\Omega = 10\,V$$

$$E_{R2} = I_{R2} \times R_2$$
$$E_{R2} = 1A \times 10\,\Omega = 10\,V$$

$$E_{R3} = I_2 \times R_3$$
$$E_{R3} = 1A \times 20\,\Omega = 20\,V$$

Figure 12S.12

The final solution with polarities is shown in Figure 12S.13.

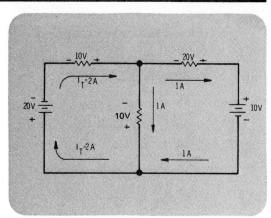

Figure 12S.13

Where Useful

This method is useful when solving for a particular voltage at a node, or branch point, without having to solve for complete current and voltage values within the circuit.

Definitions and Methods of Use

A node is a connection of two or more circuit components. A *principal node* has three or more connections (such as points A and B in Figure 12S.14). The first step in utilizing the method of node voltage analysis is to select *one* principal node in the circuit to act as the *reference node* (point B in Figure 12S.14). Then a current equation is written for each of the other principal nodes in the circuit. With this method, there will be one less node equation written than there are principal nodes. (Figure 12S.14 would then require only one equation.)

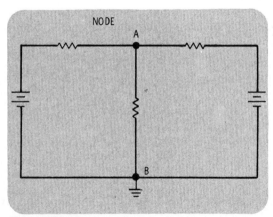

Figure 12S.14

To actually write the equation, begin by *choosing a current convention* (in the circuit shown in Figure 12S.15 electron current will be assumed), and by *labeling all* of the *currents* flowing in the circuit *with an assumed direction*. Then, using Kirchhoff's current law, *write a Kirchhoff's current law equation for every node except the reference node*. For the circuit shown in Figure 12S.15, you would have only one node equation: $I_3 = I_1 + I_2$.

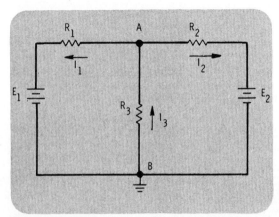

Figure 12S.15

Now the real trick in using node voltage analysis is this: Each of these *currents* is now *expressed in terms of the voltages* in the circuit. As shown in Figure 12S.16, I_3 equals the voltage across R_3 divided by R_3 or E_{R3}/R_3. Similarly I_1 equals E_{R1}/R_1 and $I_2 = E_{R2}/R_2$ as shown in the figure.

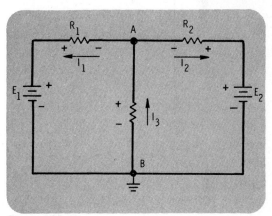

$$I_3 = I_1 + I_2$$

OR

$$\frac{E_{R3}}{R_3} = \frac{E_{R1}}{R_1} + \frac{E_{R2}}{R_2}$$

Figure 12S.16

Finally, all of the voltages can be expressed in terms of the *source voltages* and the *node voltage* E_{AB}. To do this, assign voltage polarities (Figure 12S.17) to each resistor in the circuit (using your assumed current direction). Then, keeping Kirchhoff's voltage law in mind, examine each *voltage* term in the equation

$$\frac{E_{R3}}{R_3} = \frac{E_{R1}}{R_1} + \frac{E_{R2}}{R_2}$$

and look for a way to express each in terms of the node voltage E_{AB} (which equals E_{R3}) and the applied voltages.

For example, if you traverse the left-hand loop if the circuit (Figure 12S.17) starting at point B and proceeding counterclockwise and write down the loop equation, you'd have $-I_3R_3 - I_1R_1 + E_1 = 0$. Or expressing each of these terms as a voltage directly, you'd have: $-E_{R3} - E_{R1} + E_1 = 0$. Remembering that $E_{R3} = E_{AB}$, the node voltage, this becomes $-E_{AB} - E_{R1} + E_1 = 0$. The key point is that, using this equation, E_{R1} can now be expressed in terms of the node voltage E_{AB} and the applied voltage E_1:

$$-E_{R1} = -E_1 + E_{AB}$$

or

$$E_{R1} = E_1 - E_{AB}.$$

If you do this same operation for the right-hand loop, you'll find that $E_{R2} = E_2 - E_{AB}$ and so the equation

Figure 12S.17

$$\frac{E_{R3}}{R_3} = \frac{E_{R1}}{R_1} + \frac{E_{R2}}{R_2}$$

can now be written

$$\frac{E_{AB}}{R_3} = \frac{E_1 - E_{AB}}{R_1} + \frac{E_2 - E_{AB}}{R_2}$$

The only unknown in this equation is now the *node voltage*, E_{AB}, and this equation may then be solved for it.

Problem

Solve Figure 12S.18 for the voltage across R_3. Use node voltage analysis.

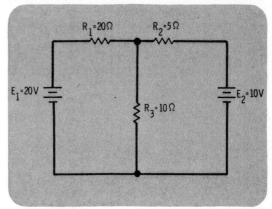

Figure 12S.18

Solution

First, identify the principal nodes and select a reference node. Here B will be the reference node (and is labeled with a ground in Figure 12S.19), and the only other principal node in the circuit is A. Then, choose a current convention (electron current will be used here) and label all the currents on the circuit assigning each a direction as shown in Figure 12S.19. Then, label voltage polarities across each resistor.

Next, write the current equation for the principal node A using Kirchhoff's current law. This gives $I_3 = I_1 + I_2$.

Then, express the node equation in terms of circuit voltages.

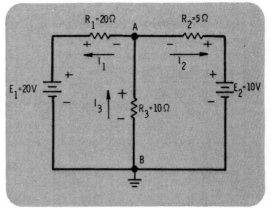

Figure 12S.19

$$I_3 = \frac{E_{R3}}{R_3}, I_1 = \frac{E_{R1}}{R_1}, \text{ and } I_2 = \frac{E_{R2}}{R_2}$$

With these terms substituted in the original current equation, you have

$$\frac{E_{R3}}{R_3} = \frac{E_{R1}}{R_1} + \frac{E_{R2}}{R_2}$$

Now, express the voltages in this equation in terms of the desired node voltage, E_{AB}, and the applied voltages E_1 and E_2. First, E_{R3} is the node voltage, so $E_{R3} = E_{AB}$.

To obtain the other voltage relations you need, first write a loop equation for the left-hand loop in this circuit (starting at point B and traversing counterclockwise):

This gives: $-E_{R3} - E_{R2} + E_1 = 0$

Then, rearranging terms:

$E_{R1} = E_1 - E_{R3}$, and substituting E_{AB} for E_{R3}, $E_{R1} = E_1 - E_{AB}$.

Following a similar procedure for the right-hand loop gives you $-E_2 + E_{R2} + E_{AB} = 0$.
Then rearranging: $E_{R2} = E_2 - E_{AB}$.

At this point, rewrite the node equation with substituted terms, and then substitute known values into the equation.

$$I_3 = I_1 + I_2$$

or

$$\frac{E_{R3}}{R_3} = \frac{E_{R1}}{R_1} + \frac{E_{R2}}{R_2}$$

or

$$\frac{E_{AB}}{R_3} = \frac{E_1 - E_{AB}}{R_1} + \frac{E_2 - E_{AB}}{R_2}$$

With known values: $E_1 = 20$ V, $E_2 = 10$ V, $R_1 = 20\,\Omega$, $R_2 = 5\,\Omega$, and $R_3 = 10\,\Omega$, the equation becomes:

$$\frac{E_{AB}}{10} = \frac{20 - E_{AB}}{20} + \frac{10 - E_{AB}}{5}$$

Then solve for E_{AB}. Multiply the equation by 20 for:

$$2E_{AB} = 20 - E_{AB} + 4(10 - E_{AB})$$

or

$$2E_{AB} = 20 - E_{AB} + 40 - 4E_{AB}$$

Combine like terms

$$2E_{AB} = 60 - 5E_{AB}$$

Transpose

$$7E_{AB} = 60$$

Divide

$$E_{AB} = \frac{60}{7} = 8.57 \text{ V}$$

Where Useful

Thevenin's theorem is useful where one particular component in a circuit must be repeatedly replaced, such as in a complex circuit where several different loads are to be applied. For example, what will the current and voltage values be for R_4 in Figure 12S.20 with R_4 equal to 20 ohms, 5 ohms, 30 ohms, and 50 ohms? Normal solution methods for this would involve four separate solutions of a complex circuit. However, Thevenin's theorem allows the rest of the circuit (without R_4) to be analyzed just once, and a very simple analysis to be made each time a new R_4 is "plugged in."

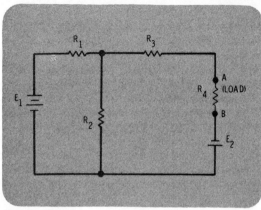

Figure 12S.20

Definitions and Methods of Use

Thevenin's theorem states that a circuit such as Figure 12S.20 above can be replaced by an equivalent series circuit consisting of one voltage (Thevenin's voltage, E_{TH}) and one series resistance (Thevenin's resistance, R_{TH}), as shown in Figure 12S.21.

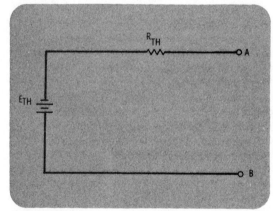

Figure 12S.21

Once E_{TH} and R_{TH} are known, various loads can be placed across points A and B (Figure 12S.22) and the desired current and voltage values for this load can then be easily determined.

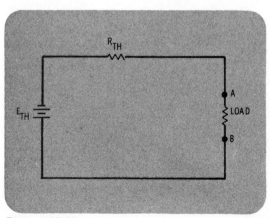

Figure 12S.22

Thevenin's voltage, E_{TH}, is found by first *removing* the load component to be tested (R_4 in Figure 12S.20) and then finding the voltage between these two open terminals (see Figure 12S.23). The voltage across the open terminals A and B in Figure 12S.23 will be E_{TH} for this circuit.

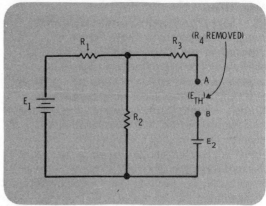

Figure 12S.23

Thevenin's resistance, R_{TH}, is found also with the test component (R_4 of Figure 12S.20) removed. R_{TH} is the resistance as measured between the *open test terminals* A and B, with all of the sources replaced by short circuits (see Figure 12S.24).

Once E_{TH} and R_{TH} are found, the circuit has been Thevenized and the equivalent circuit (Figure 12S.22) can be used to solve for the voltages across, and the current through, any load attached to the terminals A and B.

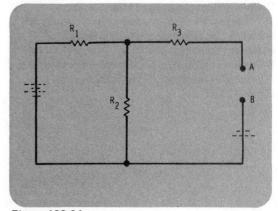

Figure 12S.24

Problem

Find the current and voltage for R_4 in Figure 12S.25 for these values of R_4: $R_4 = 20\ \Omega$ and $R_4 = 5\ \Omega$. Use Thevenin's theorem.

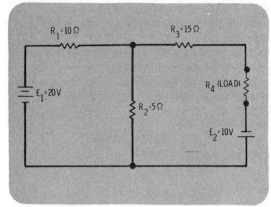

Figure 12S.25

Solution

First label the test points and remove the load to be tested (Figure 12S.26).

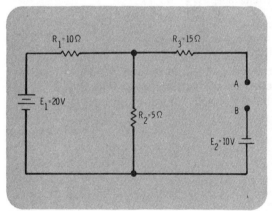

Figure 12S.26

Next, find E_{TH}, the open circuit voltage between test points A and B. In Figure 12S.27, E_{TH} will be the algebraic sum of E_{R3}, E_{R2}, and E_2. Since A and B are open, no current flows through R_3; therefore, $E_{R3} = 0$. E_{R2} can be solved using series circuit solution methods, because with A and B *open*, the only complete circuit path for current flow is the single series loop containing E_1, R_1, and R_2. Once E_{R2} is found, it can be algebraically added to E_2 to find the total voltage across points A and B (which is the Thevenin voltage).

Using electron current to analyze the problem as shown in Figure 12S.27, I_1 flows through R_2 and R_1 and gives rise to the voltage polarities

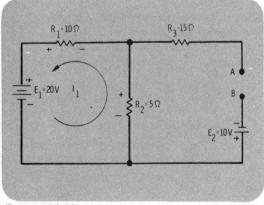

Figure 12S.27

labeled. (Note that you'd arrive at the same polarities using conventional current in your analysis.)

To find E_{R2} just calculate the total current flowing in the left-hand loop, I_1:

$$I_1 = E_1/R_T$$

where R_T just equals $R_1 + R_2$ which equals 5 + 10 = 15 Ω.

Substituting

$$I_1 = 20 \text{ V}/15 \ \Omega = 1.33 \text{ A}$$

Then E_{R2} equals I_1 times R_2 or

$$1.33 \text{ A} \times 5 \ \Omega$$

or

$$E_{R2} = 6.65 \text{ V}$$

So at this point E_{R3}, E_{R2}, and E_2 are known. E_{TH} is just the algebraic sum of these. (Remember $E_{R3} = 0$.)

$$E_{TH} = E_{R3} + E_{R2} + E_2$$
$$E_{TH} = 0 \text{ V} + 6.65 + 10 \text{ V}$$
$$E_{TH} = 16.65 \text{ V}$$

Note: The polarities of the voltages must be considered. E_{R2} and E_2 were in the same relative direction so they were added. Thus, E_{TH} is 16.65 V, negative at point B, positive at point A.

Next solve for R_{TH} by mentally replacing all source voltages with short circuits, and finding R_{TH} between the open test points A and B (Figure 12S.28) using the circuit reduction techniques you've seen earlier (Figure 12S.29).

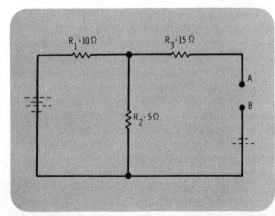

Figure 12S.28

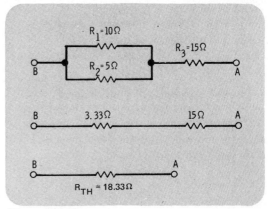

Figure 12S.29

Then draw the equivalent circuit and substitute the desired loads for solution (Figure 12S.30).

If $R_4 = 20 \, \Omega$ find I and E_{R4},

$$I = \frac{E_{TH}}{R_T} \text{ where } R_T = R_{TH} + R_4$$

$$I = \frac{16.65 \text{ V}}{18.33 + 20} = 0.434 \text{ A}$$

$$E_{R4} = I \times R_4 = 0.434 \text{ A} \times 20 = 8.68 \text{ V}$$

If $R_5 = 5$ ohms.

$$I = \frac{16.65}{18.33 + 5} = 0.714 \text{ A}$$

$$E = I \times R_4 = 0.714 \times 5 = 3.57 \text{ V}$$

Note that once a circuit is "Thevenized," it is easy to solve for E_{load} and I_{load} for any given load. A complete repeat of the circuit analysis is not required as different loads are plugged in.

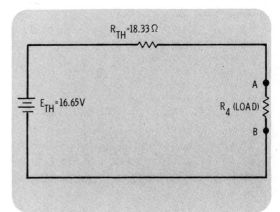

Figure 12S.30

Where Useful

As with Thevenin's theorem, Norton's theorem is useful when repeated substitutions of one component are desired in a circuit, and you have no desire to do repeated complex solutions for the circuit for each load. For example, in Figure 12S.31, Norton's theorem provides a method that will allow you to easily calculate the voltage across and current through any load placed across terminals A and B without completely analyzing the circuit for each load.

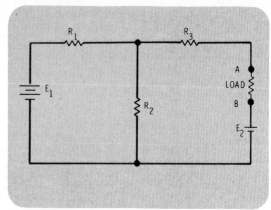

Figure 12S.31

Definitions and Method of Use

Norton's theorem states that a circuit with several sources and components can be replaced for evaluation at two test points by an equivalent circuit consisting of one *current source* (Norton current, I_N) and one parallel, or shunt, resistance (Norton resistance, R_N). See Figure 12S.32.

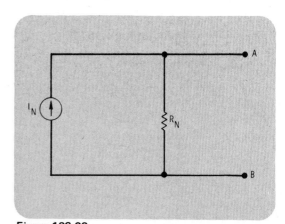

Figure 12S.32

I_N is found by mentally shorting the test terminals, A and B of Figure 12S.31, together and calculating the current that would flow through the short (see Figure 12S.33). The value of current found, I_N, will be the value of the current source in the equivalent circuit (Figure 12S.32), whose output will be shared by R_N and the load placed across A and B.

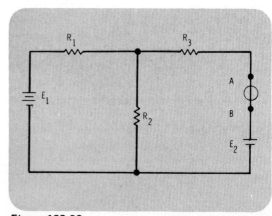

Figure 12S.33

R_N is found by shorting all voltage sources and computing the resistance from the open load terminals with the load removed (R_4), Figure 12S.34. Note that R_N and R_{TH} (Thevenin resistance) are found in an identical manner; thus, $R_N = R_{TH}$.

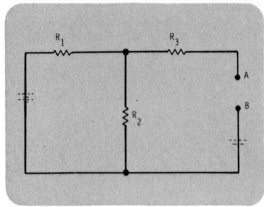

Figure 12S.34

Once I_N and R_N are known, the loads to be tested are placed across the equivalent circuit and the current and voltages calculated (Figure 12S.35) using parallel circuit analysis techniques.

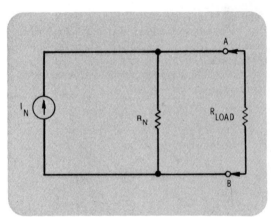

Figure 12S.35

Problem

Find the current and voltage for R_4 in Figure 12S.36 for $R_4 = 30\ \Omega$. Use Norton's theorem.

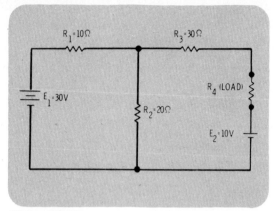

Figure 12S.36

Solution

First remove the load (R_4) and then short the load terminals and solve for I_N through the short (Figure 12S.37).

Several methods are available to solve for I_N. Here, superposition is used.

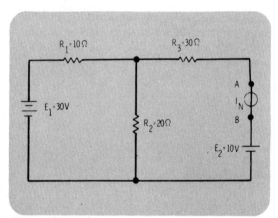

Figure 12S.37

First mentally short out E_2 and find the current produced by E_1 acting alone as shown in Figure 12S.38. To do this, find the total resistance for the circuit and then solve for the total circuit current.

$$R_T = \frac{R_2 \times R_3}{R_2 + R_3} + R_1 = \frac{30 \times 20}{30 + 20} + 10$$

$$R_T = 12 + 10 = 22\ \Omega$$

$$I_T = \frac{E}{R} = \frac{30}{22} = 1.36\ A$$

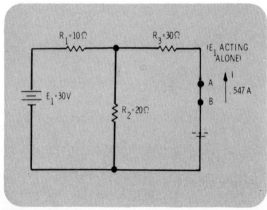

Figure 12S.38

Then solve for the voltage across R_1. This subtracted from E_1 will give you E_{R3}. Given E_{R3}, I_{R3} can be calculated, and this gives you the current past points A and B. Then

$$E_{R1} = I \times R_1 = 1.36 \text{ A} \times 10 = 13.6$$

$$E_{R3} = 30 - 13.6 = 16.4 \text{ V}$$

$$I_{R3} = \frac{E}{R3} = \frac{16.4 \text{ V}}{30 \text{ }\Omega} = 0.547 \text{ A}$$

The I through the shorted terminals for E_1 alone is 0.547 amp *entering B to A* for electron current.

Then solve again with E_2 in the circuit and E_1 shorted (Figure 12S.39). To do this, again first find R_T and solve for I_T which is through the A, B terminal short.

$$R_T = R_3 + \frac{R_1 \times R_2}{R_1 + R_2}$$

$$R_T = 30 + \frac{10 \times 20}{10 + 20} = 30 + 6.67$$

$$R_T = 36.67$$

$$I_T = \frac{10 \text{ V}}{36.67} = 0.272 \text{ A}$$

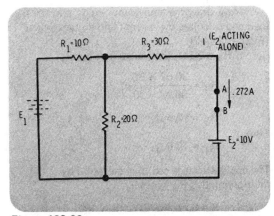

Figure 12S.39

This second current through the shorted terminal is in an opposite direction to the first; therefore, the value of I_N is the difference of the two, with the direction of the larger (Figure 12S.40). So

$$I_N = 0.547 - 0.272$$

or

$$I_N = 0.275 \text{ amp from B to A.}$$

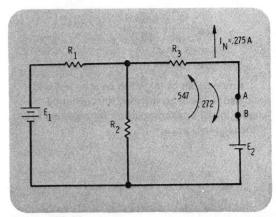

Figure 12S.40

Then to complete "Nortonizing" this circuit, find R_N, the resistance between the open load terminals A and B with all sources replaced with shorts (Figure 12S.41).

$$R_N = R_3 + \frac{R_1 \times R_2}{R_1 + R_2} = 30 + \frac{200}{30}$$

$$R_N = 36.67 \ \Omega$$

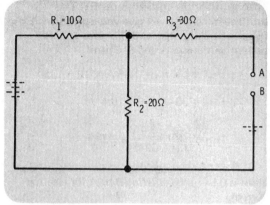

Figure 12S.41

Now, draw the equivalent circuit, add the R_4 load and calculate the current and voltage values (Figure 12S.42).

If $R_4 = 30 \ \Omega$, then solve first for R_T.

$$R_T = \frac{36.67 \times 30}{36.67 + 30}$$

$$R_T = 16.5 \ \Omega$$

Thus

$$R_T = 16.5 \ \Omega.$$

The applied voltage can then be found by multiplying this R_T times I_N.

$$E = I_N \times R_T = 0.275 \times 16.5$$

$$E = 4.54 \ V$$

Then

$$I_{R4} = \frac{E}{R4} = \frac{4.54 \ V}{30 \ \Omega}$$

$$I_{R4} = 0.15 \ A$$

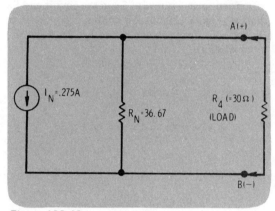

Figure 12S.42

Any number of loads can be attached and evaluated in the same manner, thus avoiding the lengthy initial problem solution for repeated values of the load resistance.

Where Useful

Millman's theorem is particularly useful in finding the common value of voltage across several parallel branches each having different voltage sources as shown in Figure 12S.43.

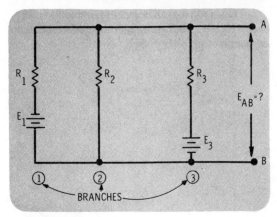

Figure 12S.43

Definitions and Methods of Use

Basically, Millman's theorem states that all parallel branches of a circuit such as Figure 12S.43 can be converted to current sources by taking each branch's total source voltage and dividing it by the total resistance of that branch. Thus for *Branch 1*

$$I_1 = \frac{E_1}{R_1}$$

for *Branch 2*

$$I_2 = \frac{E_2}{R_2}$$

and for *Branch 3*

$$I_3 = \frac{E_3}{R_3}$$

Once this is done for each branch, the results are combined by adding the currents $I_1 + I_2 + I_3$ and then dividing by the sum of the conductance of each branch, $1/R_1 + 1/R_2 + 1/R_3$. This gives a formula of this form:

$$E = \frac{I_1 + I_2 + I_3}{\dfrac{1}{R_1} + \dfrac{1}{R_2} + \dfrac{1}{R_3}}$$

And since this is simply the total current divided by total conductance, it can also be written $E = I_T/G_T$.

This I_T/G_T of the branches is actually $I_T R_T$, since $1/G = R$, thus resulting in the desired branch voltage since $I_T \times R_T = E_T$.

Millman's Formula

This can be expressed in the formula shown in Figure 12S.44. This formula will work for any number of branches and is simply extended as necessary. However, *some precautions should be observed*. First: All branches must be parallel with no series resistances between them. Second: If branches contain more than one source, or multiple resistances, combine them for totals for each branch before solving. Third: The *polarity* used in the formula *for each source* is taken to be *the polarity it is applying to the top point* (point A of Figure 12S.43) *of the branches*, with respect to point B as a reference point. For example in Figure 12S.43, E_1 is negative, E_3 is positive. Notice also that the second branch with R_2 has no source, so E_2 would simply be equal to zero for that branch.

Once these precautions are observed, it is a simple matter to plug in the known values and solve for the desired voltage. Once this voltage is known, the other voltages and currents in the circuit can easily be found.

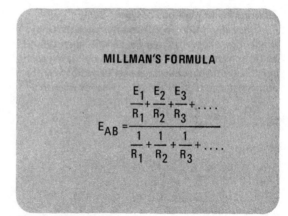

MILLMAN'S FORMULA

$$E_{AB} = \frac{\dfrac{E_1}{R_1} + \dfrac{E_2}{R_2} + \dfrac{E_3}{R_3} + \ldots}{\dfrac{1}{R_1} + \dfrac{1}{R_2} + \dfrac{1}{R_3} + \ldots}$$

Figure 12S.44

Problem

Solve for the voltage between points A and B (or E_{AB}) in Figure 12S.45. Use Millman's theorem.

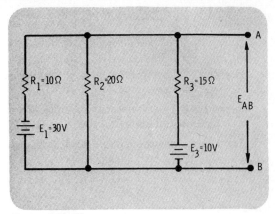

Figure 12S.45

Solution

Since there is only one source and one resistance in each parallel branch, simply place the values in the formula and solve.

$$E_{AB} = \frac{\dfrac{E_1}{R_1} + \dfrac{E_2}{R_2} + \dfrac{E_3}{R_3}}{\dfrac{1}{R_1} + \dfrac{1}{R_2} + \dfrac{1}{R_3}}$$

By substituting the voltages and resistances, you get

$$E_{AB} = \frac{\dfrac{-30}{10} + \dfrac{0}{20} + \dfrac{10}{15}}{\dfrac{1}{10} + \dfrac{1}{20} + \dfrac{1}{15}}$$

(Note the −30 for E_1, the 0 for E_2 and the +10 for E_3.) Then, simply reduce the terms and solve.

$$E_{AB} = \frac{-3 + 0 + 0.667}{0.1 + 0.05 + 0.067}$$

$$E_{AB} = \frac{-2.333}{0.217}$$

$$E_{AB} = -10.75 \text{ V}$$

Note that the minus indicates that the voltage is negative at point A with respect to point B as the reference point.

Additional Reference Material on Complex Circuit Analysis Methods

Grob, B., *Basic Electronics*, Third Edition, (New York, N.Y.: McGraw-Hill Book Co., 1971) pp. 119-166.

Herrick, C. N., *Unified Concepts of Electronics*, (Englewood Cliffs, N.J.: Prentice-Hall, 1970), pp. 187-232.

Singer, B. B., *Basic Mathematics for Electricity and Electronics*, Second Edition, (New York, N.Y.: McGraw-Hill Book Co., 1965), pp. 178-238.

Tocci, R. J., *Introduction to Electric Circuit Analysis*, (Columbus, Ohio: Charles E. Merrill Publishing Co., 1974), pp. 263-300.

Lesson 13

Capacitors and the RC Time Constant

This lesson introduces a new type of circuit component: the capacitor. Capacitor construction, units of capacitance measurement, and effects of capacitors in dc circuits are discussed. The very important concepts of the resistor-capacitor (RC) time constant and Universal Time Constant graph are introduced and discussed in detail, along with the effects of capacitor charge storage and rapid discharge.

LESSON 13. CAPACITORS AND THE RC TIME CONSTANT

● Objectives

In this lesson you will be introduced to an entirely new type of circuit component that produces some very useful effects in dc circuits: the capacitor. At the end of this lesson you should be able to:

1. *Sketch* the construction of a capacitor, *label* each of its component parts, and *draw* its schematic symbol.

2. *Write* and *show* with sketches what is meant by a charged and discharged capacitor, including a description of the procedures that may be used to charge and discharge a capacitor.

3. *Write* a definition of capacitance; explaining its relationship to the amount of charge stored and voltage across a capacitor's plates. *List* the three factors that determine the capacitance of a capacitor, and *define* the unit of capacitance; the farad.

4. For the circuit shown in Figure A below:

   a. *Calculate* the resistor-capacitor (RC) time constant.

   b. *Calculate* the voltage across the capacitor's plates one time constant after switch S is thrown to position 1.

   c. *Calculate* the time it takes for the capacitor to *fully charge.*

   d. *Sketch* a graph which describes how the voltage across the capacitor varies with time after the switch is thrown to position 1.

   e. With the capacitor in Figure A fully charged, the switch is thrown to position 2. *Calculate* the voltage on the capacitor 1 time constant later. *Calculate* the time it takes for the capacitor to fully discharge.

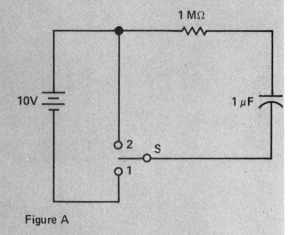

Figure A

   f. *Sketch* a graph which describes the behavior of the voltage across the capacitor with time, after the switch is thrown to position 2.

5.

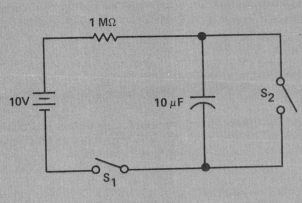

Figure B

In Figure B above, switch S_1 is closed for 5 time constants, then opened. *Write* a description of what happens when switch S_2 is closed.

LESSON 13. CAPACITORS AND THE RC TIME CONSTANT

- **Lesson Objectives**
- **Capacitor Construction**

By now you have come quite a long way in your study of dc electricity. You have probed the nature of electricity itself, learned much of the language and conventions used in describing electricity, and should be getting more and more familiar with the mathematical tools that enable you to predict and control the behavior of dc circuits. The circuits that have been discussed so far, however, have only contained combinations of voltage sources and resistors. In this lesson an entirely new type of circuit component will be introduced: the capacitor.

Lesson Objectives — To begin, it is always a good idea to set down objectives for the topics you will be studying. At the end of this lesson you should be able to sketch the construction of typical capacitors, and mathematically describe their behavior in circuits. You will also be able to define the important new concept of an "RC (resistor-capacitor) time constant," and sketch and explain what is called the *universal time constant graph*. You should also be familiar with common applications of capacitors in dc circuits, and be able to explain *why* capacitors are useful in these applications.

> **AT THE END OF THIS LESSON
> YOU SHOULD BE ABLE TO:**
>
> DRAW TYPICAL CAPACITOR
>
> DESCRIBE CAPACITOR BEHAVIOR
> MATHEMATICALLY
>
> DEFINE "RC TIME CONSTANT"
>
> USE "UNIVERSAL TIME CONSTANT GRAPH"

Figure 13.1

Capacitor Construction — In its most basic form, a capacitor consists of two conductors separated from each other by an insulator. In a capacitor, this insulator is called a *dielectric*. A sketch of a simple capacitor is shown in Figure 13.2. It basically consists of two metal plates separated by air, which is just one type of insulator or dielectric material commonly employed in capacitors. Connected to each plate is a lead or wire so that the capacitor can be connected to other circuit components. Capacitors used to be (and sometimes still are) referred to as *condensors*. The term capacitor is by far in more widespread use today.

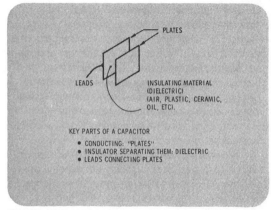

KEY PARTS OF A CAPACITOR
- CONDUCTING: "PLATES"
- INSULATOR SEPARATING THEM: DIELECTRIC
- LEADS CONNECTING PLATES

Figure 13.2

Capacitor Symbols — The schematic symbol for a capacitor is easy to associate with the real capacitor. As you see in Figure 13.3(A), the symbol shows two plates electrically separated from one another, with leads attached to each plate.

Figure 13.3(B) shows the more commonly accepted symbol for a capacitor. Notice that the major difference between this and the previous symbol is that in this symbol one plate is curved while the other plate is symbolized with a straight line. Generally, the curved line indicates the plate that should be connected to a more *negative voltage* than the other plate. Figure 13.3(C) shows some symbols you may see for *variable* capacitors or trimmer capacitors used in a variety of electronic applications that will be discussed for you in later courses in electronics.

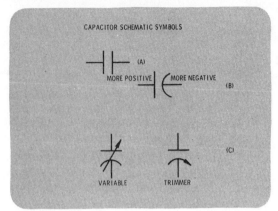

Figure 13.3

Electrolytic Capacitor Symbol — One other important capacitor symbol is shown in Figure 13.4. This symbol has a "plus" sign next to the "flat" plate. This symbol is the most common one used to signify an *electrolytic capacitor*. Electrolytic capacitors should *always* have the plate marked with the "plus sign" connected to a more positive voltage than the other plate. Electrolytic capacitors are designed to be used in dc or pulsating dc applications only.

A variety of other symbols are in use for electrolytic capacitors, and some additional ones are shown in Figure 13.4. In any of them, however, the polarity of the plates is identified; one of the plates is either labeled positive, or its shape tells

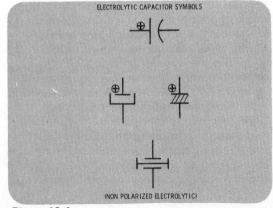

Figure 13.4

you that it is positive. (Certain specially constructed electrolytic capacitors are available that can be used in ac applications, and the symbol used to indicate these "nonpolarized" electrolytic capacitors is also shown in Figure 13.4.)

If this is your first opportunity to learn about the capacitor, you may be wondering just how the thing works in dc circuits. There is no direct conduction path for current flow through the device, so what good is it? Just what does it do? These questions about capacitors will be answered. First, however, a little discussion of the history of capacitors may be interesting to you. The fact is that the capacitor was discovered by accident in 1746 in Leyden, Holland by a physicist named Pieter Van Musschenbroek.

Leyden Jar — Pieter was doing some experiments in an attempt to "electrify" water. The water "electrification" device consisted of a large jar lined inside and out with copper foil as shown in Figure 13.5. This "Leyden jar" as it is called, has all of the elements of a capacitor. As shown in the detail view, the rod sticking through the lid of the jar had a chain on the end that hung down and supplied connection to the inside layer of foil. This formed one plate of the capacitor, the glass wall of the jar served as the dielectric, and the outer foil served as the other plate.

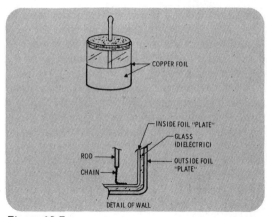

Figure 13.5

Leyden Jar Operation — As the story goes, Van Musschenbroek connected the Leyden jar to a voltage source for a period of time (Figure 13.6), then the voltage source was disconnected, and the jar was removed.

Figure 13.6

Zap — At this point Van Musschenbroek's assistant is said to have held the jar with one hand while disconnecting the high-voltage lead with the other hand. The lab assistant received an unexpected shock of considerable intensity (Figure 13.7). Unfortunately, history didn't record the words spoken by the assistant.

The shock received by the assistant points up a most important aspect of capacitors: *They are devices that can store an electrical charge.* Electrical charge, as well as electrical energy, can be stored or held in a capacitor and then released at a later time. Note that capacitors *must be given* the charge they store. A capacitor cannot produce electrical energy by itself, the way a battery does with chemical action.

Figure 13.7

Capacitor Action — How does a capacitor store charge? It does this through the action of an *electrostatic field*. To see just how this works, focus your attention on the sketch of capacitor plates shown in Figure 13.8.

As has been said, a capacitor consists of two conductors separated from each other by a layer of insulating material called a dielectric. Normally, the two metal conducting plates will have equal amounts of net positive and negative charge. As was discussed in Lesson 1, objects that contain equal amounts of positive and negative net charge are said to be *electrically neutral*. What will happen to these two plates if a potential difference is applied to them, say with a battery? The negative terminal of the battery pushes electrons out onto the negative connected plate, while the positive battery terminal draws electrons from the positive connected plate. As one plate receives a negative charge due to excess electrons building up, the other plate receives a positive charge due to the lack of electrons created. The net charges created on the capacitor's plates are *equal* and *opposite*. This is an important point. The battery takes electrons from one plate of the capacitor and essentially puts them on the other plate.

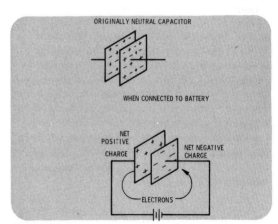

Figure 13.8

Charges Pack on Negative and Positive Plates —
Negative charges (electrons) cannot build up on the
negative plate forever. As more and more electrons
move toward the negative plate, they start getting
pushed back or repelled by the electrons already
there. After a while the battery cannot push any
more electrons onto the negative plate. The same
thing is true with the positive plate. After a while,
so many electrons have been removed from the
positive plate that the battery cannot pull any
more off.

Consequently the capacitor develops a net
positive charge on one plate and a net negative
charge on the other plate. Because of this, a
potential difference or voltage exists between the
two plates. At the point where no more charge is
flowing, the voltage across the capacitor's plates
equals the battery's voltage. Notice in Figure 13.9
that the voltage across the capacitor opposes the
battery's voltage. The capacitor's negative plate is
pushing electrons back in opposition to the push of
the negative battery terminal, and the capacitor's
positive plate attracts electrons and thus prevents
their further removal from the positive plate. At
the point where the voltage on the capacitor equals
the voltage of the battery, no more current flows,
since these voltages are in opposing directions.

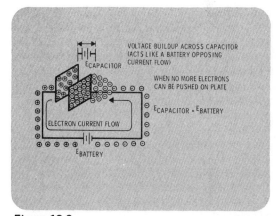

Figure 13.9

Capacitor Equivalent Circuit — As illustrated in
Figure 13.10, once the capacitor's plates have
accumulated enough negative and positive charges,
the capacitor acts like a battery wired to act
against the original battery in this circuit.

Review in your mind for a moment the
important points about a capacitor which have
been discussed. When a capacitor originally having
neutral plates is connected to a battery, *charge
flows* as a *current exists* in the circuit. Electrons
flow out of the battery's negative terminal and
build up on the negative connected plate, while
electrons are drawn from the positive connected
plate. The charge created on each plate causes a
potential difference or voltage, to build up on the

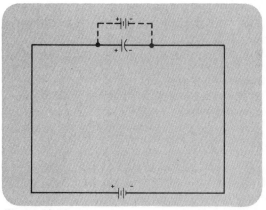

Figure 13.10

capacitor. As the charge on the capacitor's plates increases, the voltage across the plates increases, until finally the potential difference of the capacitor equals that of the battery. Since these two voltages are in opposition, when they equal each other there can be no more electron current flow. So notice: Charges flow in the *circuit*, but none flow *through* the capacitor because of the insulating gap between the two plates. Also, charge flow or *current only goes on for the short time* necessary for the voltage across the capacitor to become equal to the battery voltage.

Charged Capacitor — Whenever a net charge exists on each of the capacitor's plates, a *potential difference* or voltage exists across the capacitor. Likewise, when a voltage is *placed* across a capacitor, "charge" gets stored inside. Here's an important definition for you to remember: whenever a potential difference or voltage exists between the plates of a capacitor, the capacitor is said to be *charged*. A primary function of capacitors is this ability to store a charge. Capacitors are rated or measured by how well they perform this function.

    The factors that affect a capacitor's ability to perform the function of charge storage are the *size of its plates*, the *spacing between the plates*, and the *type of insulating material* separating the plates.

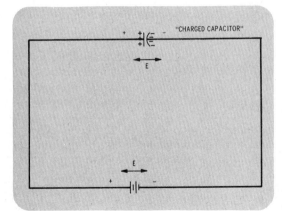

Figure 13.11

Plate Area and Capacitance — Each of these factors will be discussed in detail and you will probably find the first factor, plate size, is the most obvious thing about a capacitor that will affect its charge storage ability. As shown in Figure 13.12, if each of a capacitor's plates is made larger (that is, if the plate area is increased), and then connected to a battery, more charge will be stored on the larger plates than on the smaller plates. More charge would have to flow until the potential difference across the capacitor equals that of the battery. When current flow stops, the capacitor would be storing more charge than a capacitor with smaller plates, even though both capacitors were charged to the same voltage.

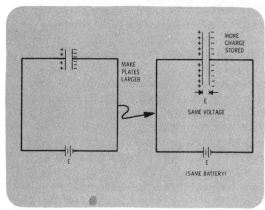

Figure 13.12

Capacitance — Since charge storage is really one of the most basic capacitor functions, a quantity is needed to describe how well a capacitor does this. This quantity is called the *capacitance* of the capacitor. As shown in Figure 13.13, *capacitance* is defined as the charge stored on a device, divided by the voltage across the device when that charge was stored.

CAPACITANCE

THE RATIO OF THE CHARGE STORED DIVIDED BY THE VOLTAGE ACROSS THE DEVICE.

Figure 13.13

Formula for Capacitance — Using the letter C to represent capacitance, Q for charge, and E for potential difference, an equation for capacitance may be written: $C = Q/E$. The capacitance of a device equals the charge it is storing divided by the voltage across it (Figure 13.14).

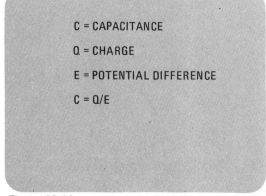

C = CAPACITANCE

Q = CHARGE

E = POTENTIAL DIFFERENCE

$C = Q/E$

Figure 13.14

- **Other Factors Affecting Capacitance**
- Dielectric Strength and Dielectric Constant

Other Factors Affecting Capacitance – It has already been shown that the capacitance of a capacitor depends on the area of its plates, the bigger the plate area, the bigger the capacitance. As has been mentioned, there are two additional factors that affect capacitance. One is the spacing between the capacitor's plates. It turns out that if the *plates* are pushed *closer together*, the *capacitance* will be *increased*. The third factor deals with the *type of dielectric* used. If different dielectrics are placed between the plates, the capacitance will vary. Details concerning *why* capacitance varies with capacitor construction will be covered in a later course, but for now just remember that the capacitance of a device depends on these three things: the spacing between plates, the area of the plates, and the dielectric material (specifically, the *dielectric constant* of the material). These three factors have been arranged into an easy-to-remember form for you in Figure 13.15. The first letters of the three factors spell SAD.

FACTORS AFFECTING CAPACITANCE

SPACING BETWEEN THE PLATES

AREA OF THE PLATES

DIELECTRIC MATERIAL USED

Figure 13.15

Dielectric Strength and Dielectric Constant – Notice that the dielectric in a capacitor is actually doing two things, First of all it is the ***insulating material*** that prevents charges from flowing from one plate to the other. Second, dielectric materials because of their makeup, actually act to ***help*** the capacitor store charge. As different dielectrics are placed between the plates of a capacitor, its capacitance will vary.

Two special quantities are used to describe how well a dielectric performs these ***two*** functions. One is ***dielectric strength***. This describes how ***resistant to breakdown*** a dielectric is. In capacitors, dielectrics consisting of very thin sheets are often subjected to very high voltages. When the voltage

DIELECTRIC STRENGTH

ABILITY OF A MATERIAL TO WITHSTAND
ELECTRICAL BREAKDOWN

MATERIAL	DIELECTRIC STRENGTH (VOLTS/MIL)
AIR	20
CERAMICS	600-1250 VARIES WITH TYPE
PYREX GLASS	330
MICA	600-1500 VARIES WITH TYPE
TEFLON	1525
OIL	375
PAPER	400-1250 VARIES WITH TYPE

Figure 13.16

applied across the dielectric becomes high enough, the dielectric will "break down," and electrons will "punch their way through" the dielectric. This creates a conducting path from one plate to the other through the dielectric and the capacitor malfunctions. *Dielectric strength* is a measure of how *resistant* a dielectric is to this type of breakdown. The dielectric strength of a material is commonly measured in volts per mil (V/mil); and some common values are listed in Figure 13.16. (One mil = 1/1000th of an inch.) These values tell you how many volts a one-mil thickness of dielectric can withstand before breaking down.

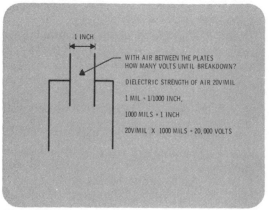

1 INCH

WITH AIR BETWEEN THE PLATES
HOW MANY VOLTS UNTIL BREAKDOWN?

DIELECTRIC STRENGTH OF AIR 20V/MIL

1 MIL = 1/1000 INCH,

1000 MILS = 1 INCH

20V/MIL X 1000 MILS = 20,000 VOLTS

Figure 13.17

Example — For example, the dielectric strength of air is 20 volts/mil. How many volts would be required to break down 1 inch of air and jump a spark through it? As 1 mil is 0.001 inch; so 1 inch is 1000 mils. Air can withstand 20 volts for each mil of thickness so 20 volts/mil times 1000 mils equals 20,000 volts. In a capacitor with 1 inch of air between its plates, 20,000 volts applied would be enough to cause breakdown, as shown in Figure 13.17.

DIELECTRIC CONSTANT

A MEASURE OF HOW WELL A DIELECTRIC HELPS
A CAPACITOR STORE CHARGE (AS COMPARED
TO AIR)

DIELECTRIC MATERIAL	DIELECTRIC CONSTANT
AIR	1
CERAMICS	80-1200 VARIES WITH TYPE
GLASS	8
MICA	3-8 VARIES WITH TYPE
TEFLON	2.1
OIL	2-5 VARIES WITH TYPE
PAPER	2-6 VARIES WITH TYPE

Figure 13.18

Dielectric Constant — The other factor of importance that describes how well a dielectric functions in a capacitor is called its *dielectric constant*. This is a measure of how well a dielectric helps a capacitor store charge. Dielectric constants for some common materials are listed in Figure 13.18. Notice that the dielectric constant for air is listed as 1. The dielectric constant for any other material tells you how much more (or less) effective a material is (as compared to air) in helping a capacitor store charge. Glass, for example, has a dielectric constant of 8. If you had a capacitor with air originally between its plates, and replaced the air with a glass slab of the same thickness, the capacitor's ability to store charge would have increased *8 times*. Just how dielectrics work to help capacitors store charge has to do with how they affect the *electric field* between the capacitor's plates. Details of exactly how this process works are beyond the scope of this text, and are covered for you in later courses.

Farad — The units used to measure capacitance are called *farads*, named after the scientist Michael *Farad*ay. Following the equation C = Q/E, a capacitor has 1 farad of capacitance when it can store 1 coulomb of charge with a 1-volt potential difference placed across it.

In reality, 1 farad turns out to be an extremely *large* unit of capacitance, so common capacitors are rated in microfarads or picofarads as listed in Figure 13.19. An important point to note is that on many capacitors the letter m is used as an abbreviation for micro. Keep in mind that on capacitors m does not mean milli or mega, it is always used to indicate micro. Also you may notice the abbreviation mmF used to indicate *micro micro farads*, which are identical to picofarads. Pico is simply a newer term.

CAPACITANCE IS MEASURED IN FARADS

$$C = Q/E$$

$$1 \text{ FARAD} = \frac{1 \text{ COULOMB}}{1 \text{ VOLT}}$$

COMMON CAPACITOR VALUES IN

MICROFARADS = μF = mF

PICOFARADS = pF = mmF

Figure 13.19

Charged Capacitor — As has been mentioned, in order to *charge* a capacitor, you must connect it to a dc supply and apply a voltage. After a very short while, current stops flowing and the plates have a potential equal to the supply voltage.

An important point: if the capacitor is *disconnected* from the battery or supply while in its charged condition, *the charge still remains on its plates*. How does the capacitor hold this charge on its plates?

Recall from an earlier lesson, the first law of electrostatics states that *unlike charges attract* each other. Look at Figure 13.20, and you can see that even with the battery disconnected, the positive charges on one plate will attract the negative charges on the other plate and hold them there. Because of these unlike charges on the plates, an *electric field* exists in the dielectric region between them. This field is actually the mechanism that holds the charge on the plates, and also while acting to do that, actually *stores energy*. Since this electric field exists in the dielectric in the absence of current flow, it is called an electro*static* field.

Since the charges on the plates cannot move to reach each other because of the insulating dielectric between the plates, in this charged condition the capacitor can store charge and energy for long periods of time. This means that a charged capacitor can be used to provide a current or do some work for you at some later time.

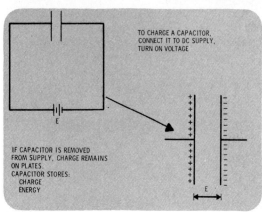

TO CHARGE A CAPACITOR, CONNECT IT TO DC SUPPLY, TURN ON VOLTAGE

IF CAPACITOR IS REMOVED FROM SUPPLY, CHARGE REMAINS ON PLATES.
CAPACITOR STORES:
CHARGE
ENERGY

Figure 13.20

Discharging a Capacitor — The energy and charge stored in a capacitor may be *recovered* if a conducting path is provided between the plates. This procedure is called *discharging* the capacitor. The excess electrons on the negative plate will flow to the positive plate, until both plates have no net excess charge, or are neutral. In this condition, the capacitor is said to be *discharged*. The current that flows is called the discharge current, and the path taken by the current is called the discharge path. A charged capacitor may be discharged by providing an appropriate conducting path between its plates as shown in Figure 13.21. An important observation for you to make should be emphasized at this point. Even though the power may be shut

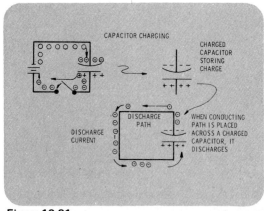

CAPACITOR CHARGING

CHARGED CAPACITOR STORING CHARGE

DISCHARGE PATH

DISCHARGE CURRENT

WHEN CONDUCTING PATH IS PLACED ACROSS A CHARGED CAPACITOR, IT DISCHARGES

Figure 13.21

off to a circuit, every capacitor in the circuit may retain its charge for a long period of time. Therefore, before working on high-voltage electronic circuitry, you should be sure to discharge all capacitors. The stored charge and energy in capacitors can discharge through you, *giving you a nasty shock*.

Charge Formula — The amount of charge present on the capacitor's plates and the voltage across the plates are related. To restate, the capacitance of a capacitor in farads equals the charge on the capacitor in coulombs divided by the voltage across it, or $C = Q/E$. This equation may now be rearranged as shown in Figure 13.22 to read $Q = CE$, or the charge stored in a capacitor (in coulombs) equals its capacitance (in farads) times the voltage between its plates. Look at this new equation carefully. It states that the amount of charge stored in a capacitor is directly related to the voltage across the capacitor's plates. The more voltage across the plates, the more charge on the plates, and vice versa. For a given voltage, the higher the capacitance, the more charge will be stored on the plates, and vice versa.

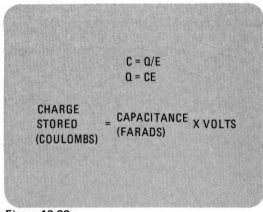

$$C = Q/E$$
$$Q = CE$$

CHARGE STORED (COULOMBS) = CAPACITANCE (FARADS) X VOLTS

Figure 13.22

Capacitor Charging: No Resistance — Up to this point, capacitor construction and charging and discharging of a capacitor have been introduced and discussed for you. When a capacitor is connected to a voltage source, current flows for a short time until the capacitor charges up to equal the source voltage. A capacitor can hold that charge for long periods of time. Consider what will happen if a *resistor* is placed in series with a capacitor as it charges and discharges.

The original circuit presented in this lesson consisted only of a capacitor and a power supply connected in series. When the battery was connected to the capacitor, electrons surged onto the negative-connected plate, and away from the positive-connected plate. When a capacitor is connected to a dc source in this way, it charges very rapidly. In fact, with virtually no resistance in the circuit, the capacitor would become fully charged almost *instantly*, as described in Figure 13.23.

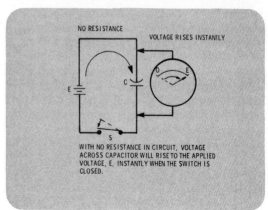

Figure 13.23

Capacitor Charging: Resistance Present — If a resistor is placed in series with this circuit, current cannot flow as freely as before. The current flowing in the circuit must "fight" its way through the resistor in order to charge the capacitor. Since there will be reduced current flow, *more time* will be required to charge the capacitor. Resistance has the important effect of causing a delay in the time required to charge a capacitor.

When the switch, S, is closed in a circuit consisting of a resistor, capacitor, and voltage source in series, the voltage across the capacitor's plates will take a *longer time* to reach the battery voltage than before (as shown in Figure 13.24). This is an important new effect. With a resistor in

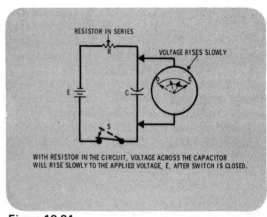

Figure 13.24

the circuit, the voltage across the capacitor rises more slowly than before, taking more time to reach the battery voltage. The current that flows in the circuit behaves in the opposite fashion. When the switch is first closed, a lot of current flows; and as the opposing voltage across the capacitor rises, less and less current flows, until finally, when the battery voltage and opposing capacitor voltage are equal, current flow ceases.

Factors Affecting Delay — This time delay required for the capacitor's voltage to reach the supply voltage, is a very useful effect — and you can control it. To enable you to *control* this delay, consider the factors that affect it.

    The time required for the capacitor to charge and the voltage across it to rise up to the supply voltage depends on *two* factors (Figure 13.25): 1) *how much resistance* is in the circuit opposing current flow, and 2) *how big* the capacitor is.

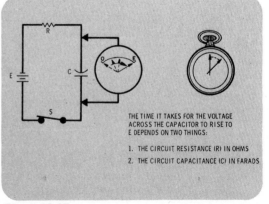

THE TIME IT TAKES FOR THE VOLTAGE ACROSS THE CAPACITOR TO RISE TO E DEPENDS ON TWO THINGS:

1. THE CIRCUIT RESISTANCE (R) IN OHMS
2. THE CIRCUIT CAPACITANCE (C) IN FARADS

Figure 13.25

Time Constant Definition — This time delay is such an important effect in electricity that engineers and technicians use a special way to describe how fast or how slow a capacitor charges. The time required for the capacitor to charge may be described in a standard way by defining what is known as the *capacitive time constant*. One time constant is defined as the *time required for a capacitor to charge up to 63% of the battery or supply voltage* (Figure 13.26). The time constant is also used to describe capacitor *discharge*, as will be discussed in a moment.

DEFINITION: TIME CONSTANT

1 TIME CONSTANT — THE TIME REQUIRED BY A CAPACITOR TO CHARGE TO 63% OF ITS FULL CHARGE VALUE (OR DISCHARGE DOWN 63% FROM FULL CHARGE).

Figure 13.26

Time Constant Formula — Why this rather unusual definition for time constant? As it turns out, this specific time interval may be easily calculated from a simple formula as shown in Figure 13.27. You can calculate the capacitive time constant using the formula $T = RC$, where T is the capacitive time constant in seconds, R is the circuit resistance in ohms, and C is the capacitance in farads. For a series resistive-capacitive (RC) circuit, you find the time constant by simply *multiplying* the circuit's resistance in ohms, times the capacitance in farads.

CAPACITIVE TIME CONSTANT FORMULA

T = RC

T = TIME CONSTANT **IN SECONDS**

R = CIRCUIT RESISTANCE **IN OHMS**

C = CAPACITANCE **IN FARADS**

Figure 13.27

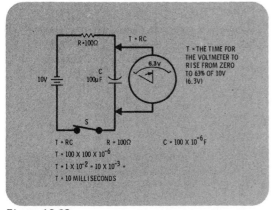

Figure 13.28

Time Constant Example — As an example, calculate the time constant for the circuit shown in Figure 13.28. In the circuit diagram the resistance is shown as 100 ohms, and the capacitance as 100 microfarads. To calculate the time constant, simply use the formula T = RC. Substituting: T = $100 \times 100 \times 10^{-6}$, which equals 1×10^{-2} or 10×10^{-3} or **10 milliseconds**. Remember, this is the time it would take for the voltage to rise from zero to 63% of the applied voltage. So in this case the voltage will rise to 63% of 10 volts or 0.63×10 or 6.3 volts in 10 milliseconds.

Second Example — Consider next the circuit of Figure 13.29. Here the resistance is a much larger 100 megohms, while the capacitance has been reduced slightly to 10 microfarads. Again, using the formula T = RC and substituting, you get T = $100 \times 10^6 \times 10 \times 10^{-6}$ or T = 1000 seconds. In this circuit you have to wait 1000 seconds (16-2/3 minutes) for the capacitor's voltage to rise very slowly to 63% of the applied voltage or 6.3 volts. The extremely large resistance really slows things up. This predictable rising voltage can be put to many uses in electricity and electronics.

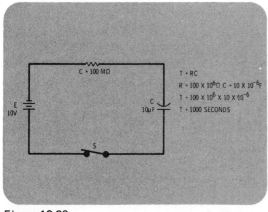

Figure 13.29

Universal Time Constant Graph — At this point you know that the time constant of a circuit such as that shown in Figure 13.29 (called an "RC" circuit) equals R times C, and that in one time constant, the capacitor's voltage rises to 63% of the applied voltage, E. Focus your attention on the details of how this voltage behaves as time goes on after switch, S, is closed. This is a new dimension in the study of dc — *time* is involved now. You know that when the switch is closed in this circuit, the voltage across the capacitor starts to rise. To describe how this happens, the behavior of the circuit may be plotted on the graph shown in Figure 13.30. The voltage across the capacitor, that is the voltmeter's reading at different times, will be plotted on the vertical axis. Time (as measured by, say, a stopwatch) will be plotted on the horizontal axis. The switch is closed and the stopwatch started. As the watch measures off one time constant, the voltage will have risen to 63% of the battery voltage. As time goes on during the next and each successive time constant, the voltage continues to rise 63% of the *remaining voltage*, until full charge is obtained. During the second time constant, the voltage rises 63% of the remaining voltage or to 86.4% of full charge, and so on. This process continues until *after five time constants* the capacitor is for all practical purposes fully charged. The line on the graph shows the voltage behavior with time. This curve is important and deserves some close examination.

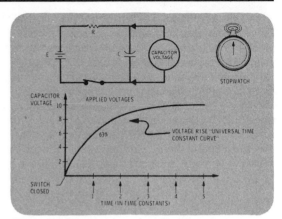

Figure 13.30

Universal Time Constant Graph (Charge) — The graph shown in Figure 13.31 is called the universal time constant graph, and may be used to describe any circuitry with a time constant type of behavior. This graph marks out in detail exactly what percentage of the applied voltage a capacitor will have charged to during specific time intervals after the voltage is applied to an RC circuit. If you know a circuit's time constant, this curve will enable you to accurately predict RC circuit behavior as time goes on. For example, suppose that you had calculated a circuit's time constant to be one second and you want to know what the capacitor's voltage will be one-half second after applying a voltage of 10 volts to it. In this case

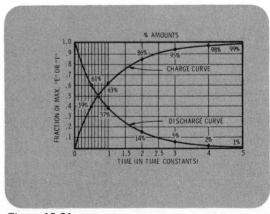

Figure 13.31

one-half second is one-half of a time constant or
0.5 time constant, so find that point on the
bottom (or horizontal axis) of the graph. Follow
the vertical line up from that point until it touches
the charge curve, and you see that after 0.5 time
constant, the capacitor would be at 39% of
10 volts, or 3.9 volts. Notice that there are *two
curves* on this graph, one is the charging curve that
has already been discussed. Now consider the
process of *discharging* a capacitor through a
resistor.

Universal Time Constant Graph (Discharge) — In
Figure 13.32, the same capacitor used before is
now fully charged but is rewired to discharge
through the resistor, R. Notice that this is an RC
circuit whose *time constant is identical* to the one
considered in Figure 13.30. This time when the
switch, S, is closed, the capacitor *discharges*, and
the behavior of the capacitor's voltage is the
reverse of that you saw previously. As the
stopwatch ticks off the first RC time constant, the
capacitor's voltage *falls 63%* of its fully charged
voltage, so that at the end of one RC time
constant, the voltage remaining on the capacitor is
37% of its fully charged value. During each
successive time constant, the voltage continues to
fall 63% of the remaining voltage value. After *five
time constants*, the voltage may be assumed to be
zero. Notice this important point: it takes five time
constants for a capacitor to either fully charge or
fully discharge. This is an important fact to
remember concerning RC circuits.

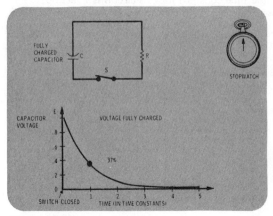

Figure 13.32

Example: RC Circuit Demonstration — To understand RC circuits more fully, a complete RC circuit example of a type you could easily build and study in the lab will now be discussed. Follow the circuit analysis as if you were actually watching the circuit perform. First, you will calculate the time constant for the circuit (shown in Figure 13.33), and then examine its behavior in detail. The circuit contains a 10-microfarad capacitor, and a 100-kilohm resistor and a switch, S, as shown. These components will be connected to a 100-volt dc supply. First calculate the time constant for this circuit, using the formula T = RC. Substituting R = 100 kΩ or 1 X 10^5 Ω and C = 10 μF or 1 X 10^{-5} farads. Multiplying 1 X 10^{+5} and 1 X 10^{-5} gives you 1 X 10^0 or 1. So the time constant for this circuit is *1 second*.

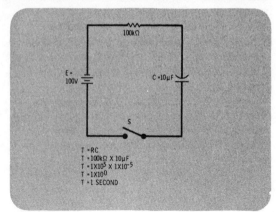

Figure 13.33

Storage Oscilloscope — To check the circuit's behavior, you could connect it to a special instrument called a *storage oscilloscope*, if one was available in your laboratory. This special instrument will actually plot a graph of the capacitor's voltage on the vertical axis, and time on the horizontal axis. On a typical storage scope you can set up the scope face so that each division on the scope's vertical axis represents 20 volts, and each one of the scope's horizontal divisions represents 1 second (Figure 13.34). When the switch, S, is closed, the oscilloscope *actually measures* the voltage across the capacitor for you, and plots it at different time intervals.

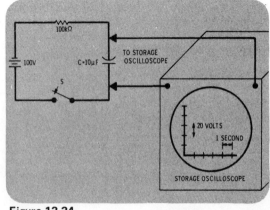

Figure 13.34

Scope Face: Capacitor Charging — When the switch in the circuit is closed, as shown in Figure 13.34, the scope face traces the patterns as shown in Figure 13.35 during the first five time constants. Figure 13.35(A) shows what the scope trace (picture) would look like at the end of 1 second. At this point, 63 volts is indicated on the scope face, which is 63% of 100 volts. In Figure 13.35(E), after five time constants or 5 seconds, the graph line has risen up to 100 volts or full charge.

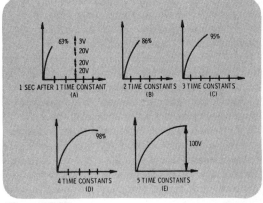

Figure 13.35

Discharge Demonstration Circuit — If the 100-volt power supply is removed from the circuit and the capacitor is connected so that it discharges through the same 100-kilohm resistor, the capacitor's discharge curve may be observed on the storage oscilloscope face (Figure 13.36), set at the same scales as before.

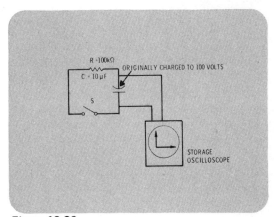

Figure 13.36

Scope Face (Discharge) — What the storage oscilloscope would display for you is shown in Figure 13.37. In Figure 13.37(A), which shows the scope face after one time constant or 1 second, the voltage has fallen from full charge of 100 volts down to about 38 volts. In the successive figures the line continues to descend until in Figure 13.37(E) at five time constants, the voltage is at zero and the capacitor is discharged.

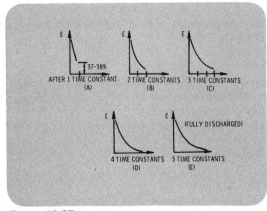

Figure 13.37

Slow Discharge of Capacitor: Low Current — At this point, you have seen what capacitors are, and how they operate in dc circuits. As has been demonstrated, a capacitor can *store charge*. This effect finds application in many dc circuits, for the following reason. When a charged capacitor is discharged, the rate at which the voltage across it drops and the amount of current that flows in the discharge circuit depend on the resistance in the discharge path. As you have seen, more resistance in the circuit causes a longer RC time constant, and hence it takes a longer time for a charged capacitor to discharge. In an RC circuit with high resistance, the amount of discharge current that flows is small. This trickle of discharge current flows for a long time, until the capacitor is discharged as shown in Figure 13.38.

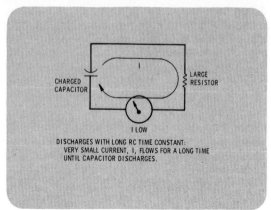

Figure 13.38

Rapid Discharge: High Current — Consider the following situation. If a charged capacitor is discharged by *shorting* its leads with a conductor, as shown in Figure 13.39, the resistance in the circuit that is formed approaches zero. The RC time constant then is also close to zero. As a result, a charged capacitor can be made to deliver an extremely large burst of current for a *very short* time. This makes the capacitor useful for providing power for special loads which require short, very high bursts of current to operate.

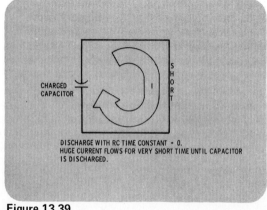

Figure 13.39

Example — One type of load that requires a high burst of current for a short time is a flash tube of the type used with many cameras. In cases such as this, a dc supply, usually having a low-current capability, may be used to gradually charge a large capacitor. This capacitor may then be discharged through the loads that require a short burst of very high current. In Figure 13.40, a high-voltage supply with a low-current capability is first used to charge a capacitor when the switch is thrown to the left. Then after the capacitor is fully charged, the switch is thrown to the right to provide a large burst of current to fire the flash tube. This, by the way, is a good time to remind you that in circuits containing capacitors, even though the power is off, there may still be a "jolt" awaiting the careless technician. Be careful.

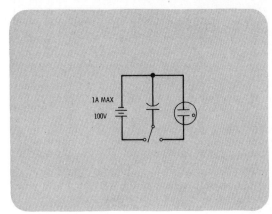

Figure 13.40

Capacitors Oppose Changes in Voltage — Now that you have seen capacitors in action, consider for a moment another of their effects that find use in a variety of applications. In all the circuits studied in the earlier lessons of this course, when power was applied to the circuit, the voltage at all parts of the circuit instantly reached its final value. No "time delays" were ever considered. In RC circuits of the type that have been discussed for you, the voltage across the capacitor rises more slowly, depending on the circuit's *time constant*. So it may be said that capacitance tends to *fight or oppose* changes of *voltage* in a circuit. As you have seen, if a rising voltage (such as a power supply which is abruptly switched on) is applied to a capacitor, the

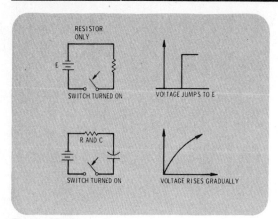

Figure 13.41

capacitor will act to *slow up* or oppose the rise of voltage, making it more gradual. Alternately, if the voltage is turned off in a circuit containing a capacitor, the capacitor will give up some of its charge and keep the voltage up longer. For this reason the capacitor finds application as a *filter* in certain circuits. In these types of circuits (Figure 13.41), a capacitor may be used to "smooth out" any abrupt voltage changes that may occur. Another common use for a capacitor is in a variety of *timing* applications. Both of these applications, as well as many others, will be covered for you in depth in later courses.

Capacitor Types — Now that you have seen some aspects of how capacitors work, consider some of the actual components you may find available in the laboratory. Figure 13.42 is a chart listing various types of capacitors that are available. Capacitors are usually *named* by the dielectric material used in them.

Because of the many applications of capacitors and the different properties of the dielectric, each capacitor you encounter will probably be labeled with a WVDC or "Working Voltage dc." This is the maximum voltage the capacitor can tolerate without its dielectric breaking down. As you apply capacitors in circuits, the voltage across them must be kept below this WVDC.

Capacitor Type	Range of Capacitance	Range of WVDC (in volts)	Range of Temperature (°C)	R_i = Insulation Resistance I_L = Leakage Current (at 25°C)	Comments
Ceramic	1 pF – 2.5 μF	20 – 200	–55 +125	R_i = 100 GΩ/μF	Small size Low cost
Paper	0.001 – 200 μF	50 – 200,000	–55 +105	R_i = 3 – 20 GΩ/μF	Low cost
Electrolytic	0.5 – 1,000,000 μF	2.5 – 700	–80 +125	I_L = 0.1 μA or more	Very small size Very low cost
Mylar	0.001 – 20 μF	50 – 1000	–55 +150	R_i = 50 GΩ/μF	Small size Relatively high cost
Air	10 – 400 pF	200/0.01 in air gap	–	–	Variable
Mica	1 pF – 1 μF	50 – 100,000	–55 +150	R_i = 10 – 100 GΩ/μF	Cap. change with age very small
Oil Filled	0.001 – 15 μF	100 – 12,500	–55 +85	R_i = 2 – 100 GΩ/μF	Low cost

Figure 13.42

- Fixed and Variable Capacitors
- Capacitor Construction
- **Electrolytic Capacitor vs Oil-Filled Capacitor**

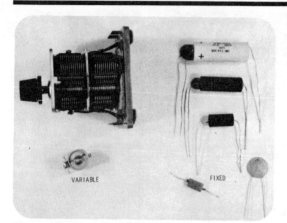

Figure 13.43

Fixed and Variable Capacitors — Capacitors are available in both fixed and variable types as shown in Figure 13.43. Several different common types including paper, mica, and ceramic capacitors are shown in the figure. As mentioned, these are named for the *dielectric material* employed in manufacturing them, and are produced in a set of preferred values, with tolerances similar to those of resistors. In these devices the capacitance is usually labeled on the device or follows a color code (listed for you in the Appendix).

Figure 13.44

Capacitor Construction — To keep the size of these devices as small as possible, a series of foil plates is usually employed with thin sheets of dielectric rolled up in between them as shown in Figure 13.44. Also shown in the figure is a typical *variable* capacitor. Variable capacitors are most often constructed so that their effective plate area can be varied by rotating a shaft connected to one set of plates. These plates can be moved between a stationary set of plates changing the capacitance value of the device.

Electrolytic Capacitor vs Oil-Filled Capacitor — Often in electronics, large amounts of capacitance may be needed in a small space. Most standard fixed capacitors are limited to values of about 1 microfarad or less due to cost and size considerations. There are capacitors available, however, that use a special chemical action to cram a large amount of capacitance into a small space. These are called *electrolytic* capacitors. An electrolytic capacitor having a capacitance of 42,000 microfarads may be about the same size as a 10 microfarad standard oil-filled capacitor (Figure 13.45). A price is paid, however, because the maximum working voltage of electrolytic capacitors is usually much lower than oil-filled or

Figure 13.45

other standard types. As mentioned, the maximum working voltage or WVDC of a capacitor is governed by its plate spacing and dielectric. If voltages on the capacitor's plates are higher than the WVDC, the dielectric will usually break down, allowing charge from one plate to pass through the dielectric to the other plate. When this happens, the capacitor is no longer properly functioning. With severe breakdown, the plates may actually short together, rendering the capacitor useless. A good rule of thumb is to choose capacitors that have a working voltage well above the highest voltage with which they will come in contact.

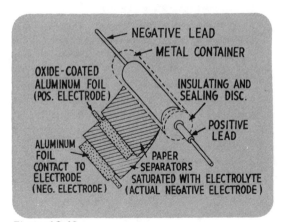

Figure 13.46

Electrolytic Construction — Most newer dry-type electrolytic capacitors commonly available are manufactured from two aluminum sheets separated by a saturated paper layer (Figure 13.46). The paper is saturated with a special chemical called an *electrolyte*, and rolled together with the aluminum sheets and packaged in a compact roll. When the capacitor is manufactured, a dc voltage is applied to it causing a thin *oxide layer* to be formed on one sheet. This oxide is *very thin* and acts as the insulating dielectric between the two plates. One of the capacitor's plates is now the aluminum sheet with the oxide deposited on it (positive plate), and the other plate is actually the electrolyte, which is connected to external circuits by the second aluminum sheet. Since the oxide layer is extremely *thin*, very large capacitances are available in electrolytic form. Since the oxide is fairly fragile, only lower voltages can normally be used with electrolytics. But most important, because of the way electrolytic capacitors are constructed, they can never be used in ac applications. They are restricted to only dc or pulsating dc applications, and must *always* be wired into dc circuits in the correct polarity, positive terminal to more positive voltages, negative to more negative voltages. If wired in the opposite direction, the electrolytic behaves as a low-value resistor, and the capacitor may actually explode due to heat generated by large amounts of current flow through it.

Leakage Resistance — It should be pointed out that even in the best capacitors, some "leakage" of charge will occur over periods of time. For good capacitors, however, significant charge can be stored almost indefinitely. Leakage occurs for a variety of reasons in capacitors and tends to be higher in electrolytic types. Some charge actually makes it through the dielectric, as if there were a very high-value resistor connected between the plates as shown in Figure 13.47. Often a minimum value of leakage resistance R_L is specified for capacitors in applications where leakage may be critical.

When discharging capacitors, it is generally *not* recommended that the leads simply be shorted together with a conductor. Although many capacitors can take this abuse, the very large internal currents created can damage them. So it is always a good idea to discharge capacitors gradually through an appropriate resistor. This keeps internal currents low and prevents damage to what are often expensive and difficult to replace parts.

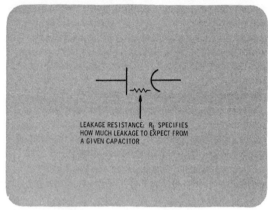

LEAKAGE RESISTANCE: R_L SPECIFIES HOW MUCH LEAKAGE TO EXPECT FROM A GIVEN CAPACITOR

Figure 13.47

Capacitor Applications — The charge storage capability of capacitors is utilized in a variety of devices and applications. These are listed in Figure 13.48. Briefly consider several of these applications: discharging capacitors can produce currents large enough to weld metals together, which is a common capacitor application in spot-welding devices. Flash tubes, as has been mentioned, also require very large, very short bursts of current to operate. A capacitor is used to provide this burst of current. Also, in capacitive discharge ignition systems, the same principle is used to get a "hotter spark" in an automobile engine. More recently, in medical emergency rooms a device known as a *defibrillator* discharges a

TYPICAL CAPACITOR APPLICATIONS

SPOT WELDER

FLASH TUBE

CAPACITOR DISCHARGE IGNITION SYSTEM

MEDICAL DEFIBRILLATOR

ELECTRONICS — TIMING, COUPLING, RESONANT CIRCUITS, ETC.

Figure 13.48

capacitor through two paddles so that a controlled amount of energy is delivered to a patient's heart, restoring normal rhythm to a heart that has stopped beating. You will find capacitors applied in a variety of ways as you continue your study of electricity.

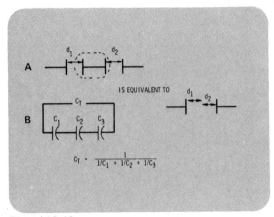

Figure 13.49

Capacitors in Series — Before leaving the topic of capacitors, one more topic should be mentioned. Often to obtain a desired value of capacitance, capacitors are combined in series and parallel connections. Consider what happens when two capacitors are connected in series as shown in Figure 13.49(A). The two inner plates of these capacitors are wired directly together so effectively that they act as one plate. The result is that two capacitors in series act as a single capacitor whose plates are separated by *both* plate separations. As you have learned, the *larger* the spacing between the plates, the *lower* the capacitance. For this reason, the total capacitance of any series combination of capacitors is less than that of any individual capacitor in the circuit. As it turns out, the total capacitance of capacitors in series is calculated in the same way as the total resistance of parallel resistors. The formula is

$$C_T = \frac{1}{1/C_1 + 1/C_2 + 1/C_3}$$

etc., as shown in Figure 13.49(B).

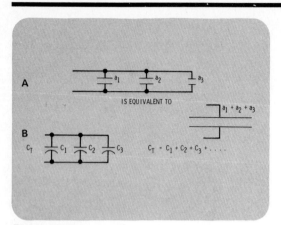

Figure 13.50

Capacitors in Parallel — When several capacitors are connected in parallel, an interesting thing happens. As shown in Figure 13.50(A), since all the *plates* are wired together in tandem, a parallel connection of capacitors acts like a single capacitor having a total plate area equal to the sum of the plate areas of the individual capacitors. Since capacitance varies directly with plate area, the total capacitance of several capacitors connected in parallel is found by adding all the individual capacitances, as with series resistors, as shown in Figure 13.50(B).

    A new and useful component has been examined in this lesson and some of the wide variety of its applications have been shown. In the next lesson, the behavior of dc in circuits containing other new components, called *inductors*, will be introduced. You will find many similarities, some big differences, and some interesting and important new effects.

LESSON 13. CAPACITORS AND THE RC TIME CONSTANT

● **Worked Through Examples**

1. Find the time constant of a circuit containing a 10-kilohm resistor in series with a 0.82-microfarad capacitor.

   To solve this problem, you must use the time constant formula $T = RC$. Substituting in the circuit values, the formula reads $T = 10 \text{ k}\Omega \times 0.82 \text{ }\mu\text{F}$. In scientific notation the values are: $T = 1.0 \times 10^4 \times 8.2 \times 10^{-7}$.

$$\begin{array}{c} 1.0 \times 10^4 \\ \underline{\times 8.2 \times 10^{-7}} \\ T = 8.2 \times 10^{-3} \text{ seconds (s) or 8.2 milliseconds (ms)} \end{array}$$

2. Find the time constant of this circuit:

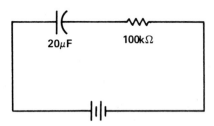

   $20\mu\text{F}$ $100\text{k}\Omega$

   Use the formula: $T = RC$. First substitute in the circuit values: $R = 100 \text{ k}\Omega$, $C = 20 \text{ }\mu\text{F}$

$$T = 100 \text{ k}\Omega \times 20 \text{ }\mu\text{F}$$
$$T = 1.0 \times 10^5 \times 2.0 \times 10^{-5}$$
$$T = 2.0 \text{ seconds}$$

3. How long will it take the capacitor in the following circuit to reach full charge?

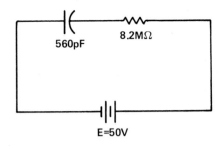

   560pF $8.2\text{M}\Omega$

   E=50V

   First, use the time constant formula $T = RC$

$$T = RC$$
$$T = 8.2 \text{ M}\Omega \times 560 \text{ pF}$$
$$T = 8.2 \times 10^6 \times 5.6 \times 10^{10}$$
$$T = 4.59 \times 10^{-3} \text{ s or 4.59 ms}$$

You must remember that the RC time constant formula you just worked gives you *one* time constant (in seconds). *Five* time constants are required for full charge. So, multiply the time constant by 5 to arrive at the correct answer.

$$4.59 \times 10^{-3}$$
$$\underline{\times \quad 5}$$
$$22.95 \times 10^{-3} \text{ or } 2.3 \times 10^{-2} \text{ seconds}$$

The capacitor will be fully charged after 2.3×10^{-2} seconds or 23 milliseconds.

4. Find the voltage across the capacitor in the circuit shown below 500 milliseconds after the switch is closed. (Use the universal time constant graph.)

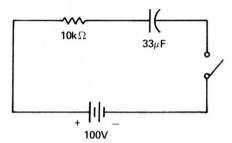

First, you should calculate the time constant of the circuit. T = RC.

$$T = RC$$
$$T = 10 \text{ k}\Omega \times 33 \text{ } \mu F$$
$$T = 1.0 \times 10^4 \times 3.3 \times 10^{-5}$$
$$T = 3.3 \times 10^{-1} \text{ or } 330 \text{ ms}$$

Now look at the universal time constant graph. Time (horizontal axis) is measured in time constants. To convert this chart to seconds, multiply 330 milliseconds by each of the time divisions. For example:

$$1 \times 330 \text{ ms} = 330 \text{ ms}$$
$$1.5 \times 330 \text{ ms} = 495 \text{ ms}$$
$$2 \times 330 \text{ ms} = 660 \text{ ms}$$
$$3 \times 330 \text{ ms} = 990 \text{ ms}$$
$$4 \times 330 \text{ ms} = 1.32 \text{ s}$$
$$5 \times 330 \text{ ms} = 1.65 \text{ s}$$

Now these values are applied to the universal time constant graph.

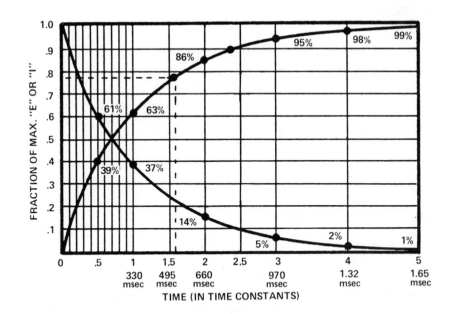

Look at the chart and locate the 500 millisecond position on the horizontal axis. Now trace directly upward (following the dotted line) and note the point on the charging curve that is reached at 500 ms. Tracing to the left from that point, across the graph, you can see that the amplitude at the intersection point is about 0.78 or 78% of the full charge voltage; 0.78 X 100 V = 78 V. So after 500 ms, the capacitor is charged to 78 volts.

5. Find the charge in coulombs of the capacitor in problem 4, at the end of 500 milliseconds.

The formula for calculating the charge stored in a capacitor is

$$Q = CE$$

where

Q = the stored charge in coulombs
C = the capacitance in farads
E = the voltage between the capacitor plates

Substituting the values of capacitance and voltage:

$$Q = 33 \, \mu F \times 78 \, V$$
$$Q = 3.3 \times 10^{-5} \times 7.8 \times 10^{1}$$
$$Q = 2.57 \times 10^{-3} \text{ coulombs (or 2.57 millicoulombs)}$$

6. Using the universal time constant graph, calculate the time required for the capacitor shown below to charge to 55 volts.

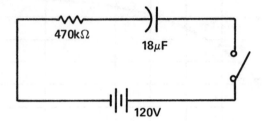

First, calculate the circuit's time constant using the formula: T = RC

$$T = RC$$
$$T = 470 \text{ k}\Omega \times 18 \ \mu F$$
$$T = 4.7 \times 10^5 \times 1.8 \times 10^{-5}$$
$$T = 8.46 \text{ s}$$

Now, the universal time constant curve may be used as follows in solving this problem. First, examine the vertical axis. On this axis the *fraction* of the maximum voltage is located. The *maximum* voltage here is 120 volts: the total applied voltage. What *fraction* of 120 volts is 55 volts? Thus, 55/120 equals 0.458. This is the *fraction* of the applied voltage 55 volts represents. Now, locate 0.458 on the vertical axis of the universal time constant graph. Trace to the right horizontally (a dotted line is drawn in for you to follow) until you intersect the charging curve.

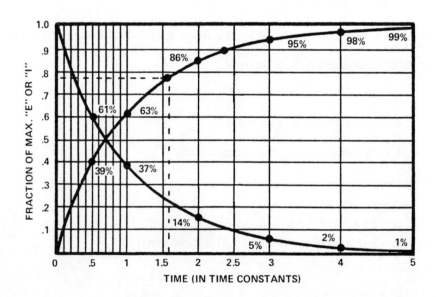

Locate that point on the curve, and then trace directly *down* to the horizontal axis. At this point you read the time elapsed: 0.6 *time constants*. You know that 1 time constant is 8.46 seconds, so the total elapsed time is 0.6 X 8.46 or 5.08 seconds.

7. A "strobe" flash attachment for a camera has a bulb that requires 0.02 coulomb of charge at 450 volts in order to flash properly. What is the minimum size capacitor that could be satisfactorily used?

    Since both the quantity of charge (Q) and voltage (E) are known, the equation C = Q/E can be used to solve this problem. Simply substitute in the capacitor values and solve for C.

    $$C = Q/E$$
    $$C = \frac{0.02\ C\ \text{(coulomb)}}{450\ V}$$
    $$C = 0.0000444\ F\ \text{or}\ 44.4\ \mu F$$

8. Find the approximate frequency of oscillation in the circuit shown here.

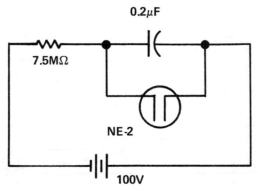

The circuit shown above is a "relaxation oscillator." It operates on the basis of its RC time constant. The bulb shown connected across the capacitor is an NE-2 neon glow lamp. These lamps require a certain voltage (called the "firing voltage") in order to light. Once lit, the voltage across the lamp must fall significantly below the firing voltage before it will turn "off." Typical "on" and "off" voltages for neon glow lamps are: 75 volts "on" and 50 volts "off." This means that the typical NE-2 will not "light" until the voltage across it reaches 75 volts, but once lit, will continue to glow until the voltage drops below 50 volts. Before the lamp lights, it has a very high resistance (essentially an open circuit). Once the lamp is on, its resistance drops to a low value.

Consider what will happen when one of these lamps is connected across a capacitor as shown in the circuit above. When power is applied to the circuit, the capacitor will begin to charge up to the source voltage. The rate of charging will be controlled by the RC time constant. When the capacitor reaches 75 volts, the neon bulb (which is connected in parallel with the capacitor) will also have 75 volts applied across it. At this instant, the bulb will light, allowing heavy current flow, and thus discharging the capacitor very quickly. As the capacitor discharges, its voltage will drop down below the 50 volts required to keep the neon bulb lit. The bulb goes out and the capacitor again charges up to the 75 volts required to fire the bulb, and the cycle is repeated again and again. As you can see, there are several factors that affect the rate of blinking (or oscillation) of the bulb: the resistor size, the size of the capacitor, the supply voltage, and the characteristics of the individual neon bulb.

To analyze this problem, first calculate the RC time constant of the circuit and plot it on a universal time constant graph.

$$T = RC$$
$$T = 7.5 \text{ M}\Omega \times 0.2 \,\mu\text{F}$$
$$T = 7.5 \times 10^6 \times 2.0 \times 10^{-7}$$
$$T = 1.5 \text{ s}$$

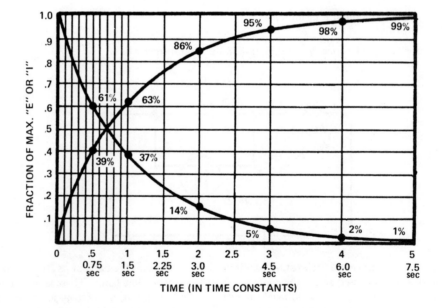

TIME (IN TIME CONSTANTS)

$$1 \times 1.5 \text{ s} = 1.5 \text{ s}$$
$$1.5 \times 1.5 \text{ s} = 2.25 \text{ s}$$
$$2 \times 1.5 \text{ s} = 3.0 \text{ s}$$
$$3 \times 1.5 \text{ s} = 4.5 \text{ s}$$
$$4 \times 1.5 \text{ s} = 6.0 \text{ s}$$
$$5 \times 1.5 \text{ s} = 7.5 \text{ s}$$

To give a clearer picture of the operation of this circuit, these values are plotted on the horizontal axis of the universal time constant graph above.

The lamp fires at 75 volts, and causes the voltage across the capacitor to rapidly drop to 50 volts so that the lamp then goes out. Voltage across the capacitor, plotted as time goes on, will appear as shown below.

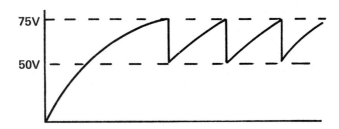

In order to find the time duration between flashes, simply look back at the Universal Time Constant graph you just filled in. Locate 75 volts and 50 volts, and measure the time elapsed between these two points. Seventy-five volts occurs at approximately 1.4 time constants or 2.1 seconds. Fifty volts occurs at 0.7 time constants or 1.05 seconds. The time elapsed is the *difference* between the two times. Subtract and you get 2.1 s − 1.05 s = 1.05 s. So the lamp will blink once every 1.05 seconds. Dividing 60 by 1.05 yields a frequency of 57 flashes per minute.

9. Calculate the total capacitance of this circuit.

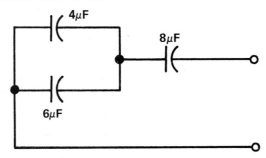

Problems of the type shown above give many students headaches because capacitors "add" just the opposite of the way resistors do. Parallel capacitors are added by using a formula similar to the series resistance formula: $C_T = C_1 + C_2 + C_3 \ldots$ Series capacitors must be added by using a formula similar to the parallel resistance formula:

$$C_T = \frac{1}{1/C_1 + 1/C_2 + 1/C_3 \ldots}$$

To solve this problem, the 4-microfarad and the 6-microfarad capacitors should be combined by using the parallel capacitance formula $C_T = C_1 + C_2 + C_3 \ldots$

$C_T = 4\,\mu F + 6\,\mu F$

$C_T = 10\,\mu F$

This 10 microfarads of capacitance must be combined with the 8 microfarads of capacitance by using the series capacitance formula.

$$C_T = \frac{1}{1/C_1 + 1/C_2 + 1/C_3}$$

$$C_T = \frac{1}{1/10 + 1/8}$$

$$C_T = \frac{1}{0.1 + 0.125}$$

$$C_T = \frac{1}{0.225}$$

$$C_T = 4.44 \, \mu F$$

10. Calculate the total capacitance of the following circuit.

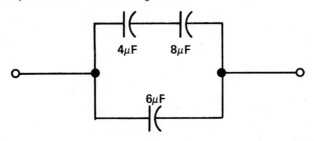

First, find the total capacitance of the upper circuit branch using the series capacitance formula:

$$C_T = \frac{1}{1/C_1 + 1/C_2 + 1/C_3}$$

$$C_T = \frac{1}{1/4 + 1/8}$$

$$C_T = \frac{1}{0.25 + 0.125}$$

$$C_T = \frac{1}{0.375}$$

$$C_T = 2.67 \, \mu F$$

Now the total capacitance may be found by combining the two parallel capacitances using the parallel capacitance formula $C_T = C_1 + C_2 + C_3 \ldots$

$$C_T = 2.67 \, \mu F + 6 \, \mu F$$

$$C_T = 8.67 \, \mu F$$

LESSON 13. CAPACITORS AND THE RC TIME CONSTANT

● **Practice Problems**

Depending upon the approach you use in solving these problems and how you round off intermediate results, your answers may vary slightly from those given here. However, any differences you may encounter should only occur in the third significant digit of your answer. If the first two significant digits of your answers do not agree with those given here, recheck your calculations.

1. Calculate the RC time constant for the following circuits. Fold over   **Fold Over**
   the page to check your answers.

a.

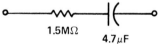

1.5MΩ   4.7μF   T = _____

b.

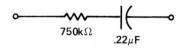

750kΩ   .22μF   T = _____

c.

43pF   110MΩ   T = _____

d.

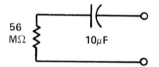

56 MΩ   10μF   T = _____

e.

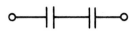

56pF   100kΩ   T = _____

2. Calculate the total capacitance in the following circuits. (All capacitors are 2 μF).

a.

C_T = _____

b.

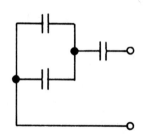

C_T = _____

Answers

1.

a.   T = 7.05 s

b.   T = 165 ms

c.   T = 4.73 ms

d.   560 s

e.   T = 5.6 μs

2.

a. C_T = 1 μF

b. C_T = 1.33 μF

c.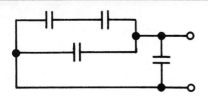

Fold Over

$C_T =$_____

d.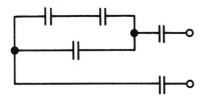

$C_T =$_____

e.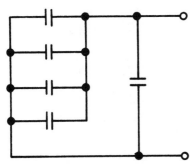

$C_T =$_____

3. Find the following unknown values using the formula Q = CE.

a.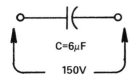

C=6μF

150V

Q = _____

b.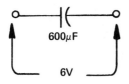

600μF

6V

Q = _____

c.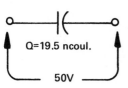

Q=19.5 ncoul.

50V

C = _____

Answers

c. $C_T = 5\,\mu F$

d. $C_T = 0.75\,\mu F$

e. $C_T = 10\,\mu F$

3.
a. $Q = 900\,\mu C$

b. $Q = 3.6\ mC$

c. $C = 390\ pF$

d.

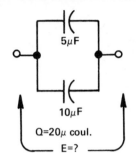

E = _____

e.

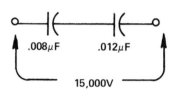

Q = _____

4. For the circuit shown below, calculate, or use the universal time constant graph to find:

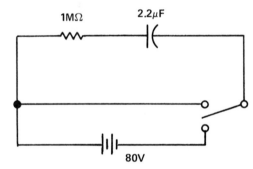

a. RC time constant.   _____

b. Time required for capacitor to charge fully.   _____

c. Voltage across the capacitor after 1.5 seconds.   _____

d. Voltage across the capacitor after 6.5 seconds.   _____

e. Time required for the capacitor to charge to 30 volts.   _____

Answers

d.   E = 1.33 V

e.   Q = 72 μC

4.
a.   T = 2.2 s

b.   11 s

c.   39.5 V

d.   75.8 V

e.   1.03 s

1. A device which, in its basic form, consists of two conductors separated from each other by an insulator is called a:
   a. Battery
   b. Generator
   c. Motor
   d. Capacitor

2. Capacitors are devices that can:
   a. Store electric charge
   b. Be charged and discharged repeatedly
   c. Hold a charge for a long time
   d. All of the above
   e. None of the above

3. When a potential difference exists between the plates of a capacitor, the capacitor is said to be:
   a. Discharged
   b. Open
   c. Charged
   d. Disconnected

4. The equation of Capacitance C in farads is _____ where Q is charge in coulombs and E is potential difference in volts.
   a. $C = QE$
   b. $C = \dfrac{Q}{E}$
   c. $C = \dfrac{E}{Q}$
   d. $C = Q^2E$

5. Capacitance is the ratio of the charge stored _____ the voltage across the capacitor.
   a. Multiplied by
   b. Divided by
   c. Added to
   d. Subtracted from

6. When a capacitor can store 1 coulomb of charge with 1 volt potential difference across it, it is said to have _____ of capacitance.
   a. 10 farads
   b. 1 farad
   c. 0.1 farad
   d. A small value

7. The charge stored in a capacitor with C farads of capacitance and E volts applied is:
   a. $Q = CE$
   b. $Q = CE^2$
   c. $Q = \dfrac{C}{E}$
   d. None of above

8. When a capacitor is placed across a battery without resistance in the circuit, the capacitor charges:
   a. Very slowly
   b. To twice the voltage
   c. Instantaneously
   d. To one-half the voltage

9. When a capacitor is placed across a battery with resistance in the circuit, the capacitor charges:
   a. Instantaneously
   b. Much slower than without resistance
   c. To twice the voltage
   d. To one-half the voltage

10. The capacitive time constant is defined by the following equation:
    a. $T = R^2C$
    b. $T = CR$
    c. $T = \dfrac{R}{C}$
    d. $T = RC$
    e. b and d above

11. When a capacitor and resistor are placed across a voltage, the time required to charge the capacitor to 63% of the applied voltage is called:
    a. The rise time
    b. The discharge curve
    c. The capacitive time constant
    d. The peak value

12. When evaluating a capacitive time constant, the R in the equation is in ohms and the C is in farads. As a result T is in:
    a. Seconds
    b. Minutes
    c. Hours
    d. Relative time

13. The capacitor is considered charged for all practical purposes after _____ time constants.
    a. 2
    b. 3
    c. 1
    d. 5

14. The discharge time constant of a capacitor is the same as the charge time constant if:

    a. The voltage is reversed
    b. The circuit is open
    c. The resistor and capacitor are the same
    d. Other paths are removed

15. What is the discharge time of a 10 microfarad capacitor charged to 10 volts shunted by a 10 K resistor?

    a. 1 sec
    b. 0.1 sec
    c. 10 sec
    d. 100 sec

Calculate the following quantities:

16.

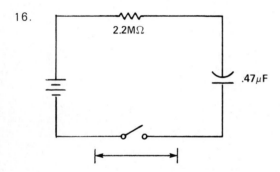

Time Constant _____ seconds

17.

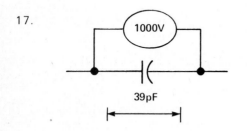

Stored charge _____ coulombs

18.

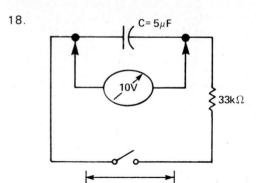

Time to discharge _____ seconds

19.

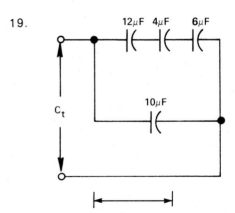

C_T, Total capacitance _____ μF

20.

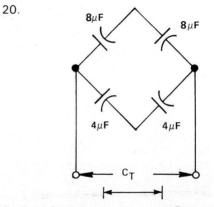

C_T, Total Capacitance _____ μF

Lesson 14

Inductors and the L/R Time Constant

In this lesson an entirely new type of circuit component — the *coil* or *inductor* — will be introduced and discussed. You will see that the behavior of coils in dc circuits may be described in a similar fashion to the action of *capacitors* discussed in the previous lesson. Coils, however, act to oppose *changes in current*, rather than voltage, and store energy in a *magnetic field*, rather than an electrostatic field. The concept of the L/R time constant used in describing circuits containing resistance (R) and inductance (L) will also be introduced.

●   Objectives

This lesson gives you a brief introduction to a new type of circuit component that you will see called by many names, such as:

Coil

Inductor

Choke

Solenoid

The behavior of coils in dc circuits can be described using many of the terms that were introduced to you in the last lesson. At the end of this lesson, you should be able to

1.   *Sketch* the construction of a typical inductor, labeling its basic component parts. *Sketch* the schematic symbols for:

   a.   Air or insulating core inductor

   b.   Iron core inductor

   c.   Powdered iron core inductor

2.   *Write* a brief description of the magnetic field created by a coil and the key effect this magnetic field has on a coil's behavior in circuits. *Sketch* the magnetic field lines around a simple current carrying coil. *State* the units used to measure the inductance.

3.   Given a schematic for any circuit of the type shown below:

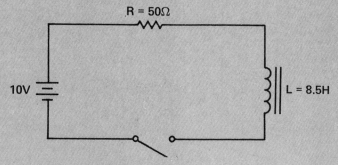

   a.   *Calculate* the circuit's time constant

   b.   *Calculate* the value of the steady-state current

   c.   *Calculate* the value of the current flowing in the circuit after one time constant

   d.   *Calculate* the time it takes for the current in this circuit to reach its steady-state value

   e.   *Sketch* a graph that shows how the current rises in this circuit from the time the switch is closed through five time constants

   f.   *Write* a description of the effect that would occur if the switch is opened after the current in this circuit has been allowed to reach its steady-state value.

LESSON 14. INDUCTORS AND THE L/R TIME CONSTANT

- **Coil Construction**
- **Iron Core Coil**

Lesson 13 covered one of the most important circuits in electronics — the resistor-capacitor (or RC) circuit. In that lesson you were shown how to predict the behavior of a circuit containing a resistor and capacitor connected in series. The characteristic behavior of these circuits, especially the characteristic time required to charge and discharge the capacitor in RC circuits was discussed. The RC time constant was discussed and it was shown how RC circuits may be used to perform many useful tasks in electronics. It was also shown how capacitors store charge and energy, and how they can be used to provide large bursts of current for short periods of time for special loads requiring such power.

Now in this lesson, *another* basic electronic component the "coil," or as it is often called, the "inductor," or "choke," will be examined. You will be seeing that the behavior of a coil, or inductor, in a circuit is, in many ways, similar to, and in some ways opposite from, the behavior of a capacitor. Puzzling? By the end of this lesson, you should have a fairly good understanding of coils and how they operate in circuits.

Coil Construction — To begin, a *coil* simply consists of wire that is wrapped or coiled around a "core." The wire may be any size, or length, and typical core material may be anything from iron to air. The most common schematic symbol for a coil looks just like several turns of wire adjacent to one another. Figure 14.1 shows the symbol used to signify an air core or insulating core coil. Typically, cores of this type are used in high-frequency ac applications.

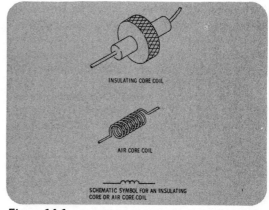

Figure 14.1

Iron Core Coil — If the coil symbol also contains two parallel lines as shown in Figure 14.2, an iron core coil is indicated. These inductors are typically used in lower frequency ac applications and dc applications. A typical iron core coil is also shown in Figure 14.2. In this type of coil, the core material may actually surround the wire and forms the most substantial component of the coil.

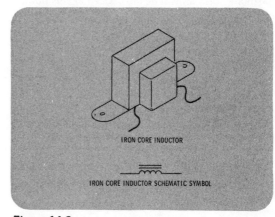

Figure 14.2

- Powdered Iron Core Coil
- Wire Inside Coils

Powdered Iron Core Coil — Figure 14.3 shows a schematic symbol indicating a coil that has an iron core indicated by dashed lines. These dashed lines indicate a coil that has a *powdered iron* core. A powdered iron core coil is used in applications where radio-frequency waves are being processed. A common use for this type of coil is for the antenna in a standard AM radio.

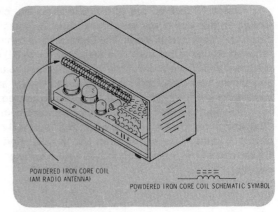

POWDERED IRON CORE COIL
(AM RADIO ANTENNA)

POWDERED IRON CORE COIL SCHEMATIC SYMBOL

Figure 14.3

Wire Inside Coils — What do coils do in dc circuits? How do they work? To get into the processes by which coils function, it is necessary to focus your attention on the individual parts of the coil one at a time (Figure 14.4). First, consider the wire making up the coil itself.

Very early in your study of dc electricity, you saw how a single strand of conducting wire contains *billions* of free electrons. Normally these electrons are moving around in the wire in random motion. You have already seen that if a potential difference is applied across the wire, electrons begin drifting from the negative to the positive potential, as discussed earlier. This is the phenomenon of electron current flow.

The action of coils depends upon a phenomenon that will be introduced at this point: the electromagnetic field.

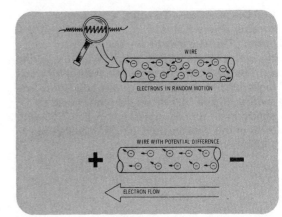

WIRE

ELECTRONS IN RANDOM MOTION

WIRE WITH POTENTIAL DIFFERENCE

ELECTRON FLOW

Figure 14.4

Electromagnetic Field — Here's a new and important fact about electrons in motion: "Whenever current flows in a conductor, a *magnetic field* is set up around the conductor" (Figure 14.5). A magnetic field is a type of field that has some properties similar to that of the electrostatic field examined earlier.

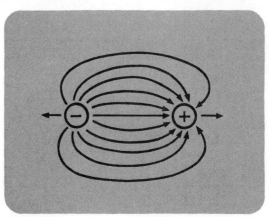

"WHENEVER CURRENT FLOWS THROUGH A CONDUCTOR, A **MAGNETIC FIELD** IS SET UP AROUND THE CONDUCTOR."

Figure 14.5

Electrostatic Field and Lines of Force — Recall that in an earlier lesson it was shown that an *electrostatic field* exists in the area around any charged body, and that this field may be visualized with electrostatic lines of force, as shown in Figure 14.6. Considerable time has been spent discussing the *effects* of the electrostatic field on electrons.

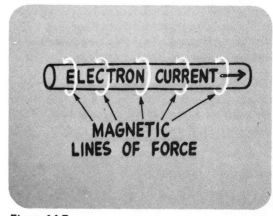

Figure 14.6

Wire with Electromagnetic Field — An *electromagnetic field* is a field which surrounds any current-carrying conductor, and can also be visualized by using what are called "magnetic lines of force." The magnetic lines of force go right around the current, and thus around the wire in little rings as shown in Figure 14.7. The center of these rings is the current itself.

ELECTRON CURRENT →

MAGNETIC LINES OF FORCE

Figure 14.7

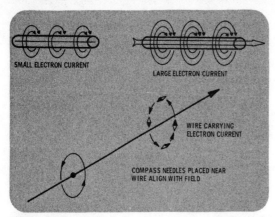

Figure 14.8

Magnetic Lines of Force — An electromagnetic field is created around any wire carrying current. The larger the current in the wire, the stronger the field will be. In a diagram such as shown in Figure 14.8, this is represented by drawing more and denser lines of force. The magnetic field around a current-carrying wire may be detected with a small compass needle placed near the wire. The compass needle, being magnetized, will line up with the magnetic lines of force near the wire. The magnetic lines of force are assigned a direction (indicated by the small arrows on them). This direction is the direction that the north pole of the compass will point when placed into the field.

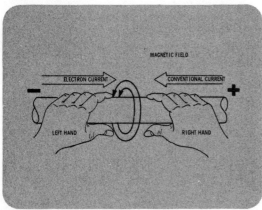

Figure 14.9

Direction of Lines of Force — To find the direction of the field lines, you can use one of the "hand rules." In Figure 14.9, a wire is shown with electron current flowing from left to right (this is equivalent and identical to conventional current flowing from right to left). To find the direction of the magnetic field, you mentally grasp the wire with your *left* hand, the thumb pointing in the direction of the *electron* current. Your fingers will curl around the wire, pointing in the direction of the magnetic field. (Do the same thing, only with your *right* hand, if you are considering *conventional* current; the magnetic field direction will be the same.)

Simple Coil — In most current-carrying wires, the surrounding magnetic field is small and goes unnoticed. However, if the wire were to be wrapped into a coil, the wire would be concentrated into a smaller area, thus concentrating the lines of force around it into a smaller area, as shown in Figure 14.10. This increases the strength of the magnetic field, especially inside the loop.

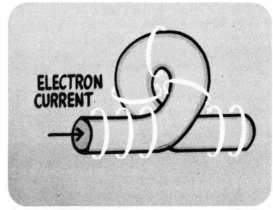

Figure 14.10

- **Multiple Loops**
- **Large Coil**
- **Bar Magnet**

Multiple Loops — If the wire is looped two or more times, an interesting and very useful effect occurs. The lines of force produced by the first loop of wire join together and reinforce the lines of force produced by the second loop, strengthening the magnetic field inside the coil as shown in Figure 14.11.

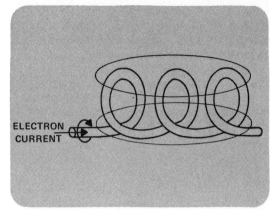

Figure 14.11

Large Coil — If many loops of current-carrying wire are wrapped into a coil, a strong electromagnetic field can be created inside the coil, as shown in the cutaway view in Figure 14.12.

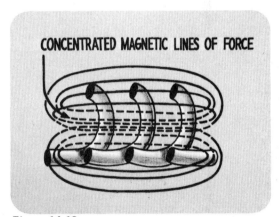

Figure 14.12

Bar Magnet — A coil, when carrying current, contains a *strong magnetic field*. Because of this, a coil carrying current will behave just like a regular bar magnet (Figure 14.13).

Most of you have probably seen magnets and are familiar with how they attract metal objects. A coil, when carrying current, does the same thing with the added advantage that it may be turned on and off. With the current turned on, a piece of steel may be drawn into the coil. If no current flows through the coil, no magnetic field exist and metal objects will be released.

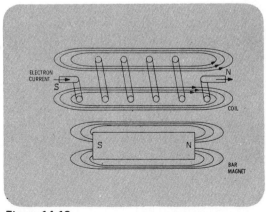

Figure 14.13

Relay — This feature makes coils useful in many applications. The device, shown in Figure 14.14, is called a "relay." In a relay, a current is used to create a magnetic field in a coil which causes the coil to draw in a metal bar connected to a switch. The current may be made to open or close the switch, as the application requires. So by using a relay, one current (called the "control current") may be used to switch an entirely different current.

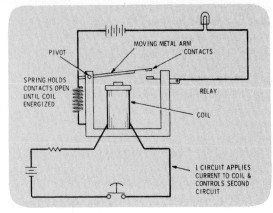

Figure 14.14

Automotive Starting System — Relays find many uses in a variety of applications in electricity and electronics. One typical situation in which they are used is where a very large *current* needs to be switched on and off from a remote location. One such situation is in the starter circuit of your automobile. The starter motor for most cars is an extremely large load: drawing currents of 100 amperes or more for short intervals. It would be impractical and undesirable to run a large length of heavy cable capable of handling 100 amps to your dashboard to a heavy-duty key switch. Instead, a special relay is actually used to switch the starter motor current. This relay is commonly called a *solenoid* (pronounced "soul-annoyed"), and may be mounted right in the casing with the starter motor or separately elsewhere in your car. The word solenoid is another one of those words that is often used to mean different things. It is used in different contexts to mean a coil, a relay, or an electromagnet, depending on what is being discussed. At any rate, the solenoid in your car works as described in Figure 14.15. When you turn your ignition key switch, it turns on only a small "control current" which activates the windings of the solenoid coil. The solenoid then completes the circuit, handling the very large current that cranks over your engine. (At the same time, another coil is at work, the spark coil, which will be discussed later in this lesson.)

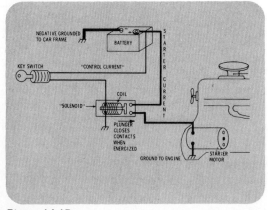

Figure 14.15

Effects of Coils in Circuits — Up to this point you have seen that when current flows through a coil, a magnetic field builds up in the coil. This magnetic field may be used for a variety of purposes, as mentioned, but its presence in the coil has some other interesting effects on the behavior of the circuit containing the coil.

Anytime you try to force current through a coil, a magnetic field must build up inside it. The key point is that it takes a certain amount of time to build up this field. Because of the mechanisms involved in this field buildup (Figure 14.16), the *rise of the current in circuits containing coils is slower. Coils act to oppose changes in current.*

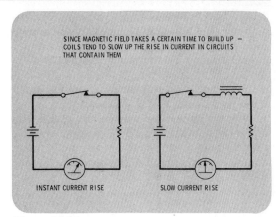

SINCE MAGNETIC FIELD TAKES A CERTAIN TIME TO BUILD UP — COILS TEND TO SLOW UP THE RISE IN CURRENT IN CIRCUITS THAT CONTAIN THEM

INSTANT CURRENT RISE   SLOW CURRENT RISE

Figure 14.16

Current Increase — Exact details of the mechanism by which coils oppose or fight changes in current will be discussed at great length in later courses. One way to quickly visualize what is going on is as follows. Magnetic fields (like electric fields) store *energy* inside them. (Energy is actually the ability to do work. In capacitors you saw how the electric field of a charged capacitor stored energy and gave up this energy when the capacitor was discharged).

A magnetic field may actually be considered to be an energy storage device that is present in a circuit whenever a coil is connected. The workings of this system may be visualized as shown in Figure 14.17. When current tries to *increase* through the coil, the coil will take some energy from the circuit and dump it into the magnetic field. The current now finds itself building up the magnetic field as it tries to start flowing through the circuit. As a result of this magnetic field buildup, *it takes the current longer to build up in the circuit with the coil present than with no coil.* For this reason, it may be said that coils fight or oppose current increasing through them. Again, it takes longer for current to build up in a circuit containing a coil than it does in a circuit with no coil.

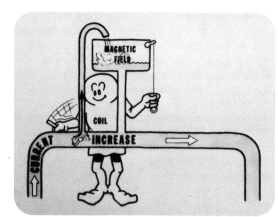

Figure 14.17

Steady Current Flow — After a period of time, the magnetic field increases to a maximum value, and the current in the circuit reaches a steady-state value. Here's another important point: *As long as the current is not changing, the magnetic field remains built up at a steady-state (Figure 14.18). When the magnetic field is in a steady-state, it has no effect on the current flowing in the circuit.* In a steady-state condition, the only thing that affects the amount of current flowing in the circuit is the total resistance present (and the applied voltage).

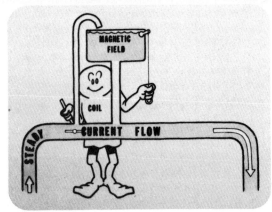

Figure 14.18

Current Decrease — If the switch is opened in a circuit containing an inductor, the circuit current will try to rapidly fall to zero. At this point, the energy stored in the coil's magnetic field gets dumped back into the circuit and tries to help keep the current flowing. It appears as if the coil saw the current trying to fall and flushed the energy into the circuit to try to keep current flowing, as visualized in Figure 14.19. So it is said that a coil *also* acts to fight or oppose any current *decrease* through it. *It takes longer for current to fall to zero in a circuit containing a coil than it does in a circuit with no coil.*

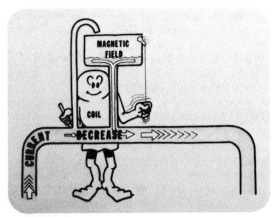

Figure 14.19

Coil Rule — Think about the actions of a coil for a moment. Through the actions of the magnetic field, a coil acts to oppose any current increase or decrease through it. *It may be said that coils act to oppose any change in current through them* (Figure 14.20).

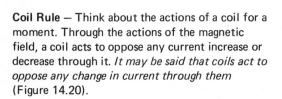

"COILS OPPOSE ANY CHANGE IN CURRENT THROUGH THEM".

Figure 14.20

- **Energy Storage**
- **Capacitor Action (Charge)**
- **Capacitor Action (Discharge)**

Energy Storage — As you will recall from the last lesson, a *capacitor stores electrons*. The imbalance of electrons (many on one plate, very few on the other) produces an electrostatic field in the region between the plates. It was stated that *energy* was stored in a capacitor in this electrostatic field. This stored energy was demonstrated to you as it was released in a capacitor discharge. Now, it has been seen that in a coil, energy is stored in the *electromagnetic field* created by a current flowing through it (Figure 14.21). Because of these similar effects, the voltage and current characteristics of coils and capacitors have an interesting interrelationship.

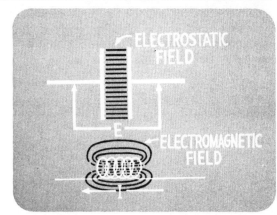

Figure 14.21

Capacitor Action (Charge) — *Capacitors* in circuits *oppose the change of voltage* across the circuit. When the switch in the circuit of Figure 14.22 is thrown to position A, a voltage is applied to the RC circuit. The capacitor takes a certain amount of time to charge up, with the voltage across it rising slowly. When the voltage across the capacitor equals the applied voltage, *no further current flow occurs in the circuit*, and the capacitor is said to be fully charged.

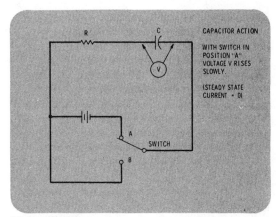

Figure 14.22

Capacitor Action (Discharge) — When the switch in this circuit is thrown to position B (Figure 14.23), the applied voltage is removed and the capacitor discharges, giving up its stored charge and energy. The voltage across the capacitor falls gradually to zero. Therefore, it may be said that a capacitor fights *changes* in *voltage* across it.

As has been discussed, a coil opposes *changes* in the *current flowing* through a circuit by means of storing and giving up the energy contained in its magnetic field. The process by which a coil fights changes in current is called electromagnetic induction. A coil sets up a voltage called an induced voltage, which actually acts to fight against changes in current. In later courses on

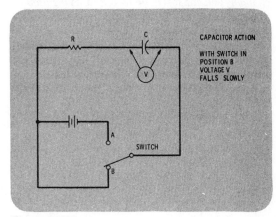

Figure 14.23

alternating current, induction and induced voltage will be discussed in more detail. For now, just remember that a coil tends to slow up or oppose changes in current through it.

Demonstration Circuit Schematic — To help you visualize how a coil operates in a dc circuit, consider what would happen in a simple dc demonstration circuit like the one shown in Figure 14.24. In this circuit schematic, you have a coil connected in series with a battery, a switch and a resistor. To simplify the analysis of the circuit at first, assume that the coil has no resistance and that all of the resistance in the circuit is supplied by the resistor, which has a labeled value of 500 ohms. The battery voltage is 6 volts.

As has been said, if the switch is thrown to position B_2, so that the battery is connected, the coil will fight or oppose the *change or buildup* in the circuit current. After the current is *finished changing* and the magnetic field is at its *maximum value, the coil will no longer oppose the current*. The coil itself opposes the current *change*, but not the current itself. So once the current has reached its final steady-state value, the only opposition to it will be the 500 ohms of resistance in the circuit.

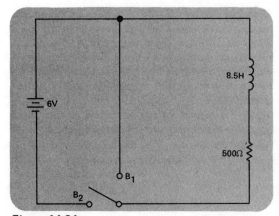

Figure 14.24

- **Ohm's Law (I)**
- **Current Rise in Inductive Circuits**
- **Current Fall in Inductive Circuits**

Ohm's Law (I) — Because the only factor controlling the circuit's steady-state current is the circuit resistance, Ohm's law can be used to find this current, as shown in Figure 14.25. After the magnetic field around the coil has reached its final value, you can use Ohm's law in the form I = E/R to calculate the current. Substituting, I = 6/500, which equals 12 milliamps. Keep in mind what will happen when the voltage is applied to the circuit of Figure 14.24. *The current will slowly rise, finally reaching this steady state value of 12 milliamps.*

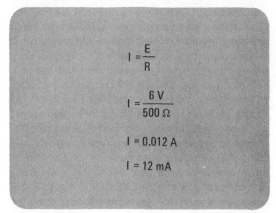

Figure 14.25

Current Rise in Inductive Circuits — Trace this process step by step. When the switch is first switched to position B_2, connecting the battery, the current jumps up from zero and starts to flow. The coil opposes this drastic change in current by taking some circuit energy and storing it in its magnetic field. Gradually, as the magnetic field gets completely built up and reaches a steady-state, the resistor provides the only opposition to current flow in the circuit.

At that point, the circuit current has reached a steady-state and the final current flowing is 12 milliamps as shown in Figure 14.26.

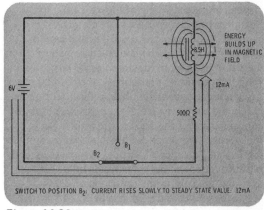

Figure 14.26

Current Fall in Inductive Circuits — If the switch in this circuit is instantly switched to position B_1, the circuit current tries to immediately change back to zero. The coil will fight this change by dumping energy back into the circuit from its magnetic field, thus the current falls *gradually* to zero as seen in Figure 14.27. Coils act to oppose any *changes in current.*

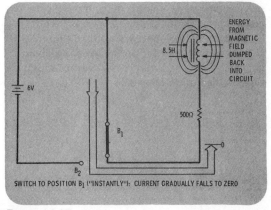

Figure 14.27

14-15

Demonstration Circuit — Figure 14.28 is a line drawing of what an actual circuit of the type being examined would look like in your laboratory. When the switch is closed, current flows from the power supply through the switch, through the 100-ohm resistor, through the coil, and then back to the battery. You will notice back in Figure 14.27, the schematic diagram indicated that a *500-ohm* resistor was connected in series with the coil. In an actual circuit of this type, probably about 400 ohms of resistance would be contained in the large amount of wire wound inside the coil. The coil probably contains about 100 feet of fairly small gage wire, so it is reasonable to expect the coil to have a considerable resistance. Keep this in mind. You must remember to include the internal resistance of coils when analyzing practical circuits. As seen in the schematic of Figure 14.27, *the resistance of a coil appears to be connected in series with the coil itself.* So the actual circuit represented by this schematic diagram would consist of a 100-ohm resistor in series with the coil for a total equivalent series resistance of 500 ohms.

In order to actually observe the effect of the coil in the circuit, a "storage oscilloscope" may be connected across the 100-ohm resistor. If you remember, the storage oscilloscope was used in the last lesson. This type of "scope" will measure and plot the voltage across this resistor on its vertical axis, and time on its horizontal axis.

Stop and consider what is being performed, keeping in mind Ohm's law. The oscilloscope will now plot the *voltage* across the resistor versus time. The voltage across any resistor, however, is directly proportional to the current flowing through it. The graph plotted by the storage scope will, in effect, be a picture of how the *current* in the circuit behaves as time goes on. On the screen will be a graph representing *current* on the vertical axis, and *time* on the horizontal axis.

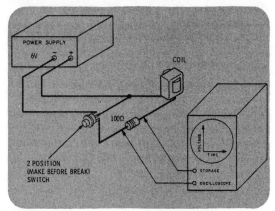

Figure 14.28

Current Rise in L/R Circuit — When the switch is thrown so that the 6-volt power source is applied to the circuit, the waveform shown in Figure 14.29 is produced. Immediately the shape of this curve should look somewhat familiar to you. It is a universal time constant graph, similar to the graph produced by an RC circuit, with some exceptions. Figure 14.29 is a plot of *current* flowing in the circuit versus time, instead of voltage as it was when you were examining a capacitor. A capacitor opposes change in voltage across a circuit, where a coil opposes any change in current through a circuit. In series circuits with inductance and resistance, the *current* rises with a characteristic "time constant" type of behavior.

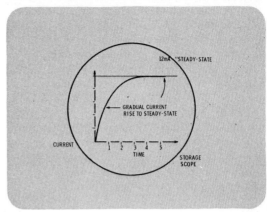

Figure 14.29

Factors Affecting Time Constant in Circuits Containing Coils — Circuits containing inductance demonstrate a time-constant type of behavior similar to the action you have seen for RC circuits. Previously it was seen that the amount of resistance in the circuit and the size of the capacitor (in farads) were the two factors that affected the time constant for an RC circuit. In a circuit containing a resistor and a coil, you will see that the time constant behavior again depends on how much resistance is present in the circuit, and also on how well the coil does its job of slowing up current changes (Figure 14.30). Focus your attention on each of these factors one at a time.

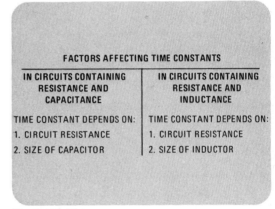

Figure 14.30

FACTORS AFFECTING TIME CONSTANTS	
IN CIRCUITS CONTAINING RESISTANCE AND CAPACITANCE	IN CIRCUITS CONTAINING RESISTANCE AND INDUCTANCE
TIME CONSTANT DEPENDS ON:	TIME CONSTANT DEPENDS ON:
1. CIRCUIT RESISTANCE	1. CIRCUIT RESISTANCE
2. SIZE OF CAPACITOR	2. SIZE OF INDUCTOR

Ohm's Law — Consider the effect of resistance on the time constant. If more resistance is added to this circuit, what will be the result?

You know from Ohm's law that more resistance in a circuit causes *less* total circuit current to flow (Figure 14.31).

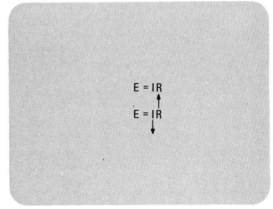

$$E = IR$$
$$E = IR$$

Figure 14.31

Time Constant and Resistance — With increased resistance in the circuit, when the power supply is turned on, the change from zero current to maximum current will be less as shown in Figure 14.32. Since the *change in current flowing* is less, the coil will offer less opposition to the change in current.

If the coil offers less opposition, the current flowing in the circuit will reach its maximum value in a shorter period of time, thus providing a shorter time constant. So the *higher the circuit's series resistance* in a circuit containing a resistor and a coil, the *shorter* the time constant will be.

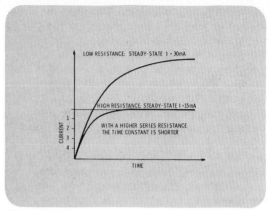

Figure 14.32

Inductance — What other factor will affect the current buildup? The better a coil is at storing energy, the more it will be able to oppose changes in current through a circuit.

The ability of a coil to store energy and fight changes in current flowing through it is specified by what is called its *"inductance."* It is for this reason that coils are often called inductors. The symbol you will see used to indicate inductance in formulas is "L." The units used to measure inductance are called "henries," named after the American scientist Joseph Henry. The unit *henry* is abbreviated "H". So the mathematical statement: "The inductance of this coil is 10 henries" may be written in shorthand form L = 10 H, as shown in Figure 14.33. The more henries of inductance a coil has, the more energy it will store in its magnetic field. As a result, the coil will be better able to oppose current changes through it. The inductance of a coil depends on how it is constructed. In general, the more turns of wire a coil has, the bigger the cross-sectional area, and the shorter its length is, the bigger the coil's inductance will be. The core material used in its construction also drastically affects coil inductance. In general, coils that employ an iron core can pack a lot more inductance in a smaller space than is possible with air or insulating cores. More details on inductance and the mechanics of how coils fight current change are really part of an ac course, and therefore will not be covered in this course.

INDUCTANCE — "L" MEASURED IN HENRIES

ABBREVIATION . . . H

"THE INDUCTANCE IS 10 HENRIES" MAY BE ABBREVIATED: "L = 10H"

Figure 14.33

In circuits containing coils and resistors (often termed RL or LR circuits), the time constant will depend directly on the number of henries of inductance present. The more inductance, the longer the time constant, and vice versa. As you have already seen, the time constant depends *inversely* on the resistance. The *more series resistance* in the circuit, the *shorter* the time constant will be, and vice versa.

L/R Time Constant Formula — In formula form, the time constant for an RL circuit can be expressed as: T = L/R. In the formula, T represents time constant in seconds; L is inductance in henries, and R is the series resistance in ohms, as shown in Figure 14.34.

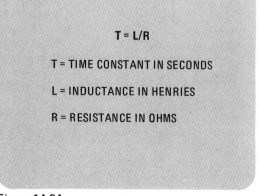

$$T = L/R$$

T = TIME CONSTANT IN SECONDS

L = INDUCTANCE IN HENRIES

R = RESISTANCE IN OHMS

Figure 14.34

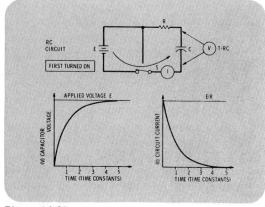

Figure 14.35

RC Circuit: First Turned On — You have now seen that inductive and capacitive dc circuits behave in somewhat similar ways, with a time constant associated with each. As a review, examine the behavior of both of these type circuits, during the various phases of their operation. First, consider what happens to the circuits in Figures 14.35 and 14.36 when power is first applied. In the RC circuit, the voltage across the capacitor starts at zero and gradually rises until it reaches the applied voltage. The current in the RC circuit flows only a short time necessary to charge the capacitor (five time constants).

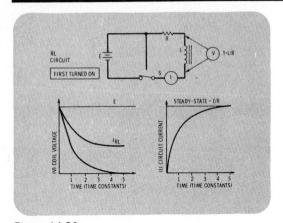

Figure 14.36

RL Circuit: First Turned On — In the RL circuit (Figure 14.36), the current in the circuit starts at zero and rises slowly to the steady-state value E/R. The voltage across the coil behaves in an interesting fashion. It starts out equal to the applied voltage when the switch is first closed, and gradually falls as the steady-state is reached. If the coil had no internal resistance, the coil voltage would gradually fall to zero. If the internal resistance of the coil is R_L, the coil's voltage will fall to a value equal to the steady-state current times R_L or $I \times R_L$.

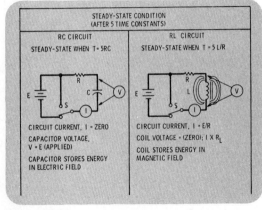

Figure 14.37

RC and RL Circuits: Steady-State — As has been seen, when the steady-state is reached in the RC circuit (after five time constants as shown in Figure 14.37), no more current is flowing in the circuit. At this point, the voltage across the capacitor equals the applied voltage. The capacitor stores energy in an electrostatic field between its plates. In the RL circuit, after five time constants, the circuit current has risen to a value equal to E/R. The coil's voltage would be zero *if the coil had no internal resistance*. Actually, the voltage across the coil will be equal to the circuit current times the coil's internal resistance, R_L. The coil stores energy in the steady-state in its *magnetic field*.

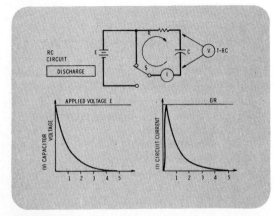

Figure 14.38

Voltage Fall: RC Circuits — If the switch in the RC circuit shown in Figure 14.38 is thrown after the capacitor has been allowed to reach steady-state, the capacitor will give up its stored energy. A burst of current will flow for a short while, and the capacitor's voltage will gradually fall to zero.

- Current Fall: RL Circuits
- **Example: Time Constant Calculation**
- **Current Rise**

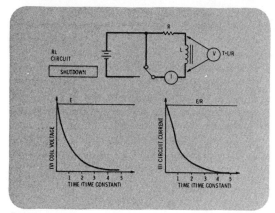

Figure 14.39

Current Fall: RL Circuits — If the RL circuit shown in Figure 14.39 has been allowed to reach the steady-state and then the switch is *instantly* switched to the short position, the current will fall *gradually* to zero. The coil voltage will jump to E and then fall off. Notice that the switch must be "instantly" switched to do this. In practice, if you built a circuit such as this and attempted to switch the current off, a *spark* would probably appear at the switch as soon as the circuit was broken. A special "make before break" switch can be used to avoid this. This "sparking" effect will be explained later in the lesson.

Example: Time Constant Calculation — You are fairly familiar with the RC time constant and how it is calculated. Some examples of how the L/R time constant of a circuit such as that shown in Figure 14.40 can be calculated.

The time constant of this circuit can be found by substituting the circuit values into the time-constant formula, $T = L/R$. "L" is equal to 8.5 henries and R is 500 ohms. Performing the necessary calculations, T, the circuit time constant, is equal to 0.017 second, or 17 milliseconds. This means that it takes 17 milliseconds for the *circuit current to rise to 63% of its steady-state value.*

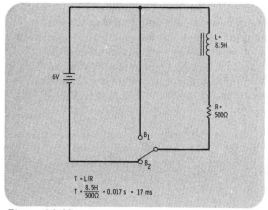

Figure 14.40

Current Rise — To calculate the steady-state value of current in this circuit, use Ohm's law in the form $I = E/R$, as mentioned before. Here $I = E/R =$ 6 V/500 Ω = 12 mA. So, the circuit current rises as shown in Figure 14.41.

As in RC circuits, this L/R circuit also requires five time constants before the "steady-state" current value is reached, and no further change is taking place in the circuit. In this case, five time constants equal 5 X 17 milliseconds or 85 milliseconds. The graph shown in Figure 14.41 is a picture of the way a storage oscilloscope trace would illustrate the behavior of this circuit.

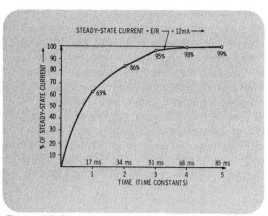

Figure 14.41

Current Fall with Resistor — If the switch is *instantaneously* moved to position B_1 as shown in Figure 14.42, the coil is connected in series with 500 ohms of resistance, and the current source is removed. When this is done, it will be seen that the current follows a reverse time constant behavior, and gradually falls to zero, after five time constants. After one time constant, the current has fallen to 37% of its final value.

Once again, when the drop in current is discussed, it is carefully specified that the switch must be "instantaneously" switched to position B_1. This is because most switches would *open* the circuit for a little while before the contacts reached position B_1. Opening a current-carrying circuit containing a coil produces an interesting, important, and somewhat drastic effect. As has been said before, a coil opposes changes in the current through a circuit. The more drastic the current change is, the greater the coil's opposition to it will be. When a circuit containing a coil is broken or opened, an infinite resistance is placed in the circuit. Stop for a moment and think about the effect this has on the time constant of the circuit.

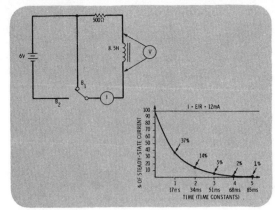

Figure 14.42

L/R Time Constant for an Open Circuit — Using the time constant formula $T = L/R$, notice that when R becomes equal to infinity, then the time constant for the circuit will become zero, because *any number divided by infinity is zero*, as shown in Figure 14.43. Now think: with a *zero time constant*, the current in the circuit will try to fall to zero *instantly*. This is a very drastic current change, and as you recall, a coil will oppose *any current change* and attempt to keep current flow constant through a circuit. The more drastic the change, the greater the coil will try to oppose it.

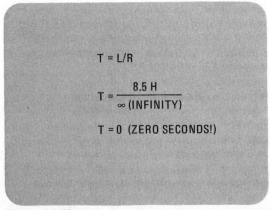

$$T = L/R$$

$$T = \frac{8.5\ H}{\infty\ (INFINITY)}$$

$$T = 0\ (ZERO\ SECONDS!)$$

Figure 14.43

Ohm's Law in Open Inductive Circuit — However, the open circuit has an *infinite resistance*. In order to maintain any value of current at a constant level through an infinitely large resistance, the coil would need an *infinitely large voltage*. In theory, the coil should produce an infinite voltage as shown in Figure 14.44. In practice, *that is just what the coil tries to do*. The current flowing in the circuit before the switch was opened was 17 milliamps. When the circuit is opened, the coil will attempt to maintain current flow at 17 milliamps. In order to do this across an infinite resistance, the energy contained in the coil's magnetic field is converted to a very high voltage, perhaps thousands of volts. The result is usually a spark somewhere in the circuit, probably at the switch. This phenomenon is affectionately called the "kick" of a coil. (If you should ever have you body connected across a current-carrying inductor when the circuit is opened, you will know how it got that name.)

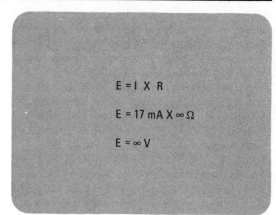

$$E = I \times R$$

$$E = 17 \, mA \times \infty \, \Omega$$

$$E = \infty \, V$$

Figure 14.44

Inductive Kickback Circuit — To illustrate this high voltage "kick," a circuit may be connected as shown in Figure 14.45. Two special neon glow lamps are now connected across the inductor and resistor in a series RL circuit, just like the one investigated previously. This circuit will demonstrate the high voltage produced by a current-carrying coil when the circuit is opened. Each of the neon lamps requires about 70 to 75 volts in order to turn "on" and flash. Together they will require 140 volts or more to light at all. Until the voltage across the lamps reaches at least 140 volts, these two lamps will not operate. Notice that the power supply in this circuit is the same one considered earlier. It has an output of only

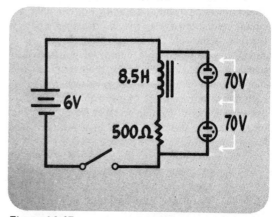

Figure 14.45

6 volts. When the switch is closed, 6 volts are applied to the circuit. Current flow gradually builds up until after five time constants, 17 milliamps of steady-state current flows in the circuit. At this point, the magnetic field has completely built up in the coil. The bulbs cannot light as yet because up to this point there is no voltage in the circuit anywhere near 140 volts. However, if the switch in the circuit is *opened*, energy from the coil's magnetic field will be abruptly dumped back into the circuit in the form of a very high voltage. Both of the neon bulbs will flash, indicating that a voltage of at least 140 volts was produced. In actuality, a much higher voltage than that is produced in this circuit.

Coil and Capacitor "Special Effects" — The high voltage "kick" effect is very important and finds many applications in electricity and electronics. You have now seen that a *coil* can produce a burst of *very high voltage* for a short period of time. This is similar to the effect shown in the preceding lesson, where a *capacitor* was used to produce a *large burst of current* for a short time (Figure 14.46).

Both of these effects have many uses in electrical applications. Coils are used to produce high voltages for firing fluorescent lights, in electric fences, and as was seen in an earlier lesson, for firing the spark plugs in an automotive ignition system. As a technician, it is necessary for you to keep in mind that sometimes the high-voltage "kick" a coil produces can be an *undesirable side effect* of having a coil in a circuit. For example, a circuit containing a relay may be damaged when the relay is shut off and a high-voltage burst appears across it. For this reason, relays often have special protective devices across their terminals to short-out and eliminate this high voltage. More than one expensive piece of electronic gear has been "wiped out" by high voltages from relays and other inductive devices.

A COIL: CAN PRODUCE A BURST OF VERY HIGH **VOLTAGE.**

A CAPACITOR: CAN PRODUCE A LARGE BURST OF **CURRENT.**

Figure 14.46

Automotive Ignition System — You have seen a coil in action in an earlier lesson in this course; a portion of an automobile electrical system was shown. Your ignition system contains an "ignition coil" which produces the spark for your spark plugs, as well as an undesirable side effect. Figure 14.47 shows the basics of the circuit. The "points" in your ignition system perform the same function as the switch that provides current to this coil.

When the "points" in an automobile engine close, current flows through the ignition coil and builds up the magnetic field inside it. The coil is designed so that when the points open, this magnetic field inside the coil collapses, producing a high-voltage arc or spark that appears across the spark plug. In addition, since the points are part of an inductive current-carrying circuit, when they open the circuit, an electrical discharge or spark can be expected to appear across the points also. Unfortunately, this is not a good or desirable effect and can cause the points to become burned and age rapidly.

To stop the arcing across the points, a component that *opposes a change in voltage* can be inserted across them and control the action of the spark. This is the reason a *capacitor* is found in an automobile distributor connected directly across the points. The capacitor prevents the coil's induced voltage from damaging the points. This capacitor is often called a "buffer" capacitor, and capacitors of this sort are often employed where a switch or a "set of points" is used to interrupt the magnetic field produced by a coil.

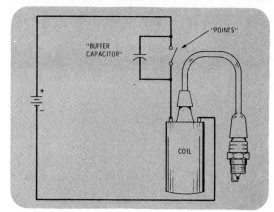

Figure 14.47

Summary — In this lesson a new electrical component, the coil, and the L/R time constant have been introduced and discussed. You have seen what coils are and examined the key factors that affect their behavior in dc circuits. In many ways coils are seen to operate in "reverse" fashion to capacitors. As will be seen in any ac circuits course, and later electronics courses, capacitors and inductors are very useful and integral parts of almost every electronic circuit. In your ac courses, you will explore the reasons why inductors behave the way they do in greater detail.

LESSON 14. INDUCTORS AND THE L/R TIME CONSTANT

● **Worked Through Examples**

---

1. Describe the magnetic field around a simple coil of the type shown in the figure below. What is the key effect of a coil's magnetic field on the behavior of coils in dc circuits?

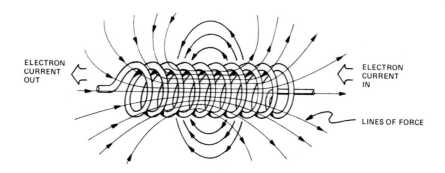

Solution: A magnetic field surrounds any wire carrying current. When this wire is wound into a coil, the magnetic field is concentrated inside the coil as shown by the magnetic lines of force drawn in the figure. This concentrated magnetic field is in effect an energy storage reservoir. Energy is stored when current attempts to increase through the coil, and this energy is released back into the circuit when current attempts to decrease through the coil. For this reason, coils are said to *oppose changes in current* in circuits.

2. Find the following values for the circuit shown below:
    a. Time constant
    b. Maximum steady-state current
    c. Voltage across the resistor after two time constants

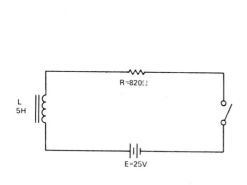

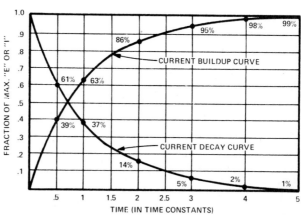

The time constant for this circuit may be found by using the inductive time constant formula, $T = L/R$. In this circuit, L is equal to 5 henries and R is equal to 820 ohms. 5/820 = 0.0061 second, or 6.1 milliseconds. This is one time constant for this circuit. Five time constants are required for the circuit to reach its steady-state condition. The maximum steady-state current in an inductive circuit is determined by using Ohm's law. The total voltage, E (here 25 volts), must be divided by the total circuit resistance R_T to give you the steady-state current. In this circuit, the total resistance is taken to be 820 ohms, the value of the resistor performing the calculation: 25 V/820 Ω = 30.5 mA. This value of current will be flowing in the circuit after five time constants.

The value of current flowing after only two time constants may be found by using the universal time constant graph. First, locate the two time constant mark on the horizontal line. Trace the graph line up until it intersects the "current buildup" curve. The intersection point is labeled 86%. This means that at this point, the circuit current is at 86% of the steady-state value. So, the current value at 2 time constants may be found by multiplying 0.86 X 30.5 mA. The current flowing after two time constants is equal to 26.2 mA. The value of the current at any time constant point may be determined by using the universal time constant graph in the manner just presented. To find the voltage across the resistor at the end of two time constants, multiply the current at that point (26.2 milliamps), times the resistance (820 ohms), to get your answer (21.5 volts).

3. Find the following values for the circuit shown below:

   a. Time constant

   b. Maximum steady-state current

   c. Voltage across the resistor after 2 milliseconds (2 ms).

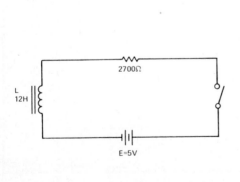

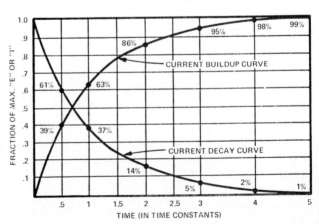

Solution:

a.   T = L/R

   T = 12/2700

   T = 4.44 ms

b. $E_T/R_T = I_T$

   5/2700 = 1.85 mA = steady-state current

c.   To find the circuit current at 2 milliseconds, the first thing to do is locate 2 milliseconds on the horizontal axis of the time constant graph. This axis of the graph is measured out in terms of time constants. You must get the chart to read out in seconds. This may be done by dividing 2 milliseconds by 4.44 milliseconds, to determine the exact percentage 2 milliseconds is as compared to 4.44 milliseconds. Two ms/4.44 ms = 0.45. In terms of time constants, 2 milliseconds is equal to 0.45 (or 45%) of one time constant. Locate 0.45 on the horizontal axis of the graph. Trace upward until that graph line intersects the current buildup curve. The intersection occurs at approximately 37%. This indicates that the current flowing at this point is 37% of the steady-state current, or 0.37 X 1.85 mA which is equal to 0.68 mA. To find the voltage across the resistor, multiply this current (0.68 milliamps) times the resistance (2700 ohms) to yield the voltage (1.84 volts).

LESSON 14. INDUCTORS AND THE L/R TIME CONSTANT

● **Practice Problems**

Solve the following problems related to inductance and the L/R time constant, using the time constant formula and the universal time constant graph given below. Fold over the sheet to check your answers.
Depending upon the approach you use in solving these problems and how you round off intermediate results, your answers may vary slightly from those given here. However, any differences you encounter should only occur in the third significant digit of your answer. If the first two significant digits of your answers do not agree with those given here, recheck your calculations.

Fold Over

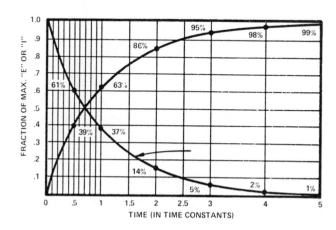

TIME (IN TIME CONSTANTS)

1.

L = 25mH

150kΩ

10V

Circuit time constant = _____

I_{max} = _____

Voltage across the 150-kilohm resistor

after two time constants = _____

2.

16H

150Ω

5V

Circuit time constant = _____

I_{max} = _____

Voltage across the 150-ohm resistor

after 50 milliseconds = _____

Answers

1. Circuit time constant = 167 nanoseconds

    I_{max} = 66.7 microamps

    Voltage across the 150-kilohm resistor

    after two time constants = 8.6 volts

2. Circuit time constant = 107 milliseconds

    I_{max} = 33.3 milliamps

    Voltage across the 150-ohm resistor

    after 50 milliseconds = 1.86 volts

Fold Over

3.

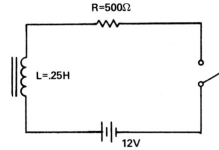

R=500Ω

L=.25H

12V

Circuit time constant = _____

I_{max} = _____

Voltage across the 500-ohm resistor

after 1 millisecond = _____

4.

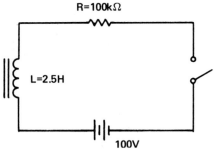

R=100kΩ

L=2.5H

100V

Circuit time constant = _____

I_{max} = _____

Voltage across the 100-kilohm resistor

after three time constants = _____

5.

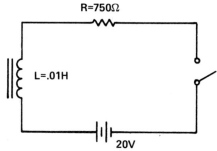

R=750Ω

L=.01H

20V

Circuit time constant = _____

I_{max} = _____

Voltage across the 750-ohm resistor

after 25 microseconds = _____

- **Practice Problems**

Answers

3. Circuit time constant = 500 microseconds

   I_{max} = 24 milliamps

   Voltage across the 500-ohm resistor

   after 1 millisecond = 10.3 volts

4. Circuit time constant = 25 microseconds

   I_{max} = 1 milliamp

   Voltage across the 100-kilohm resistor

   after three time constants = 95 volts

5. Circuit time constant = 13.3 microseconds

   I_{max} = 26.7 milliamps

   Voltage across the 750-ohm resistor

   after 25 microseconds = 17 volts

1. Coils or inductors are commonly used at high frequency when they:
   a. Have an iron core
   b. Have an air core
   c. When they are wound backwards
   d. None of the above

2. Coils oppose:
   a. Changes in current
   b. Changes in voltage
   c. Changes in resistance
   d. Changes in direction

3. Whenever current flows in a conductor, a _____ field is set up around the conductor.
   a. Voltage
   b. Current
   c. Electrostatic
   d. Magnetic

4. When many loops of current-carrying wire are wrapped into a coil, a strong _____ is created inside the coil.
   a. Current
   b. Voltage
   c. Electromagnetic field
   d. Bond

5. Magnetic fields store_____
   a. Coulombs
   b. Current
   c. Voltage
   d. Energy

6. It takes_____ for current to build up in a series circuit containing a coil than it does in a circuit with no coil.
   a. A shorter time
   b. Longer
   c. Forever
   d. None of above

7. In a coil with steady-state current flowing, the magnetic field has _____ on the current.
   a. Lots of effect
   b. No effect
   c. A new dependence
   d. None of the above

8. In a circuit with a coil, a resistor, a battery and an open switch, when the switch is closed the current rises:
   a. Instantaneously
   b. In a step function
   c. Gradually
   d. The same as the voltage

9. The final value of current in the circuit of question 8 is determined by:
   a. Ohm's Law
   b. The L and R ratio
   c. LR
   d. The amount of inductance

10. The time it takes for the current to attain 63% of its final value in a circuit containing inductance and resistance (besides a battery) is called:
   a. The capacitive time constant
   b. A coulomb
   c. The first step
   d. The L/R time constant

11. Inductance is measured in units called:
   a. Millivolts
   b. Henries
   c. Nanoseconds
   d. Gigahertz

12. When a switch opens the circuit with a steady-state current flowing through a coil, the coil will cause_____ voltage to appear across the infinite resistance of the open circuit.
   a. Twice the battery
   b. A very large
   c. A very small
   d. 3 Millivolts of

13. A coil can produce a burst of:
   a. Very high voltage
   b. Very low voltage
   c. Very large current
   d. Very small current

14. A capacitor can produce a burst of:
   a. Very high voltage
   b. Very low voltage
   c. Very large current
   d. Very small current

15. If more series resistance is added to a series R-L circuit, its time constant:

    a. Increases
    b. Stays the same
    c. Decreases
    d. None of above

16. The time constant for an R-L circuit is 25 milliseconds. Steady-state current is flowing. When the current is turned off it reaches approximately zero in:

    a. 25 Milliseconds
    b. 50 Milliseconds
    c. 75 Milliseconds
    d. 125 Milliseconds

Calculate the indicated unknown quantities:

17.   L = 2.5 H   R = 600 Ω

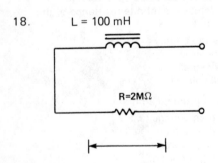

Time Constant _____ sec

18.   L = 100 mH

R=2MΩ

Time Constant _____ sec

19.   L = 15 H

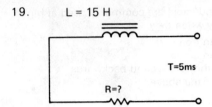

T=5ms

R=?

R _____ ohms

20.   L = ?

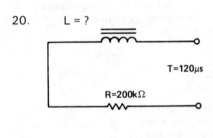

T=120μs

R=200kΩ

L _____ Henries

Appendices

APPENDIX 1. SPECIFIC RESISTANCES OF VARIOUS MATERIALS

(Listed in Increasing Values)

Symbol	Name	Specific Resistance* (ohms)	Symbol	Name	Specific Resistance* (ohms)
Ag	Silver	9.9	Pd	Palladium	61.4
Cu	Copper	10.4	Pt	Platinum	65.9
Au	Gold	14.7	Rb	Rubidium	78.2
Cr	Chromium	15.6	Sn	Tin	78.2
Al	Aluminum	17.0	Ta	Tantalum	87.8
Ti	Titanium	19.2	Tl	Thallium	105.9
Na	Sodium	25.9	Cs	Cesium	114.3
Mg	Magnesium	26.2	Pb	Lead	122.7
Ca	Calcium	27.6	Sr	Strontium	141.4
Rh	Rhodium	28.2	As	Arsenic	210.5
W	Tungsten	30.1	Sb	Antimony	234.6
Mn	Manganese	30.1	Ga	Gallium	318.8
Mo	Molybdenum	31.9	Os	Osmium	336.9
Zn	Zinc	34.6	Hg	Mercury	565.9
Ir	Iridium	36.7	Bi	Bismuth	661.7
K	Potassium	36.7	—	Graphite	4812.3
Ni	Nickel	41.7	C	Carbon	18,046.2
Cd	Cadmium	42.3	Te	Tellurium	1.2×10^6
In	Indium	50.3	P	Phosphorus	6.02×10^{12}
Li	Lithium	51.4	B	Boron	4.81×10^{13}
Fe	Iron	52.9	Se	Selenium	6.02×10^{13}
Co	Cobalt	54.1	S	Sulfur	6.02×10^{17}

*Resistance of a wire that is one foot long, one mil diameter, at 20°C, made of the material listed.

APPENDIX 2. SPECIFIC RESISTANCE AND RESISTIVITY

1. For materials used in making wire of various sorts, the specific resistance (ohm-circ mil/ft) is specified.

The *specific resistance* of a material is the resistance of a piece of wire made from the material that is:

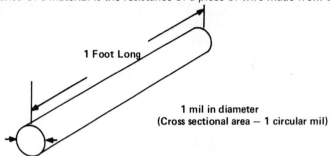

1 Foot Long

1 mil in diameter
(Cross sectional area — 1 circular mil)

A wire that is 1 mil in diameter has a cross-sectional area of *1 circular mil*. The circular mil is a special unit used in specifying the cross-sectional area of wires. The number of *circular mils* of cross-sectional area of a wire is defined as its *diameter squared*.

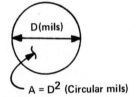

D(mils)

A = D² (Circular mils)

So if a wire is D mils in diameter, its cross-sectional area is D^2 circular mils.

(Do not confuse the circular mil with the *square mil*. 1 circular mil = $\pi/4$ *square mils*.

To find the resistance of a piece of wire, given its diameter in mils and its length in feet:

    a. Look up its *specific resistance* in the table.

    b. Use the formula:

$$R \text{ (ohms)} = \frac{\text{(Specific Resistance)}}{\text{(ohms circ mil/foot)}} \times \frac{\text{Length (Feet)}}{\text{(Diameter in mils)}^2}$$

2. For materials not used in making wires, the *resistivity* in ohm-meters is commonly specified.

The *resistivity* of a material is generally expressed as the resistance that would be measured between the faces of a 1-meter cube made of the material.

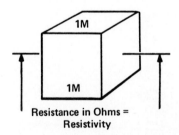

1M

1M

Resistance in Ohms =
Resistivity

To find the resistance of a slab of material of any given dimension, use the formula:

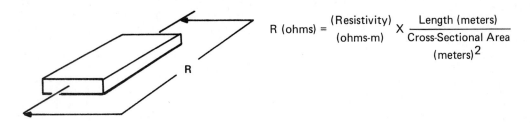

$$R \text{ (ohms)} = \frac{\text{(Resistivity)}}{\text{(ohms-m)}} \times \frac{\text{Length (meters)}}{\text{Cross-Sectional Area (meters)}^2}$$

Note: A problem that exists here is that in various textbooks the Greek letter "rho" (ρ) is used to mean *both* specific resistance and resistivity. Actually the only difference between the two are their units of measure.

Resistivity (ohm-meter) = Specific Resistance (ohms-circ mil/ft) Times 1.662×10^{-9}

Specific Resistance (ohms-cir mil/ft) = Resistivity (ohm-meter) Times (6.015×10^8)

WIRE TABLE, STANDARD ANNEALED COPPER
American Wire Gauge (B. & S.) English Units

Gauge No.	Diameter in mils at 20°C	Cross section at 20°C		Ohms per 1000 feet*			
		Circular mils	Sq. Inches	0°C (32°F)	20°C (68°F)	50°C (122°F)	75°C (167°F)
0000	460.0	211600	0.1662	0.04516	0.04901	0.05479	0.05961
000	409.6	167800	.1318	.05695	.06180	.06909	.07516
00	364.8	133100	.1045	.07181	.07793	.08712	.09478
0	324.9	105500	.08289	.09055	.09827	.1099	.1195
1	289.3	83690	.06573	.1142	.1239	.1385	.1507
2	257.6	66370	.05213	.1440	.1563	.1747	.1900
3	229.4	52640	.04134	.1816	.1970	.2203	.2396
4	204.3	41740	.03278	.2289	.2485	.2778	.3022
5	181.9	33100	.02600	.2887	.3133	.3502	.3810
6	162.0	26250	.02062	.3640	.3951	.4416	.4805
7	144.3	20820	.01635	.4590	.4982	.5569	.6059
8	128.5	16510	.01297	.5788	.6282	.7023	.7640
9	114.4	13090	.01028	.7299	.7921	.8855	.9633
10	101.9	10380	.008155	.9203	.9989	1.117	1.215
11	90.74	8234	.006467	1.161	1.260	1.408	1.532
12	80.81	6530	.005129	1.463	1.588	1.775	1.931
13	71.96	5178	.004067	1.845	2.003	2.239	2.436
14	64.08	4107	.003225	2.327	2.525	2.823	3.071
15	57.07	3257	.002558	2.934	3.184	3.560	3.873
16	50.82	2583	.002028	3.700	4.016	4.489	4.884
17	45.26	2048	.001609	4.666	5.064	5.660	6.158
18	40.30	1624	.001276	5.883	6.385	7.138	7.765
19	35.89	1288	.001012	7.418	8.051	9.001	9.792
20	31.96	1022	.0008023	9.355	10.15	11.35	12.35
21	28.45	810.1	.0006363	11.80	12.80	14.31	15.57
22	25.35	642.4	.0005046	14.87	16.14	18.05	19.63
23	22.57	509.5	.0004002	18.76	20.36	22.76	24.76
24	20.10	404.0	.0003173	23.65	25.67	28.70	31.22
25	17.90	320.4	.0002517	29.82	32.37	36.18	39.36
26	15.94	254.1	.0001996	37.61	40.81	45.63	49.64
27	14.20	201.5	.0001583	47.42	51.47	57.53	62.59
28	12.64	159.8	.0001255	59.80	64.90	72.55	78.93
29	11.26	126.7	.00009953	75.40	81.83	91.48	99.52
30	10.03	100.5	.00007894	95.08	103.2	115.4	125.5
31	8.928	79.70	.00006260	119.9	130.1	145.5	158.2
32	7.950	63.21	.00004964	151.2	164.1	183.4	199.5
33	7.080	50.13	.00003937	190.6	206.9	231.3	251.6
34	6.305	39.75	.00003122	240.4	260.9	291.7	317.3
35	5.615	31.52	.00002476	303.1	329.0	367.8	400.1
36	5.000	25.00	.00001964	382.2	414.8	463.7	504.5
37	4.453	19.83	.00001557	482.0	523.1	584.8	636.2
38	3.965	15.72	.00001235	607.8	659.6	737.4	802.2
39	3.531	12.47	.000009793	766.4	831.8	929.8	1012
40	3.145	9.888	.000007766	966.5	1049	1173	1276

*Resistance at the stated temperatures of a wire whose length is 1000 feet at 20°C.

APPENDIX 4. INTERPRETING THE RESISTOR COLOR CODE

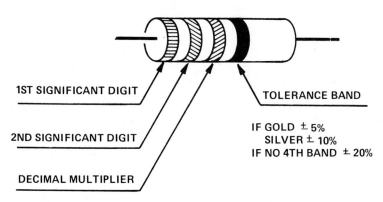

1ST SIGNIFICANT DIGIT

2ND SIGNIFICANT DIGIT

DECIMAL MULTIPLIER

(# OF ZEROS TO PLACE AFTER FIRST TWO DIGITS)

TOLERANCE BAND

IF GOLD ± 5%
SILVER ± 10%
IF NO 4TH BAND ± 20%

	Significant Digit	Decimal Multiplier (Put These Zeros Be-hind First Two Digits)	(Power of Ten)
Black	0	1	10^0
Brown	1	1 0	10^1
Red	2	1 00	10^2
Orange	3	1 000	10^3
Yellow	4	1 0000	10^4
Green	5	1 00000	10^5
Blue	6	1 000000	10^6
Violet	7	1 0000000	10^7
Gray	8	1 00000000	10^8
White	9	1 000000000	10^9
Gold	—	Multiply by 0.1	10^{-1}
Silver	—	Multiply by 0.01	10^{-2}

APPENDIX 5. PREFERRED VALUES FOR RESISTORS AND CAPACITORS

The numbers listed in the chart below, and *decimal multiples* of these numbers, are the commonly available resistor values at 5%, 10%, and 20% tolerance.

20% Tolerance (*No* 4th Band)	10% Tolerance (*Silver* 4th Band)	5% Tolerance (*Gold* 4th Band)
10*	10	10
		11
	12	12
		13
15	15	15
		16
	18	18
		20
22	22	22
		24
	27	27
		30
33	33	33
		36
	39	39
47	47	47
		51
	56	56
		62
68	68	68
		75
	82	82
		91
100	100	100

APPENDIX 6. RESISTOR SIZE COMPARISON

(By Wattage Rating)

$\dfrac{1}{10}$ watt

$\dfrac{1}{4}$ watt

$\dfrac{1}{2}$ watt

1 watt

2 watt

APPENDIX 7. SCIENTIFIC NOTATION AND THE METRIC PREFIXES

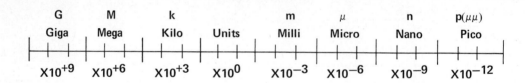

$$1 \text{ unit} = 1 \, .$$

$$. \, 0 \, 0 \, 1 = 1 \text{ milli}$$

$$1 \text{ kilo} = 1 \, 0 \, 0 \, 0 \, .$$

$$. \, 0 \, 0 \, 0 \, 0 \, 0 \, 1 = 1 \text{ micro}$$

$$1 \text{ mega} = 1 \, 0 \, 0 \, 0 \, 0 \, 0 \, 0 \, .$$

$$. \, 0 \, 0 \, 0 \, 0 \, 0 \, 0 \, 0 \, 0 \, 1 = 1 \text{ nano}$$

$$1 \text{ giga} = 1 \, 0 \, 0 \, 0 \, 0 \, 0 \, 0 \, 0 \, 0 \, 0 \, .$$

$$. \, 0 \, 0 \, 0 \, 0 \, 0 \, 0 \, 0 \, 0 \, 0 \, 0 \, 0 \, 1 = 1 \text{ pico}$$

STANDARD FORM: X.XX $\times$ $10^{+\text{exponent}}$			
Symbol	Prefix	Value	Power of 10
G	giga	1,0 0 0, 0 0 0, 0 0 0 .	$\times 10^{+9}$
M	mega	1, 0 0 0, 0 0 0 .	$\times 10^{+6}$
k	kilo	1, 0 0 0 .	$\times 10^{+3}$
—	(units)	1 .	$\times 10^{0}$
m	milli	.0 0 1	$\times 10^{-3}$
μ	micro	.0 0 0 0 0 1	$\times 10^{-6}$
n	nano	.0 0 0 0 0 0 0 0 1	$\times 10^{-9}$
p $(\mu\mu)$	pico	.0 0 0 0 0 0 0 0 0 0 0 1	$\times 10^{-12}$

APPENDIX 8. BASIC SCHEMATIC SYMBOLS

Symbol	Device	Symbol	Device
⊣⊢⊢⊢	Battery or DC Power Supply	⊥	Push Button Normally Open (PBNO)
—⋀⋀⋀—	Resistor	⊙⊥⊙	Push Button Normally Closed (PBNC)
—⋀⋀⋀— (arrow)	Potentiometer	⏚	Earth Ground
—⋀⋀⋀— (arrow)	Rheostat	⏢	Chassis Ground
—⋀⋀⋀— (tapped)	Tapped Resistor	⊥⊤	Capacitor
(V) (A) (mA)	Meters — Symbol to Indicate Function	⊥⊤ +	Capacitor, Polarized (Electrolytic)
⊐	Lamp	—⌒⌒⌒—	Coil, Air Core
⟋⟍	Switch SPST	≡⌒⌒⌒	Coil, Iron Core
∘ ⟋⟍ ∘	Switch SPDT	—⌒—	Fuse
⟋⟍ (dual)	Switch DPST	Conductor, General No Connection	
⟋⟍ (dual)	Switch DPDT	Connection	

The following table can be used to find the square root or square of most any number. Numbers from 1 to 120 can be read directly from the table. But what about a number such as 150? How can its square or square root be found? The secret to the use of this table is in the understanding of *factoring*. Factoring a number means to break the original number up into two smaller numbers, that, when multiplied together, give you back the original. For example, 150 is equal to 10 times 15. Ten and 15 are said to be *factors* of 150. If 10 times 15 is equal to 150, then the square root of 10 times the square root of 15 is equal to the square root of 150. Both 10 and 15 are listed on the square and square root table. The square root of 10 from the table is equal to 3.162. The square root of 15 is equal to 3.873; 3.162 times 3.873 is equal to 12.246426, which should be the square root of 150. You can test this number by multiplying it by itself. Thus, 12.246426 squared is equal to 149.97, etc., — very close to 150. (Small errors due to rounding will normally occur when using the tables.) The factoring procedure written out mathematically would then be:

$$150 = 10 \times 15$$

$$\sqrt{150} = \sqrt{10} \times \sqrt{15} \quad (\text{Look up } \sqrt{10}, \sqrt{15} \text{ in tables})$$

$$\sqrt{150} = 3.162 \times 3.873$$

$$\sqrt{150} = 12.246 \ldots \ldots$$

Try another number now, say, 350. First, factor 350:

$$350 = 35 \times 10$$

The square root of 350 must equal the square root of 35 times the square root of 10.

$$\sqrt{350} = \sqrt{35} \times \sqrt{10}$$

Go to the tables and look up the square roots of 10 and 35:

$$\sqrt{350} = 5.9161 \times 3.162$$

Multiply the square roots of 10 and 35, and you have found the square root of 350.

$$\sqrt{350} = 18.706 \ldots \ldots$$

To check the accuracy of your calculations, multiply 18.706 by itself.

$$18.706^2 = 349.91$$

Again, very close to the original number.

Try one more number, this time 1150.

First, factor 1150.

$$1150 = 115 \times 10$$

The square root of 1150 must equal the square root of 115 times the square root of 10.

$$\sqrt{1150} = \sqrt{115} \times \sqrt{10}$$

Look up the square roots of 115 and 10 from the tables.

$$\sqrt{1150} = 10.7238 \times 3.162$$

Multiply the square roots of 115 and 10, and you have the square root of 1150.

$$\sqrt{1150} = 33.908$$

To check the validity of this number, square it. It should be very close to 1150.

N	$\sqrt{N}$	N^2	N	$\sqrt{N}$	N^2	N	$\sqrt{N}$	N^2
1	1.000	1	41	6.4031	1681	81	9.0000	6561
2	1.414	4	42	6.4807	1764	82	9.0554	6724
3	1.732	9	43	6.5574	1849	83	9.1104	6889
4	2.000	16	44	6.6332	1936	84	9.1652	7056
5	2.236	25	45	6.7082	2025	85	9.2195	7225
6	2.449	36	46	6.7823	2116	86	9.2736	7396
7	2.646	49	47	6.8557	2209	87	9.3274	7569
8	2.828	64	48	6.9282	2304	88	9.3808	7744
9	3.000	81	49	7.0000	2401	89	9.4340	7921
10	3.162	100	50	7.0711	2500	90	9.4868	8100
11	3.3166	121	51	7.1414	2601	91	9.5394	8281
12	3.4641	144	52	7.2111	2704	92	9.5917	8464
13	3.6056	169	53	7.2801	2809	93	9.6437	8649
14	3.7417	196	54	7.3485	2916	94	9.6954	8836
15	3.8730	225	55	7.4162	3025	95	9.7468	9025
16	4.0000	256	56	7.4833	3136	96	9.7980	9216
17	4.1231	289	57	7.5498	3249	97	9.8489	9409
18	4.2426	324	58	7.6158	3364	98	9.8995	9604
19	4.3589	361	59	7.6811	3481	99	9.9499	9801
20	4.4721	400	60	7.7460	3600	100	10.0000	10000
21	4.5826	441	61	7.8102	3721	101	10.0499	10201
22	4.6904	484	62	7.8740	3844	102	10.0995	10404
23	4.7958	529	63	7.9373	3969	103	10.1489	10609
24	4.8990	576	64	8.0000	4096	104	10.1980	10816
25	5.0000	625	65	8.0623	4225	105	10.2470	11025
26	5.0990	676	66	8.1240	4356	106	10.2956	11236
27	5.1962	729	67	8.1854	4489	107	10.3441	11449
28	5.2915	784	68	8.2462	4624	108	10.3923	11664
29	5.3852	841	69	8.3066	4761	109	10.4403	11881
30	5.4772	900	70	8.3666	4900	110	10.4881	12100
31	5.5678	961	71	8.4261	5041	111	10.5357	12321
32	5.6569	1024	72	8.4853	5184	112	10.5830	12544
33	5.7446	1089	73	8.5440	5329	113	10.6301	12769
34	5.8310	1156	74	8.6023	5476	114	10.6771	12996
35	5.9161	1225	75	8.6603	5625	115	10.7238	13225
36	6.0000	1296	76	8.7178	5776	116	10.7703	13456
37	6.0828	1369	77	8.7750	5929	117	10.8167	13689
38	6.1644	1444	78	8.8318	6084	118	10.8628	13924
39	6.2450	1521	79	8.8882	6241	119	10.9087	14161
40	6.3246	1600	80	8.9443	6400	120	10.9545	14400

APPENDIX 10. HOW TO EXTRACT SQUARE ROOTS MANUALLY

This procedure outlines, step by step, how to extract square roots manually.

Problem

Compute $\sqrt{4139}$

Solution

Step 1: Begin at the decimal point (which is to the right of the last digit) and divide the number into two-digit groups (underlining indicates the groups).

$$4\underline{1}\underline{39}.\underline{00}$$

Step 2: Place the decimal point for the square root directly above the decimal point that appears under the radical sign.

$$\sqrt{4139.00}$$

Step 3: Find the largest number that *when multiplied by itself will give a product equal to or less than* the first pair of digits. In this case, 6 X 6 = 36, which is the largest perfect square that does not exceed 41. Place 6 on the radical sign above 41.

$$\begin{array}{c} 6 \quad . \\ \overline{\sqrt{4139.00}} \end{array}$$

Step 4: Square 6 to obtain 36 and place it below the first two digits (41). Subtract 36 from 41 to obtain 5. Bring down the next pair of digits (39).

$$\begin{array}{r} 6 \quad . \\ \sqrt{4139.00} \\ 36 \quad\quad\quad \\ \hline 539 \quad\quad \end{array}$$

Step 5: Double the first digit of the answer, 6, to obtain a trial divisor of 12. Place 12 to the left of 539 as shown.

$$\begin{array}{r} 6 \quad . \\ \sqrt{4139.00} \\ 36 \quad\quad\quad \\ \hline 12 \,|\, 539 \quad\quad \end{array}$$

Step 6: Divide the trial divisor (12) into *all but the last digit* of the modified remainder 539. It will divide into 53 four times. This will be the next digit of the answer. *Place the 4 above the second pair of digits and also place the 4 to the right of the trial divisor.* The completed divisor is 124. Multiply 124 by 4 to obtain 496. Subtract 496 from 539 to obtain 43. Bring down the next pair of digits (.00).

```
        6  4.
      √4139.00
        36
   124 | 539
         496
         4300
```

Step 7: Double the first two digits of the answer (64) to obtain the new trial divisor (128). Place 128 to the left of 4300 as shown.

```
        6  4.
      √4139.00
        36
   124 | 539
         496
   128 | 4300
```

Step 8: Divide the trial divisor 128 into all but the last digit of the modified remainder 4300. It will go into 430 three times. This will be the next digit of the answer. Place the 3 on the radical sign over the next pair of numbers. Also, place the 3 to the right of the trial divisor 128. The completed trial divisor is 1283. Multiply 1283 by 3 to obtain 3849. Subtract 3849 from 4300. The remainder is 4.51.

```
        6  4.3
      √4139.00
        36
   124 | 539
         496
  1283 | 4300
         3849
          451
```

Note: If greater accuracy is required, the number may be carried out by adding more pairs of zeros to the right of the decimal place and performing Steps 7 and 8 until the desired accuracy is obtained.

Step 9. The answer may be checked by multiplying the answer by itself and adding the remainder from the last step, 64.3 times 64.3 plus 4.51 is equal to 4139.

$$64.3 \times 64.3 = 4134.49 + 4.51 = 4139$$

APPENDIX 10. HOW TO EXTRACT SQUARE ROOTS MANUALLY

Problem

Find the square root of 240.25

Solution

Step 1: Begin at the decimal point and divide the number into digit groups in both directions. (Notice that a zero was added in front of the two in order to make a pair.)

$$\sqrt{\overline{02}\,\overline{40}.\overline{25}}$$

Step 2: Place the decimal point for the square root directly above the decimal point that appears under the radical sign.

$$\sqrt{02\overline{40}.\overline{25}}$$

Step 3: Determine the largest number that when multiplied by itself will give a product equal to or less than the first pair of digits, 02. The number 1 is the only number that meets these requirements. Place 1 over the first pair of digits.

$$\overset{1\quad.}{\sqrt{0240.25}}$$

Step 4: Square 1 to obtain 1. Place this number below the first two digits, 02. Subtract 1 from 02 to obtain 1. Bring down the next pair of digits (40).

$$\begin{array}{r} 1\quad.\ \\ \sqrt{02\overline{40}.\overline{25}} \\ \underline{1} \\ 140 \end{array}$$

Step 5: Double the first digit of the answer 1 to obtain a trial divisor of 2. Place the 2 to the left of 140 as shown.

$$\begin{array}{r} 1\quad.\ \\ \sqrt{02\overline{40}.\overline{25}} \\ \underline{1} \\ 2\,\big|\,140 \end{array}$$

Step 6: Divide the trial divisor (2) into all but the last digit of the modified remainder 140. Two will divide into 14, seven times. This will be the next digit of the answer. Place the 7 on the radical sign above the second pair of digits and also place a 7 to the right of the 2 in the trial divisor. The completed trial divisor is now 27. Multiply 27 by 7 to obtain 189. Oops! 189 will not subtract from 140. Now you know why the *trial* divisor is so named! At this point go back to the trial divisor, subtract one from it, and try again. Notice that one was also subtracted from the answer on the radical sign. You now have a trial divisor of 26, and the last digit of the answer is 6; 26 times 6 is equal to 156. Unfortunately, 156 will not subtract from 140. Take 1 from both the trial divisor and the last digit of the answer. You now have a trial divisor of 25 and the last digit of the answer is 5. Multiply 25 by 5 to get 125. Fortunately, 125 *will* subtract from 140 to leave a remainder of 15. Bring down the next pair of digits (25).

```
             15 .
          √ 0240.25
             1
     25 |  140
          125
             1525
```

Step 7:   Double the first two digits of the answer 15, to obtain the new trial divisor of 30. Place 30 to the left of 1525 as shown.

```
             15 .
          √ 0240.25
             1
     25 |  140
          125
     30 |  1525
```

Step 8:   Divide the trial divisor, 30, into all but the last digit of the modified remainder 1525; 152 divided by 30 is equal to 5. This will be the next digit of the answer. Place the 5 above the next pair of digits in the number whose square root is being extracted. Also, place the 5 to the right of the 30 in the trial divisor. The new trial divisor is 305. Multiply 305 by 5 to obtain 1525. Subtract 1525 from 1525 to obtain zero. The square root of 240.25 is 15.5.

```
             15 .5
          √ 0240.25
             1
     25 |  140
          125
    305 |  1525
          1525
             0
```

Step 9:   Check your answer by multiplying 15.5 by 15.5 and adding the remainder, if any, to the product.

$$15.5 \times 15.5 = 240.25$$

APPENDIX 11. THE UNIVERSAL TIME CONSTANT GRAPH

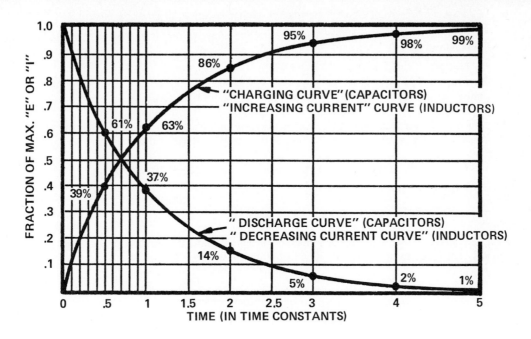

How to Use This Chart

This chart can be used to graphically determine the voltage or current at any point in time for an RC or L/R circuit, during charging (or current buildup), or discharge (or current collapse).

The examples shown below illustrate the use of the chart.

1. Find the voltage across the capacitor shown in the circuit below, 1 second after the switch is thrown.

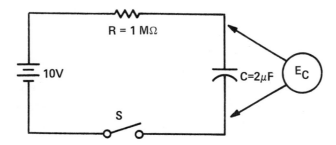

Solution

a. First find the circuit time constant

$$T = RC$$

$$T = (1 \times 10^6) \times (2 \times 10^{-6}) = 2 \text{ seconds}$$

b.   Express the time (t) at which the capacitor voltage is desired in *time constants*.

Here you want the voltage after 1 second and the time constant is 2 seconds, so t = 1/2 (the time constant)

or t = 0.5T

c.   Look at the chart, on the horizontal axis and locate 0.5 time constants.

d.   Move up the vertical line until it reaches the appropriate curve (in this case the charging curve). Read from the vertical axis the fraction of the applied voltage at the time (here 39%).

e.   At t = 1 second, the voltage across the capacitor equals 39% of 10 volts or

E_C = 0.39 X 10

E_C = 3.9 volts

2.   Find the voltage across the capacitor shown in the circuit below 2 seconds after the switch, S, is thrown. The capacitor is charged to 20 volts before the switch is thrown.

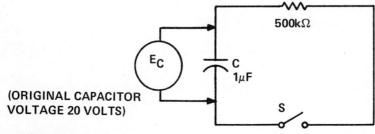

(ORIGINAL CAPACITOR VOLTAGE 20 VOLTS)

Solution

a.   Find the circuit time constant

T = RC

T = (500 X 10^{+3}) X (1 X 10^{-6})

T = 0.5 seconds

b.   Express the time at which the capacitors voltage is desired in time constants. Here, 2 seconds divided by 0.5 seconds is 4; 2 seconds is 4 time constants for this circuit.

t = 4T

c.   Look at the chart, locate 4 time constants on the horizontal axis.

d.   Move up the vertical line until it reaches the appropriate curve (the discharge curve). Read the fraction of the original voltage from the vertical axis (2%).

e.   AT t = 2 seconds, the voltage across the capacitor is at 2% of the original voltage or is at 2% of 20 volts.

$$E_C = 0.02 \times 20$$

$$E_C = 0.40 \text{ volts}$$

Remember that 5 time constants is required for a 100% charge (full charge or discharge for RC circuits, maximum or zero current for L/R circuits).

The voltage at any point along a charge or discharge curve may be calculated by using one of these two mathematical formulas:

Charge: e (at time t) $= E_{app} (1 - \epsilon^{-t/RC})$

Discharge: e (at time t) $= E_{app} (\epsilon^{-t/RC})$

The scientific calculator greatly reduces the degree of difficulty in the solution of problems of this type. For example, the *charge formula* shown above may be solved by using a calculator such as the SR-50, and this procedure:

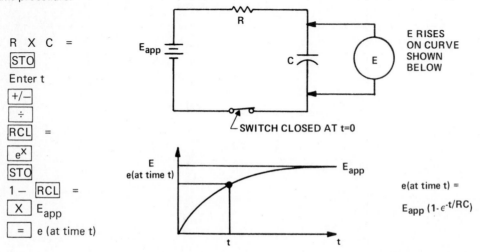

The decay of a capacitor's charge (discharge) may be calculated by using this calculator sequence:

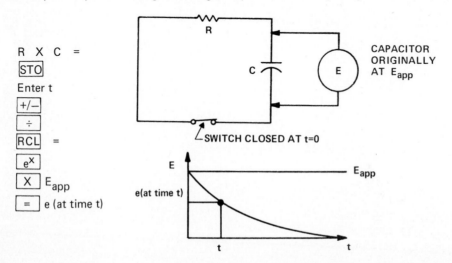

CAPACITOR COLOR CODE CHART

Color	Significant Digit	Decimal Multiplier	Tolerance in %	Voltage Rating
Black	0	1	20	
Brown	1	10	1	100
Red	2	$\cdot 10^2$	2	200
Orange	3	10^3	3	300
Yellow	4	10^4	4	400
Green	5	10^5	5	500
Blue	6	10^6	6	600
Violet	7	10^7	7	700
Gray	8	10^8	8	800
White	9	10^9	9	900
Gold		0.1		1000
Silver		0.01		2000
No Color			20	500

APPENDIX 13. CAPACITOR COLOR CODES

- **Mica Capacitors**

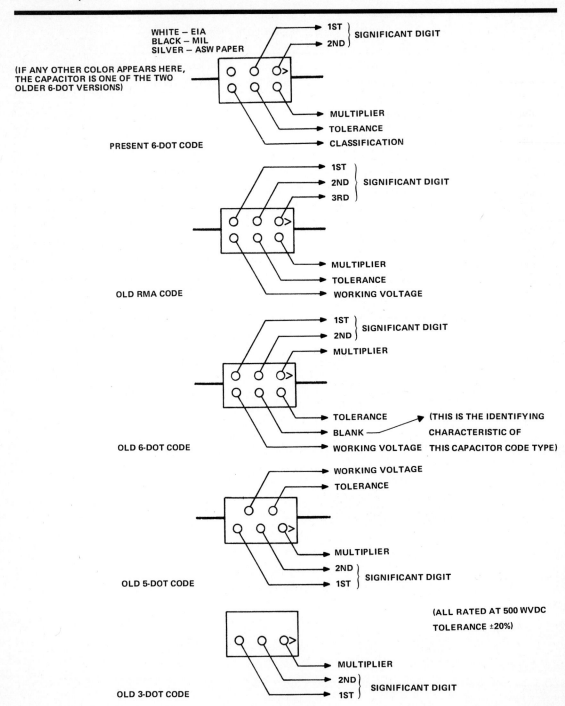

WHITE — EIA
BLACK — MIL
SILVER — ASW PAPER

(IF ANY OTHER COLOR APPEARS HERE,
THE CAPACITOR IS ONE OF THE TWO
OLDER 6-DOT VERSIONS)

1ST
2ND } SIGNIFICANT DIGIT

MULTIPLIER
TOLERANCE
CLASSIFICATION

PRESENT 6-DOT CODE

1ST
2ND } SIGNIFICANT DIGIT
3RD

MULTIPLIER
TOLERANCE
WORKING VOLTAGE

OLD RMA CODE

1ST
2ND } SIGNIFICANT DIGIT

MULTIPLIER

TOLERANCE
BLANK ———— (THIS IS THE IDENTIFYING
WORKING VOLTAGE   CHARACTERISTIC OF
                  THIS CAPACITOR CODE TYPE)

OLD 6-DOT CODE

WORKING VOLTAGE
TOLERANCE

MULTIPLIER
2ND }
1ST } SIGNIFICANT DIGIT

OLD 5-DOT CODE

(ALL RATED AT 500 WVDC
TOLERANCE ±20%)

MULTIPLIER
2ND }
1ST } SIGNIFICANT DIGIT

OLD 3-DOT CODE

A-24

● Ceramic Capacitors

CERAMIC CAPACITORS

All Values Are Read in Picofarads

Color	Significant Digit	Decimal Multiplier	Tolerance		Temperature Coefficient ppm/°C
			Above 10 pF (in %)	Below 10 pF (in pF)	
Black	0	1	20	2.0	0
Brown	1	10	1		−30
Red	2	100	2		−80
Orange	3	1000			−150
Yellow	4				−220
Green	5		5	0.5	−330
Blue	6				−470
Violet	7				−750
Gray	8	0.01		0.25	30
White	9	0.1	10	1.0	500

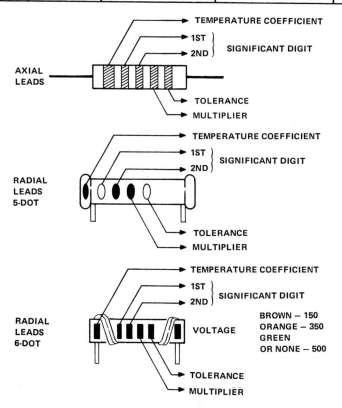

APPENDIX 13. CAPACITOR COLOR CODES

- **Ceramic Capacitors**

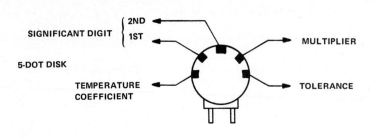

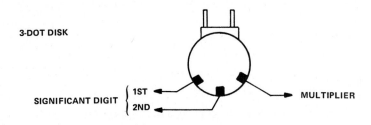

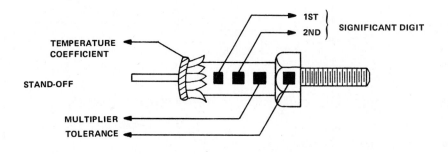

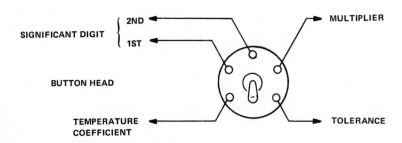

The following charts are listed to give a convenient method for comparing various common English and metric units to allow easy conversion from one unit to another. These comparisons are for common values of lengths, areas, volume, speed, and electric resistivity. Included also is a listing of several other miscellaneous unit comparisons.

Length Comparisons

To use this chart to compare (and thus convert) one unit to another, find the existing measurement in the From column and then find the desired unit in the vertical headings (TO). Where these two intersect will give you the conversion of *one* existing unit (From) into *one* new unit (To). For example, if you have *one inch* and you need this *in centimeters*; find "1 inch" in the From column (4th line down) and go over to the vertical column labeled cm; and you find that 1 inch = 2.54 cm. Then, if you wanted to convert 25 inches (or any value of inches) into centimeters you would simply multiply 25 (or any given number of inches) by 2.54 for 63.5 centimeters.

Length Comparisons

To / From	cm	meter	km	in	ft	mile	naut. mile
1 Centimeter	1	1×10^{-2}	1×10^{-5}	0.3937	3.281×10^{-2}	6.214×10^{-6}	5.40×10^{-6}
1 Meter	100	1	1×10^{-3}	39.37	3.281	6.214×10^{-4}	5.40×10^{-4}
1 Kilometer	1×10^5	1×10^3	1	3.937×10^4	3281	0.6214	0.540
1 Inch	2.54	2.54×10^{-2}	2.54×10^{-5}	1	8.333×10^{-2}	1.578×10^{-5}	1.371×10^{-5}
1 Foot	30.48	0.3048	3.048×10^{-4}	12	1	1.894×10^{-4}	1.646×10^{-4}
1 Statute Mile	1.609×10^5	1609	1.609	6.336×10^4	5280	1	0.8670
1 Nautical Mile	1.852×10^5	1852	1.852	7.293×10^4	6076.1	1.1508	1

The charts that follow are used in the same manner as the length comparison chart with the "From" in the left column and the "To" conversions listed in the following vertical columns.

Area Comparison

From \ To	meter^2	cm^2	ft^2	in^2	circ mil
1 square meter	1	1×10^4	10.76	1550	1.974×10^9
1 square centimeter	1×10^{-4}	1	1.076×10^{-3}	0.1550	1.974×10^5
1 square foot	9.290×10^{-2}	929.0	1	144	1.833×10^8
1 square inch	6.452×10^{-4}	6.452	6.944×10^{-3}	1	1.273×10^6
1 circular mil	5.067×10^{-10}	5.067×10^{-6}	5.454×10^{-9}	7.854×10^{-7}	1

Volume Comparison

From \ To	meter^3	cm^3	1	ft^3	in^3
1 cubic meter	1	1×10^6	1000	35.31	6.102×10^4
1 cubic centimeter	1×10^{-6}	1	$1. \times 10^{-3}$	3.531×10^{-5}	6.102×10^{-2}
1 liter	1.000×10^{-3}	1000	1	3.531×10^{-2}	61.02
1 cubic foot	2.832×10^{-2}	2.832×10^4	28.32	1	1728
1 cubic inch	1.639×10^{-5}	16.39	1.639×10^{-2}	5.787×10^{-4}	1

Speed Comparison

From \ To	ft/sec	km/hr	meter/sec	miles/hr	cm/sec	knot
1 foot per second	1	1.097	0.3048	0.6818	30.48	0.5925
1 kilometer per hour	0.9113	1	0.2778	0.6214	27.78	0.540
1 meter per second	3.281	3.6	1	2.237	100	1.944
1 mile per hour	1.467	1.609	0.4470	1	44.70	0.8689
1 centimeter per second	3.281×10^{-2}	3.6×10^{-2}	0.01	2.237×10^{-2}	1	1.944×10^{-2}
1 knot	1.688	1.852	0.5144	1.151	51.44	1

Electric Resistivity Comparison

From \ To	μohm-cm	ohm-cm	ohm-m	ohm-circ mil/ft
1 micro-ohm-centimeter	1	1×10^{-6}	1×10^{-8}	6.015
1 ohm-centimeter	1×10^{6}	1	0.01	6.015×10^{6}
1 ohm-meter	1×10^{8}	100	1	6.015×10^{8}
1 ohm-circular mil per foot	0.1662	1.662×10^{-7}	1.662×10^{-9}	1

Miscellaneous Unit Comparisons

1 fathom = 1 ft
1 yard = 3 ft
1 rod = 16.5 ft
1 U.S. gallon = 4 U.S. fluid quarts
1 U.S. quart = 2 U.S. pints
1 U.S. pint = 16 U.S. fluid ounces
1 U.S. gallon = 0.8327 British imperial gallon
1 British imperial gallon = 1.2 U.S. gallons

1 liter = 1000 cm^3
1 knot = 1 nautical mile/hr
1 mile/min = 88 ft/sec = 60 miles/hr
1 meter = 39.4 in = 3.28 ft
1 inch = 2.54 cm
1 mile = 5280 ft = 1.61 km
1 angstrom unit = 10^{-10} meters
1 horsepower = 550 ft-lb/sec = 746 watts

The Greek Alphabet

(Including common use of symbols in basic electricity)

Letter	Capital	Common Use of Symbol	Lower	Common Use of Symbol
Alpha	A		α	
Beta	B		β	
Gamma	Γ		γ	
Delta	Δ	change in	δ	change in
Epsilon	E		ϵ	base of natural logs
Zeta	Z		ζ	
Eta	H		η	
Theta	Θ		θ, ϑ	angle (phase angle)
Iota	I		ι	
Kappa	K		κ	dielectric constant
Lambda	Λ		λ	wavelength
Mu	M		μ	micro
Nu	N		ν	frequency
Xi	Ξ		ξ	
Omicron	O		o	
Pi	Π		π	3.14159
Rho	P		ρ	specific resistance, resistivity
Sigma	Σ	sum of terms	σ, ς	
Tau	T		τ	
Upsilon	Υ		υ	
Phi	Φ		ϕ, φ	magnetic flux
Chi	X		χ	
Psi	Ψ		ψ	
Omega	Ω	ohms	ω	angular frequency
(Reversed Omega)	$(\mho)$	mho		

APPENDIX 16. BASIC EQUATIONS OF BASIC ELECTRICITY

TERM	UNIT	SYMBOL	FORMULA SERIES	FORMULA PARALLEL
Charge	Coulomb	Q	1 coulomb = 6.28 X 10^{18} electrons	
Voltage (Potential difference, EMF)	Volt (V)	E	$E_T = E_1 + E_2 + E_3 + \ldots$ $E = IR$	$E_T = E_1 = E_2 = E_3 \ldots$
Current (Flow of charge)	Ampere (Amp) (A)	I	$I_T = I_1 = I_2 = I_3 \ldots$ $I = E/R$	$I_T = I_1 + I_2 + I_3 \ldots$
Resistance	Ohm (Ω)	R	$R_T = R_1 + R_2 + R_3 + \ldots$ $R = E/I$ $R = 1/G$	$R_T = \dfrac{1}{1/R_1 + 1/R_2 + 1/R_3 + .}$ $R_T = \dfrac{R_1 R_2}{R_1 + R_2}$ $R_T = \dfrac{R_s}{N}$
Conductance	Mho ($\mho$)	G	$G_T = 1/R_T$ $G = 1/R$	$G_T = G_1 + G_2 + G_3 \ldots$
Power	Watt (W)	P	$P = IE$ $P = E^2/R$ $P = I^2 R$	$P = IE$ $P = E^2/R$ $P = I^2 R$
Capacitance	Farad (F)	C	$C_T = \dfrac{1}{1/C_1 + 1/C_2 + 1/C_3 + \ldots}$ $C = Q/E$ $T = RC$	$C_T = C_1 + C_2 + C_3 + \ldots$
Inductance	Henry (H)	L	$L_T = L_1 + L_2 + L_3 + \ldots$ $T = L/R$	$L_T = \dfrac{1}{1/L_1 + 1/L_2 + 1/L_3 + .}$

Bibliography

Adams, J. E., *Electrical Principles and Practices*, New York, N.Y.: McGraw-Hill Book Co., 1963.

Bureau of Naval Personnel, *Basic Electricity*, New York, N.Y.: Dover Publications, Inc., 1970.

Cooke, N. M., *Basic Mathematics for Electronics*, Second Edition, New York, N.Y.: McGraw-Hill Book Co., 1960.

DeFrance, J. J., *Electrical Fundamentals*, Englewood Cliffs, New Jersey: Prentice-Hall, Inc., 1969.

Doyle, J. M., *An Introduction to Electrical Wiring*, Reston, Virginia: Reston Publishing Co., Inc., 1975.

Fiske, K. A., and Harter, J. H., *Direct Current Circuit Analysis Through Experimentation*, Third Edition, Seal Beach, California: The Technical Education Press, 1970.

Gillie, A. C., *Electrical Principles of Electronics*, Second Edition, New York, N.Y.: McGraw-Hill Book Co., 1969.

Graf, R. F., *Modern Dictionary of Electronics*, Indianapolis, Indiana: Howard W. Sams and Co., Inc., 1968.

Graham, K. C., *Fundamentals of Electricity*, Fifth Edition, Chicago, Illinois: American Technical Society, 1968.

Grob, B., *Basic Electronics*, Second Edition, New York, N.Y.: McGraw-Hill Book Co., 1971.

Halliday, D., and Resnick, R., *Physics for Students of Science and Engineering, Part II*, New York, N.Y.: John Wiley and Sons, 1960.

Harris, N. C., and Hemmerling, E. M., *Introductory Applied Physics*, Third Edition, New York, N.Y.: McGraw-Hill Book Co., 1972.

Herrick, C. M., *Unified Concepts of Electronics*, Englewood Cliffs, N.J.: Prentice-Hall, Inc., 1970.

Mileaf, H., *Electricity One-Seven*, New York, N.Y.: Hayden Book Co., Inc., 1966.

Mileaf, H., *Electronics One-Seven*, Rochelle, Park, N.J.: Hayden Book Co., 1967.

Naval Air Technical Training Command, *Basic Electricity*, Memphis, Tennessee: NavAirTech Training Center, 1967.

Oppenheimer, S. L., Hess, F. R., Borchers, J. P., *Direct and Alternating Currents*, Second Edition, New York, N.Y.: McGraw-Hill Book Co., 1973.

Philco Education Operations, *Basic Concepts and DC Circuits*, Fort Washington, Pennsylvania: Philco Corporation, 1960.

Shrader, R. L., *Electrical Fundamentals for Technicians*, New York, N.Y.: McGraw-Hill Book Co., 1969.

Singer, B. B., *Basic Mathematics for Electricity and Electronics*, Second Edition, New York, N.Y.: McGraw Hill Book Co., 1965.

Siskind, C. S., *Electrical Circuits, Direct and Alternating Current*, Second Edition, New York, N.Y.: McGraw-Hill Book Co., 1965.

Staff of Buck Engineering Co., Inc. (DeVito, M. J., Project Supervisor), *Introduction to Electricity and Electronics*, Farmingdale, New Jersey: Buck Engineering Co., Inc., 1971.

Staff of Electrical Technology Department, New York Institute of Technology, Schure, A., Project Director, *A Programmed Course in Basic Electricity*, New York, N.Y.: McGraw-Hill Book Co., 1970.

Suffern, M. G., *Basic Electrical and Electronic Principles*, Third Edition, New York, N.Y.: McGraw-Hill Book Co., 1962.

Timbie, W. H., *Essentials of Electricity*, Third Edition, New York, N.Y.: John Wiley and Sons, Inc., 1963.

Tocci, R. J., *Introduction to Electric Circuit Analysis*, Columbus, Ohio: Charles E. Merrill Publishing Co., 1974.

Trejo, P.E., *DC Circuits*, Palo Alto, California: Westinghouse Learning Press, 1972.

Turner, R. P., *Basic Electricity*, Second Edition, New York, N.Y.: Holt, Rinehart, and Wilson, 1963.

Weick, C. B., *Principles of Electronic Technology*, New York, N.Y.: McGraw-Hill Book Co., 1969.

Wellman, W. R., *Elementary Electricity*, Second Edition, New York, N.Y.: Van Nostrand Reinhold Company, 1971.

Glossary

algebraic sum
All positive quantities in an expression added together and each negative quantity subtracted from that result.

alternating current
Current which continuously reverses direction, usually in periodic fashion.

ammeter
A meter connected in series with a circuit, branch, or component which measures the current flowing through that circuit, branch, or component.

ampere
The unit of measure for current flow which equals 1 coulomb of electrons passing one point in a circuit in 1 second.

atom
The smallest part into which an element can be divided and still retain the behavior of the element.

battery
A device which maintains a potential difference between its terminals by chemical action.

bleeder current
Current flow through the bleeder resistor in a voltage divider circuit. Used to stabilize the output voltage of the voltage divider.

bleeder resistor
A resistor found in a voltage divider circuit or power supply and used to stabilize the output voltage.

branch
Path for current flow in a circuit.

bridge circuit
A special type of parallel-series circuit in which the voltages in each branch may be balanced by adjustment of one component. A special version called a Wheatstone bridge may be used to accurately measure resistance.

capacitance
How well a capacitor stores charge. Equal to the quantity of stored charge (Q) divided by the voltage (E) across the device when that charge was stored. Unit of measurement is the farad.

capacitor
A device that can store a charge on conducting plates through the action of an electrostatic field between the plates.

cell
A single unit device which converts chemical energy into electrical energy.

chassis
> A metal frame used to secure and house electrical components and associated circuitry.

choke coil
> An inductor used to oppose changes in current flow.

circuit
> A complete path for current flow from one terminal to the other of a source such as a battery or a power supply.

circuit analysis
> A technique of examining components in circuits to determine various values of voltage, current, resistance, power, etc.

circuit breaker
> An automatic device which, under abnormal conditions, will open a current-carrying circuit to help prevent unnecessary damage. Unlike a fuse, a circuit breaker may be reset to reconnect the circuit.

circuit reduction
> A technique of circuit analysis whereby a complex resistive circuit is replaced by a single equivalent resistance.

circuit sense
> An ability to recognize series or parallel portions of complex circuits to apply series and parallel circuit rules to those portions of the circuit for circuit analysis.

coil
> A number of turns of wire that is wrapped around a core. (Also called an inductor.)

combining like terms
> Algebraic addition of parts of an equation that each contain the same unknown quantity.

common point
> A voltage reference point in a circuit. A point which is "common" to many components in the circuit.

condenser
> See capacitor.

conductance
> The ability to conduct or carry current. Conductance is equivalent to the reciprocal of (or one over) the resistance.

conductor
> A material with many free electrons that will carry current.

constant current source
> An idealized source whose output current does not change with changes in the load, but whose output voltage varies with the load connected to it.

coulomb
A large quantity of electrons that is convenient when working with electricity and equals 6.25 billion, billion electrons (or 6.25 X 10^{18} electrons).

current
"Electron current" is the flow of electrons through a material from negative to positive. "Conventional current" is the flow of positive charges from positive to negative. "Current flow" is a general term often used to mean either of the above. Symbol is I, unit is ampere.

dielectric
An insulating material with properties that enable its use between the two plates of a capacitor.

dielectric breakdown (in a capacitor)
Failure of an insulator to prevent current flow from one plate of a capacitor through the insulator to the other plate. Often causes permanent damage to the capacitor.

dielectric constant
A factor which indicates how much more effective a material is as compared to air in helping a capacitor store a charge (when used as the insulating material between the capacitor's plates). The abbreviation for dielectric constant is K; air has a K of 1.

dielectric strength
A factor which indicates how well a dielectric resists breakdown under high voltages.

direct current
Current that flows in only one direction.

direct relationship
One in which two quantities both increase or both decrease while other factors remain constant.

direct short
A circuit situation in which a conductor with little or no resistance is placed across a battery or power supply. Results in very high current which will damage the source if a protective device such as a fuse or circuit breaker is not included in the circuit.

dropping resistor
A resistor used to decrease a given voltage by an amount equal to the voltage dropped across the resistor.

dry cell
A cell with a paste-like electrolyte which can be used in any position as compared to a wet cell which must be used in an upright position.

earth ground
A point that is at the potential of the earth or something that is in direct electrical connection with the earth such as water pipes.

electricity

The flow of electrons through simple materials and devices.

electrolyte

A chemical (liquid or paste) which reacts with metals in a cell to produce electricity.

electromagnetic field

A field of force produced around a conductor whenever there is current flowing through it. This field can be visualized with magnetic *lines of force*.

electron

Negatively charged particles surrounding the nucleus of an atom which determine chemical and electrical properties of the atom.

electron shells

The specific paths of rotation, or orbits, which electrons follow as they revolve around the nucleus of the atom.

electrostatic field

A field of force that surrounds any charged object. This field can be visualized with electrostatic lines of force.

electrostatic force

A force which exists between any two charged objects. If the two objects each have the *same* type of charge, the force is a *repulsion*. If the two objects each have *different* types of charge, the force is an *attraction*.

element

One of the 106 different substances which is the basic building block of all matter, and cannot be divided into simpler substances by chemical means.

energy

The ability to do work. Unit commonly used in measuring energy is the joule; equal to the energy supplied by a 1-watt power source in 1 second.

equivalent resistance

The value of one single resistor that can be used to replace a more complex connection of several resistors.

exponent

A number written above and to the right of another number called the base. Example: 10^2, 10 is the base, 2 is the exponent. A number which indicates how many times the base is multiplied by itself. $10^2 = 10 \times 10 = 100$.

farad

The unit of capacitance. A capacitor has 1 farad of capacitance when it can store 1 coulomb of charge with a 1-volt potential difference placed across it.

filament
A wire in an electric light bulb or electron tube which is heated by passing a current through it in order to make it glow or emit light, and/or electrons.

free electrons
Electrons which are not bound to a particular atom but circulate among the atoms of the substance.

fuse
A protective device usually containing a material with a low melting point which melts when current through it exceeds the ampere value for which it is rated and opens the circuit stopping the current flow.

giga
The metric prefix meaning one billion or 10^9. Abbreviated G.

ground
A voltage reference point in a circuit which may be connected to earth ground.

henry
Unit of measure for inductance. A 1-henry coil produces 1 volt when the current through it is changing at a rate of 1 ampere per second. Abbreviated H.

horsepower
A measure of power. One horsepower equals 746 watts. Abbreviated hp.

hot wire
A wire which is connected to a source of voltage or current and is not grounded.

inductance
The ability of a coil to store energy and oppose changes in current flowing through it.

inductor
A number of turns of wire wrapped around a core used to provide inductance in a circuit. (Also called a coil.)

insulator
A material with very few free electrons. A nonconductor.

inverse relationship
A relationship between two quantities in which an increase in one quantity causes a decrease in the other quantity while other factors are held constant.

jumper cables
Heavy wire conductors (usually stranded) used as an aid in starting a car with a weak battery.

junction
A connection common to more than two components in a circuit. A node.

kilo

A metric prefix meaning 1000 or 10^3. Abbreviated k.

Kirchhoff's current law

One of many tools of circuit analysis which states that the sum of the currents arriving at any point in a circuit must equal the sum of the currents leaving that point.

Kirchhoff's voltage law

Another tool of circuit analysis which states that the algebraic sum of all the voltages encountered in any loop equals zero.

leakage resistance

The normally high resistance of an insulator such as a dielectric between the plates of a capacitor.

load

A device such as a resistor which receives electrical energy from a source and that draws current and/or has resistance, requires voltage, or dissipates power.

loop

A closed path for current flow in a circuit.

loop equation

The algebraic sum of all the voltages in a loop set equal to zero.

main line current

The total current in a parallel circuit.

mega

A metric prefix meaning one million or 1,000,000 or 10^6. Abbreviated M.

mesh

The simplest form of a loop, resembling a single window pane in a circuit.

mesh current

The current flowing in a mesh. Usually assumed to be flowing in a particular direction.

mho

The unit of conductance. Symbol ℧ .

micro

A metric prefix meaning one millionth or 1/1,000,000 or 10^{-6}. Abbreviated with the Greek letter mu (μ).

milli

A metric prefix meaning one thousandth or 1/1000 or 10^{-3}. Abbreviated m.

Millman's theorem

A tool of circuit analysis which states that the voltage across several branches of a multisource parallel circuit that has no series resistance between the branches equals the sum of the branch currents divided by the total conductance of the circuit.

nano

A metric prefix meaning one billionth or 1/1,000,000,000 or 10^{-9}. Abbreviated n.

negative ion

An atom which has gained one or more electrons.

node

A junction. A connection common to more than two components in a circuit.

node current equations

A mathematical expression of Kirchhoff's current law at a junction or node.

node voltage

The voltage at a node with respect to some reference point in the circuit.

ohm

The unit of resistance. Symbol Ω.

ohmmeter

An instrument used to measure resistance.

Ohm's law

A basic tool of circuit analysis which states that in simple materials, the amount of current through the material varies directly with the applied voltage and varies inversely with the resistance of the material. Gives rise to three common equations for use in circuit analysis: $E = IR$, $R = E/I$, $I = E/R$.

open circuit

A circuit interruption that causes an incomplete path for current flow.

oscilloscope

An instrument that can visually display rapidly varying quantities as a function of time. Often used to measure voltage.

parallel circuit

A circuit that has two or more paths (or branches) for current flow.

parallel-series circuit

A circuit with several branches wired in parallel. Each branch contains one or more components connected in series, but no single component carries the total circuit current.

partial short

A path with essentially 0 ohms of resistance connected across part of a circuit but not connected directly across the source.

percent
A ratio of one part to the total amount. One part of a hundred.

pico
A metric prefix meaning one million millionth or 10^{-12}. Abbreviated p.

polarity of voltage
A means of describing a voltage with respect to some reference point, either positive or negative.

positive ion
An atom that has lost one or more electrons.

potential difference
A measure of force produced between charged objects that moves free electrons. Also called voltage or electromotive force. Symbol is E, unit is the volt (abbreviated V).

potentiometer
A resistance element with a sliding wiper contact used in applications where a division of resistance is required. (A *three*-terminal adjustable resistive divider.)

power
The rate at which work is done or the rate at which heat is generated (abbreviated P). The unit of power is the *watt* (abbreviated W), which is equal to 1 joule per second.

power dissipated
Power which escapes from a resistance in the form of heat by the convection of air moving around the component.

power rating of a resistor
How much power a resistor can dissipate (give off) safely in the form of heat in watts.

power supply
A device which is usually plugged into a wall outlet and can replace a battery in many applications by providing a known potential difference between two convenient terminals.

protons
Particles in the nucleus of the atom which has a positive charge equal to the electron's negative charge.

recall
Function key on a calculator which when pressed causes the calculator to display the contents of the memory.

reciprocal
Mathematical "inverse". The reciprocal of any number is simply that number divided into one.

resistance
Opposition to current flow which is a lot like friction because it opposes electron motion and generates heat. Symbol R. Unit is the ohm (Ω).

reference node
> A junction in a circuit from which all voltages are measured.

reference point
> An arbitrarily chosen point in a circuit to which all other points in the circuit are compared, usually when measuring voltages.

regulated voltage
> The output voltage of a power supply which contains special circuitry to keep the output voltage constant, even if the current drain on it is changing.

relay
> A switch (or combination of switches) activated by an electromagnetic coil.

rheostat
> An adjustable device with two terminals which can be used to produce a variable resistance in a circuit.

rounding off
> A procedure by which a number with many digits can be reduced to a number with only three significant digits. The first three significant digits are kept, and the fourth examined. If the fourth digit is 5 or greater, the third significant digit is raised by one. If the fourth digit is 4 or less, the first three digits are kept unchanged.

scientific notation
> A type of shorthand used to keep track of decimal places which utilizes powers of the number 10. Standard form for scientific notation is $D.DD \times 10^E$, where D represents each of the first 3 significant digits, and E represents the exponent, or power of ten.

series circuit
> A circuit with only one path through which current can flow.

series-parallel circuit
> A group of series and parallel components in which at least one circuit element lies in the path of the total current.

short circuit
> A path with little or no resistance connected across the terminals of a circuit element.

shunt
> Another term which means parallel. Often also refers to the low value of parallel resistance used in an ammeter for determining or changing the "range" of the meter.

significant digits
> Those digits within a number which have the greatest weight. In the decimal system digits to the left of any designated digit are more significant than those to the right.

sign of a voltage

A notation, either positive (+) or negative (−), in front of a voltage. (Important in solving loop equations and depends on whether the voltage aids or opposes current flow in a circuit.)

simultaneous equations

A series of equations which contain the same unknown quantities, and which can be manipulated to solve for each of the unknowns.

solenoid

A term used to mean coil or inductor, also used to mean a type of relay such as that used to switch the starter current in an automobile.

source

A device, such as a battery or dc power supply, which supplies the potential difference and electrical energy to the circuit.

specific resistance

The resistance of any material in a particular size and shape. The resistance of a piece of a substance that is 1 foot long and 1 mil in diameter at 20°C measured in ohms.

square root of a number

Another number which must be multiplied by itself to obtain the original number.

square of a number

That number multiplied by itself.

store

A calculator operation where the number in the display is transferred to the memory where it is held until it is recalled.

substitute

To replace one part of a formula or equation with another quantity which is its equal.

superposition theorem

A tool of complex circuit analysis which states that in a network with two or more sources the current or voltage for any component is equal to the algebraic sum of the effects produced by each source acting separately.

switch

A device that is used to open or close circuits, thereby stopping or allowing current flow in a circuit or through a component.

terminal

A connection point on a device or component.

Thevenin's theorem

A tool of circuit analysis which states that a complex circuit can be reduced to an equivalent series circuit with a single voltage source and a single series resistance, as long as all components are linear.

time constant

The time it takes in seconds for a capacitor to charge up to 63 percent of the applied voltage or the time it takes for a fully charged capacitor to discharge from 100 percent down to 37 percent of full charge. Equal to the product of R (in ohms) times C (in farads) in a resistive-capacitive circuit. Also a measure of the current rise and fall in inductive circuits. (Equal to the quotient of L/R in resistive-inductive circuits, L in henries, R in ohms.)

transposing

Moving a quantity from one side of an equation across the equal sign to the other side of the equation and changing its sign.

traverse

Move around a circuit or across a component. Usually refers to a mental process used for keeping track of various voltages encountered in a loop.

troubleshooting

A technique used to locate a problem in a circuit.

valence electrons

Those electrons in the outermost or valence shell of an atom.

valence shell

The outermost shell of an atom.

volt

The unit of voltage or potential difference. Abbreviated V.

voltage

A measure of the push on each electron which makes it move. Symbol E. Unit is the volt.

voltage divider

A type of circuitry that provides an economical way to obtain one or several lower voltages from a single higher voltage supply.

voltage drop

Change in voltage available between points in a circuit produced by current flow through resistors. Also called an IR drop. Unit is the volt.

voltmeter

An instrument used to measure voltage between two points in a circuit.

watt

The unit of power. Abbreviated W. Equal to 1 joule per second.

Wheatstone bridge
A specialized circuit, sometimes housed as an instrument, for measuring resistance very accurately, by comparison to a standard resistance.

working voltage
The recommended maximum voltage at which a capacitor should be operated.

Index

QUIZ ANSWERS — CHAPTER 1 THROUGH 7

Lesson 1	Lesson 2	Lesson 3	Lesson 4
1. h	1. e	1. e	1. d
2. f	2. g	2. a	2. c
3. f	3. b	3. c	3. b
4. f	4. d	4. b	4. a
5. g	5. c	5. e	5. f
6. a	6. b	6. e	6. b
7. c	7. a	7. b	7. a
8. a	8. c	8. c	8. f
9. d	9. e	9. d	9. d
10. b	10. a	10. a	10. a
11. f	11. f	11. a	11. a
12. f	12. b	12. e	12. e
13. a	13. e	13. c	13. c
14. c	14. c	14. c	14. a
15. e	15. b	15. b	15. c
16. d	16. a	16. d	16. d
17. a	17. b	17. a	17. b
18. e	18. d	18. d	18. b
19. d	19. b	19. c	19. e
20. c	20. e	20. a	20. d

Lesson 5	Lesson 6	Lesson 7
1. d	1. d	1. b
2. c	2. d	2. a
3. b	3. c	3. d
4. a	4. a	4. b
5. d	5. c	5. a
6. c	6. e	6. c
7. a	7. d	7. c
8. c	8. c	8. b
9. a	9. b	9. d
10. 1175	10. c	10. b
11. 11,110	11. a	11. b
12. 200,500	12. e	12. a
13. 30Ω	13. 410Ω	13. a
14. 820Ω	14. 7.5K	14. c
15. b	15. 11.37K	15. e
16. b	16. 1.0 MΩ	16. c
17. a	17. 100Ω	17. a
18. e	18. 60Ω	18. b
19. d	19. 1K	19. a
20. c	20. b	20. d

QUIZ ANSWERS — CHAPTERS 8-14

Lesson 8	Lesson 9	Lesson 10	Lesson 11
1. SP	1. SP	1. b	1. b
2. P	2. P	2. c	2. d
3. PS	3. SP	3. a	3. c
4. S	4. S	4. b	4. d
5. SP	5. SP	5. d	5. e
6. PS	6. 54K	6. 160mW	6. d
7. SP	7. 13.9	7. 400W	7. e
8. SP	8. 45.3	8. 25mA	8. a
9. PS	9. 84.6K	9. 9.6mW	9. b
10. PS	10. −1mA	10. 112V	10. d
11. 95Ω	11. −0.5mA	11. b	11. d
12. 123Ω	12. 15.1V	12. a	12. b
13. 133Ω	13. 222μA	13. b	13. a
14. 16.9Ω	14. 6mA	14. d	14. c
15. 24.5Ω	15. −820Ω	15. c	15. d
16. 1	16. −7.2V	16. c	16. a
17. R_4	17. −2mA	17. e	17. a
18. 3.15K	18. 19.7V	18. b	18. c
19. 10V	19. d		19. e
20. 1mA	20. d		20. d

Lesson 12	Lesson 13	Lesson 14
1. a	1. d	1. b
2. a	2. d	2. a
3. a	3. c	3. d
4. d	4. b	4. c
5. b	5. b	5. d
6. +320mA	6. b	6. b
7. +3.0A	7. a	7. b
8. −80mA	8. c	8. c
9. +30.5mA	9. b	9. a
10. −5mA	10. e	10. d
11. c	11. c	11. b
12. a	12. a	12. b
13. f	13. d	13. a
14. d	14. c	14. c
15. c	15. b	15. c
16. b	16. 1.03sec	16. d
17. b	17. 39×10^{-9}	17. 4msec
18. a	18. 0.825sec	18. 50nsec
19. c	19. 12μF	19. 3K
20. b	20. 6μF	20. 24H